THE BIOLOGY OF MARSUPIALS

BIOLOGY AND ENVIRONMENT
General Editor: Bernard Stonehouse

THE BIOLOGY OF PENGUINS
Edited by BERNARD STONEHOUSE

THE BIOLOGY OF MARSUPIALS
Edited by BERNARD STONEHOUSE
and DESMOND GILMORE

THE BIOLOGY OF MARSUPIALS

Edited by
BERNARD STONEHOUSE

School of Environmental Science
University of Bradford, England

and

DESMOND GILMORE

Department of Physiology
University of Glasgow, Scotland

UNIVERSITY PARK PRESS
Baltimore • London • Tokyo

First published 1977 by
THE MACMILLAN PRESS LTD
London and Basingstoke

Published in North America by
UNIVERSITY PARK PRESS
Chamber of Commerce Building
Baltimore, Maryland 21202

Printed in Great Britain

Library of Congress Cataloging in Publication Data

Main entry under title:

The Biology of marsupials.

   (Biology and environment)
   Includes indexes.
   1. Marsupialia. I. Stonehouse, Bernard.
II. Gilmore, Desmond.
QL737.M3B5   1976   599'.2   76 - 10944
ISBN 0 - 8391 - 0852 - 4

Contributors

ARUNDEL, J. H – Veterinary Clinical Centre, University of Melbourne, Werribee, Victoria 3030, Australia

BARBOUR, R. A. – Department of Anatomy and Histology, University of Adelaide, South Australia 5001

BARKER, I. K. – Ontario Veterinary College, Guelph, Canada

BEVERIDGE, I. – Veterinary Clinical Centre, University of Melbourne, Werribee, Victoria 3030, Australia

BRADLEY, A. J. – Department of Physiology, Monash University, Clayton, Victoria 3168, Australia

BRAITHWAITE, R. W. – Department of Physiology, Monash University, Clayton, Victoria 3168, Australia

BUCHMANN, O. L. K. – Department of Zoology, University of Tasmania, Hobart, Tasmania

CALABY, J. H. – Division of Wildlife Research, C.S.I.R.O., Lyneham, ACT, Australia

CLEMENS, W. A. – Department of Paleontology, University of California, Berkeley, California 94720, USA

CROMPTON, A. W. – Museum of Comparative Zoology, Harvard University, Cambridge, Mass., USA

GILMORE, D. P. – Institute of Physiology, University of Glasgow, Glasgow, UK

GUILER, E. R. – Department of Zoology, University of Tasmania, Hobart, Tasmania

HAYMAN, D. L. – Department of Genetics, University of Adelaide, South Australia 5001

HEARN, J. P. – MRC Unit of Reproductive Biology, Forrest Road, Edinburgh, UK

HIIEMAE, K. – Unit of Anatomy with special reference to Dentistry, Guy's Hospital Medical School, London, UK

KEAST, A. – Department of Biology, Queen's University, Kingston, Ontario, Canada

KIRKBY, R. J. – Department of Behavioural Sciences, Lincoln Institute, Carebon, Victoria 3053, Australia

KIRSCH, J. A. W. – Peabody Museum of Natural History, Yale University, New Haven, Conn., USA

LEE, A. K. – Department of Physiology, Monash University, Clayton, Victoria 3168, Australia

McDONALD, I. R. – Department of Zoology, Monash University, Clayton, Victoria 3168, Australia

MARTIN, P. G. – Department of Botany, University of Adelaide, South Australia 5001

NEWSOME, A. E. – Division of Wildlife Research, C.S.I.R.O., Lyneham, ACT, Australia

PACKER, W. C. — Department of Zoology, University of Western Australia, Nedlands, Western Australia 6009

PARKER, P. — Museum of Comparative Zoology, Harvard University, Cambridge, Mass., USA

REID, I. A. — Department of Physiology, School of Medicine, University of California, San Francisco 94143, USA

SALLIS, J. D. — Department of Biochemistry, University of Tasmania, Hobart, Tasmania

SETCHELL, B. — ARC Institute of Animal Physiology, Babraham, Cambridge, UK

STODART, Eleanor — 38 James Street, Curtin, ACT, Australia

STONEHOUSE, B. — School of Environmental Science, University of Bradford, Bradford, UK

THEXTON, A. J. — Royal Dental Hospital School of Dental Surgery, London, UK

WEBB, S. — Department of Zoology, La Trobe University, Melbourne, Victoria, Australia

WOOLLEY, P. — Department of Zoology, La Trobe University, Melbourne, Victoria, Australia

ZIEGLER, A. C. — Bishop Museum, Honolulu, Hawaii 96818, USA

Contents

Contents

1 Introduction: The Marsupials

B. Stonehouse*

The marsupials or pouched mammals form a distinctive group within the Class Mammalia. Though clearly possessing the diagnostic features of typical mammals, including high and stable body temperature, furry pelt, simple lower jaw and mammary glands, they have for long been accepted as standing apart from the main body of the Class. Generally thought of as 'primitive' (that is, displaying archetypal features which more advanced mammals have lost), they are considered to stand closer to placental mammals than to egg-laying monotremes or to the earlier forms – triconodonts, docodonts and multituberculates – known only as Mesozoic or early Tertiary fossils. This recognition is reflected in their taxonomic position. Though there is considerable taxonomic disagreement over the number of subclasses into which the Class Mammalia may properly be divided, most of the argument concerns the status of the early fossil groups; Simpson's (1945) classification, which includes the marsupials and placentals in a single subclass Theria, is generally accepted. Marsupials form the Infraclass Metatheria, placentals the Infraclass Eutheria. Both groups made their first appearance in the Cretaceous, derived from pantotherian (trituberculate) stock of the Upper Jurassic.

Within their infraclass, marsupials stand alone, traditionally grouped in the single Order Marsupialia, and thus comparable in status with each of the 17 or 18 living and 14 fossil orders of placental mammals. This arrangement stresses the basic uniformity of marsupial structure and the anatomical differences which separate them from other mammals, and is to some degree justified by the relative sizes of the two groups: about 140 living and fossil genera of marsupials are recognised, compared with over 2600 genera of placental mammals. However, it gives inadequate recognition both to the striking ecological diversity to be found within the group, and to the plasticity of form which has given rise to animals so apparently dissimilar as the Virginia opossum *Didelphis virginiana*, the Red Kangaroo *Macropus rufus* and the Marsupial mole *Notoryctes typhlops*. Recent field and laboratory studies have tended to stress diversity rather than uniformity, and modern taxonomists (for example, Ride, 1964; Kirsch, 1968, Van Valen, 1971) appear willing to recognise two or more orders within the infraclass, based on well-defined and generally agreed family groupings. Kirsch and Calaby in this volume draw attention in their first table to three different arrangements of higher category classifications of living marsupials, involving one, three and four orders. There is substantial agreement at family level, though Kirsch's own classification, based largely on serological studies, involves more families than the others. These

*Dr. Bernard Stonehouse was formerly a Reader in Zoology in the University of Canterbury, Christchurch, New Zealand, where his research included studies of introduced marsupials in forest and grassland habitats.

are to some extent different ways of expressing the general feeling among those
who work with marsupials that — whatever taxonomic rank may be ascribed —
marsupials are didelphoid, dasyuroid, perameloid, caenolestoid or phalangeroid.
Kirsch and Calaby's contribution, based on an annotated checklist, recognises
some 250 living species of marsupials, contained in 75 genera and sixteen families.
Except where specifically mentioned in the text, it has been accepted by all
other contributors to this volume.

The fossil record of marsupials is slight compared with that of placentals, and
heavily weighted in favour of American species. All fossil material so far ascribed
to the infraclass is included in extant orders, much of it in extant families, though
some eight to ten entirely fossil families have also been proposed. The phylogeny
and fossil history of marsupials are reviewed by Clemens in this volume.

Marsupials are mammals in which the young are born at an early stage of
development and nurtured during a further period of development and growth
outside the mother's body, attached to teats of abdominal mammary glands
which are often but not always enclosed in a pouch. Anal and urogenital openings of
the female are enclosed in a common sphincter; there are two lateral vaginae which
open jointly into a median birth-passage or separately into a urogenital canal.
The penis may be bifurcate, and in some species the testes lie anterior to it. The
egg is yolky and surrounded by a firm membrane. It may draw extra nourishment
from uterine secretions, but only in the bandicoots (Peramelidae) is there a well-
established and briefly functional placenta. Gestation extends between eight and
about 40 days according to species, and pouch life usually lasts nine weeks or
more. Male marsupials are normally pouchless and play little part in rearing the
young. The special features of marsupial gestation and early development are
traditionally regarded as primitive, representing an early, archaic stage in mammalian evolution from which the placentals advanced, while the marsupials did
not. While it may well be true that development of a placenta gave advantages
which ultimately allowed the eutherian mammals to diversify and spread to a
greater degree than marsupials, patterns of modern marsupial reproduction cannot
be taken to represent the pre-placental condition of ancestral mammals. Nor can
it be assumed that modern marsupials in their present environments are in any
way at disadvantage or hazard because of their lack of a functional placenta.
Parker's contribution to this volume draws attention to some of the advantages
which marsupials have retained in pursuing their independent evolutionary course,
and to their present reproductive success in a variety of ecologically unrewarding
habitats.

The marsupial skull is clearly mammalian, though again different from that of
eutherian mammals. The braincase is small; median and posterior crests, jugal
arches, the inturned ventral posterior border of the lower jaw (classically though
rather confusingly described as the 'inflected angle') and other muscle attachments, are often comparatively large, giving the marsupial skull unmistakable
flowing lines which contrast with the functional austerity of the placental skull.
There is no post-orbital bar; at the hinge of the jaw the jugal bone is always
involved in the glenoid cavity, and the palate is often incomplete or perforate
posteriorly. Teeth vary in form as much as those of placentals, and are usually
more numerous; there may be up to 10 incisors in the upper jaw and six in the
lower, with typically three premolars and four molars on either side. The tym-

panic chamber may be open, or covered by the alisphenoid bone. The axial skeleton is similar to that of placentals and almost as variable. One most striking difference is the presence of epipubic bones attached to the front of the pelvic girdle and lying in the abdominal musculature; their function is nowhere clearly stated, though often related to the presence of the pouch. Other peculiarities of both skeletal and soft parts of the marsupial body are detailed in Barbour's contribution to this volume, which reviews an extensive literature.

Marsupials are currently found in two widely separated regions of the world – Australasia and the Americas. Fossil remains are known also from Europe, but have not yet been reported from Africa, Asia or Antarctica. Many attempts have been made to explain this curious geographical distribution, though most of the earlier accounts are now outmoded by recently developed concepts of plate tectonics. As major continental movement has occurred within the time-span of mammalian evolution, adaptive radiation and dispersion, biogeographical models which fail to take it into account are unlikely to be satisfactory. Contributions to this symposium by Keast and Martin discuss the biogeography and adaptive radiation of marsupials in the context of the most recent findings about late Mesozoic and Tertiary movements of continental masses, over a time scale of 100 million years or more.

Three families – the Didelphidae, Microbiotheriidae and Caenolestidae – inhabit South and Central America, one species of didelphid, the Virginia opossum, extending across North America to beyond the Canadian border. The American marsupials alive today are mostly small, ranging in size from mouse to rabbit; generally either carnivorous or omnivorous, many live in forests and feed mainly on insects. Didelphids, which form the most versatile group, have given rise to some 70 species. They include the ubiquitous Virginia opossum, which has been intensively studied in both the forests and the laboratories of North America, several lesser-known species of similar form and ecology but narrower geographical limits, small murine opossums (*Marmosa, Monodelphis*), frugivorous woolly opossums (*Caluromys*), and semi-aquatic water opossums (*Chironectes*) with webbed hind feet and waterproof pouch.

Marsupials were formerly more plentiful in North America, making up a considerable proportion of the mammalian fauna in Upper Cretaceous times (Tyndale-Biscoe, 1973) though declining later (Clements, this volume). Many were didelphoid, referrable either to the Didelphidae or to closely related families now extinct. The single genus of European marsupials *Peratherium*, which flourished until the Miocene, is also didelphoid, and presumably of North American origin. Marsupials were plentiful throughout South America from the early Tertiary onward, diversifying to occupy a range of ecological niches. The major family Borhyaenoidea included a selection of large carnivores which resembled wolves, mustelids, foxes, sabre-tooth tigers and hyaenas. Other families included the mole-like *Necrolestes* and a number of rat-like forms, both herbivorous and carnivorous. The caenolestids, represented today by three genera of small woodland insectivores, appear in the fossil record from the early Eocene onward.

South American marsupials flourished alongside placental mammals through-out the early and middle Tertiary, while their continent remained isolated from other land masses. They declined rapidly, together with several of the placental groups, after the establishment of the Panama bridge during the Pliocene epoch, and following a new influx of placentals from North America. Their decline is usually considered the result of competition from the newly arriving species. This is perhaps a näive view, argued by analogy from the disappearance of several Australian marsupial species after the arrival of European man with his fires, his agriculture, and his menagerie of placental attendants. Modern mammalogists are more impressed than their predecessors with the efficiency of marsupials on their home ground, and with their adaptability in changing environments. Though the legend that pouched mammals are in some ways 'second-class' or 'primitive' dies hard, recent studies of marsupials in their native environment, in environments modified by man, and as introduced aliens, leave us little reason to believe that they must inevitably decline when faced with competition from placentals (see papers by Newsome, Stodart, Guiler and Sallis, and Gilmore in this volume). Reasons for the late Tertiary decline of South America's indigenous marsupials and placentals bear further investigation; climatic and ecological changes may well have contributed more to their downfall than direct competition for niches.

Some 172 living species of marsupials, contained in 13 families, occur in Australia and its offshore islands, and in Tasmania, New Guinea, Celebes and Timor. They include the kangaroos, wallabies, wombats, koala, bandicoots, Tasmanian devils and possums (similar but only distantly related to the American opossums), and many lesser-known forms down to the size-range of small mice. Fossil remains of Australasian marsupials — particularly those of early Tertiary forms — are comparatively rare, but their record indicates that a major adaptive radiation occurred during the Palaeocene and Eocene, giving rise to most of the existing families and to some now extinct. This was followed by a series of minor radiations which established the late-Tertiary and present day abundance of species. Both Australia and New Guinea were centres of marsupial evolution, at times linked to each other and at times separated by shallow but isolating sea. Ziegler in this volume draws attention to the importance of New Guinea as a centre of marsupial evolution separate from mainland Australia. While marsupials of the mainland are primarily animals of grassland and scrub, most of New Guinea's 42 endemic species and nearly half of the species shared with Australia are forest-adapted forms. Ziegler's account includes both a survey of the little-known New Guinea fauna, and an analysis of the mid-to late-Tertiary changes in sea level and flora which produced it.

Modern Australasian marsupials show a very wide range of environmental adaptation and specialisation. Many are curiously similar in form to placentals which occupy comparable niches in other parts of the world. Though analogies are sometimes strained, there are strong resemblances between wolves (*Canis*) and thylacines (*Thylacinus*), civets (*Civettictis*) and native cats (*Dasyurus*), coypus (*Myocastor*) and wombats (*Vombatus*), hares (*Lepus*) and hare wallabies (*Lagorchestes, Lagostrophus*), moles (*Talpa*) and marsupial moles (*Notoryctes*), jerboas (*Dipus, Allactaga*) and jerboa marsupials (*Antechinomys*), flying squirrels (*Glaucomys*) and flying phalangers and gliders (*Acrobates, Petaurus, Schoinobates*). On a broader scale the kangaroos and wallabies show several points of convergence with grazing and browsing placentals of the suborder Ruminantia, including grinding molars and a ruminant form of digestion, and many kinds of small marsupials occupy the range of niches which in other continents are occupied by placentals of the orders Insectivora and Rodentia. There are notable omissions from the marsupial range. No bat-like, seal-like or whale-like forms have appeared, perhaps not surprisingly in animals with the marsupial pattern of reproduction. No marsupial seems ever to have reached elephant size, though some very large herbivores, including giant kangaroos, disappeared from the Australian mainland late in the Pleistocene, possibly destroyed by early man (Tyndale-Biscoe, 1973). A large phalangeroid carnivore, *Thylacoleo*, with claws and shearing premolars, may have shared this fate, leaving no carnivorous marsupial larger than the wolf-like thylacine.

This volume summarises recent research, in several fields, in which marsupials have been involved for their special qualities as non-placental mammals. Most of the studies are Australasian, reflecting the healthy interest now shown by Australian biologists in their native mammals, and to some degree balancing the weight of earlier marsupial studies which, American based, tended to be centred on the Virginia opossum. Though by no means exhaustive, the selection includes contributions from many areas in which work is still in progress; the editors hope that it will help biologists of the several disciplines involved — taxonomy, cyto-genetics, evolution, biogeography, population ecology, behaviour, comparative anatomy, and cellular, endocrine and metabolic physiology — to discover what others are doing and so broaden their own interests in the biology of marsupials.

The editors thank Mrs Heather French for her valuable work as Editorial Assistant; and Miss Hebe Jerrold for her indexing.

Bradford, 1976   B.S.
                                                                           D.P.G.

References

Kirsch, J. A. W. (1968). Prodromus of the comparative serology of Marsupialia. *Nature*, 217, 418-420

Ride, W. D. L. (1964). A review of Australian fossil marsupials. *J. Proc. R. Soc. West. Australia,* 47, 97-131

Simpson, G. G. (1945). The principles of classification and a classification of mammals. *Bull. Amer. Mus. Nat. Hist.,* 85, 1-350

Tyndale-Biscoe, C. H. (1973). *Life of Marsupials.* Edward Arnold, London, 1-254

Van Valen, L. (1971). Adaptive zones and the orders of mammals. *Evolution,* 25, 420-428

Section 1 The species and their chromosomes

Since marsupials first came to the attention of European naturalists in the early sixteenth century, their relationships with more orthodox eutherian mammals and with each other have provided topics of perennial interest to taxonomists. Early classifiers, struggling to pigeon-hole the mass of new species brought home by explorers, tended to place their new finds wherever possible in existing groups, selecting such features as the shapes of claws, limbs and teeth as indicators of similar ways of life, and therefore of natural affinities. The significance of the unique method of reproduction — common to all marsupials — was overlooked; this is not surprising, for it is anatomically apparent only in well-preserved female specimens, and far less indicative of way of life than the features which we now call adaptive. Several marsupials accordingly found themselves grouped with their placental analogues — native cats, for example, close to the Carnivora, and kangaroos and wombats among the Rodentia. As the concept of common ancestry came to replace natural affinity as a basis for grouping, the fundamental characteristics came into their own. Now there are no remaining doubts as to which animals are marsupials and which are not; the taxonomic relationships of marsupials to monotremes and eutherians have been established on criteria which satisfy anatomists, palaeontologists, serologists and cytogeneticists, and only the questions of relationships within the infraclass remain. Some of these questions are discussed in the opening chapter by two biologists — Dr Kirsch of America and Dr Calaby of Australia — whose experience covers the two geographical areas in which marsupials are found. Their checklist of living species should be useful to biologists for many years to come.

Detailed studies of chromosome number and structure have thrown much light on taxonomic relationships in many diverse taxa. As Sharman, Martin, Hayman and others have shown, marsupials tend to distinguish themselves from other mammals by having notably fewer and larger chromosomes. Where placentals may show as many as 78 in their karyotype, with a modal figure of 48, the greatest number recorded so far in marsupials is 32; almost half the known species have a karyotype of 14, and over 90% lie within the range 14–22. The small number of chromosomes and the clarity with which they can be demonstrated have facilitated comparative studies within the families of marsupials, especially among the many species — diverse both geographically and taxonomically — which share a karyotype of 14. Dr David Hayman's chapter discusses some of the implications of these studies, and their significance in the general field of cytogenetics.

2 The species of living marsupials: an annotated list

John A. W. Kirsch* and John H. Calaby†

Introduction

As all first-year zoology students well know, the Australian marsupials represent
an adaptive radiation recalling that of eutherian mammals elsewhere. Australian
marsupials include a wide range of herbivores, insectivores and carnivores; they
are found in virtually every habitat from alpine to desert to rain forest; and
the group lacks only conspicuous rodent-like and bat-like forms – significantly,
relatively recently immigrant placental mammals fill these roles.

The radiation of marsupials in the Australian region may or may not be due
to isolation. There is no fossil evidence one way or the other, but it is generally
assumed that marsupials are competitively inferior to placentals, and could have
radiated so exuberantly *only* if left to themselves.

The American marsupials, on the other hand, are more limited in ecological
range, though their extensive fossil record attests that this range was once greater.
Thus extant American marsupials represent only three families, compared with
the six or more of Australia. The narrower extent of the radiation in America
reflects a different situation from that possibly obtaining in Australia. The
earliest and isolated Tertiary mammal fauna of South America consisted of a
mixture of marsupials and placentals, the first placentals including edentates
and notoungulates, later enriched by several rodent and primate taxa, and
ultimately by a number of northern groups in the late Pilocene. Marsupials and
placentals thus evolved together in South America, the marsupials providing the
bulk of the carnivores and insectivores. The extant marsupials are mainly small
to medium-sized carnivores, some of which became established in Middle America
at the end of the Tertiary.

The ecological breadth of the marsupial radiations is reflected in the taxonomic diversity of marsupials. We estimate that there are about 250 living species
in all, of which two-thirds are Australasian. Not surprisingly in a group so diverse,
the taxonomic validity of named species varies greatly, reflecting very great
differences in intensity of study of the several genera. Moreover, the geographical
disjunction of Australasian and American marsupials contributes to the inequality:
few workers on one geographical moiety of marsupials have worked as extensively
on the other.

*John A. W. Kirsch is Assistant Professor of Biology at Yale University, where he also
curates the Mammal section of the Peabody Museum Of Natural History. He has worked on
the comparative serology of marsupials under W. D. L. Ride at the University of Western
Australia and spent fourteen months collecting marsupials and rodents in South America. He
is currently working on the functional morphology of marsupials, and the DNA of New
World rodents.

†John H. Calaby is a Senior Principal Research Scientist of C.S.I.R.O. Division of Wildlife
Research, Canberra, A.C.T., Australia.

9

The object of this chapter is to present a reasonable and usable list of the species of living marsupials. We have not listed subspecies: aside from the philosophical question of the reality of subspecies, few Australasian marsupials, and even fewer American ones, are so well known as to justify taxonomy of this detail. The multiplicity of named New Guinean and South American taxa is particularly marked, no doubt partially because of speciation due to the diversity and stability of habitats in the tropical regions, but also reflecting the difficulty of collecting geographically adequate samples there.

Higher-category classification of marsupials

Higher-category classification, while important to those of us who practise it, contributes little to the predictive function of a classification. Given the tenuous nature of affinities at these levels, it functions to encode rather few data about the included organisms. Thus this discussion of classification above the family level is intentionally brief. In any event, the subject of the major classification of marsupials has been treated elsewhere (Kirsch, 1968, 1976; Ride, 1964; Sharman, 1974; Simpson, 1945, 1970), and discussion must properly include the fossil taxa, which this paper does not do.

Taxonomically, the marsupial families have shown a remarkable stability in composition, and have, for the most part, been universally recognised as such for a century. Moreover, certain of these families group easily into a number of superfamilies, a fact which Simpson formalised in his classifications of 1930 and 1945. However, Simpson insisted that further groupings of the superfamilies would be inappropriate and based on inevitably convergent single characters. The problem is lucidly discussed in Simpson (1945). Briefly, the difficulty is that perameloids (bandicoots) and phalangeroids (in Simpson's sense, phalangers, kangaroos, the Koala and wombats) are distinguished by syndactyly of the second and third toes of the pes, while all other marsupials have the five digits free from each other. Phalangeroids also show reduction of the lower incisors to a single functional pair, which are enlarged and procumbent; this dental feature is a character also of the caenolestoids (shrew- opossums). However, caenolestoids, like the didelphoids (opossums) and dasyuroids (Australian native cats and mice, etc.), do not have syndactyl feet.

The incompatibility of the two taxonomic systems is obvious, and almost certainly requires that either or both the shared pes and dental features be convergently evolved in the groups possessing them. Most zoologists now assume that the 'diprotodont' incisors are convergent in phalangeroids and caenolestoids, and this is supported by serological studies (Hayman *et al.*, 1971) as well as by the geographical separation of the two groups. Moreover, the Australasian diprotodonts (phalangeroids) have a number of other distinguishing features, including peculiarities of the brain (Abbie, 1937, 1941) and possession of a superficial thymus gland (Yadav, 1973), which set them apart from the American caenolestoids. On the other hand, Australian zoologists at least tend to assume (with Jones, 1923) that syndactyly does indicate a special *phylogenetic* relationship between the perameloids and phalangeroids, the bandicoots being seen as a non-diprotodont but syndactylous intermediate between the non-diprotodont, non-syndactylous dasyuroids and the diprotodont, syndactylous phalangeroids.

Diprotodonty must therefore be convergently evolved in Caenolestoidea.

Be that as it may, a number of alternate schemes for grouping the super-families *not* based on pes or dentition have been proposed since 1945. Because of the ecological breadth, antiquity and taxonomic diversity of marsupials, and because the historical accident of late discovery resulted in their being classified in a single order, Ride (1964) suggested the recognition of several orders of marsupials in an arrangement that combines some of the superfamilies of

Table 2.1 Comparison of three higher-category classifications of the living marsupial families. Modified and expanded after Sharman (1973)

(Kirsch, 1976)	(Ride, 1964)	(Simpson, 1945)
Superorder Marsupialia	Superorder Marsupialia	
Order Polyprotodontia	Order Marsupicarnivora	Order Marsupialia
Suborder Didelphi-morphia		
Superfamily Didel-phoidea	Superfamily Didel-phoidea	Superfamily Didelphoidea
1 Family Didelphidae	Family Didelphidae (implied)	Family Didelphidae
2 Family Micro-biotheriidae		
3 Family Thylacinidae		
Suborder Dasyuromorphia		
Superfamily Dasyuroidea	Superfamily Dasy-uroidea	Superfamily Dasyuroidea
4 Family Dasyuridae	Family Dasyuridae	Family Dasyuridae
5 Family Myrmecobiidae		
	Family Thylacinidae	
		Family Notoryctidae
	Order Peramelina	
Suborder Peramelemorphia		
Superfamily Perameloidea		Superfamily Perameloidea
6 Family Peramelidae	Family Peramelidae	Family Peramelidae
7 Family Thylacomyidae		
Suborder Notoryctemorphia		
Superfamily Notoryctoidea		
8 Family Notoryctidae		
Order Paucituberculata	Order Paucituberculata	
Superfamily Caenolestoidea		Superfamily Caenolestoidea
9 Family Caenolestidae	Family Caenolestidae	Family Caenolestidae
Order Diprotodonia	Order Diprotodonia	
Superfamily Phalan-geroidea		Superfamily Phalangeroidea
10 Family Phalangeridae	Family Phalangeridae	Family Phalangeridae
11 Family Petauridae		
12 Family Burramyidae		
13 Family Macropodidae	Family Macropodidae	Family Macropodidae
Superfamily Vombatoidea		
14 Family Vombatidae	Family Vombatidae	Family Phascolomidae (= Vombatidae)
15 Family Phascolarctidae		
Superfamily Tarsipedoidea		
16 Family Tarsipedidae		
	Marsupialia *incertae sedis*	
	Family Notoryctidae	

Simpson. Kirsch (1968, 1976) later modified Ride's scheme and made a number
of detailed rearrangements within the orders, basing much of this revision on
serological data. These two classifications are compared with Simpson's in table 2.1.

Later Simpson (1970) half-seriously suggested grouping marsupials into two
suborders based on their distribution; his objections to the ordinal schemes
involved the problematic positions of several extinct marsupial groups. Van Valen
(1971) recognised only two orders, Diprotodontia (the phalangeroids) and Poly-
protodontia (all other marsupials, including the caenolestoids); while Turnbull's
provocative reclassification of all mammals (1971) follows Ride's ordinal arrangement of marsupials, but places them as a cohort within the infraclass Eutheria:
the zalambdodont marsupials are removed to another cohort of mammals
including placentals with similar molar morphology.

Familial groupings

It is certainly useful to have some indication of familial affinities, however, and
we list the species under their families in Table 2.2. In addition to three American
families, our list shows 13 families of Australasian marsupials, as opposed to
Simpson's six, and it is necessary to explain the additions. At present the familial
groupings of phalangeroids (Simpson's sense) are those of Kirsch (1968), which
were founded primarily on serological data; a few genera not tested by him (e.g.
Dactylopsila, Distoechurus) were classified on the basis of morphology. Thus most
of the new families are Australasian diprotodonts.

The first additional family, Phascolarctidae, is that of the Koala, traditionally
classified with the phalangers; its resemblances to some of those animals seem
certainly convergent or retentions of primitive characters, and the weight of the
evidence (see Troughton, 1967, for a list) favours affinities with the wombats,
although at a familially distinct level.

Three other additions are the Tarsipedidae, Petauridae and Burramyidae, now
separated from the diverse Phalangeridae. The new arrangement seems to have
met with general approval (e.g. Ride, 1970; Tyndale-Biscoe, 1973), and is
followed here.

Two more additions are Thylacinidae, containing the genus *Thylacinus*, and
Myrmecobiidae, containing *Myrmecobius*. The Tasmanian tiger or Thylacine,
has frequently been considered a foreign element in the Australian fauna, and
Ride (1964) regarded it as at least familially distinct from the Dasyuridae. We
accept and follow this separation, as we do that of Archer and Kirsch (in press),
who argue for a return to the practice of recognising the Numbat, *Myrmecobius*,
at a familial level.

Those authors also suggest familial recognition for the Rabbit-eared bandicoots,
Macrotis, and their fossil relatives. Thus two families are distinguished among
the Perameloidea.

Generic and specific groupings

It is apparent from the taxonomic literature that marsupials, like most mammalian
groups, have undergone periods of lumping and splitting, reflecting both fashion
and intensity of study. The present consensus as to the genera strikes us as fairly

well balanced, and no major alterations will be found in our list. A surprisingly large number of new species have been described recently, however, either as a result of revision of problematical groups (e.g. the Grey kangaroos and certain dasyurids), or outright discovery of unknown animals (e.g. several dasyurids and didelphids). Subspecies of Australian marsupials are virtually meaningless and in our opinion (see above) even less real in the case of New Guinean and American marsupials. We do not list subspecies here.

The strategy of the list

The most recent standard listings of marsupial species are those of Ride (1970: the Australian mainland and associated continental islands), Laurie and Hill (1954: New Guinea and Sulawesi) and Cabrera (1957: South America). The degrees of certainty with which one can accept the names and concepts are unequal: we give most assurance to Ride's list. As noted above, many South American and New Guinean marsupial species have been collected from few, often widely separated, localities. Subspecies and species therefore proliferate, and the taxonomy of many of these forms is in disarray. What we have done in the following list is to work from the three main references and ancillary publications, making such changes as are necessary consequent on later work, new discoveries and partial revisions — sometimes including our own opinions. In notes for each family listing we comment on the discrepancies, noting the groups presently being worked on (and by whom) and indicating which genera are most in need of revision.

Marsupials from the New Guinea region are indicated by an asterisk; species occurring also in Australia are given a double asterisk. We prefer Ride's names for genera common to Australia and New Guinea where these differ from names used by Laurie and Hill. *Dasyurus, Isoodon, Cercartetus* and *Macropus* are thus used in place of *Satanellus, Thylacis, Eudromicia* and *Protemnodon,* respectively.

The species are arranged alphabetically within genera, which in turn are ordered according to what we imagine to be their phylogenetic relationships, the most primitive genera being placed first.

Species of living marsupials

Family Didelphidae: opossums (70 species)

The earliest workers on opossums placed virtually all members of this family in the genus *Didelphis* (along with some of the Australasian marsupials as well!). Even with the narrowing of generic concepts later in the nineteenth century, the generic nomenclature of opossums remained unstable. This is particularly true of the allocation of *Philander*, which has been misapplied to the woolly opossums (*Caluromys*) and is now generally used for the Grey and the Black four-eyed opossums. Recently Pine (1973) argued that *Philander* was meant to apply to the Brown four-eyed opossum. *Metachirus nudicaudatus,* but Hershkovitz (1976) maintains that Pine was in error; we retain the commoner usage here. Thus, *Philander opossum* is used for the Grey four-eyed opossum; *Philander mcillhenyi* is a recently described species (Gardner and Patton, 1972) which differs from the more common species in pelage, which is darker in *P. mcillhenyi,* and in details of dentition and the cranium.

Table 2.2 A list of the species of living marsupials, arranged by families

1. FAMILY DIDELPHIDAE: Opossums
    (70 species)

Marmosa (Marmosa) alstoni (J. A. Allen, 1900)
 canescens (J. A. Allen, 1893)
 cinerea (Temminck, 1824)
 constantiae Thomas, 1904
 domina Thomas, 1920
 emiliae Thomas, 1909
 fuscata Thomas, 1896
 germana Thomas, 1904
 impavida (Tschudi, 1844)
 incana (Lund, 1841)
 invicta Goldman, 1912
 juninensis Tate, 1931
 lepida (Thomas, 1888)
 leucastra Thomas, 1927
 mapiriensis Tate, 1931
 mexicana Merriam, 1897
 murina (Linné, 1758)
 noctivaga (Tschudi, 1844)
 ocellata Tate, 1931
 parvidens Tate, 1931
 phaea Thomas, 1899
 quichua Thomas, 1899
 rapposa Thomas, 1899
 regina Thomas, 1898
 robinsoni Bangs, 1898
 rubra Tate, 1931
 scapulata (Burmeister, 1856)
 tyleriana Tate, 1931
 yungasensis Tate, 1931

Marmosa (Stegomarmosa) andersoni Pine, 1972
Marmosa (Thylamys) aceramarcae Tate, 1931
 agilis (Burmeister, 1854)
 agricolai Moojen, 1943
 dryas Thomas, 1898
 elegans (Waterhouse, 1839)
 formosa Shamel, 1930
 grisea (Desmarest, 1827)
 karimii Petter, 1968
 marica Thomas, 1898
 microtarsus (Wagner, 1842)
 pusilla (Desmarest, 1804)
 tatei Handley, 1956
 unduaviensis Tate, 1931
 velutina (Wagner, 1842)
Monodelphis adusta (Thomas, 1897)
 americana (Müller, 1776)
 brevicaudata (Erxleben, 1777)
 dimidiata (Wagner, 1847)
 domestica (Wagner, 1842)
 henseli (Thomas, 1888)
 orinoci (Thomas, 1899)
 scalops (Thomas, 1888)
 sorex (Hensel, 1872)
 touan (Shaw, 1800)
 unistriata (Wagner, 1842)
Lestodelphys halli (Thomas, 1921)
Metachirus nudicaudatus (Geoffroy, 1803)
Didelphis albiventris Lund, 1841
 marsupialis Linné, 1758
 virginiana Kerr, 1792
Philander mcillhenyi Gardener and Patton, 1972
 opossum (Linné, 1758)
Lutreolina crassicaudata (Desmarest, 1804)
Chironectes minimus (Zimmermann, 1780)

Table 2.2 (*continued*)

Caluromys derbianus (Waterhouse,
    1841)
    lanatus (Illiger, 1815)
    philander (Linné, 1758)
Caluromysiops irrupta Sanborn, 1951
Glironia criniger Anthony, 1926
    venusta Thomas, 1912

2. FAMILY MICROBIOTHERIIDAE:
    Monito del monte (1 species)

    Dromiciops australis (Philippi, 1894)

3. FAMILY CAENOLESTIDAE: Shrew-
    opossums (7 species)

    Caenolestes caniventer Anthony, 1921
        convelatus Anthony, 1924
        fuliginosus (Tomes, 1863)
        obscurus Thomas, 1895
        tatei Anthony, 1923
    Lestoros inca (Thomas, 1917)
    Rhyncholestes raphanurus Osgood,
        1924

4. FAMILY DASYURIDAE: Native cats,
    marsupial mice, Tasmanian devil (49
    species)

    Murexia longicaudata (Schlegel, 1866)*
        rothschildi (Tate, 1938)*
    Neophascogale lorentzii (Jentink,
        1911)*
    Phascolosorex doriae (Thomas, 1886)*
        dorsalis (Peters and
            Doria, 1876)*
    Myoictis melas (Müller, 1840)*
    Planigale gilesi Aitken, 1972
        ingrami (Thomas, 1906)
        maculatus (Gould, 1851)
        novaeguineae Tate and
            Archbold,
            1941*
        tenuirostris Troughton,
            1928
    Antechinus apicalis (Gray, 1842)
        bellus (Thomas, 1904)
        bilarni Johnson, 1954
        flavipes (Waterhouse,
            1838)
        godmani (Thomas, 1923)
        macdonnellensis (Spencer,
            1896)

    melanurus (Thomas,
        1899)*
    minimus (Geoffroy,
        1803)
    naso (Jentink, 1911)*
    rosamondae Ride, 1964
    stuartii Macleay, 1841
    swainsonii (Waterhouse,
        1840)
    wilhelmina Tate, 1947*
Phascogale calura Gould, 1844
    tapoatafa (Meyer, 1793)
Dasycercus cristicauda (Krefft, 1867)
Dasyurus albopunctatus Schlegel,
        1880*
    geoffroii Gould, 1841
    hallucatus Gould, 1842
    maculatus (Kerr, 1792)
    viverrinus (Shaw, 1800)
Sarcophilus harrisii (Boitard, 1841)
Dasyuroides byrnei Spencer, 1896
Ningaui ridei Archer, 1975
    timealeyi Archer, 1975
Sminthopsis butleri Archer, 1977
    crassicaudata (Gould,
        1844)
    granulipes Troughton,
        1932
    hirtipes Thomas, 1898
    leucopus (Gray, 1842)
    longicaudata Spencer,
        1909
    macroura (Gould, 1845)
    murina (Waterhouse,
        1838)
    ooldea Troughton, 1965
    psammophila Spencer, 1895
    douglasi Archer, 1977
    virginiae De Tarragon,
        1847**
Antechinomys laniger (Gould, 1856)

5. FAMILY MYRMECOBIIDAE: Numbat
    (1 species)
    Myrmecobius fasciatus Waterhouse, 1836

6. FAMILY THYLACINIDAE: Tasmanian
    tiger or Thylacine (1 species)

    Thylacinus cynocephalus (Harris, 1808)

7. FAMILY NOTORYCTIDAE: Marsupial
    mole (1 species)

    Notoryctes typhlops (Stirling, 1889)

(continues)

Table 2.2 (*continued*)

8.   FAMILY PERAMELIDAE: Bandicoots
     (16 species)

     Peroryctes longicauda (Peters and
                         Doria, 1876)*
          papuensis Laurie, 1952*
          raffrayanus (Milne-Edwards,
                         1878)*
     Microperoryctes murina Stein, 1932*
     Perameles bougainville Quoy and
                         Gaimard, 1824
          eremiana Spencer, 1897
          gunnii Gray, 1838
          nasuta Geoffroy, 1804
     Echymipera clara Stein, 1932*
          kalubu (Lesson, 1828)*
          rufescens (Peters and
                         Doria, 1875)**
     Rhynchomeles prattorum Thomas,
                         1920*
     Isoodon auratus (Ramsay, 1887)
          macrourus (Gould, 1842)**
          obesulus (Shaw, 1797)
     Chaeropus ecaudatus (Ogilby, 1838)

9.   FAMILY THYLACOMYIDAE: Rabbit-
     eared bandicoots (2 species)

     Macrotis lagotis (Reid, 1837)
          leucura (Thomas, 1887)

10. FAMILY PHALANGERIDAE: Phalangers
     (11 species)

     Trichosurus arnhemensis Collett, 1897
          caninus Ogilby, 1836
          vulpecula (Kerr, 1792)
     Wyulda squamicaudata Alexander,
                         1919
     Phalanger atrimaculatus Tate, 1945*
          celebensis (Gray, 1858)*
          gymnotis (Peters and Doria,
                         1875)*
          maculatus (Desmarest,
                         1818)**
          orientalis (Pallas, 1766)**
          ursinus (Temminck, 1824)*
          vestitus (Milne-Edwards,
                         1877)*

11. FAMILY BURRAMYIDAE: Pygmy
     phalangers (7 species)

     Cercartetus caudatus (Milne-Edwards,
                         1877)**

          concinnus (Gould, 1845)
          lepidus (Thomas, 1888)
          nanus (Desmarest, 1818)
     Distoechurus pennatus (Peters, 1874)*
     Acrobates pygmaeus (Shaw, 1793)
     Burramys parvus Broom, 1895

12. FAMILY PETAURIDAE: Ringtails,
     gliding phalangers (22 species)

     Gymnobelideus leadbeateri McCoy,
                         1867
     Petaurus australis Shaw, 1791
          breviceps Waterhouse, 1839**
          norfolcensis (Kerr, 1792)
     Pseudocheirus
     (Pseudocheirus) canescens (Waterhouse,
                         1847)*
          caroli Thomas, 1921*
          forbesi Thomas, 1887*
          herbertensis (Collett,
                         1884)
          mayeri Rothschild and
                         Dollman, 1932*
          peregrinus (Boddaert,
                         1785)
          schlegeli Jentink, 1884*
     Pseudocheirus
     (Pseudochirops) albertisii (Peters,
                         1874*)
          archeri (Collett, 1884)
          corinnae Thomas,
                         1897*
          cupreus Thomas,
                         1897*
          dahli Collett, 1895
     Pseudocheirus
     (Hemibelideus) lemuroides (Collett,
                         1884)
     Schoinobates volans (Kerr, 1792)
     Dactylopsila
     (Dactylopsila) megalura Rothschild
                         and Dollman,
                         1932*
          tatei Laurie, 1952*
          trivirgata Gray, 1858**
     Dactylopsila
     (Dactylonax) palpator Milne- Edwards,
                         1888*

Table 2.2 *(continued)*

13. FAMILY MACROPODIDAE: Kangaroos
     (56 species)

Hypsiprymnodon moschatus Ramsay,
          1876
Potorous platyops (Gould, 1844)
          tridactylus (Kerr, 1792)
Bettongia gaimardi (Desmarest, 1822)
          lesueur (Quoy and Gaimard,
          1824)
          penicillata Gray, 1837
          tropica Wakefield, 1967
Aepyprymnus rufescens (Gray, 1837)
Caloprymnus campestris (Gould, 1843)
Thylogale billardierii (Desmarest, 1822)
          brunii (Schreber, 1778)*
          stigmatica Gould, 1860**
          thetis (Lesson, 1827)
Petrogale brachyotis Gould, 1841
          godmani Thomas, 1923
          penicillata (Griffith, 1827)
          purpureicollis Le Souef,
          1924
          rothschildi Thomas, 1904
          xanthopus Gray, 1855
Peradorcas concinna (Gould, 1842)
Lagorchestes asomatus Finlayson,
          1943
          conspicillatus Gould,
          1842
          hirsutus (Gould, 1844)
          leporides (Gould, 1841)
Setonix brachyurus (Quoy and
          Gaimard, 1830)
Lagostrophus fasciatus (Péron, 1807)
Macropus agilis (Gould, 1842)**
          antilopinus (Gould, 1842)
          bernardus (Rothschild,
          1904)
          dorsalis (Gray, 1837)
          eugenii (Desmarest, 1817)
          fuliginosus (Desmarest,
          1817)
          giganteus Shaw, 1790
          greyi (Waterhouse, 1846)
          irma (Jourdan, 1837)
          parma Waterhouse, 1846
          parryi (Bennett, 1835)
          robustus Gould, 1841
          rufogriseus (Desmarest,
          1817)
          rufus (Desmarest, 1822)
Wallabia bicolor (Desmarest, 1804)

Onychogalea fraenata (Gould, 1841)
          lunata (Gould, 1840)
          unguifera (Gould, 1841)
Dendrolagus bennettianus De Vis,
          1887
          dorianus Ramsay, 1883*
          goodfellowi Thomas,
          1908*
          inustus Müller, 1840*
          lumholtzi Collett, 1844
          matschiei Förster and
          Rothschild,
          1907*
          ursinus Müller, 1840*
Dorcopsis atrata Van Deusen, 1957*
          hageni Heller, 1897*
          veterum (Lesson, 1827)*
Dorcopsulus macleayi (Miklouho-
          Maclay, 1885)*
          vanheurni (Thomas,
          1922)*

14. FAMILY PHASCOLARCTIDAE: Koala
     (1 species)

Phascolarctos cinereus (Goldfuss, 1817)

15. FAMILY VOMBATIDAE: Wombats
     (3 species)

Vombatus ursinus (Shaw, 1800)
Lasiorhinus krefftii (Owen, 1872)
          latifrons (Owen, 1845)

16. FAMILY TARSIPEDIDAE: Honey
     possum (1 species)

Tarsipes spencerae Gray, 1842

The extremely polytypic *Marmosa* has sometimes been considered more than one genus, and acquaintance with even a few living species certainly confirms this judgement; Pine (1972) suggests a formal splitting into several genera. In this light, the great number of species presently placed in *Marmosa* does not seem so extreme. Cabrera (1957) lists 37 species of South American *Marmosa*, a reduction to 36 from Tate's 47 South American species (Tate, 1933), with the addition of *M. (Thylamys) agricolai,* which Cabrera states was originally described as part of a subspecies of *M. (Thylamys) agilis.* Cabrera's sub-generic groupings employ Gray's (1821, 1843) generic names *Marmosa* and *Thylamys*, a procedure followed in Cabrera's earlier work (1919) and later by Gilmore (1941).

Species we have added to Cabrera's list include *M. tatei,* which was described by Handley (1956) just prior to publication of Cabrera's catalogue, and is a member of the subgenus *Thylamys*; *M. (Thylamys) karimii,* described by Petter (1968); M. *(Stegomarmosa) andersoni,* the new subgenus and species described by Pine (1972); and the three exclusively North American species, *M. (Marmosa) canescens, M. (Marmosa) invicta* and *M (Marmosa) mexicana* (see Hall and Kelson, 1959). Finally, we include *M. (Thylamys) formosa,* synonymised by Cabrera with *M. (Thylamys) velutina* but given as a full species in Pine's provisional list of the species of *Marmosa* (Collins, 1973). That list, which inadvertently omits *M. (Stegomarmosa) andersoni,* differs slightly from the one printed here in that it does not include *M. (Marmosa) alstoni* and *M. (Marmosa) juninensis,* which Pine regarded as synonyms of *M. (Marmosa) cinerea* and *M. (Marmosa) parvidens,* respectively. Pine (personal communication) now believes that both *M. (Marmosa) juninensis* and *M. (Marmosa) parvidens* are synonyms of *M. (Thylamys) emiliae,* which he removes to the subgenus *Marmosa* (where we have also placed it) from *Thylamys*. Our list is thus longer by two species than Pine's.

Our listings for other polytypic genera also differ from those of Cabrera. *Monodelphis* has received even less attention than *Marmosa* but Reig *et al.* (in preparation) recognise *M. orinoci* as a distinct species: it is a llanos (open plains) animal as opposed to *M. brevicaudata,* the darker rain forest species with which it is usually synonymised. On the other hand, Reig (1964) showed that *M. fosteri* specimens were simply juveniles of *M. dimidiata*, and *M. fosteri* is accordingly omitted. Similarly, we follow Handley (1966) in considering the Central American *M. melanops* (Goldman, 1912) synonymous with *M. adusta.* Pine (personal communication) is at present describing a new species of *Monodelphis,* and he believes that several species synonymised by Cabrera should be recognised. With Cabrera's assignments indicated in parentheses, these are:

Monodelphis iheringi (Thomas, 1888) (Cabrera: in *M. americana*)
           maraxina Thomas, 1923 (in *M. domestica*)
           osgoodi Doutt, 1938 (in *M. adusta*)
           theresa Thomas, 1921 (in *M. americana*)
           umbristriata (Miranda Ribeiro, 1936) (in *M. americana*)

Thus there may be six or more full species additional to those we list. We do not use the subgenera *Monodelphis* and *Minuania*.

With regard to *Didelphis* itself, Hershkovitz (1969) indicates that the correct name for the White-eared opossum is *D. albiventris* (not *D. azarae*); and Gardner (1973) presents a convincing case that *D. marsupialis* and *D. virginiana* are distinct species which overlap but do not interbreed in Middle America.

The last three genera listed are sometimes grouped in a separate subfamily with *Dromiciops* following Reig's paper of 1955, in which he points out their similarities to the extinct Microbiotheriidae. It now appears that *Dromiciops* only is a living microbiothere, and this genus alone is listed under Microbiotheriidae.

Finally, we are inclined, as was Cabrera, to regard *Glironia criniger* as a colour variant of *G. venusta,* but it is listed here pending supplementation of the few available specimens.

In summary, the opossums are, despite their central importance as the animals most nearly representative of the earliest marsupials, perhaps the ones least well understood taxonomically. This list can only be regarded as very tentative, pending further work by Pine and others who are currently interested in opossum classification.

Family Microbiotheriidae: Monito del monte (1 species)

The evidence from serology (Kirsch, 1976) is that *Dromiciops* is very distinct from all other opossums, and Reig (personal communication) has independently concluded that the dental similarities between *Dromiciops* and the other putative microbiotheres are convergent. Although both island and mainland species of *Dromiciops* have been described, we concur with Cabrera that they are synonymous.

Family Caenolestidae: shrew-opossums (7 species)

Cabrera's arrangement of the shrew-opossums is followed exactly here, but the three genera are very similar and probably some of the species of *Caenolestes* should be combined: *C. tatei* and *C. caniventer* are both found at somewhat lower elevations than the others, and, like *C. convelatus,* are known only from a single locality each. The type species, *C. fuliginosus,* and *C. obscurus,* are found disjunctively throughout the Andes from Venezuela to Peru (*C. obscurus* north of Ecuador and *C. fuliginosus* south) and are almost certainly conspecific: only the Ecuadorian border separates them.

Family Dasyuridae: native cats, marsupial mice, Tasmanian devil (49 species)

The chief areas of uncertainty in Dasyuridae remain the species, and even genera, of the smaller forms, including most of those other than species of *Dasyurus, Sarcophilus, Phascogale* and *Antechinomys.* The New Guinean species are particularly poorly known, and it is apparent from continuing studies of Archer that there are unknown species — and even genera — in the Australian fauna as well.

These comments notwithstanding, Archer's work (1975, 1977) has done much to clarify the species and their relationships in *Sminthopsis, Antechinomys* and Archer's new genus, *Ningaui.* With Ride (1970) we abandon the several additional genera Tate (1947) used for some species of *Antechinus;* but we reinstate A.

bilarni, not listed by Ride, which in the opinion of one of us (J.H.C.) is distinct from *A. macdonnellensis.* We also include the recently described *Planigale gilesi* of Aitken (1972). Technically, *Sminthopsis butleri* and *S. douglasi* must at present be considered *nomina nuda.*

Family Myrmecobiidae: Numbat (1 species)

As noted above, we accept the argument of Archer and Kirsch (in press) that recognition of the Numbat at this level is more consistent with contemporary usage of the family category in Marsupialia.

Family Thylacinidae: Tasmanian tiger, or Thylacine (1 species)

The Tasmanian tiger's undeniable similarities to some of the extinct South American carnivores led Sinclair (1906) to place these and *Thylacinus* in the same family. Whether this is reasonable (Simpson, 1941, and later, maintains not), we follow Ride (1964, 1970) in placing the Thylacine in a distinct family.

Family Notoryctidae: Marsupial mole (1 species)

The relationships of this animal remain mysterious, in spite of cytological studies which show that the chromosomes have a strong resemblance to those of some phalangeroids (Calaby *et al.* 1974), and serological tests which hint at dasyuroid-perameloid affinities (Kirsch, in prep.). There is a single, variable species.

Family Peramelidae: bandicoots (16 species)

The New Guinean bandicoots are especially in need of revision, as are the species of *Isoodon. Chaeropus* has not been seen for about 70 years.

Family Thylacomyidae: Rabbit-eared bandicoots (2 species)

These distinctive bandicoots are elevated to a full family by Archer and Kirsch (in press). The family name derives from a preoccupied generic name, *Thylacomys,* which Bensley (1903) used to form a subfamily name. By the rules of the International Code of Zoological Nomenclature, Thylacomyinae must continue to be used as the basis of the family name.

Family Phalangeridae: phalangers (11 species)

Systematics of this family are fairly secure: there are strong ecological as well as morphological differences among the species of *Trichosurus,* while *Wyulda* is known only from a restricted area of north-west Australia and is certainly mono-typic. *Phalanger,* on the other hand, is a mostly New Guinean genus that is poorly understood. Graeme G. George in Papua New Guinea is currently working on *Phalanger,* and considers that more species should be recognised than those listed.

Family Burramyidae: pigmy phalangers (7 species)

Based on general considerations and the morphology of the chromosomes, the pigmy phalangers may represent the most primitive of the Australasian diprotodont marsupials. The New Guinean *Distoechurus* has not been examined serologically or cytologically, but belongs in the family as judged by morphological features.

Cercartetus is a much broader genus than the others (Turnbull and Schram, 1973), which are all monotypic: *Acrobates pulchellus* Rothschild, 1892 is dropped from this list for reasons given by Van Deusen (1960). However, Calaby, Dimpel and Cowan (1971) pointed out that the type specimen of *Burramys parvus* is a Pleistocene fossil, and since the discovery of *Burramys* as a living animal (anon., 1966) no critical comparison of the type material of *B. parvus*, fossil specimens from other localities and modern specimens had been made. This remark still holds and the specific name given above to living populations remains an assumption.

Family Petauridae: ringtails, gliding phalangers (22 species)

Dactylopsila (including *Dactylonax* as a subgenus) is tentatively placed in this family of marsupials, some of which (*Gymnobelideus, Petaurus*) it resembles, especially in pelage. *Petaurus* contains only three species, the named New Guinean and north-west Australian forms being very similar to *P. breviceps*. The most difficult problems concern the ringtails, here all considered members of a single genus, *Pseudocheirus*, with three subgenera. *P. (Pseudochirops) dahli* is often considered to represent a fourth subgenus, or even a full genus, *Petropseudes*, but we are of the opinion that it is not that different. Again, the New Guinean species are the most uncertain.

Family Macropodidae: kangaroos (56 species)

The first five genera are conventionally referred to a separate subfamily of 'rat-kangaroos', although the peculiarities of *Hypsiprymnodon moschatus* (a simple stomach, retention of the hallux and especially pronounced plagiaulacoid premolars) argue for a third subfamily for this species. The only departure from Ride's work in our listing for the other four genera is the elimination of the Tasmanian and Victorian *Potorous apicalis* (Gould, 1851). Johnston (1973) has shown that the characters by which this 'species' was separated from *P. tridactylus* were insufficient and inconsistent with information from isozymes, karyotypes, morphometric analyses and cross-breeding experiments. *P. platyops*, the other species recognised, has not been collected for a century, and *Caloprymnus* was last recorded in 1935.

Ride (1962) remarked that the species of true kangaroos (the remaining 13 genera) are fairly certain but the genera are not. Nevertheless, most of the genera are now reasonably clear-cut. *Peradorcas, Setonix* and *Lagostrophus* are monotypic but sufficiently distinct to warrant that status. The tree-kangaroos *Dendrolagus* of New Guinea and north-eastern Australia are in need of some revision, as are the New Guinean forest wallabies *Dorcopsis* and *Dorcopsulus*. The latter two genera do not differ so much from each other that they could not be accommodated reasonably comfortably in a single genus that would take the name *Dorcopsis*. *Dorcopsulus vanheurni* is probably conspecific with *D. macleayi*. Use of the name *Dorcopsis veterum* in place of *D. muelleri* (Schlegel, 1866) and the spelling of the specific name of *Thylogale brunii* follow Husson (1955). Troughton and Le Souef's (1936) two species *Dendrolagus deltae* and *D. spadix* are dropped from the list; they should be synonymised with *D. matschiei* and *D. goodfellowi*, respectively. *D. spadix* is based on a shrunken badly prepared adult skin (without skull) and a juvenile skin and skull. In Calaby's personal copy of Troughton and

Le Souef's paper there is a manuscript note in Troughton's handwriting, that he
regarded *D. deltae* as 'doubtful' as there was an 'error of locality provided by Le
Souef from zoo specimens'.

Petrogale is perhaps the most perplexing of all the Australian kangaroos, with
its abundance of island and mainland nominal species and subspecies. In the few
populations sampled cytologically chromosome numbers range from 18 to 22.
Three tropical forms have been examined to date; all have 18 chromosomes, but
there are considerable differences in chromosome morphology (Hayman and
Martin, 1974; Sharman and Calaby, unpublished data). There is no doubt that
more species will eventually be recognised than those listed. The species of
Lagorchestes are also chromosomally diverse. *L. asomatus* is known from a single
unsexed skull, and *L. leporides* has not been seen for more than a century. *Onychogalea lunata* was last collected in 1964, and *O. fraenata* is known from a single
remnant population.

The 14 species here placed in *Macropus* and the Swamp wallaby (*Wallabia
bicolor*), the typical kangaroos and large wallabies and two smaller wallabies, are
a taxonomically difficult group with inconsistent similarities and distinctions
between various species or species-groups. Calaby (1966) recommended that the
species with 16 chromosomes be referred to *Macropus* until such time as a clear
picture of their relationships emerges, and this convention has been followed by
most authors. He suggested also that the cytologically unique *W. bicolor* be placed
in a monotypic genus; it differs in its breeding biology and behaviour from other
macropodids (Sharman, Calaby and Poole, 1966); and its massive permanent
premolar and certain features of the molars, such as low links and relatively low
lophs, are reminiscent of the Pliocene and Pleistocene *Protemnodon* spp. We agree
that its monotypic status should be maintained.

The Red kangaroo, *Macropus rufus* ($2n = 20$), has been placed in the monotypic genus *Megaleia* by virtually all authors for many years. Recent studies
(unpublished) by Sharman and Calaby on chromosome morphology of the Red
kangaroo and the wallaroo-euro group (*M. antilopinus, M. bernardus, M. robustus*)
show that *M. bernardus* is chromosomally distinct from the remaining wallaroos
and euros and intermediate between these and the Red kangaroo. It appears that
the species within this group, including *M. rufus,* are closer to each other than
any is to other members of *Macropus s. l.* It seems logical to abandon *Megaleia*
and place the Red kangaroo in *Macropus.* The serological reationships of
Macropus species are also consistent with this arrangement (Kirsch, 1968). If
this grouping were to be recognised nomenclaturally, perhaps at subgeneric level,
the name *Osphranter* is available. It is noted that all of these species (and only
these species) have been placed in *Osphranter* by one authority or another in
the past.

Most species of *Macropus* and *W. bicolor* are common animals in their respective habitats. *M. parma* was believed to be very rare or even extinct for many
years but is now known to occur in a considerable area of forest in north-eastern
New South Wales. The last record of the Toolache wallaby, *M. greyi,* was in 1927.

Family Phascolarctidae: Koala (1 species)

The family is certainly monotypic, and probably closest to Vombatidae; with the
wombats, it forms a distinct group of Australian diprotodonts.

Family Vombatidae: wombats (3 species)

The nomenclature of wombats has an arbitrary element and the number of species recognised by Ride is a distinct reduction from earlier authors (Iredale and Troughton, 1934). Ride followed Crowcroft (1967) in recognising three species of *Lasiorhinus* – *barnardi* Longman, 1939, *gillespiei* (De Vis, 1900) and *latifrons*. Material available in collections of *L. barnardi* and *L. gillespiei* is extremely limited. The taxa are morphologically close but differ markedly in size, although this may be an expression, at least in part, of age differences in the available specimens. Their status is a matter of subjective judgement and we prefer to lump them. Wilkinson (in Merrilees, 1973) has shown that *L. krefftii* is the correct name for *L. gillespiei*. *L. latifrons* is still common but only a single remnant population of *L. krefftii* remains, at the type locality of *L. barnardi*.

Family Tarsipedidae: Honey possum (1 species)

The single species is the most divergent, serologically as well as morphologically, of all the living Australasian diprotodonts; but its chromosomes ($2n = 24$) are at least similar to those of *Trichosurus*. The species is, and was, limited to south-west Australia.

Acknowledgements

Many of our co-workers in marsupial taxonomy read parts or all of this paper and contributed ideas and often new information. We are especially grateful to Michael Archer and Ronald Pine for their criticisms of the sections on Dasyuridae and Didelphidae, respectively, and for permitting us to quote from their unpublished work. Similarly, we thank Des Cooper, Graeme George, Peter Johnston, Geoffrey Sharman and Osvaldo Reig for their information and advice, and Pamela Parker for critical reading of the manuscript. This paper was written while Kirsch held a Yale College Junior Faculty Fellowship, 1974 - 1975.

References

Abbie, A. A. (1937). Some observations on the major subdivisions of the Marsupialia with especial reference to the position of the Peramelidae and Caenolestidae. *J. Anat.*, **71**, 429 - 36.

Abbie, A. A. (1941). Marsupials and the evolution of mammals. *Aust. J. Sci*, **44**, 77 - 92.

Aitken, P. F. (1972). *Planigale gilesi* (Marsupialia: Dasyuridae): a new species from the interior of eastern Australia. *Rec. S. Aust. Mus. (Adelaide)*, 16 1 - 14.

Anon. (1966). A relict marsupial. *Nature*, **212**, 225.

Archer, M. (1975). *Ningaui*, a new genus of tiny dasyurids (Marsupialia) and two new species, *N. timealeyi* and *N. ridei*, from arid Western Australia. *Mem. Qd. Mus.*, **17** (2), 237 - 49.

Archer, M. (1977). Revision of the marsupial dasyurid genus *Sminthopsis* Thomas, a description of two new species from northern Australia, *S. butleri* and *S. douglasi*, and a consideration of arid-adapted characters. *Bull. Am. Mus. Nat. Hist.* (in press).

Archer, M. A. and Kirsch, J. A. W. (1976). The case for Thylacomyidae and
    Myrmecobiidae, or why are marsupial families so extended? *J. Linn. Soc.
    N. S. W.* (in press).
Bensley, B. A. (1903). On the evolution of the Australian Marsupialia; with remarks
    on the relationships of the marsupials in general. *Trans. Linn. Soc. Lond.
    Zool.,* (2) **9**, 83 - 217.
Cabrera, A.(1919), *Genera Mammalium. Monotremata, Marsupialia,* Museo
    Nacional de Ciencias Naturales, Madrid.
Cabrera, A. (1957). *Catalogo de los Mamiferos de America del Sur,* Museo
    Argentino de Ciencias Naturales 'Bernardino Rivadavia,' Buenos Aires.
Calaby, J. H. (1966). Mammals of the upper Richmond and Clarence Rivers, New
    South Wales. *C.S.I.R.O. Division of Wildlife Research Technical Paper No.
    10,* 1 - 55.
Calaby, J. H., Corbett, L. K., Sharman, G. B. and Johnston, P. G. (1974). The
    chromosomes and systematic position of the Marsupial Mole, *Notoryctes
    typhlops. Aust. J. Biol. Sci.,* **27**, 529 - 32.
Calaby, J. H., Dimpel, H. and Cowan, I. McT.(1971). The mountain pigmy possum,
    Burramys parvus Broom (Marsupialia) in the Kosciusko National Park, New
    South Wales. *C.S.I.R.O. Division of Wildlife Research Technical Paper No.
    23,* 1 - 11.
Collins, L. R. (1973). *Monotremes and Marsupials: a Reference for Zoological
    Institutions.* Smithsonian Institution Press, Washington.
Crowcroft, P. (1967). Studies on the hairy-nosed wombat *Lasiorhinus latifrons*
    (Owen 1845) I. Measurements and taxonomy. *Rec. S. Aust. Mus.,* **15**,
    383 - 98.
Gardner, A. L. (1973). The systematics of the genus *Didelphis* (Marsupialia:
    Didelphidae) in North and Middle America. *Special Publications The
    Museum Texas Tech University No. 4,* 1 - 81.
Gardner, A. L. and Patton, J. L. (1972). New species of *Philander* (Marsupialia:
    Didelphidae) and *Mimon* (Chiroptera: Phyllostomidae) from Peru. *Occ.
    Papers Mus. Zool. Louisiana State Univ. No. 43,* 1 - 12.
Gilmore, R. M. (1941). Zoology: pp. 314 - 19 in Bugher, J. C. *et al.* The
    susceptibility to yellow fever of the vertebrates of eastern Colombia. *Am.
    J. Trop. Med.,* **21** (2).
Gray, J. E. (1821). On the natural arrangement of vertebrose animals. *London
    Med. Repository,* **15**, 296 - 310.
Gray, J. E. (1843). *List of the Specimens of Mammalia in the Collection of the
    British Museum,* British Museum, London,
Hall. E. R. and Kelson, K. R. (1959). *The Mammals of North America,* vol. I,
    Ronald Press, New York.
Handley, C. O., Jr. (1956). A new species of murine opossum (genus *Marmosa*)
    from Peru. *J. Wash. Acad. Sci.,* **46**, 402 - 404.
Handley, C. O., Jr. (1966). Checklist of the mammals of Panama, in *Ectoparasites
    of Panama* (ed. R. L. Wenzel and V. J. Tipton), Field Museum of Natural
    History, Chicago, pp. 753 - 95.
Hayman, D. L., Kirsch, J. A. W., Martin, P. G. and Waller, P. F. (1971). Chromo-
    somal and serological studies of the Caenolestidae and their implications
    for marsupial evolution. *Nature,* **231**, 194 - 195.

Hayman, D. L. and Martin, P. G. (1974). *Animal Cytogenetics,* vol. 4: *Chordata 4 Mammalia I: Monotremata and Marsupialia,* Gebrüder Borntraeger, Berlin - Stuttgart.

Hershkovitz, P. (1969). The evolution of mammals on southern continents. VI. The Recent mammals of the Neotropical Region: a zoogeographic and ecological review. *Quart. Rev. Biol.,* **44**, 1 - 70.

Hershkovitz, P. (1976). Comments on generic uaiues of four-eyed opossums. *Proc. Biol. Soc. Wash.* (in press).

Husson, A. M. (1955). Notes on the mammals collected by the Swedish New Guinea Expedition 1948 - 1949. *Nova Guinea,* n. s. **6**, 283 - 306.

Iredale, T. and Troughton, E. Le G. (1934). *A Check-list of the Mammals Recorded from Australia,* The Australian Museum, Sydney.

Johnston, P. G. (1973). 'Variation in island and mainland populations of *Potorous tridactylus* and *Macropus rufogriseus* (Marsupialia)', unpublished Ph. D. thesis, University of New South Wales, Sydney.

Jones, F. W. (1923). *The Mammals of South Australia,* part I, Government Printer, Adelaide.

Kirsch, J. A. W. (1968). Prodromus of the comparative serology of Marsupialia. *Nature,* **217**, 418 - 20.

Kirsch, J. A. W. (1976). The classification of marsupials with special reference to karyotypes and serum proteins, in *The Biology of Marsupials* (ed. D. Hunsaker), Academic Press, New York.

Laurie, E. M. O. and Hill, J. E. (1954). *List of Land Mammals of New Guinea, Celebes and Adjacent Islands 1758 - 1952,* British Museum (Natural History), London.

Merrilees, D. (1973). Fossiliferous deposits at Lake Tandou, New South Wales, Australia. *Mem. Natl. Mus. Victoria,* **34**, 177 - 82.

Petter, F. (1968). Une sarigue nouvelle du norde-est du Bresil, *Marmosa karimii* sp. nov. (Marsupiaux, Didelphides). *Mammalia,* **32**, 313 - 17.

Pine, R. H. (1972). A new subgenus and species of murine opossum (genus *Marmosa*) from Peru. *J. Mamm.,* **53**, 279 - 82.

Pine, R. H. (1973). Anatomical and nomenclatural notes on opossums. *Proc. Biol. Soc. Wash.,* **86**, 391 - 402.

Reig, O. A. (1955). Noticia preliminar sobre la presencia de microbiotherinos vivientes en la fauna Sudamericana. *Investnes. Zool. Chil.,* **2**, 121 - 30.

Reig. O. A. (1964). Roedores y marsupiales del partido de General Pueyrredón y regiones adyacentes (Provincia de Buenos Aires, Argentina). *Publs. Museo de Ciencias Nat. de Mar del Plata,* **1**, 203 - 24.

Reig, O. A., Gardner, A. L., Patton, J. L. and Bianchi, N. O. The chromosomes of the Didelphidae and their systematic and evolutionary significance. (Manuscript in preparation).

Ride, W. D. L. (1962). On the use of generic names for kangaroos and wallabies (subfamily Macropodinae). *Aust. J. Sci.,* **24**, 367 - 72.

Ride, W. D. L. (1964). A review of Australian fossil marsupials, presidential address, 1963. *J. Proc. R. Soc. West. Aust.,* **47**, 97 - 131.

Ride, W. D. L. (1973). *A Guide to the Native Mammals of Australia,* Oxford University Press, Melbourne.

Sharman, G. B. (1974). Marsupial taxonomy and phylogeny. *Aust. Mammalogy,* 1, 137 – 54.

Sharman, G. B., Calaby, J. H. and Poole,. W. E. (1966). Patterns of reproduction in female diprotodont marsupials. *Symp. Zool. Soc. Lond. No. 15,* 205 – 32.

Simpson, G. G. (1930). Post-Mesozoic Marsupialia, in *Fossilium catalogus* 1: *Animalia,* pt. 47. W. Junk, Berlin, pp 1 - 87.

Simpson, G. G. (1941). The affinities of the Borhyaenidae. *Am. Mus. Novitates, No. 1118,* 1 -6.

Simpson, G. G. (1945). The principles of classification and a classification of mammals. *Bull. Am. Mus. Nat. Hist.,* 85, vi – xvi, 1 –350.

Simpson, G. G. (1970). The Argyrolagidae, extinct South American marsupials. *Bull. Mus. Comp. Zool., Harvard,* 139, 1 – 86.

Sinclair, W. J. (1906). Mammalia of the Santa Cruz beds. Marsupialia. *Rep. Princeton Univ. Exped. Patagonia,* 4, pt. 3, 333 –459.

Tate, G. H. H. (1933). A systematic revision of the marsupial genus *Marmosa. Bull. Am. Mus. Nat. Hist.,* 66, 1 – 250.

Tate, G. H. H. (1947). Results of the Archbold Expeditions. No. 56 On the anatomy and classification of the Dasyuridae (Marsupialia). *Bull. Am. Mus. Nat. Hist.,* 88, 97 – 155.

Troughton, E. LeG. (1967). *Furred Animals of Australia,* 9th edn, Angus and Robertson, Sydney.

Troughton, E. LeG. and Le Souef, A. S. (1936). Two new tree kangaroos from Papua, with notes on allied forms. *Aust. Zoologist,* 8, 193 – 7.

Turnbull, W. D. (1971). The Trinity therians: their bearing on evolution in marsupials and other therians, in *Dental Morphology and Evolution* (ed. A. A. Dahlberg), University Press, Chicago pp, 151 - 79.

Turnbull, W. D. and Schram, F. R. (1973). Broom Cave *Cercartetus:* with observations of pigmy possum dental morphology, variation, and taxonomy. *Rec. Aust. Mus. Sydney,* 28, 437 – 64.

Tyndale-Biscoe, H. (1973). *Life of Marsupials,* American Elsevier, New York.

Van Deusen, H. M. (1960). Notes on the marsupial feather-tailed glider of Australia. *J. Mamm.,* 41, 263 - 4.

Van Valen, L. (1971). Adaptive zones and the orders of mammals. *Evolution,* 25, 420 - 8.

Yadav, M. (1973). The presence of the cervical and thoracic thymus lobes in marsupials. *Aust. J. Zool.,* 21, 285 – 301.

3 *Chromosome number – constancy and variation*

D. L. Hayman*

Introduction

As with any large group of organisms, the cytogenetic features of the marsupials possess aspects of particular or specialised interest because of the nature and diversity of the group, and they also possess features of fundamental biological interest. This chapter discusses two aspects of chromosome number in marsupials, one relating to conservation and the other relating to mosaicism. In neither situation is the matter finalised, so that implications for further research are indicated.

Chromosome number and its origin – one view

The basic cytological information about marsupial chromosome numbers is very well known, and marsupials must be among the best-known groups of organisms in this regard. Opinions differ as to the number of species which should be recognised, but if one takes the figure of some 260 given by Kirsch and Calaby's paper in this volume, then it is possible to say that some 50 per cent of the species have been examined cytologically, some in considerable detail. In addition, the nuclear DNA contents of a large number of diverse species of Australian marsupials have been measured. This provides an added dimension to the formal chromosome studies. Given this intensive study, it is possible to make a number of generalisations about marsupial cytogenetics.

As discussed by Martin in Chapter 6, marsupials are found today in North and South America, Australia and New Guinea. It is believed that they evolved in what is now North America, and that they diverged from the Eutheria some 130 million years ago (Air *et al.*, 1971). As is shown by haemoglobin chain sequence differences, the present-day distribution represents an extremely old separation between the two groups, which have been isolated and evolving independently ever since (Stenzel, 1974). There are no comparable parallels to this disjunction in the Eutheria. A biological feature of importance when combined with this disjunction is the comparison between the American and Australian marsupials in relation to the presence of competing Eutherian species. The American species of marsupials have had to compete with Eutherian species. The Australian species have had no such experience on the Australian mainland – with the exception of rodents in the last 3 million years or so and Man and his dog in the last 50 thousand years. Consequently a considerable range of adaptation and diversity of types

*Dr D. L. Hayman is Reader in Genetics at the University of Adelaide.

27

The biology of marsupials

has been reached by marsupials in Australia in the absence of competition from
other mammals which may have limited the range of habitats open to American
marsupials.

The marsupials with their ancient disjunction between two major continents,
and their different histories of adaptation with and without competition from
Eutheria, present an attractive opportunity for comparative study of the role of
numerical and morphological changes in the chromosomes associated with
evolutionary divergence.

The role of chromosome number in evolution is really not well understood.
Clearly the chromosomes act as a mechanical conveyance at meiosis and mitosis for
the genetic material. However, there are at least two other features which chromosome number affects. Firstly, the number of chromosomes reflects the number of
independently assorting units of inheritance and thus has implications for the
release of variability; secondly, changes in number are always associated with
changes in structure and these may produce consequent changes in genetic
expression due to 'position effect' phenomena. That is, the physical organisation
of the chromosome is related to the expression of the genetic material the
chromosome carries. John and Lewis (1968) recognise this complex role by
giving the name 'endophenotype' to the chromosome number and morphology.

The chromosome numbers found in the marsupials are listed in Table 3.1. The
marsupials are organised into different taxonomic groupings known as superfamilies which reflect not only taxonomic affinities but also the present-day
distribution of marsupials. The superfamilies Caenolestoidea and Didelphoidea
are exclusively American, while the other six superfamilies are exclusively
Australian. Two of these are monospecific superfamilies and therefore do not
give as much information as do the others. The major features of chromosome
number can be identified as follows (see Figure 3.1). Firstly, the marsupials are
characterised by very low chromosome numbers. The lowest described is $2n = 10$
(Pseudocheirus cupreus) and the highest $2n = 32$ *(Aepyprymnus rufescens)*.
Secondly, the distribution of chromosome numbers is bimodal with one mode at
$2n = 14$ and the other $2n = 20$ or 22. Thirdly, this bimodality is reflected in
both the American and the Australian species.

So far as the range and frequency of chromosome number is concerned, there
is little difference between the two groups of marsupials. However chromosome
number is not the only basis on which the chromosomes of different animals may
be compared. The gross morphology of the chromosomes is even more important.
A simple comparison may be made of the different $2n = 14$ complements from
the six large superfamilies. Representative chromosome complements from each
of these superfamilies are shown in Figure 3.2. It can be seen that each autosomal
complement has a number of points of similarity. The autosomes consist of three
pairs of large metacentric or sub-metacentric chromosomes, one pair of medium-
size metacentric chromosomes, and two pairs of smaller chromosomes one of
which may have a satellited short arm. This general form of the complement is
found among all $2n = 14$ species of the two American superfamilies, in all but
one of the Australian Dasyuroidea and Perameloidea, and in some of the $2n = 14$
species of the Phalangeroidea and Vombatoidea, notably those of the Burramyidae
and Vombatidae. It appears to be a most remarkably constant feature among
diverse and disjunct groups of species. Hayman and Martin (1969, 1974) have

Table 3.1

	$2n$	Sex chromosomes	Reference

SUPERFAMILY DIDELPHOIDEA (GRAY)

Family Didelphidae Gray

	$2n$	Sex chromosomes	Reference
Didelphis virginiana Kerr, 1792	22	XY/XX	Sharman (1961)
D. marsupialis L., 1758	22	XY/XX	Sharman (1961)
D. albiventris Lund, 1841	22	XY/XX	Sharman (1961)
Lutreolina crassicaudata (Desmarest, 1804)	22	XY/XX	Sharman (1961)
Chironectes minimus (Zimm., 1780)	22	XY/XX	Hayman and Martin (1974)
Philander opossum L., 1758	22	XY/XX	Biggers *et al.* (1965)
Monodelphis brevicaudata (Erxleben, 1777)	18	XY/XX	Reig and Bianchi (1969)
M. dimidiata (Wagner, 1874)	18	XY/XX	Reig and Bianchi (1969)
Metachirus nudicaudatus (E. Geoffroy St-Hilaire, 1803)	14	XY/XX	Hayman and Martin (1974)
Marmosa mexicana Merriam, 1897	14	XY/XX	Biggers *et al.* (1965)
M. marina (L., 1758)	14	XY/XX	Hayman and Martin (1974)
M. cinerea (Temminck, 1824)	14	XY/XX	Hayman and Martin (1974)
M. robinsoni Bangs, 1898	14	XY/XX	Reig (1968)
M. fuscata Thomas, 1896	14	XY/XX	Reig and Sonneschein (1970)
M. elegans (Waterhouse, 1839)	14	XY/XX	Reig *et al.* (1972)
Caluromys derbianus (Waterhouse, 1841)	14	XY/XX	Biggers *et al.* (1965)
C. lanatus (Illiger, 1815)	14	XY/XX	Hayman and Martin (1974)
C. philander (L., 1758)	14	XY/XX	Hayman and Martin (1974)
Dromiciops australis (F. Phillipi, 1894)	14	XY/XX	Hayman and Martin (1974)

SUPERFAMILY CAENOLESTOIDEA (TROUESSART)

Family Caenolestidae Thomas

	$2n$	Sex chromosomes	Reference
Caenolestes obscurus Thomas, 1895	14	XY/XX	Hayman *et al.* (1971)
C. fuliginosus (Thomas, 1863)	14	XY/XX	Hayman *et al.* (1971)
C. convelatus Anthony, 1921	14	XY/XX	Hayman *et al.* (1971)
Lestoros inca (Thomas, 1917)	14	XY/XX	Hayman *et al.* (1971)

SUPERFAMILY DASYUROIDEA (GOLDFUSS)

Family Dasyuridae Goldfuss

	$2n$	Sex chromosomes	Reference
Dasyurus maculatus (Kerr, 1792)	14	XY/XX	Sharman (1961)
D. viverrinus (Shaw, 1800)	14	XY/XX	Sharman (1961)
D. geoffroii (Gould, 1841)	14	XY/XX	Martin and Hayman (1967)
D. hallucatus (Gould, 1842)	14	XY/XX	Sharman (1961)
Sarcophilus harrisii (Boitard, 1841)	14	XY/XX	Sharman (1961)
Dasyuroides byrnei Spencer, 1896	14	XY/XX	Hayman and Martin (1969)
Dasycercus cristicauda (Krefft, 1867)	14	XY/XX	Martin and Hayman (1967)
Antechinus stuartii Macleay, 1841	14	XY/XX	Sharman (1961)
A. flavipes (Waterhouse, 1838)	14	XY/XX	Hayman and Martin (1974)
Sminthopsis crassicaudata (Gould, 1844)	14	XY/XX	Sharman (1961)
S. leucopus (Gray, 1842)	14	XY/XX	Cooper in Sharman (1974b)
S. macroura (Gould, 1845)	14	XY/XX	Sharman (1961)
S. psammophila Spencer, 1895	14	XY/XX	Hayman and Martin (1974)
Antechinomys laniger (Gould, 1856)	14	XY/XX	Hayman and Martin (1969)
Myrmecobius fasciatus Waterhouse, 1836	14	XY/XX	Sharman (1961)
Ningaui sp.	14	XY/XX	Rofe and Baverstock (unpublished)

SUPERFAMILY NOTORYCTOIDEA OGILBY 1892

Family Notoryctidae Ogilby

	$2n$	Sex chromosomes	Reference
Notoryctes typhlops (Stirling, 1889)	20	XY/XX	Calaby *et al.* (1974)

SUPERFAMILY PERAMELOIDEA (WATERHOUSE)
Family Peramelidae Waterhouse

	2n	Sex chromosomes	Reference
Isoodon obesulus (Shaw, 1797)	14	XY/XX	Sharman (1961)
I. macrourus (Gould, 1842)	14	XY/XX	Sharman (1961)
Perameles nasuta Geoffroy, 1804	14	XY/XX	Sharman (1961)
P. gunnii Gray, 1838	14	XY/XX	Sharman (1961)
P. bougainville Quoy and Gaimard, 1824	14	XY/XX	Sharman (1961)
Echymipera kalabu (Lesson, 1828)	14 + 0 – 5 supernumeraries	XY/XX	Hayman, Martin and Waller (1969)
E. rufescens (Peters and Doria, 1875)	14 + 1 – 3 supernumeraries	XX	Sharman (pers. comm., 1973)
Peroryctes longicauda (Peters and Doria, 1875)	14	XX	Hayman, Martin and Waller (1969)
Macrotis lagotis (Reid, 1837)	19♂,18♀	XY_1/Y_2/XX	Martin and Hayman (1967)

SUPERFAMILY VOMBATOIDEA IREDALE AND TROUGHTON
Family Vombatidae Iredale and Troughton

	2n	Sex chromosomes	Reference
Vombatus ursinus (Shaw, 1800)	14	XY/XX	Sharman (1961)
Lasiorhinus latifrons (Owen, 1845)	14	XY/XX	Sharman (1961)

Family Phascolarctidae Owen

	2n	Sex chromosomes	Reference
Phascolarctos cinereus (Goldfuss, 1817)	16	XY/XX	Sharman (1961)

SUPERFAMILY PHALANGEROIDEA THOMAS
Family Burramyidae Broom

	2n	Sex chromosomes	Reference
Burramys parvus Broom, 1896	14	XX/XY	Gunson, Thompson and Sharman (1968)
Acrobates pygmaeus (Shaw, 1793)	14	XY/XX	Gunson, Thompson and Sharman (1968)
Cercartetus concinnus (Gould, 1845)	14	XY/XX	Sharman (1961)
C. nanus (Desmarest, 1818)	14	XY/XX	Martin and Hayman (1967)
C. lepidus (Thomas, 1888)	14	XY/XX	Sharman (1961)
C. caudatus (Milne-Edwards, 1877)	14	XY/XX	Hayman and Martin (1974)

Family Phalangeridae Thomas

	2n	Sex chromosomes	Reference
Trichosurus vulpecula (Kerr, 1792)	20	XY/XX	Sharman (1961)
T. arnhemensis Collett, 1897	20	–/XX	Hayman and Martin (1969)
T. caninus Ogilby, 1836	20	XY/XX	Sharman (1961)
Phalanger vestitus (Milne-Edwards, 1877)	14	XY/XX	Hayman and Martin (1974)
P. gymnotis (Peters and Doria, 1875)	14	XY/XX	Hayman and Martin (1974)

Family Petauridae Gill

	2n	Sex chromosomes	Reference
Pseudocheirus peregrinus (Boddaert, 1785)	20	XY/XX	Sharman (1961)
P. canescens (Waterhouse, 1847)	18	XY/XX	Hayman and Martin (1974)
P. corinnae Thomas, 1897	16	–/XX	Hayman and Martin (1974)
P. archeri (Collett, 1884)	16	–/XX	Hayman and Martin (1969)
P. cupreus Thomas, 1897	10	XY/XX	Hayman and Martin (1974)
Hemibelideus lemuroides (Collett, 1884)	20	–/XX	Hayman and Martin (1969)
Petaurus breviceps Waterhouse, 1839	22	XY/XX	Sharman (1961)
Schoinobates volans (Kerr, 1792)	22 + 2 – 6 supernumeraries	XY/XX	Hayman and Martin (1965)

Table 3.1 (*Continued*)	$2n$	Sex chromosomes	Reference
Dactylopsila trivirgata Gray, 1856	18	$-/XX$	Hayman and Martin (1974)
D. palpator Milne-Edwards, 1888	18	$-/XX$	Hayman and Martin (1974)

Family Macropodidae Gray

	$2n$	Sex chromosomes	Reference
Hypsiprymnodon moschatus Ramsay, 1876	22	XY/XX	Sharman (1961)
Potorous tridactylus (Kerr, 1792)	$13\delta12\female$	XY_1Y_2/XX	Sharman (1961)
Bettongia penicillata Gray, 1837	22	XY/XX	Sharman (1961)
B. gaimardi (Desmarest, 1822)	22	XY/XX	Sharman (1961)
B. lesueur (Quoy and Gaimard, 1824)	22	XY/XX	Sharman (1961)
Aepyprymnus rufescens (Gray, 1837)	32	XY/XX	Hayman and Martin (1969)
Lagostrophus fasciatus (Peron, 1807)	24	XY/XX	Sharman (1961)
L. hirsutus (Gould, 1844)	22	XY/XX	Sharman (1961)
L. conspicillatus Gould, 1842	$15\delta 16\female$ (also 24δ)	$X_1X_2Y/X_1X_1X_2X_2$	Martin and Hayman (1966)
Onychogalea unguifer (Gould, 1841)	20	XY/XX	Sharman (1961)
Petrogale rothschildi Thomas, 1904	22	XY/XX	Hayman and Martin (1969)
P. penicillata (Griffith, 1827)			
var. *pearsoni* Thomas, 1922	22	XY/XX	Sharman (1961)
var. *hacketti* Thomas, 1905	20	XY/XX	Hayman and Martin (1969)
var. *inornata* Gould, 1842	18	XY/XX	Hayman and Martin (1969)
P. brachyotis Gould, 1841	18	XY/XX	Hayman and Martin (1969)
P. xanthopus Gray, 1855	22	XY/XX	Martin and Hayman (1966)
Thylogale billardieri (Desmarest, 1822)	22	XY/XX	Sharman (1961)
T. thetis (Lesson, 1827)	22	XY/XX	Sharman (1961)
T. stigmatica Gould, 1860	22	XY/XX	Sharman (1961)
T. bruijni (Schreber, 1778)	22	XY/XX	Hayman and Martin (1974)
Setonix brachyurus (Quoy and Gaimard, 1830)	22	XY/XX	Sharman (1961)
Dorcopsulus macleayi (Miklouho-Maclay, 1885)	18	$-/XX$	Hayman and Martin (1974)
Dendrolagus lumholtzi Collett, 1844	14	$XY/-$	Hayman and Martin (1969)
D. goodfellowi Thomas, 1908	14	XY/XX	Hayman and Martin (1969)
D. matschei Forster and Rothschild, 1907	14	$-/XX$	Hayman and Martin (1969)
D. ursinus Muller, 1840	12	$XY/-$	Hayman and Martin (1969)
D. doriamus Ramsay, 1883	12	$-/XX$	Hayman and Martin (1974)
Wallabia bicolor (Desmarest, 1804)	$11\delta10\female$	XY_1Y_2/XX	Sharman (1961)
Macropus rufus (Desmarest, 1822)	20	XY/XX	Sharman (1961)
M. giganteus Shaw, 1790	16	XY/XX	Sharman (1961)
M. fuliginosus (Desmarest, 1817)	16	XY/XX	Sharman (1961)
M. robustus Gould, 1841	16	XY/XX	Sharman (1961)
M. antilopinus (Gould, 1842)	16	XY/XX	Sharman (1961)
M. agilis (Gould, 1842)	16	XY/XX	Sharman (1961)
M. rufogriseus (Desmarest, 1817)	16	XY/XX	Sharman (1961)
M. dorsalis (Gray, 1837)	16	XY/XX	Sharman (1961)
M. parryi (Bennett, 1835)	16	XY/XX	Sharman (1961)
M. irma (Jourdan, 1837)	16	XY/XX	Sharman (1961)
M. euginii (Desmarest, 1817)	16	XY/XX	Sharman (1961)
M. parma Waterhouse, 1846	16	XY/XX	Hayman and Martin (1969)

SUPERFAMILY TARSIPEDOIDEA

Family Tarsipedidea Gill

	$2n$	Sex chromosomes	Reference
Tarsipes spencerae Gray	24	$-/XX$	Hayman and Martin (1969)

The species names used in this Table are those used by Kirsch and Calaby in this volume. The varieties of *Petrogale penicillanta* are included in this species by Ride (1970). These would appear to justify specific status on their cytology but a complete taxonomic study remains to be performed.

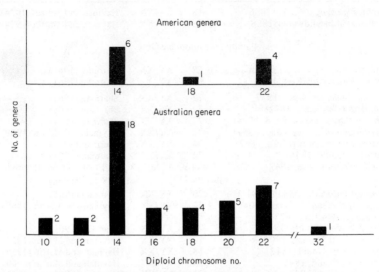

Figure 3.1 Histogram of the diploid chromosome numbers found in Australian and American genera.

presented a detailed argument for considering that the presence of this common form of chromosome complement constitutes strong evidence for these being features of the original chromosome complement preserved in a relatively un-modified state.

The evidence for this view is naturally circumstantial, since a direct test is not possible. Firstly, there is the diversity of species and groups in which this com-plement is found, and the representative range of species studied in all groups, which would argue that the information available represents a representative sample of the marsupials as a whole. Secondly, certain approximations can be made which allow a comparison to be made between the chromosomes of all the different species with the postulated basic complement of $2n = 14$. These approxi-mations are shown in Tables 3.2 and 3.3, which list the 'average' chromosome complement for each of the $2n = 14$ species in each superfamily. If the autosomes alone are considered, there is considerable similarity as measured by the sum of differences from the average of chromosome arms, and more similarity if whole chromosomes are considered. Inclusion of the sex chromosomes in the analysis produces more variation and the reasons for this are discussed later.

Properly, comparisons between chromosomes of different species should only be made when the nuclear DNA values are known for each species (Martin and Hayman ,1965). The nuclear DNA values for many of the Australian species are known; those of the South American species are not known. The similarity of the average chromosomes as a whole suggests that the variations in nuclear DNA content are due to the DNA content of all arms varying in a largely proportional manner. It follows from this that the sort of variations in morphology which do occur between the autosomes of the $2n = 14$ group are due to changes largely in arm ratio as a result of centromere shifts.

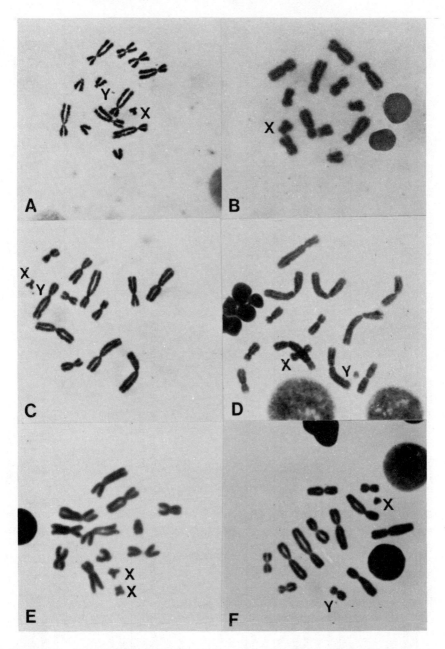

Figure 3.2   Somatic divisions in representative species of each large superfamily. These are species with the postulated 'basic' complement. A, *Acrobates pygmaeus.* B, *Isoodon obesulus* (this species is XO in somatic cells; see text). C, *Dasyurus geoffroii* D, *Lasiorhinus latifrons.* E, *Marmosa robinsoni.* F, *Caenolestes obscurus.*

Table 3.2 Comparison of mean per cent lengths of chromosome arms in hypothetical average animals, all with $2n = 14$ from six superfamilies

Superfamily	1L	1S	2L	2S	3L	3S	4L	4S	5L	5S	6L	6S	XL	XS	Sum of differences from average	Sum of differences for whole chromosomes
Didelphoidea	7.28	4.20	6.37	3.73	6.24	3.64	3.94	3.70	3.47	1.13	3.64	0.56	1.86	0.44	4.32	1.35
Caenolestoidea	6.86	4.18	6.50	4.14	6.31	3.85	3.95	3.62	2.71	1.65	2.27	2.27	1.60	0.09	4.65	1.99
Dasyuroidea	7.00	4.26	5.26	5.26	6.60	3.07	4.55	2.86	3.45	1.61	2.08	2.08	1.64	0.32	7.25	1.47
Perameloidea	7.49	3.15	6.49	3.04	5.78	3.53	3.58	3.58	3.15	1.63	3.01	1.39	2.47	1.75	5.15	3.85
Wombatoidea	6.44	3.79	5.71	3.97	6.46	2.92	3.62	3.62	2.70	2.07	2.86	1.74	2.55	1.48	3.86	3.96
Phalangeroidea	7.98	3.21	6.57	3.30	6.54	3.48	3.87	3.68	3.20	1.47	2.80	1.18	1.92	0.75	3.14	1.76
Average marsupial	7.17	3.80	6.15	3.90	6.31	3.41	3.92	3.51	3.10	1.59	2.77	1.54	2.00	0.80		

Table 3.3 Comparison of mean per cent lengths of chromosome arms in hypothetical average animals with $2n = 14$, whose complements are believed not to be the basic $2n = 14$. The differences are from the 'average marsupial' in table 3.2

Genus	1L	1S	2L	2S	3L	3S	4L	4S	5L	5S	6L	6S	XL	XS	Sum of differences from average	Sum of differences for whole chromosomes
Potorous	10.35	2.46	7.91	4.01	6.59	3.03	2.41	2.41	4.16	Tr.*	1.37	1.37	2.55	1.20	17.8	9.84
Dendrolagus	9.81	1.08	5.86	3.46	5.32	1.00	4.59	1.51	4.35	1.74	2.65	2.32	1.79	0.30	14.3	10.11
Phalanger	8.01	8.01	4.51	4.51	4.17	2.74	3.78	2.56	6.36	Tr.*	3.93	0.20	2.10	0.18	15.8	12.64

*Tr. = Trace – a small arm which is not measurable.

The comparison between the sex chromosomes of the different species shows that there are great differences. This is particularly obvious for the Y-chromosomes (see Figure 3.2) – although there is at least a factor of 2 in the size difference between the X-chromosome of the Dasyuroidea and the X-chromosome of the Perameloidea.

This hypothesis of the retention of the basic chromosome complement and chromosome form is one of the major areas of interest in marsupial cytogenetics at present. Consequently it should be examined in somewhat more detail as to its merits and its implications.

Not all the $2n = 14$ complements have what may be called the 'basic' karyotype; *Phalanger* spp and some *Dendrolagus* spp have $2n = 14$ but these chromosomes do not fit the 'basic' complement (Table 3.3). Some members of the species *Lagorchestes conspicallutus* have $2n = 14$ (Martin and Hayman, 1966), but this is the result of the presence of a polymorphism, and the sex chromosome system in this species is a complex multiple one and indicates evolutionary reduction in number from a higher number. The species *Potorous tridactylus* ($2n = 12$ XX/XY$_1$Y$_2$) is obviously derived from a $2n = 14$ species with an XX/XY system (Sharman and Barber, 1952), and a postulated complement for this species can be constructed (Table 3.3). It is possible to construct a similar analysis for some of these species to that conducted in Table 3.2 (see Table 3.3). Such an analysis shows that there is a very great divergence from the 'average marsupial' based on the postulated basic $2n = 14$ complement.

The presence of $2n = 14$ 'non-basic' complements does not invalidate the hypothesis. In each of these cases it is argued that the $2n = 14$ complement is derived from a higher chromosome number by the process of chromosome fusion.

The generalised view of the evolution of chromosome number in the marsupials is that of an increase in chromosome number, from the basic $2n = 14$. In certain groups such as the Caenolestoidea and Dasyuroidea there are no present examples of this, while in the Phalangeroidea and Vombatoidea many of the present-day species have a derived complement.

Thus the explanation offered for the second mode in the distribution of chromosome numbers at $2n = 20$ and $2n = 22$ is that this represents a derived mode from $2n = 14$. Given a complement of $2n = 14$ made up of 12 metacentric autosomes, then it would be possible to obtain a maximum of 24 acrocentric autosomes by the usually proposed methods of increasing chromosome number. However, different combinations of chromosome morphologies and numbers are possible as a result of pericentric inversions, and these can account for the different types of $2n = 20$ and $2n = 22$ complements that are found (Figure 3.3). In order to obtain $2n = 32$, found in *Aepyprymnus rufescens*, it is necessary to postulate a second cycle of increase from a complement with more than $2n = 14$ and with metacentric or sub-metacentric chromosomes.

There has been a curious reluctance to accept the idea of an increase in chromosome number in evolution, due to the combined problem of postulating and demonstrating a mechanism by which this could have taken place. There are two mechanisms suggested in most discussions on the matter. One of these is to suggest that 'fission' of the centromere takes place – that is, a metacentric yields two stable telocentrics. This has been attacked as an unlikely suggestion because it is argued that the telocentrics so produced are unstable. There are certain situations described, however, which may well be examples of such a mechanism

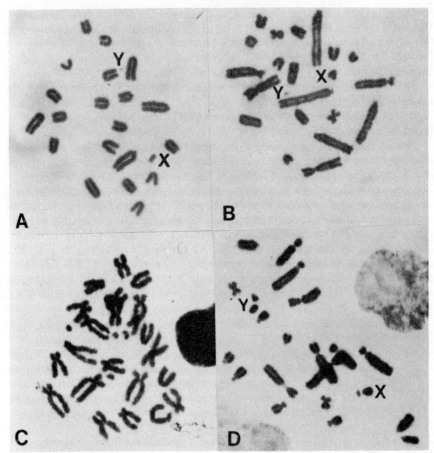

Figure 3.3 Chromosome complements from species with $2n = 22$. A, *Philander opossum* ♂. B, *Petrogale xanthopus* ♂. C. *Schoinobates* ♂ *volans* ♂ $2n = 22 + 3$ supernumeraries. The Y is one of the four smallest chromosomes — the other three are supernumerary chromosomes. D, *Setonix brachyurus* ♂

(Southern, 1969; Webster, Hall and Williams, 1972). The second suggestion is to postulate a centromere and telomere donor which is incorporated into the chromosome complement by breakage and reunion. The most likely candidates for such donor chromosomes are the supernumerary chromosomes — and as these have been described in marsupials (Hayman and Martin, 1965b; Hayman, Martin and Waller, 1969), centromere and telomere donation must not be thought of as an impossible mechanism. However, these supernumerary chromosomes are usually present in small numbers in any individual presumably as a result of the operation of natural selection. To rely on a number of them being incorporated into the chromosome complement requires that they lose their selective disadvantage.

It is possible to suggest another mechanism which does not seem to have the formal difficulties of the first two, and which requires only the occurrence of unlikely events. Such events probably occur with the frequency of the mutation rate — and therefore cannot be dismissed. If breaking and healing occur adjacent to the centromere in different arms of sister chromatids at C_2, followed by non-disjunction at anaphase, this would result in the substitution by two acrocentric chromosomes of a metacentric chromosome by using perfectly acceptable mechanisms in the light of present knowledge.

It is unlikely that a complete picture of the different ways in which chromosomes may alter their morphology exists at present. There is good evidence that regions of the chromosome may increase and decrease in amount in a dramatic fashion. Such a mechanism is necessary to explain the alterations in the repeated sequences of DNA (e.g. Masremas and Hatch, 1972; Kurnit and Maio, 1974). Precisely how such changes are brought about and what their implications are for changes in chromosome morphology and number are not known yet.

Unless it is argued that there are never any situations in which the chromosome number is increased, mechanisms for increase must have operated. What is perhaps unusual, but what ought not to be unacceptable, is that the argument is advanced for the marsupials that an increase in number occurred from a basic chromosome complement of $2n = 14$ and that relics of this basic complement are found today in representations of very different types of marsupials.

Chromosome number and its origin — another view

What are the alternative arguments to the hypothesis of conservation of a basic complement of $2n = 14$? One approach would be to argue that whatever was the basic complement, it is no longer present and that the similarities postulated between the $2n = 14$ complements are not correctly interpreted as relics. A further approach would be to postulate another chromosome number as basic. In the first extensive account published of marsupial chromosomes Sharman (1961) suggested that evolution in marsupials may have been accompanied by a general reduction in chromosome number, and suggested that $2n = 22$ may have been the basic number. He further postulated (1974a, b) that a similar argument may be made for the American species, and showed that this would involve a complement containing nearly all telocentrics to derive *Caluromys* and *Marmosa* species complements.

Since at least one of the processes of chromosome number increase, namely Robertsonian changes, is potentially reciprocal, it is theoretically correct to argue for a basic number other than $2n = 14$. However, given this possibility, it is not clear that the number has to be $2n = 22$. If the choice of this number is based only on the frequency with which $2n = 22$ occurs in the marsupials, then of course $2n = 14$ occurs more frequently and in a more diverse series of genera than any other number. Further, while there is a mode at $2n = 22$, this similarity of number does not convey anything like the uniformity of complements present in species with $2n = 14$. The $2n = 22$ complements range from complements with predominantly acrocentric chromosomes (*Petrogale*) to complements with predominantly sub- or metacentric chromosomes (*Schoinobates*). (See figure 3.3.)

It may be argued that $2n = 14$ can be derived as a complement from $2n = 22$ by more acceptable, i.e. demonstrable, processes than $2n = 22$ can be derived from $2n = 14$. However, there are at least two examples of chromosome complements greater than $2n = 22$. Both of these would then require processes of increase of chromosome number to explain their origin as Sharman notes, thus

denying this basis for the choice of $2n = 22$. On a hypothesis of some number
other than $2n = 14$ being the basic complement, it would be necessary to argue
that the features in common to the $2n = 14$ complements are derived in each of
the superfamilies by chromosome fusion and the whole situation is a remarkable
example of parallel evolution. Such a mechanism is of course possible and it
could be justified in selective terms by postulating that the same ends are served
as those which it is argued have led to the conservation of the complement.

In support of the hypothesis for decreasing chromosome numbers in marsupial
evolution Sharman argues that there are 'many separate demonstrations of the
occurrence of chromosome number reduction in various marsupials'. The only
ones of these which are unequivocal are those relating to the evolution of multiple
sex chromosome systems in *Potorous tridactylus* (Sharman and Barber, 1952),
Wallabia (Protemnodon) *bicolor* (Sharman, 1961), *Lagorchestes conspicillatus*
(Martin and Hayman, 1966) and *Macrotis lagotis* (Martin and Hayman, 1967). In
these species autosomal chromosomes have been incorporated into sex chromosomes, and a centromere and telomeres presumably lost. It is probable but certainly
not proven that among the rock wallabies (*Petrogale* spp) there are also examples
of a reduction in number by chromosome fusions (Hayman and Martin, 1969,
1974), and Hayman and Martin (1974) suggest that among the Petauridae, the
Phalangeridae and Macropodidae there are also instances of chromosome fusion.

While these latter interpretations are conjectural, there has never been any
doubt that 'fusion' can occur and has been important in marsupial chromosome
evolution. What is at issue is whether the $2n = 14$ complements are conserved or
derived in the present day marsupials. It is possible that no answer will be found
to this question.

Matthey (1974) has criticised the hypothesis of $2n = 14$ as the primitive chromosome complement of marsupials on a number of separate grounds.

Matthey does not deny that both fission and fusion have taken place in the
evolution of eutherian mammals. He argues that the primitive number of eutherian
mammals lies in the modal range from $2n = 40$ to $2n = 56$, and that numbers
greater than this must be derived by fission. There are fewer species with numbers
higher than $2n = 56$ and consequently Matthey argues that fission occurs less
frequently. This argument is suspect since there can be only one primitive number
and if it is $2n = 40$ then the relative proportions of species with numbers arrived
at by fission would be very much greater than those with numbers derived by
fusion. Matthey acknowledges that (even with the inherent error of his procedure)
there are some groups of the Artiodactyla and the Primates in which fission has
occurred more frequently than fusion.

Present day Didelphis species have $2n = 22$ and do not differ greatly in skeletal
features from very old fossils and Matthey suggests that the primitive chromosome
number was also $2n = 22$. This would be in accord with his view, which many may
find surprising, 'that it seems doubtful that the diploid number is a character
subject to natural selection by environmental factors'.

Matthey points out that parallel evolution has taken place in the certain lizard
groups which have very similar derived complements, and that consequently a
similar situation can occur in the marsupials. This does not of itself do more than
show that the process which could have given rise to the present day distribution
of chromosome numbers in marsupials could also have given rise to the first

marsupial form from which the present day species are derived. It simply restates the problem of deciding which of the two hypotheses is correct.

Because there is evidence that complex sex-chromosome systems are found among eutherian species with a very low chromosome number, Matthey argues that the four examples of such systems in marsupials being found in species with numbers less than $2n = 22$ constitute evidence of the primitive number being $2n = 22$. However it has been argued (Hayman and Martin, 1974) that in each case this reduction followed an initial increase in number from $2n = 14$, which is also acceptable. What would be absolutely destructive of the hypothesis that $2n = 14$ is the primitive complement, would be to find an example of this number and form with a complex sex chromosome system. Such has not been reported so far.

The nuclear DNA contents quoted by Matthey as being in accord with the $2n = 22$ hypothesis are only a small fraction of the information available. The extensive DNA data referred to in Hayman and Martin (1969, 1974) have not been mentioned by Matthey, and as discussed earlier in this paper and more extensively elsewhere, do not allow a distinction between the two hypotheses. It is an unjustified conclusion by Matthey to suppose that these arguments must lead to the conclusion that the primitive formula in marsupials is $2n = 22$.

The hypothesis favoured in this chapter is that of conservation. It is suggested that the solution to the problem of chromosome number in the majority of present day marsupials was arrived at many millions of years ago. The evolutionary distinction between the present day superfamilies is a very substantial one and does not only reflect the geographic separation. Kirsch (in press) finds that the serological differences between the Caenolestoidea and the Didelphoidea are as great as those between either group and the Australian superfamilies. This makes even less likely the hypothesis that parallel evolution has arrived at a similar solution to the problem of chromosome number in such different and disjunct groups of species.

The hypothesis of the conservation of this basic complement of $2n = 14$ seems more likely given our present knowledge than the hypothesis that in two highly disjunct populations containing diverse superfamilies chromosome fusion has arrived at a numerically and morphologically similar solution to the problems of chromosome number.

The function of chromosome number

The presence of the same chromosome number and arguably similar chromosome form in so many different species of marsupials with $2n = 14$ raises again the problem of the role of chromosome number and form in evolution. The species concerned here live quite different lives in terms of the number of reproductive cycles during their life time, the number of young per litter and the mobility of the species. They range from the very large, burrowing, long-lived herbivores with one young per conception and irregular breeding like the wombat, to the small, social, active, short-lived dasyurids with predictably large numbers of offspring per conception. *Sarcophilus harrisii* and *Antechinus stuartii* are examples of this diversity and are discussed elsewhere in this volume.

These different features of the life cycle are all aspects of the genetic system of the organisms; that is, they affect the release of variability, and the nature of variability in the population. Since the chromosome number is the same, as is the chromosome form, one variable that remains is chiasma frequency. The marsupials with $2n = 14$ allow an excellent opportunity to make a comparative study of chiasma frequency between representatives of these different sorts of genetic systems. One of the problems of studies of this sort is that the data must be obtained from both sexes, since differences in chiasma formation and in crossing-over have been found between the sexes in mouse, and there are theoretical reasons for supposing that one of the variations which can occur in a genetic system is a difference in recombination between the sexes (Owen, 1953).

A simple model which could be tested would be that chiasmata frequency is highest in those small marsupials with a rapid generation turnover and lowest in animals that are larger and live a long time. This would have some parallel to the situation described in plants by Stebbins (1958), Rees and Ahmad (1963) and others. Few data have been published as yet. Gunson *et al.* (1968) have collected some information and this and unpublished data of the author are shown in Table 3.4. These data indicate the range that is found in chiasma frequency in the male, and suggest that the examination of other species, and a study of the situation in the female would be a profitable area of study. Such a study should include not only the analysis of total chiasmata per cell, but also an analysis based on the distribution of chiasmata within the three size groups of autosomes.

What can be made of the groups where chromosome numbers have changed? The greatest range of chromosome number is found in the Macropodidae – the kangaroos. Many of them are wide ranging in their distribution like the red kangaroo ($2n = 20$) and euro ($2n = 16$) and the swamp wallaby ($2n = 10/11$), and this is apparently independent of chromosome number. Others, such as many of the different rock wallabies ($2n = 22$ to $2n = 18$), may exist in relatively isolated pockets, suggesting restriction to specialised disjunct habitats after adaptation from a more generalised less specific ancestor. This process is sometimes associated with a change in chromosome number. The genus *Pseudocheirus* shows a range in chromosome number sometimes associated with speciation, whereas the genus *Trichosurus,* with a similar if not more extensive distribution, does not. While chromosome number must still be important in all these groups, the relationship of the variation in number to what is known of the biology of the different species is not apparent. Further, the obvious comparisons which suggest themselves in the $2n = 14$ group have no immediate parallel in this group.

Table 3.4 Average number of chiasmata per cell in males of species
with the postulated basic $2n = 14$ complement

Species	Number of cells	Average number of chiasmata per cell	Authority
Dasyurus sp.	24	18.7	Drummond (1938)
Acrobates pygmaeus	10	14.8	Gunson, Thompson and Sharman (1968)
Burramys parvus	10	22.6	Gunson, Thompson and Sharman (1968)
Isoodon obesulus	20	12.9	Hayman (unpublished)

If constancy is postulated for the form of the autosomes, then what is the explanation for the absence of constancy for the sex chromosomes? The absence of constancy in morphology is not a novel result. Hayman and Martin (1969, 1974) have shown the extreme range in size of sex chromosomes which exists in the family Macropodidae. A possibly greater range exists in the Cricetidae and in species of the genus *Mus*. The explanation for the variation appears to be in local increases in DNA presumably of the repetitious kind; the material usually behaves as constitutive heterochromatin, and stains after C-banding techniques are applied (Hayman and Martin, 1974). However, it is not clear why considerable variation may occur in the sex chromosomes and is apparently absent from the autosomes. Two differences between these types of chromosomes are known already. The sex chromosomes lie in the middle of a hollow spindle at cell division; and the nature of the association between X- and Y-chromosomes at meiosis does not necessarily involve genetic exchange at the chiasmata, whereas that between autosomes does. Perhaps either of these two features is associated with the absence of the uniformity that is found in the groups of autosomes of the $2n = 14$ species.

Certainly there do appear to be limits to the capacity of both the autosomes and the sex chromosomes to increase in size. The largest chromosomes seen in the marsupials are approximately equal to the large autosome of the Red kangaroo, *M. rufa $2n = 20$*, and the largest autosome in *W. bicolor* $2n = 10♀/11♂$, a species with a complement believed to be derived by a process of fusion from a much higher number. However, the postulated process of fusion has not produced chromosomes of excessive length in this species or in any others. This suggests that selection favours such a limitation and it has been suggested that two of the factors affecting such selection would be the size of the spindle and the size of the cell (Hayman and Martin, 1969).

The role of new techniques

The development of 'banding' techniques (e.g. Sumner, Evans and Buckland, 1971) has greatly increased the capacity of cytologists to determine relationships between different species. The chromosomes of the sheep and the goat (Evans, Buckland and Sumner, 1973) show considerable banding similarities, as do those of Man and his primate relatives (de Grouchy *et al.*, 1973). While the precise basis of the banding patterns is not yet known and, hence, their taxonomic significance cannot be assessed, it is quite clear that it relates to an aspect of chromosome organisation – possibly to do with compaction of the chromosome at nuclear division. It also does appear to have merit as a taxonomic index. On both grounds this technique is an appropriate one to apply to the $2n = 14$ species of marsupials. If, as is argued, the $2n = 14$ complement is a conserved one, then there may be aspects of chromosome organisation which are maintained with the conserved complement and which may be detected using the banding techniques. In addition, the extreme diversity of the animals possessing the $2n = 14$ complement argues for some change in banding as a consequence of evolutionary diversification. Thus there may well be evidences of both the processes of differentiation and of diversification revealed by a study of closely related and distantly related species with $2n = 14$. This problem is being examined at present, and encouraging results have been achieved (Figure 3.4).

This banding technique has also considerable application in detecting relationships between species of *Macropus, Megaleia* and *Wallabia*. On the basis of their measurements of the chromosomes, and from what is known of the capacity of certain species to form hybrids in captivity, Hayman and Martin (1974) have

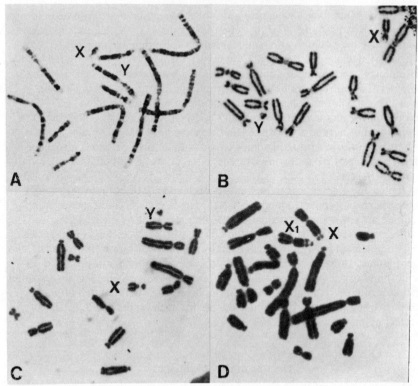

Figure 3.4 A, *Sminthopsis crassicaudata*♂ showing banding patterns after trypsin treatment (kindly supplied by R. Rofe). B, *Pseudocheirus peregrinus*♂ – a montage from a
widely spread cell showing a characteristic appearance of the centromere regions found
in this and some other species (kindly supplied by R. Rofe). C, *Macropus fuliginosus*♂.
D, *Megaleia rufa*♀ showing a normal and a 'mutant' X-chromosome X_1. This animal had
one male progeny carrying the 'mutant' X. The achromomatic region on the X-chromosome of this species and of other macropods is thought to be the nucleolar organiser
region.

suggested a pattern of relationship (see Figure 3.5). This group of species has a
range of chromosome numbers which Hayman and Martin have suggested are
derived by chromosome fusion from an ancestor with $2n = 20$ or $2n = 22$. An
analysis of the banding patterns of the chromosomes of these species would
enable another parameteter to be used in analysing their relationships.

An additional approach to examining relationships at the chromosome level
follows from what has been initiated by Hungerford *et al.* (1971) in examining
pachytene chromosome morphology in Man. Taylor (unpublished) has been
examining the pachytene morphology of Australian and American $2n = 14$
species in order to see if the chromomere pattern reveals any similarities.

The newest parameters which have to be applied to the marsupial chromosomes are the measurement of nuclear DNA contents and the examination of the
DNA. A great deal is known of the relative DNA contents of Australian marsupials
(Martin and Hayman, 1967; Hayman and Martin, 1974). There is little variation
within the Dasyuroidea and the Perameloidea and a great deal within the Phalangeroidea, where there is a range from 86 to 150. The petaurids have the highest

values. The range of nuclear DNA contents in marsupials is not as great as that found in Amphibia and is of the order of that found in the Eutheria (Bachmann, 1972).

When a comparison is made between the different $2n = 14$ groups, the range encountered is some 40%. The postulated preservation of the same essential form of chromosome complement requires a proportional change in each chromosome rather than a change in one chromosome. Indeed, that the same chromosome form should be preserved with extreme change in nuclear DNA content is yet another demonstration of the significance of this endophenotype. The increase in nuclear DNA content is not wholly ascribable to the increase in the size of the chromosomes. An additional explanation for this increase must be sought in the new discovery that a proportion of the eukaryote genome consists of so-called families of sequences of DNA multiplied many thousands – even millions – of times. This is the so-called repetitious DNA (Britten and Kohne, 1968). It may exist in different amounts in different species and it is believed to be present in all chromosomes. There is a parallelism between the so-called C-bands (Arrighi and Hsu, 1971) and the location of regions richest in repeated sequences. There is a positive correlation between the amount of C-banding material as a percentage of the total chromosomes and increasing amounts of nuclear DNA in a number of marsupials. *Sminthopsis* spp (DNA value = 86) has very small amounts compared with wombats (DNA value = 103), which have less than bandicoots (DNA value = 116). The presence of repeated DNA has been demonstrated in marsupials using hydroxyapatite chromotography (Hayman, unpublished).

The implications of the study of repeated sequences upon the different hypotheses of the evolution of chromosome number in marsupials remain to be determined. Again the possible comparisons which may be made between the different species with $2n = 14$ and the same and different DNA contents offer the most attractive areas of investigation.

Where there is circumstantial evidence of relationships such as that indicated in Figure 3.5 between certain species of macropods, the analysis of the repeated DNA sequences by techniques such as Southern's (1970) could be most rewarding. For example, *W. bicolor* is believed to have a chromosome complement derived by fusion from species with higher numbers. It has obvious affinities with *M. agilis,* with which it hybridises. Late-replicating regions found along the chromosome arms (Hayman and Martin, 1965b) of the large chromosomes of *W. bicolor* may represent the residual centromeric regions of chromosomes incorporated by fusion into the present *W. - bicolor* complement. If a satellite DNA which was specific for those regions could be identified, then it would be of some interest to see if this DNA would hybridise with the centromeric regions of *M. agilis* or any other macropod. Affinities may also be found between satellite sequences of *M. agilis, M. robustus, M. fuliginosus, M. giganteus and M. parryi* and *Megaleia rufa,* which are all closely related.

Mosaicism for chromosome number

The previous discussion has emphasised how the evolution of chromosome numbers in marsupials may be a remarkable example of constancy. The marsupials also possess one of the few examples in mammals of the variation of chromosome numbers between tissues of the same organism. This bizarre situation occurs in the Peramelidae and is found in all species of that group so far examined except

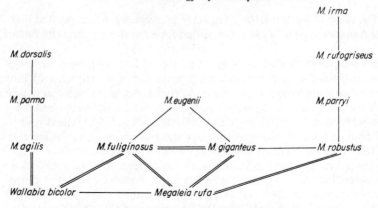

Figure 3.5 Diagram of the postulated relationships between species of the genera *Macropus*, *Wallabia* and *Megaleia*. Redrawn and corrected from Hayman and Martin (1974).

Macrotis lagotis. This species is also the only peramelid which has a chromosome number other than the basic $2n = 14$.

The essence of the situation is shown in Table 3.5. In adult, i.e. free-living, peramelids there are two forms of mosaicism for the soma and germ line. In *Isoodon* species those somatic cells in which divisions have been observed are $2n = 13$ XO irrespective of sex, while the germ line is apparently $2n = 14$ XX or XY. In *Perameles* some somatic tissues are $2n = 13$ XO, others are $2n = 14$ XX or XY and the germ line is $2n = 14$ XX or XY. *Peroryctes* spp and *Echymipera* spp are

Table 3.5 Sex chromosome mosaicism in peramelids

Genus	Tissue	$2n$	Sex chromosomes
Isoodon	Testes	14	XY
	Ovarian tissue	14	XX
	Thymus, liver, bonemarrow, leucocyte culture, spleen, corneal epithelium, intestinal epithelium	13	XO
Perameles Peroryctes Echymipera	Testes	14	XY
	Ovarian tissue	14	XX
	Bonemarrow, leucocyte culture spleen	13	XO
	Corneal epithelium intestinal epithelium	14	XX/XY

assumed to follow the same pattern as *Perameles* spp, although they have not been examined to the same extent.

The mosaicism is believed to be a developmental phenomenon, since X- and

Y-bearing sperm are produced and it is believed that only X-bearing eggs are involved in fertilisation. Subsequent to fertilisation different tissues lose either an X- or a Y-chromosome. There is evidence of the loss during development in the data of Walton (1971) and Hayman and Martin (1974) which shows the development of XO as the constitution of the tissue from either an XX or XY constitution or one in which XX or XY was the predominant condition. Different tissues become XO at different ages during development – the corneal epithelium is much later than the marrow and the spleen. There is a suggestion in the data that those tissues which are late in becoming XO are those which are XO in *Isoodon* but XX or XY in the *Perameles* type of mosaicism.

The mechanical basis of loss is not known. However, it has been suggested that the eliminated chromosome may be impaired in its replication and that consequently the chromosome fails to function at anaphase (Hayman and Martin, 1965a). This suggestion is based on the fact that the net result of elimination is to produce a tissue which is chromosomally identical irrespective of sex. Thus, so far as the X-chromosome is concerned, the elimination replaces the well-known inactivation of the X-chromosome present in mammals (Lyon, 1961). Since the inactivated X-chromosome of mammals is late-replicating, it is reasonable to assume that elimination may therefore be an extension or exaggeration of the same phenomena. It is assumed that one of the two X-chromosomes present in some of the somatic tissues of adult *Perameles* spp is late-replicating.

However, this explanation does not formally account for the elimination of the Y-chromosome, since this would not be expected on the basis of the X-inactivation mechanism. Obviously the Y-chromosome in males is characterised in the same way as is one of the X-chromosomes. It is not known whether the X-chromosome which is eliminated is a randomly chosen X-chromosome or whether it is the paternal X-chromosome, as is apparently the situation in the inactivated X-chromosome of macropods (Cooper *et al.*, 1971). The other examples of sex chromosome mosaicism, e.g. *Rattus rattus* (Young, 1971) *Choleopus hoffmani* (Corin-Frederic, 1969), *Akadon azarae* (Bianchi and Contreras, 1967) and *Microtus oregoni* (Ohno *et al.*, 1963, 1966), all involve Eutherian mammals and all involve only the X-chromosomes. It is possible that the involvement of the Y-chromosome and the X-chromosome in marsupials is a reflection of the variation between the random inactivation of the X-chromosome in Eutherian mammals and the inactivation of the paternal X-chromosome in macropods, and possibly other marsupials (Hayman and Martin, 1974).

A curious extension of the mosaicism for sex chromosomes has been found in *Echymipera kalabu* (Hayman *et al.*, 1969). This species contains varying numbers of additional or supernumerary chromosomes to its normal complement of $2n = 14$. Individuals with up to 5 supernumerary chromosomes have been identified. However, in those tissues in which the X- or Y-chromosome has been eliminated the supernumerary chromosomes have also been eliminated. Presumably the mechanism of elimination is the same, but it is not clear why the supernumerary chromosomes should be subject to the same system of regulation as the sex chromosomes. Possibly they are innocent victims of a pre-adapted system which recognises some features, which they share with the sex chromosomes. It is also possible that the supernumerary chromosomes were derived originally from the sex chromosomes and that this common origin is the basis for their recognition and their similarity in behaviour to the sex chromosomes.

The few data available show a significant excess of supernumerary chromosomes in the females as opposed to the males. Should elimination be only of the paternally derived supernumerary chromosomes, on the parallel of X-inactivation in the macropods, then additional modifications to the hypothesis would be required in order to explain this bias. Regrettably this species is not readily available and the extremely interesting implications of this system may have to wait some time before investigation.

References

Air, G. M., Thompson, E. O. P., Richardson, B. J. and Sharman, G. B. (1971). Amino-acid sequences of kangaroo myoglobin and haemoglobin and the date of marsupial Eutherian divergence. *Nature*, 229, 391 - 4.

Arrighi, F. E. and Hsu, T. C. (1971). Localisation of heterochromatin in human chromosomes. *Cytogenetics*, 10, 81.

Bianchi, N. O. and Contreras, J. R. (1967). The chromosomes of the field mouse *Akadon azarae* (Cricetidae, Rodentia) with special reference to sex chromosome anomalies. *Cytogenetics*, 6, 306 - 13.

Biggers, J. D., Fricky, H. I., Hall, W. C. D. and McFeely, R. A. (1965). Chromosomes of American marsupials. *Science*, 148, 1602 - 3.

Britten, R. J. and Kohne, D. E. (1968). Repeated Sequences in DNA. *Science*, 161, 529 - 40.

Calaby, J. H., Corbett, L. K., Sharman, G. B. and Johnston, P. G. (1974). The chromosomes and systematic position of the marsupial mole, *Notoryctes typhlops. Aust. J. Biol. Sci.*, 27, 529 - 32.

Cooper, D. W., Vanderberg, J. L., Sharman, G. B. and Poole, W. E. (1971). Phosphoglycerate kinase polymorphism in kangaroos provides further evidence for paternal X-inactivation. *Nature New Biology*, 230, 155 - 7.

Corin-Frederic, Janine (1969). Les formules gonosomiques dites aberrantes chez les mammiferes Eutheriens. *Chromosoma*, 27, 268 - 87.

de Grouchy, J., Turleau, C., Roubin, M. and Chavin-Colin, F. Chromosomal evolution of Man and the primates (*Pan aroglodykes, Gorilla gorilla, Pongo pygmaeus*). *Nobel Symposium No. 23 Stockholm, Sept. 25 -27, 1972,* 124 - 31. Almqvist and Wiksell.

Drummond, F. H. (1938). Meiosis in *Dasyurus viverrinus. Cytologia, Tokyo*, 8, 343 - 52.

Evans, H. J., Buckland, R. A. and Sumner, A. T. (1973). Chromosome homology and heterochromatin in goat, sheep and ox studied by banding techniques. *Chromosoma*, 42, 388 - 402.

Gunson, M. M., Thompson, J. A. and Sharman, G. B. (1968). The affinities of *Burramys* (Marsupialia: Phalangeroidea) as revealed by a study of its chromosomes. *Aust. J. Sci.*, 31, 40.

Hayman, D. L., Kirsch, J. A. W., Martin, P. G. and Waller, P. F. (1971). Chromosomal and serological studies of the Caenolestidae and their implications for marsupial evolution. *Nature*, 231, 194 - 5.

Hayman, D. L. and Martin, P. G. (1965a). Sex chromosome mosaicism in the marsupial genera *Isoodon* and *Perameles. Genetics*, 52, 1201 - 6.

Hayman, D. L. and Martin, P. G. (1965b). Supernumerary chromosomes in the marsupial *Schoinobates volans* Kerr. *Aust. J. Biol. Sci.,* 18, 1981 – 2.

Hayman, D. L. and Martin, P. G. (1969). Cytogenetics of marsupials, in *Comparative Mammalian Cytogenetics* (ed. K. Benirschke), Springer, New York, pp. 191 – 217.

Hayman, D. L. and Martin, P. G. (1974). *Monotremata and Marsupialia* vol. 4 *Chordata in Animal Cytogenetics, Gebrüder Borntraeger,* Berlin – Stuttgart.

Hayman, D. L., Martin, P. G. and Waller, P. F. (1969). Parallel mosaicism of supernumerary chromosomes and sex chromosomes in *Echymipera kalabu* (Marsupialia). *Chromosoma,* 27, 371 – 80.

Hungerford, D. A., La Bodie, Gundula U., Balahan, Gloria B., Messatzzia, Linda R., Haller, Gail and Miller, Alice E. (1971). Chromosome structure and function in Man IV. Provisional maps of the three long acrocentric autosomes (chromosomes 13, 14 and 15) at pachytene in the male. *Ann. Genet.,* 14, 157 – 260.

John, B. and Lewis, K. (1968). The chromosome complement. *Protoplasmologia.* Springer Verlag, Vienna.

Kirsch, John A. W. (in press). The classification of marsupials, in *Biology of Marsupials,* vol. II (ed. D. Hunseker). Academic Press, New York.

Kurnit, David M. and Maio, Joseph J. (1974). Variable satellite DNA's in the African green monkey. *Chromosoma,* 45, 387 – 400.

Lyon, Mary F. (1961). Gene action in the X-chromosome of the mouse (*Mus musculus* L). *Nature,* 190, 372 – 3.

Martin, P. G. and Hayman D. L. (1965). A quantitative method of comparing the karyotypes of related species. *Evolution,* 19, 157.

Martin, P. G. and Hayman, D. L. (1966). A complex sex-chromosome system in the hare wallaby *Lagorchestes conspicillatus* Gould. *Chromosoma,* 19, 159 – 75.

Martin P. G. and Hayman, D. L. (1967). Quantitative comparisons between the karyotypes of Australian marsupials from three different superfamilies. *Chromosoma,* 20, 290 – 310.

Matthey, Robert (1974). The chromosome formulae of eutherian mammals. *Cytoxonomy and Vertebrate Evolution,* Academic Press, London, pp. 531 – 616

Masremas, J. A. and Hatch, F. T. (1972). A possible relationship between satellite DNA and the evolution of kangaroo rat species (genus *Dipodomys*). *Nature New Biology,* 240, 102 – 5.

Ohno, S., Jainchill, J. and Stennius, C. (1963). The creeping vole (*Microtus oregoni*) as a genosomic mosaic. 1 – the OX/XY constitution of the male. *Cytogenetics,* 2, 232 – 9.

Ohno, S., Stennius, C. and Christian, L. (1966) The XO as the normal female of the creeping vole (*Microtus oregonii*), in *Chromosomes Today,* vol. I. (ed. C. D. Darlington and K. R. Lewis). Oliver and Boyd, Edinburgh and London.

Owen, A. R. G. (1953). A genetical system admitting of two distinct equilibria under natural selection. *Heredity,* 7, 97 – 102.

Rees, H. and Ahmad, K. (1963). Chiasma frequencies in *Lolium* populations. *Evolution,* 17, 575 – 9.

Reig, O. A. (1968). The chromosomes of the Didelphid marsupial *Marmosa robinsonni* Bangs. *Experientia,* 25, 185 – 6.

Reig, O. A. and Bianchi, N. O. (1969). The occurrence of an intermediate Didelphid karyotype in the short tailed opossum (genus *Monodelphis*). *Experientia,* 25, 1210 - 11.

Reig, O. A. Taul, F. D. and Sportono, O. A. (1972). *Sonderdruck aus Z. f. Saugertierkunde Bd.,* 37, HI, S37 − 42,

Reig, O. A. and Sonneschein, C. (1970). The chromosomes of *Marmosa fuscata* Thomas from Northern Venezuela (Marsupialia, Didelphidae). *Experientia,* 26, 199 - 200.

Ride, W. D. L. (1970). *A Guide to the Native Mammals of Australia.* Oxford University Press.

Sharman, G. B. (1961). The mitotic chromosomes of marsupials and their bearing on taxonomy and phylogeny. *Aust. J. Zool.,* 9, 38 - 60.

Sharman, G. B. (1974a). Marsupial taxonomy and phylogeny. *Aust. Mammology,* 1, 2, 137 - 54.

Sharman, G. B. (1974b). The chromosomes of non-eutherian mammals, in *Cytotaxonomy and Vertebrate Evolution,* Academic Press, London, pp. 480 - 530.

Sharman, G. B. and Barber, H. N. (1952). Multiple sex-chromosomes in the marsupial, *Potorous. Heredity,* 6, 345 - 55.

Southern, David I. (1969). Stable telocentric chromosomes produced following centric misdivision in *Myrmeleotettix maculatus* (Thunb.). *Chromosoma,* 26, 140 - 7.

Southern, E. M. (1970). Base sequence and evolution of guinea-pig satellite. *Nature,* 227, 794 - 8.

Stenzel, Peter (1974). Opossum Hb chain sequence and neutral mutation theory. *Nature,* 252, 62.

Sumner, A. T., Evans, H. J. and Buckland, R. A. (1971). New techniques for distinguishing between human chromosomes. *Nature New Biology,* 232, 31 - 2.

Walton, Shirley M. (1971). Sex-chromosome mosaicism in pouch young of marsupials *Perameles* and *Isoodon. Cytogenetics,* 10, 115 - 20

Webster, T. Preston, Hall, William P. and Williams, Ernest E. (1972). Fission in the evolution of a lizard karyotype. *Science,* 177, 611 - 13.

Young, H. S. (1971). Presumptive x-monosomy in Black Rats from Malaya. *Nature,* 232, 484 - 5.

Section 2   Origins and evolution of marsupials

Marsupials are currently found in two major regions of the world – South and Central America (from whence a single species has spread relatively recently across North America), and Australasia. Fossil remains show them to have been formerly more widespread in North America, and far more diverse in South America; they established little more than a token appearance in Europe, and have not yet been shown to have spread to South Africa or Antarctica. Chapter 4, by Professor William Clemens, reviews the early history and comparative anatomy of the group, as disclosed by a patchy fossil record, tracing their origin to the Americas – probably North America – in Cretaceous times.

Chapters 5 and 6 consider more fully the dynamics of marsupial evolution and zoogeography, especially in relation to recently gathered evidence from the field of plate tectonics. Professor Allen Keast is concerned especially with the validity of the possible Antarctic route between South America and Australia; his very broad review of recent literature will be of value in relation to many other zoogeographical problems of the southern hemisphere. Professor Peter Martin, though generally concurring with Keast's account, draws attention to recent evidence of tectonic activity in the Pacific area during the Cretaceous. This raises the possibility of additional migration routes across the Pacific which, now entirely lost by subduction, might help to explain a number of biogeographical anomalies – including the discontinuous distribution of marsupials – about the Pacific rim. Dr. Alan Ziegler (Chapter 7) provides a model for the adaptive radiation of an important but little-known group of marsupials in an extensive but little-known mountain forest environment. Where the familiar marsupials of Australia evolved primarily in open grassland and semi-desert, the 53 species of New Guinea – mostly endemic – are mainly forest-adapted; Ziegler's contribution includes an annotated check-list of this fauna, and an account of its relationships with the marsupial fauna of northern Australia.

4   *Phylogeny of the marsupials*

W. A. Clemens*

Introduction

The history of study of the Marsupialia begins early in the sixteenth century with discovery of the opossum by Spanish explorers of South America. As further accounts and specimens of the American and, soon thereafter, Australian marsupials reached Europe, biologists allied these new animals with a variety of eutherian (placental) mammals. Linnaeus classified the opossum with pigs, armadillos, hedgehogs and shrews. Kangaroos and wombats were thought to be related to rodents. During the nineteenth century de Blainville emphasised differences of reproductive systems in development of his classification and distinctly separated Didelphia, the marsupials, from the placentals or Monodelphia. It remained for subsequent biologists to recognise the evolutionary significance of this division.

During the latter years of the nineteenth and the twentieth centuries marsupials have been the object of a wide variety of biological investigations. The pace of this research activity ebbed and flowed with chance discoveries, carefully considered choices of research projects, or changes in emphases and interests of the scientific community. Currently it is at a high point, which appears to be the result of the interaction of at least three factors.

One contributing factor is certainly the recent, rapid and continuing expansion of our knowledge of the fossil record of marsupials. In 1953, when the late Professor R. A. Stirton began his research in Australia, the Tertiary record of marsupial evolution on that continent was minimal, almost non-existent. Knowledge of Quaternary history was heavily biased in favour of Late Pleistocene faunas. Through Stirton's efforts, work of American associates and now the varied researches of an increasing number of Australian palaeontologists, a general outline of the evolution of Australian terrestrial faunas from the Late Oligocene onward is beginning to emerge (Stirton, Tedford and Woodburne, 1968; Tedford, 1974; Ride, 1964).

In Tertiary South American faunas marsupials were the dominant carnivores, abundant omnivores and represented in niches for small herbivores. For many years our knowledge of these marsupials was essentially limited to fossils from Patagonia. Contributions by Pascual, Patterson, Paulo Couto, Simpson and others continue to increase understanding of these southern South American faunas. Also, as a result of the interests of Professor Stirton and his colleagues, as

*At the University of California Professor Clemens is a member of the Department of Paleontology and chairman of the interdisciplinary programme in Human Evolution, Prehistory and Paleoenvironments. His research interests range from investigations of Australian Tertiary faunas through studies of American and European Mesozoic mammals and Early Palaeocene faunas.

well as the research of Professors Hoffstetter, Patterson and others, a picture of the evolution of marsupials in the northern part of South America is beginning to emerge (Patterson and Pascual, 1972; Simpson, 1971).

This expansion of information has not been limited to the Southern Hemisphere. In Europe the long-known Tertiary marsupial *Peratherium* was the subject of thorough systematic revisions by von Koenigswald (1970) and Crochet (1969). Through improvement of collecting techniques the Late Cretaceous record of mammalian evolution in North America is being augmented and now documents a major adaptive radiation of marsupials (Clemens, 1971; Fox, 1971).

The wealth of new information from the fossil record is but one of the factors contributing to the resurgence of interest in evolution of marsupials. A second stems from the juxtaposition of investigations of insular situations as a focus for studies of evolutionary mechanisms, and the current revolution in the geological sciences spawned by demonstration that the composition and relative positions of continents and oceanic basins have been changing through the history of the Earth. This physical evolution is not to be viewed just in terms of opening or closing of routes of distribution by the joining or sundering of continental blocks. Change in positions of continents has modified their environments through shifts of latitude, and changing the patterns of circulation of marine currents, while tectonic activity concomitant to movement has influenced the development of their topography. These contributions from the geological sciences have broadened the dynamic, historical dimensions of biogeography.

Marsupials are a group of considerable antiquity, their first records coming from rocks of Late Cretaceous age. In comparison with the equally ancient eutherian mammals they were not as successful in either their dispersal to various continents or their evolutionary diversification. The impacts of changing configurations of continents and insular isolation are more vividly illustrated in their evolutionary history, which has become both a renewed focus of research and an example illuminating most of the flood of new books on biogeography.

Finally, 'there was a time when marsupials were regarded as second class mammals and the main justification for studying them was for the light they might shed on the evolutionary processes leading to higher mammals' (Tyndale-Biscoe, 1973). As the author of this comment and contributors to this volume thoroughly demonstrate, kangaroos and bandicoots, brush possums and opossums, and many other kinds of marsupials are now subjects of ecological, behavioural, physiological and functional studies warranted by their intrinsic scientific interest and sometimes stimulated by practical, economic necessity. The results of these researches require evaluation in light of the fossil record of evolution of marsupials and, reciprocally, are improving understanding of their phylogeny and evolutionary relationship to placentals.

Fossil record

Only a small proportion of the animals and plants that made up prehistoric biotas have left traces in the fossil record and only a minor proportion of this record has been sampled. As with all other major groups of organisms, the available documentation of the evolution of marsupials is incomplete, and is biased by both the processes of preservation and vagaries of sampling and study (Simpson, 1960).

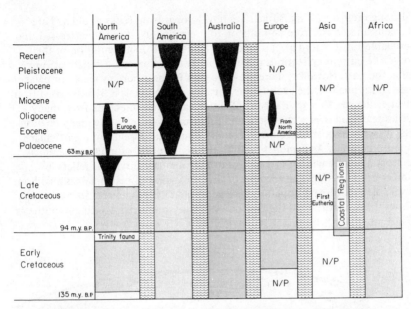

Figure 4.1 Chronological and geographical distribution of fossil and Recent records of marsupials shown by black shapes. Within each continent changes in diversity of marsupials through time are approximately represented by changes in width of shape. Different scales are used for each continent, so intercontinental comparisons cannot be made. Description of other symbols given in text.

Figure 4.1 gives a résumé of the current fossil record of the Marsupialia. The records for six continents are shown in vertical columns. Europe and Asia are represented as separate continents for, with some interruptions, during most of the Cretaceous and Early Cainozoic they were separated by a seaway (the Obik) crossing what is now the western Siberian lowlands to link the Arctic Ocean with the southern Tethys Sea (Figure 4.2). The wave pattern between the vertical columns indicates the presence of water barriers to free interchange of terrestrial vertebrates between continents. Some segments of these bars indicate the presence of wide oceanic basins; others record narrower, island-dotted straits that might have been crossed by a few terrestrial vertebrates. Where relevant, additional data are given in the following text. Finally, I have attempted to introduce explanations for absence of marsupials from parts of the currently available fossil records of these continents. The lightly shaded areas of the vertical columns indicate times and places where there are no fossil records of terrestrial mammals. Here the 'absence' of marsupials could be, and in some cases is certainly, an artifact of the available record. Unshaded sectors of the chart have a fossil record of terrestrial mammals. Sections of these are designated N/P when marsupials have not been found in well-sampled, local faunas from several areas within the continent. Here there is some basis but usually not unequivocal evidence for considering that the absence of marsupials is real and not an artifact of the available fossil record.

Although remains of marsupials are now known from rocks as old as Late Cretaceous and derive from deposits on four continents, it must be stressed that the available fossil record gives a woefully inadequate documentation of their history. The current collections record their distribution and diversity at some times in the past. However, there are large gaps in the record available to palaeontologists and much of the evolutionary history of mammals is hypothesised from indirect evidence. Also, even when collections are large and come from many localities scattered over a continent and presumably representing many environments, it is still impossible to prove beyond doubt the absence of a particular group of organisms. As demonstrated by the distribution of modern mammals, once geographically widespread species can tenaciously survive as relicts in limited areas. The following account of the phylogeny of marsupials is based on the available record, and hopefully makes adequate allowance for its many gaps; however, it is no more than the author's currently preferred working hypothesis.

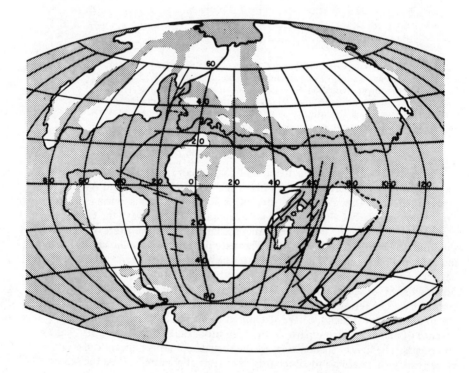

Figure 4.2 Reconstruction to show positions of continents and extent
of seas during the Maestrichtian (latest Cretaceous). In preparing
this map, surfaces of continental blocks or plates were assumed to
be above sea level unless sediments of marine origin are present.
Therefore the map depicts a minimum extent of Maestrichtian seas.

Mammalian origins

Most reconstructions of the relative positions of continents during the Mesozoic suggest that in the Triassic and Jurassic most of the modern continents were parts of one or two supercontinents. Although movement along some lines of fracture began earlier, the Cretaceous apparently was the time of major fragmentation and origin of broad oceanic barriers — for example, the southern Atlantic Ocean — that impeded or prevented dispersal of terrestrial vertebrates (see Hallam, 1973).

Kuehneotherium from the Late Triassic or Early Jurassic of southern Wales provides the oldest record of therian mammals and is considered to represent the radical of the stock that provided the common ancestors of marsupials and placentals (Kermack, Kermack and Mussett, 1968; Crompton, 1971). Occurring in the same complex of deposits in Britain are remains of morganucodontids, probably the oldest non-therian mammals (Prototheria *sensu* Hopson, 1970), a group represented today only by the monotremes. Also present are bones and teeth of tritylodontids, a group of mammal-like reptiles.

Kuehneotherium is known only from Wales but the morganucodontids are recorded in deposits in continental Europe, China, and South Africa. The tritylodontids have an even wider range, including Europe, Africa and the New World. Thus there is reason to suspect that though currently recorded only from Wales, Late Triassic or early Jurassic therians might have been inhabitants of most of the continents of the time.

During the Jurassic, epicontinental seas were widespread and terrestrial areas fragmented. In North America and Europe an evolutionary radiation of therians produced groups which are now classified in six families, divided between the two orders Symmetrodonta and Eupantotheria. At the close of the Jurassic and beginning of the Cretaceous, regression of the epicontinental seas appears to have provided another limited period for dispersal of terrestrial faunas. But soon thereafter this interchange was terminated, by the fragmentation and separation of continental blocks, and also flooding by extensive epicontinental seas (Lillegraven, 1974).

Triassic and Jurassic therians are known from only a few sites that, with one exception, are now in the Northern Hemisphere. However, data from the distribution of other terrestrial vertebrates, plus geological interpretations of the distribution of continents and development of epicontinental seas suggest that therians could have gained an essentially world-wide distribution at either the beginning or the end of the Jurassic. Most of the Early Cretaceous terrestrial regions of the Earth could have been inhabited by primitive therians, and are possible candidates for areas of origin of marsupials and placentals.

Cretaceous biotas

The Cretaceous was a time of great modification of terrestrial biotas, probably the most extensive in their entire history. Angiosperms are not certainly represented in Jurassic deposits. The first undoubted records are in sediments of Early Cretaceous age in middle latitudes of the Northern Hemisphere. From this area they spread poleward and diversified, so that by early in the Late Cretaceous angiosperms dominated terrestrial floras as far north as Alaska (Smiley, 1966; Baker and Hurd, 1968).

This revolutionary change in composition of terrestrial floras had a profound effect upon the evolution of terrestrial vertebrate herbivores. Among the lineages of dinosaurs, in the mid-Cretaceous the hadrosaurid, 'duck-billed' dinosaurs diversified from an iguanodontid radical and underwent a major evolutionary radiation. First records of the ceratopsians are of Late Cretaceous age, and in North America fossils document rapid diversification and occupation of a variety of niches for large herbivores. The extinction of stegosaurs, the origin and diversification of ankylosaurs and the modification in geographic ranges of sauropods were probably linked with change in floral composition.

Of the mammalian lineages known from the Late Jurassic only one group, the non-therian plagiaulacoid multituberculates, had dentitions appropriate for a herbivorous mode of life. The teeth of these multituberculates and their poorly known Early Cretaceous descendants do not show wide departures from what is assumed to be the primitive plagiaulacoid pattern. Then, beginning in the mid-Cretaceous, fossils record a major diversification in their dental morphology signalling the expansion of multituberculates into a variety of ecological niches (Kielan-Jaworowska, 1974).

The causal link between evolution of the angiosperms and origin of marsupials and placentals is suggested to be the changes in the terrestrial invertebrate fauna. It has been postulated that the association between plants and insects attracted to the plants' reproductive organs as a source of food began long before angiosperms appear in the fossil record (see Baker and Hurd, 1968). These authors note that a few members of the Cycadales and many of the Bennettitales show morphological evidence of being insect-pollinated. Also, they conclude that, probably, primitive angiosperms were insect-pollinated.

What caused the origin and rapid mid-Cretaceous dispersal is still a matter of speculation. However, it appears most likely that evolution of insects and other terrestrial invertebrates was closely interrelated with the change in floral composition and could have resulted in many new species of prey for insectivorous mammals. Increase in diversity, and possibly biomass, of terrestrial invertebrate faunas cannot be substantiated with evidence from their fossil record, which leaves more to be desired than that of mammalian evolution. But the few Cretaceous terrestrial invertebrates recovered so far provide some support. They include morphologically primitive members of the Lepidoptera (Mackay, 1970) and myrmecoid ants not far removed from tiphiid wasps (Wilson, Carpenter and Brown, 1966).

Additional evidence supporting this hypothesis can be drawn from studies of the function of mammalian dentitions. Therians of the Jurassic are characterised by the presence of molariform teeth made up of a series of shearing blades. Though mechanisms were present to prevent overclosure — driving the lower teeth into the palate — the teeth of these early therians lacked broad occluding surfaces with which to crush their food. The prey of these small, carnivorous mammals was either cut by the shearing blades or impaled and torn by the high cusps at the ends of the blades (see Crompton, 1971). Late in the Jurassic or early in the Cretaceous (the first record is provided by the Early Cretaceous therian *Aegialodon*; Kermack, Lees and Mussett, 1965) the tribosphenic type of molar evolved in therians probably derived from peramurid eupantotheres (Clemens and Mills, 1971). This type of molar differed from those of its ancestors in the addition of new shearing blades and evolution of a protocone on the upper molars and an occluding talonid basin on the lowers. The piston-like protocone functioned to crush food in the sharply delimited talonid basin. With the evolution of the tribosphenic molar the therian dentition had for the first time structures that could both cut and crush foodstuffs concentrated in a region of the jaws where maximum muscular forces could be applied. There is strong evidence that the last common ancestors of marsupials and placentals had molar teeth of the tribosphenic pattern.

Modern insectivorous mammals can be broadly classified in two groups according to their mode of feeding. The first group, probably the more derived, includes the non-therian Echidna, the Marsupial banded anteater (*Myrmecobius*) and the eutherian anteaters of South America and the Old World Pangolin, which catch large numbers of prey on their sticky tongues and swallow them with little or no preparation in the mouth. These insectivorous mammals are characterised by reduction or loss of the dentition. The second group, including other insectivorous marsupials (for example, dasyurids) and the eutherian shrews (Soricidae), capture individual insects and masticate them with molars that still clearly show the tribosphenic morphology.

Further support for the view that the tribosphenic dentition was particularly advantageous to insectivorous mammals comes from the Cretaceous histories of other lineages of apparently insectivorous therians and non-therians. Of the families represented in the Late Jurassic, paurodonts are unknown in Cretaceous faunas. Last records of the dryolestids are of Early Cretaceous age. Spalacotheriid symmetrodonts and triconodonts are present in the oldest known, North American, Late Cretaceous local fauna but were rare elements, in contrast to the abundant marsupials and placentals. In summary, evolution of the therian mammals with tribosphenic dentitions that were ancestral to marsupials and placentals appears to have been linked either with the origin of angiosperms or with their rapid mid-Cretaceous spread and diversification, and the coevolution of terrestrial invertebrates.

Marsupial origins

Rarely if ever have the time and place of origin of an order or class of vertebrates been determined with great precision or reliability. In part this stems from the nature of the event being sought. It is the differentiation of a species or a group of closely related lineages that might not have stood out as remarkably different from contemporaneous species but whose descendants form a group worthy of taxonomic recognition. Inadequacy of the available fossil record — both the frequency of stratigraphic (chronological) gaps separating collections, and incomplete sampling of contemporaneous faunas documenting the range of environmental variation — contribute to the lack of precision.

Evidence garnered from comparative studies of Late Cretaceous and Cainozoic marsupials and placentals indicates that they were derived from therian mammals with tribosphenic molars. None of the known Jurassic therians have this type of molar. An isolated tooth, the type and only specimen of *Aegialodon* from sediments of early Early Cretaceous age in England (Kermack *et al.*, 1965), provides the first record of a mammal with tribosphenic molars. The first sample of mammals with this advanced type of dentition that gives some hint of their initial diversity is derived from the Trinity Sandstone of Texas, which is of late Early Cretaceous age (Patterson, 1956; Slaughter, 1971).

The Trinity fauna is now represented by collections from six localities and is made up of a hundred or more specimens. Most are isolated, fragmentary teeth but a few partial maxillaries and mandibles were discovered. They include specimens of multituberculates, triconodonts and symmetrodonts, as well as those representing therians with tribosphenic dentitions and, possible, a more primitive species (Trinity molar type 6; Slaughter, 1971). Some of these therian teeth have been interpreted as representing a eutherian dubbed *Pappotherium;* others are the basis of the genus *Holoclemensia*, which has been allocated to the marsupial family Didelphidae (Slaughter, 1971). Although at one time I favoured these infraclass allocations, review of the evidence has caused me to revise my opinion.

The tribosphenic molar evolved at a time of great biotic change. Its acquisition was a major functional modification; for the first time each molar tooth could act to both cut and crush foodstuffs. A survey of the Tertiary record of mammalian evolution shows that similar major modifications in the dentitions of therian mammals – for example, the acquisition of enlarged ever-growing incisors by rodents or evolution of high-crowned (hypsodont) cheek teeth in equids – were followed by adaptive radiations. These produced many new 'experimental models' sharing the new character, which were quickly winnowed. The survivors gave rise to dominant, long-ranging lineages.

Using these Tertiary events as models, we can expect that with acquisition of the tribosphenic dentition therians underwent an adaptive radiation. Also, it is to be expected that many lineages might exhibit dental characters formerly thought to be diagnostic of marsupials or placentals, but may not have been involved in the ancestry of these groups. *Deltatheridium* from Mongolia (Butler and Kielan-Jaworowska, 1973) and *Potamotelses* from Canada (Fox, 1972), both of Late Cretaceous age, appear to be relics of this radiation, with the former clearly exhibiting a mixture of characters once thought to be diagnostic of marsupials and placentals.

Many of the morphological similarities between the Trinity therian teeth and those of marsupials and placentals are the result of shared, primitive characters. A few noted by Slaughter (1971) might be shared derived characters. It cannot be denied that additional evidence could demonstrate the metatherian or eutherian affinities of some of the Trinity therians, but until that is done I suggest a return to Patterson's (1956) view that these Trinity therians are of metatherian or eutherian grade but uncertain infraclass affinities. Turnbull (1971) came to a similar conclusion and proposed the Order Tribosphena to include these therians and some possibly ancestral forms, e.g. *Aegialodon* and *Peramus*.

At present the oldest records of unequivocal marsupials come from the Milk River Formation of Alberta, Canada (Fox, 1971). This formation was deposited during the Campanian Age of the Late Cretaceous (approximately 70 – 80 million years B.P.). The oldest published records of placental mammals, based on excellent collections including skulls and partial skeletons, come from the Djadokhta Formation of Mongolia (Kielan-Jaworowska, 1969). The age of these deposits is not as well substantiated as that of the Milk River Formation.

Currently the Djadokhta Formation is thought to be Late Cretaceous and some what older than the Milk River Formation.

Thus the fossil record suggests that the time of origin of the Marsupialia was pre-Campanian; some time in the later part of the Early Cretaceous or early in the Late Cretaceous seems likely. The Americas are usually regarded as the area of origin of the marsupials but, as will be discussed below, there is argument as to whether it was most likely North or South America. Finally, considerable research based on comparative studies of modern marsupials and placentals is illuminating their phylogenetic relationships, particularly the evolution of their reproductive systems, which markedly differentiate the two groups (Lillegraven, 1969; Parker, this volume).

Holarctic marsupials

During most of the Late Cretaceous North America was divided into eastern and western continents by the Western Interior Sea, an epicontinental sea linking the Gulf of Mexico with the Arctic Ocean (figure 4.2). Streams draining the eastern flanks of the western continent deposited a great thickness of sediments on the coastal lowlands in what are now parts of Alberta, Montana, Wyoming, Colorado, Utah and New Mexico. These deposits have been the source of many collections of fossil vertebrates that document the course of marsupial evolution during the last 20 million years of the Cretaceous, approximately from 80 to 63 million years B.P. (Clemens, 1966; Fox, 1971; Lillegraven, 1969, 1974; Sahni, 1972).

The oldest of these western North American local faunas that from the Milk River Formation (Fox, 1971), contains at least seven species of marsupials representing five genera. After the multituberculates they are the second most diverse group of mammals in the local fauna; eutherians are rare and represented by only two species. The mammals of the latest Cretaceous Lance local fauna show roughly the same taxonomic distribution. Marsupials, represented by twelve species from four genera, are slightly more diverse than the multituberculates, and both groups greatly outnumber the rare eutherians.

The Cretaceous marsupials occupied a variety of niches for mammals ranging from the size of shrews up to that of a small terrier. Their dentitions show few major departures from the primitive tribosphenic pattern, suggesting that they were primarily omnivorous or carnivorous animals. There are a few departures. The highly modified premolars and enlargement of some of the molar shearing crests in the dentition of *Didelphodon* might be related to a scavenging habit or a diet based on fresh-water invertebrates. A small marsupial, *Glasbius,* has teeth on which the cusps have a low, rounded relief and other modifications suggesting a frugivorous diet. It is the only marsupial in these faunas showing any indication of dental modifications that might be linked to a herbivorous diet.

In comparison with Cainozoic evolutionary radiations of marsupials, their diversity in the Cretaceous was not great. However, during the Late Cretaceous, probably because of the competition of dinosaurs and other reptiles, mammals were able to occupy only a limited number of niches. In the oldest North American local faunas marsupials filled nearly half of these and maintained, if not slightly increased, their adaptive diversity until the end of the Cretaceous. Judged in the context of Cretaceous mammalian evolution, marsupials underwent a significant Late Cretaceous radiation in North America.

Recently the determination of the area of origin of marsupials has been mooted (Tedford, 1974). The following evidence favours the hypothesis of North American origins. A suitable ancestral stock represented by the Trinity therians of metatherian – eutherian grade was present. The oldest records of the marsupials are provided by morphologically primitive species in a local fauna on the western Late Cretaceous, North American land mass. Finally, this area provides the records of the first adaptive radiation of marsupials. Although the possibility of origin in South America cannot be ruled out (see below), the hypothesis of northern origins still remains viable.

At the close of the Cretaceous North American marsupials almost shared the fate of the dinosaurs. All but members of the genus *Alphadon* became extinct. In the Early Palaeocene North American faunas were dominated by multituberculates and placentals; marsupials are represented by only one monotypic genus, *Thylacodon* (probably a synonym of *Peralestes*). Through the rest of the early Cainozoic until the end of the Early Miocene they remained persistent, though minority members of the North American fauna. In the United States and Canada many local faunas of Middle Miocene through Middle Pleistocene age have been sampled with techniques specially designed to recover records of mammals of small size. No remains of marsupials have been discovered. The possibility that they survived in Central America cannot be excluded. Late in the Pleistocene the opossum, *Didelphis*, which appears to have evolved in South America, invaded eastern North America (Gardner, 1973).

Jardine and McKenzie (1972) argued that North America and Eurasia were 'effectively a single continent in the Cretaceous', so it would follow that the occurrence of marsupials in Late Cretaceous deposits of North America strongly suggests that they also occurred in Eurasia. This conclusion reflects a poorly appreciated danger in the use of maps prepared by geologists and geophysists showing the distribution of continental plates as a basis for biogeographic studies. The position of these plates certainly controls the distribution of land masses; however, it is the position of the shore lines of the seas that surrounded and frequently submerged parts of continental plates that directly controls the distribution of terrestrial animals. Taking evidence of the distribution of seas into account, Lillegraven (1974) concludes that the Cretaceous distribution of terrestrial faunas was characterised by fragmentation of continental areas, endemic evolution and limited opportunities for intercontinental dispersal. The occurrence of marsupials in Eurasia cannot simply be deduced from the distribution of continental plates.

Only one mammalian fossil, the molar of a eutherian, is known from the Late Cretaceous of Europe and fossils of Early Palaeocene age have yet to be described. However, several local faunas of Middle and Late Palaeocene age are known, some from extensive samples obtained with modern collecting techniques (Russell, 1975). The absence of marsupials from these collections most likely reflects their absence from the European Palaeocene fauna. At the beginning of the Eocene there was a period of remarkably free dispersal of mammals from North America to Europe resulting in common occurrence of over 50 per cent of the genera on the two continents (Russell, 1975). The didelphid marsupial *Peratherium* was one of the mammals that dispersed to Europe at this time and survived there well into Miocene time (von Koenigswald, 1970).

There is no evidence that *Peratherium* reached Asia from Europe during the intermittent Eocene regressions or the final retreat of the Obik Sea that drained the Turgai Straits (Russell, 1975). The possibility of dispersal of marsupials into Asia via a route across the region of the modern Bering Straits has been suggested (Clemens, 1968). Two questions have become involved here. Could marsupials have been part of the Cretaceous or early Tertiary faunas of Asia? If they were, could they have been the stock that was ancestral to the marsupial fauna of Australia?

The second question can be directly and negatively answered. The current reconstructions of the distribution of continents in the Cretaceous and early Cainozoic (Tedford, 1974) place Australia well to the south of its present position. A major oceanic barrier, probably too wide to be breached even by chance, sweepstakes routes of dispersal (McKenna, 1973), separated Australia and Asia. Only late in the Cainozoic, long after marsupials became part of the Australian fauna, was this barrier sufficiently narrowed by the northward movement of Australia to be successfully breached by non-volant, terrestrial mammals (murid rodents).

Several years ago most of the collections of Late Cretaceous and Early Tertiary terrestrial mammals of Asia came from sites in central Asia, particularly Mongolia, and all lacked evidence of the presence of marsupials. A few eutherians of Eocene age discovered at sites on the Pacific coast of China show close resemblances to North American mammals. These fossils suggest that early in the Tertiary mammals could have dispersed from North America to Asia along the Pacific coast but were limited in their distribution to coastal lowlands, with environmental barriers preventing them from reaching central parts of Asia.

Collections of Late Cretaceous and Early Tertiary Asian mammals have been greatly increased in recent years without discovery of marsupials. Reconstruction of the positions of the continents, indicating that the Bering Region was at very high latitudes at this time, suggests another barrier to possible dispersal of North American marsupials (Szalay and McKenna, 1971). However, the similarities of Eocene faunas of coastal China and North America remain. Until these Chinese sites are thoroughly sampled and studied, the possibility of the presence of *Peratherium* or another vagrant marsupial in Asia should not be excluded.

In summary, the available fossil record strongly suggests that marsupials originated in the Americas some time in the Cretaceous. Apparently, oceanic barriers resulting from the origin of new oceanic basins and spread of epicontinental seas restricted their Holarctic distribution to the New World until later in the Cainozoic. Many lineages of North American marsupials became extinct at the close of the Cretaceous; however, didelphids survived until the end of the Early Miocene. North American species of *Didelphis* are very recent immigrants from South America. The didelphid *Peratherium* reached Europe at the beginning of the Eocene and survived there into the Miocene. Marsupials might have reached the Pacific coasts of Asia in the Early Tertiary. This possibility cannot be eliminated. In view of the current reconstruction of the distributions of continents based on geological and geophysical data, these marsupials, if present in Asia, were probably not the ancestors of the Australian marsupial fauna.

South American marsupials

Although heavily biased in favour of localities in Argentina, the fossil record from
the Late Palaeocene through the Pleistocene now provides a broad outline a South
American mammalian evolution. In the Late Palaeocene three groups of mammals
were part of the South American fauna: the eutherian edentates and condylarths,
and the marsupials. The earliest South American marsupials appear to have been
opossum-like omnivores. Through most of the Cainozoic some of their descendants
filled niches for carnivores until they were replaced by phororhacoid ground birds
(Marshall, 1976) or by eutherians dispersing from North America. Most of the
niches for herbivores, particularly those available to animals of large body size,
were occupied by edentates and descendants of the early condylarths. However,
several lineages of marsupials appear to have converged with the eutherian rodents.
This radiation of dominantly omnivorous and carnivorous marsupials is of the
same magnitude as that of their Australian ecological vicars. It has been the subject
of several recent surveys (Patterson and Pascual, 1972: Simpson, 1971). Rather
than attempt to summarise these excellent articles, I wish to comment on three
related and contentious problems.

Attempts to correlate the sequence of South American terrestrial faunas with
their Holarctic counterparts have been fraught with difficulties. Until late in the
Cainozoic, the South American fauna was largely composed of endemic animals;
few intercontinental dispersals link the faunal sequences. Also, only a few instances
of interdigitations of fossiliferous marine and non-marine strata have been discovered. These provide some basis for application of Lyellian epochs (Palaeocene,
Eocene, etc.) to the South American time scale. Currently only two radiometric
age determinations have been obtained from rocks that can be related to the
terrestrial faunal sequence.

A small sample of a local fauna found near Laguna Umayo in Peru has been
regarded by its describers as being of Late Cretaceous age (see Sigé, 1972). This age
determination is based on studies of charophytes and fragments of egg shells
thought to be dinosaurian. A recent survey of the geology of the area (Portugal,
1974) adds support for the correlation but also illustrates its tenuous foundation.
The Laguna Umayo local fauna might be as old as Late Cretaceous, but even this
level of refinement does not seem warranted by the limited evidence available.

However, this local fauna does appear to provide the oldest record of mammals
in South America. These include a primitive condylarth, *Perutherium*. and several
kinds of didelphid marsupials; one species is allocated to the otherwise North
American Cretaceous genus *Alphadon*. Sigé (1972) cautiously identified other
fragmentary teeth from Laguna Umayo as possibly representing a second group
of marsupials hitherto only known from the North American Cretaceous, the
Pediomyidae. The characters cited to support this identification are certainly
pediomyid-like. I doubt whether they are diagnostic, and still hold to the view
that all South American marsupials could be derived from the didelphid *Alphadon*,
or a closely related genus, that reached South America sometime in the Cretaceous.

The hypothesis of North American origin of the Marsupialia has been challenged
by Tedford (1974). He quite rightly argues that the absence in South America of
samples of Cretaceous local faunas, older than that from Laguna Umayo, which
might contain marsupials, reduces the value of the priority of their first appearance

and adaptive radiation in North America. In the absence of a fossil record, he has developed a hypothetical phylogeny of South American Cretaceous marsupials, with chance dispersals to explain their occurrence in North America.

Tedford (1974) goes on to argue that the scope of the adaptive radiation of marsupials recorded in the Laguna Umayo and other local faunas of unequivocal Tertiary age, 'suggests that marsupials may have been on that continent (South America) for a long time, possibly from early in the Cretaceous'. This assessment cannot be disproven. However, considering the scope of the adaptive radiations of the marsupials in Australia (which Tedford suggests reached that continent in the Late Cretaceous or Palaeocene) or those of Primates or the Condylarthra in the North American Palaeocene, his argument loses force. Currently the available evidence favours the hypothesis of origin of marsupials in the Americas, although origin in Antarctica (or even possibly, but unlikely, Australia) cannot be lightly dismissed. Of the working hypotheses I still prefer that of North American origins, which is based on a Cretaceous fossil record. Only when comparable samples of the South American Cretaceous fauna become available will there be a basis for choice.

Finally, I turn to the American Opossum (*Didelphis*), which is gaining popularity among the mammals employed in biological research and is often termed a 'living fossil'. The factual bases for this claim warrant special attention. It has been shown that the dentition of *Didelphis* differs in several significant features from the dentitions of species of the Cretaceous didelphid *Alphadon,* which on various grounds are thought to show the primitive marsupial dental morphology (Clemens, 1968).

Recently Gardner (1973) demonstrated that populations of modern *Didelphis* in northern and middle America represent two species. *D. virginiana* occurs in tropical to temperate habitats extending from southwestern Nicaragua and Costa Rica northward to Canada. Many laboratory colonies in the United States consist of members of this species. *D. marsupialis* is confined to tropical habitats extending from southern Tamaulipas, Mexico, to the northern border of Argentina. Gardner (1973) reports a variety of differences in morphology, patterns of environmental tolerances and karotype separating the two species.

The oldest records of *Didelphis* come from South American deposits of approximately Pliocene age. *Didelphis marsupialis* or its ancestor appears to have dispersed into tropical Central America by an island-hopping or sweepstakes route some time prior to full emergence of the Panamanian isthmus (Hershkovitz, 1969; Gardner, 1973). The latter author postulates that *D. virginiana* differentiated from *D. marsupialis* late in the Pleistocene and, because of its tolerance to temperate climates, has been able to spread northward into the United States and Canada (Guilday, 1958).

Phrases such as 'living fossil' must be taken in the context of the well-established concept of mosaic evolution. Relative to other organisms some species are characterised by the retention or relatively limited modification of many primitive characters. Admittedly, in other characters these animals can be highly derived. Current investigations suggest that in many features *Didelphis virginiana* is distinctly derived relative to other Recent and prehistoric didelphids. Data derived from studies of *D. virginiana* cannot be regarded *ipso facto* as representative of the primitive characters of didelphids or the Marsupialia.

Australia

Origin and evolution of the Australian terrestrial fauna have been the subject of or discussed in a wide variety of recent publications. The following are mentioned because of their inherent interest, diversity of viewpoints and excellent biblio- graphies: Cox, 1974; Cracraft, 1974; Hoffstetter, 1972; Jardine and McKenzie, 1972; Tedford, 1974; and Tyndale-Biscoe, 1973.

Clearly one of the major contributions towards unravelling the ancestry of the Australian marsupials comes from geological and geophysical studies. After years of inconclusive debate based upon biological data, studies of the physical evolution of the Earth's crust are making positive contributions. They show that through the Cretaceous and Cainozoic Antarctica maintained a south polar position and until some time in the Cretaceous Australia was part of this southern continent. After some Late Cretaceous and Palaeocene fracturing, associated with formation of epicontinental seas, Australia broke free from Antarctica and since the beginning of the Eocene has moved northward over more than 20° of latitude (particularly see Tedford, 1974, for discussion and references).

Using just the results of these geological studies, it becomes apparent that terrestrial vertebrates might have dispersed from Antarctica into Australia during the Cretaceous and Palaeocene. By possibly some time in the Oligocene the seaway separating the two continents had become wide enough to prohibit dis- persal even by the sweepstake route. To the north of the Australian continent during the Cretaceous and most of the Tertiary a wide sea separated it from south-eastern Asia. Only late in the Tertiary had Australia moved far enough north and sufficiently narrowed this seaway to permit dispersal via a sweepstake route. To turn to the biological evidence: the oldest marsupials known from the Australian continent are of Late Oligocene age and their diversity suggests that marsupials had been parts of the Australian biota for a considerable time (Tedford *et al.*, 1975). Therefore, if marsupials originated in the Americas, the route of dispersal into Australia must have been from South America via Antarctica.

Evidence from a variety of sources depicts something of the nature of this Antarctic route of dispersal. The oft-cited similarities between South American borhyaenids and the Tasmanian 'wolf' (*Thylacinus*) or between the South American caenolestoids and Australian phalangeroids are still best regarded as excellent examples of parallel evolution. All lineages of Australian marsupials are probably descendants of a didelphid immigrant, the only group of mammals known to have dispersed across Antarctica during the Late Cretaceous or Early Tertiary. Apparently the route was not utilised by other groups of South American marsupials or the eutherian edentates and condylarths, nor did mono- tremes disperse across it into South America. This pattern of chance dispersal is typical of a sweepstakes route. Geological information is in accord, for it suggests that part of this dispersal route was via a volcanic island chain, rather than along a continuous dry land connection (see Tedford, 1974). Early Tertiary antarctic floras were characterised by *Nothofagus* and araucarian pines suggesting temperate climates at least in coastal situations (see Raven and Axelrod, 1972); but it was an area under the influence of a polar daylight regime. Thus probably stringent environmental filters barred many animals from even being candidates for sweepstakes dispersal.

Finally, during the past two decades many concentrations of Late Tertiary and Pleistocene fossils have been discovered in Australia, and provide a general outline of marsupial evolution during this period (Stirton *et al.*, 1968; Tedford, 1974). Although a few have not yet appeared in the Tertiary fossil record, most of the families represented in the Pleistocene – Recent Australian fauna are also known in Miocene faunas. For some — for example, the kangaroos and extinct diprotodontids — major evolutionary radiations occurred during the Late Tertiary and Pleistocene. In earlier Miocene and Late Oligocene faunas there are several animals that cannot be classified in modern families (*Wynyardia*), or that represent highly derived phylogenetic side branches (*Ektopodon*). These mammals could well be relics of an earlier evolutionary radiation.

Conclusions

Currently research on the Marsupialia is going forward on a variety of fronts. Data from new collections of fossil marsupials, an increasingly broad spectrum of studies of modern forms and better understanding of the physical evolution of the Earth's crust have contributed to a better understanding of the phylogeny of the marsupials. It must be stressed, however, that the phylogenetic hypotheses are based in part on incomplete palaeontological documentation, and future discoveries could cause radical modifications.

The author favours the following phylogenetic working hypothesis. Origin of the marsupials was one aspect of major terrestrial biotic changes during the mid-Cretaceous. Most likely, marsupials differentiated from a primitive therian stock somewhere in the Americas, probably North America. In the Eocene one group of didelphids reached Europe from North America and survived there well into the Miocene. The possibility that didelphids reached Pacific coastal regions of Asia in the Late Cretaceous or Early Tertiary cannot be excluded. However, the distribution of continents at that time suggests that marsupials reached Australia via Antarctica, not from Asia.

References

Baker. H. G. and Hurd, P. D. (1968). Intrafloral ecology. *Ann. Rev. Entomol.* **13**, 385 – 414.

Butler, P. M. and Kielan-Jaworowska, Z. (1973). Is *Deltatheridium* a marsupial? *Nature*, **245**, 105 – 106.

Clemens, W. A. (1966). Fossil mammals of the type Lance Formation, Wyoming. Part II. Marsupialia. *Univ. Calif. Publ. Geol. Sci.*, **62**, 1- 122.

Clemens, W. A. (1968). Origin and early evolution of marsupials. *Evolution*, 22, 1 - 18.

Clemens W. A. (1971). Mammalian evolution in the Cretaceous, in *Early Mammals* (ed. D. M. and K. A. Kermack), *J. Linn. Soc. (Zool).*, *suppl. vol. 50, pp. 165 - 80.*

Clemens, W. A. and Mills, J. R. E. (1971). Review of *Peramus tenuirostris* Owen (Eupantotheria: Mammalia). *Bull. Br. Mus. (Nat. Hist.) Geol.*, **20**, 89 - 113.

Cox, C. B. (1974). Vertebrate palaeodistributional patterns and continental drift. *J. Biogeography*, **1**, 75 – 94.

Cracraft, J. (1974). Continental drift and vertebrate distribution. *Ann. Rev. Ecol. Systemat.*, **5**, 215 – 61.

Crochet, J. Y. (1969). Revision du genre *Peratherium* Aymard 1849 (Marsupialia). *C. R. Acad. Sci. Paris,* 268, 2038 -41.

Crompton, A. W. (1971). The origin of the tribosphenic molar, in *Early Mammals* (ed. D. M. and K. A. Kermack), *J. Linn. Soc. (Zool).,Suppl.,* vol. 50, pp. 65 - 88.

Fox, R. C. (1971). Marsupial mammals from the early Campanian, Milk River Formation, Alberta, Canada, in *Early Mammals* (ed. D. M. and K. A. Kermack), *J. Linn. Soc. (Zool.), Suppl.* vol. 50, pp. 145 - 64.

Fox, R. C. (1972). A primitive therian mammal from the Upper Cretaceous of Alberta. *Can. J. Earth Sci.,* 9, 1479 - 94.

Gardner, A. L. (1973). The systematics of the genus *Didelphis* (Marsupialia: Didelphidae) in North and Middle America. *Sp. Publ., Mus., Texas Tech. Univ.,* No. 4.

Guilday, J. E. (1958). The prehistoric distribution of the opossum. *J. Mammalogy,* 39, 39 - 43.

Hallam, A. (1973). *A Revolution in the Earth Sciences, from Continental Drift to Plate Tectonics,* Oxford University Press, Oxford.

Hershkovitz, P. H. (1969).The evolution of mammals on southern continents. VI. The Recent mammals of the Neotropical Region: a zoogeographic and ecological review. *Quart. Rev. Biol.,* 44, 1- 70.

Hoffstetter, R. (1972). Données et hypothese concernant l'origine et l'histoire biogeographique des Marsupiaux. *C. R. Acad. Sci. Paris,* 274, 2635 - 8.

Hopson, J. A. (1970). The classification of nontherian mammals. *J. Mammalogy,* 51, 1- 9.

Jardine, N. and McKenzie, D. (1972). Continental drift and the dispersal and evolution of organisms. *Nature,* 217, 418 - 20.

Kermack, D. M., Kermack, K. A. and Mussett, F. (1968). The Welsh pantothere *Kuehneotherium praecursoris. J. Linn. Soc. (Zool.),* 47, 407 - 23

Kermack, K. A., Lees, P. M., and Mussett, F. (1965). *Aegialodon dawsoni,* a new trituberculosectorial tooth from the Lower Wealden. *Proc. R. Soc. (Lond.) Ser. B.,* 162, 535 - 54.

Kielan-Jaworowska, Z. (1969). Results of the Polish-Mongolian paleontological expeditions. Pt. I. Preliminary data on the Upper Cretaceous eutherian mammals from Bayn Dzak, Gobi Desert. *Palaeontol. Polon.,* 19, 171 - 91.

Kielan-Jaworowska, Z. (1974). Migrations of the Multituberculata and the Late Cretaceous connections between Asia and North America. *Ann. S. Afr. Mus.,* 64, 231 - 43.

von Koehigswald, W. (1970). *Peratherium* (Marsupialia) im Ober-Oligozanund Miozän von Europa. *Bayer. Akad. Wiss., Math.-Naturwiss. Klasse, Abhl. N. F.,* 144, 1- 79.

Lillegraven, J. A. (1969). Latest Cretaceous mammals of upper part of Edmonton Formation of Alberta, Canada, and review of marsupial-placental dichotomy in mammalian evolution. *Paleont. Contrib., Univ. Kansas,* Art. 50 (Vert. 12), 1-122.

Lillegraven, J. A. (1974). Biogeographical considerations of the marsupial-placental dichotomy. *Ann. Rev. Ecol. Systemat.,* 5, 74 - 94.

Mackay, M. R. (1970). Lepidoptera in Cretaceous amber. *Science,* 167, 379 - 80.

McKenna, M. C. (1973). Sweepstakes, filters, corridors, Noah's arks, and beached Viking funeral ships in palaeogeography, in *Implications of Continental Drift to the Earth Sciences* (ed. Tarling, D. H. and Runcorn, S. K.), Academic Press, London and New York, pp. 295 - 308.

Marshall, L. G. (1976). Evolution of the Borhyaenidae, extract South American predacious marsupials'. Ph. D. dissertation, University of California.

Patterson, B. (1956). Early Cretaceous mammals and the evolution of mammalian molar teeth. *Fieldiana, Geol.* 13, 1 - 105.

Patterson, B. and Pascual, R. (1972). The fossil mammal fauna of South America, in *Evolution, Mammals, and Southern Continents* (ed. Keast, A., Erk, F. C. and Glass, B.) pp. 247 - 309.

Portugal, J. A. (1974). Mesozoic and Cenozoic stratigraphy and tectonic events of Puno-Santo Lucia area, Department of Puno, Peru. *Amer. Assoc. Petrol. Geol. Bull.,* 58, 982 - 99.

Raven, P. H. and Axelrod, D. I. (1972). Plate tectonics and Australasian paleobiogeography. *Science,* 176, 1379 - 86.

Ride, W. D. L. (1964). A review of Australian fossil marsupials. *J. Proc. R. Soc. West Aust.,* 47, 97 - 131.

Russell, D. E. (1975). Paleoecology of the Paleocene-Eocene transition in Europe, in Szalay, F. S., 'Approaches to Primate Paleobiology', *Cont. Primate.,* 5, 28 - 61.

Sahni, A. (1972). The vertebrate fauna of the Judith River Formation, Montana, *Bull. Amer. Mus. Nat. Hist.,* 147, 321 - 412.

Sigé, B. (1972). La faunule de mammifères du Crétacé supérieur de Laguna Umayo (Andes peruviennes). *Bull. Mus. Nation. d'Hist. Nat. Paris,* 3e sér. No. 99, Sci. Terr., 19, 375 - 409.

Simpson, G. G. (1960). The history of life, in *Evolution after Darwin* (ed. Tax, S.), vol. 1, University of Chicago Press, Chicago, pp. 117 - 80.

Simpson, G. G. (1971). The evolution of marsupials in South America. *An. Acad. Brasil Acad. (Suplemento),* 43, 103 - 18.

Slaughter, B. H. (1971). Mid-Cretaceous (Albian) therians of the Butler Farm local fauna, Texas, in *Early Mammals* (ed. Kermack, D. M. and K. A), *J. Linn. Soc. (Zool), Suppl,* vol. 50, pp. 131 - 43.

Smiley, C. J. (1966). Cretaceous floras from Kuk River area, Alaska: stratigraphic and climatic interpretations. *Geol. Soc. Amer. Bull.,* 77, 1 - 14

Stirton, R. A., Tedford, R. H. and Woodburne, M. O. (1968). Australian Tertiary deposits containing terrestrial mammals. *Univ. Calif. Publ. Geol. Sci.,* 77, 1 - 30.

Szalay, F. S. and McKenna, M. C. (1971). Beginning of the age of mammals in Asia. the Late Paleocene Gashato fauna, Mongolia. *Bull. Amer. Mus. Nat. Hist.,* 144, 271 - 317.

Tedford, R. H. (1974). Marsupials and the new paleogeography, in *Paleogeographic Provinces and Provinciality* (ed. C. A. Ross), *Soc. Econ. Paleont. Mineral. Sp. Publ.* 21, pp. 109 - 26.

Tedford, R. H., Banks, M. R., Kemp, N. R., McDougall, I, and Sutherland, F. L. (1975). Recognition of the oldest known fossil marsupials from Australia. *Nature,* 255, 141 - 2.

Turnbull, W. D. (1971). The Trinity therians: their bearing on evolution in marsupials and other therians, in *Dental Morphology and Evolution* (ed. Dahlberg, A. A.), University of Chicago Press, Chicago, pp. 151 - 80.

Tyndale-Biscoe, H. (1973). *Life of Marsupials,* Edward Arnold, London.

Wilson, E. O., Carpenter, F. M. and Brown, W. L Jr. (1966). The first Mesozoic ants, with the description of a new subfamily. *Psyche,* 74, 1 – 19.

5 Historical biogeography of the marsupials

Allen Keast*

Demonstration that the continents have moved extensively in geological time has led to a mass of palaeogeographic reassessments and reinterpretations of the early history of the marsupials, and of other animal and plant groups: see Keast (1971; republished as 1972a), Jardine and McKenzie (1972), Fooden (1972), Smith (1972), Raven and Axelrod (1972, 1974), Cracraft (1973, 1974), Cox (1973), Lillegraven (1974), Tedford (1974) and Rich (1975). Shorter discussions of specific aspects of the dispersal history of marsupials are contained in Lillegraven (1969), Hoffstetter (1970), Martin (1970) and Cox (1970). Several of the above writers include comprehensive reviews of the newer geological data on the changing palaeopositions of the continents. Tedford (1974) and Howden (1974) incorporate epicontinental seas into their reconstructions, and Cracraft, Raven and Axelrod, and Rich incorporate some data on palaeoclimates. The result is that we now have a much sounder basis for making deductions about marsupial palaeo-biogeography. Many of our assertions and conclusions remain, however, highly tentative. There still exists in the literature 'rampant contradictions and mis-interpretations' (Lillegraven, 1974: p. 263).

Six different theories have been advanced in recent years as to the early dispersive history of the marsupials.

(1)   Marsupials arose, and/or differentiated, in North America (Clemens, 1968; Lillegraven, 1969), and from there dispersed across a broad water-gap to South America in the Late Cretaceous (Simpson, 1950; Patterson and Pascual, 1968, 1972) (see figure 5.1). Eocene dispersal to Europe occurred across the north Atlantic, possibly by the 'De Geer route' of McKenna (1972). This history is based on solid palaeontological data. In North America the only continent from which there is a good Cretaceous fossil record, the marsupials had differentiated into three families and at least five genera and 13 species by the end of that period (Clemens, 1968; Lillegraven, 1974).

(2)   Marsupials arose in the South American part of Gondwanaland and dispersed northwards from there across the water-gap to North America, and thence to Europe (Tedford, 1974). They reached Australia via Antarctica. The

Professor Allen Keast is a zoogeographer working on evolutionary aspects of ecology and speciation in vertebrates, with a particular interest in the evolution of vertebrate faunas of the southern continents. A graduate of the University of Sydney, he was awarded a Ph.D. at Harvard University. Formerly Curator of Birds, Reptiles and Amphibians at the Australian Museum, he is now Professor of Biology at Queen's University, Kingston, Ontario.

Cretaceous marsupial record for South America, however, is very inadequate and, in contrast to North America, where fossils extend back to the Albian (100 m.y. B.P.), is limited to the latest part of the period. Two forms are known: one belongs to the 'North American' genus *Alphodon* and the other has been dubiously identified as a member of the North American family Pediomyidae (Sige, 1972). Tedford, however, places considerable weight on the spectacular diversity of the Palaeocene marsupial fauna of South America, relative to the somewhat limited one of the North American Cretaceous, as demonstrating a very long evolutionary history here.

(3)   Marsupials dispersed from the Americas to Australia by way of Asia. This was the accepted theory of the 'static continents' period — see Simpson (1940, 1961), Darlington (1965), and Keast (1972b). It is invalidated by the far southern position of Australia in the Cretaceous (Figure 5.6).

(4)   Marsupials reached Australia by way of South America and Antarctica — see Harrison (1924), Hoffstetter (1970), Lillegraven, Tedford, etc. There are strong arguments in favour of this route — see later.

(5)   Marsupials reached Australia by way of Africa (Cox, 1970). This theory, which received no acceptance and was subsequently dropped by its author (Cox, 1973), has recently been rejuvenated by Raven and Axelrod (1974), based on the reconstruction of Smith, Briden and Drewry (1973) which suggests that Africa may have remained in contact with India, Antarctica and Australia until about Albian times (100 m.y. B.P.) (see Figure 5.2) These authors argue that since the theory required a southward passage of no more than 35 degrees, Darlington's (1965) criticism that the South America - Antarctica route was too cold for the marsupials would be invalid. Serious objections can be raised to this theory — see later.

(6)   Marsupials arose in the central Pacific in a former landmass, the 'Darwin Rise', and from there were rafted to the Americas and Australia, respectively (Martin, 1970). The idea has received no support. See following paper in this volume for a later reassessment of this hypothesis by its author.

Finally, Fooden (1972) suggested that the distribution of the three 'evolutionary grades' of mammals - monotremes, marsupials and placentals - and of primitive placentals in relation to advanced ones, reflects the order in which these groups evolved and 'samples' became isolated historically. While there is no denying that isolation has 'protected' the primitive mammals of Australia and Madagascar, various objections can be raised to the theory. Thus, marsupials and placentals arose simultaneously, not successively; the two enjoyed long periods of cohabitation in both North America and South America; there is no evidence that monotremes have ever occurred anywhere other than Australia (see also Lillegraven, 1974).

The origin and history of the early marsupial mammals

Jurassic fossil mammals are a somewhat diversified assemblage belonging to four or five major groups, and having dental specialisations channelling them towards herbivore, carnivore and insectivore roles (Clemens, 1970). The Cretaceous saw marked evolutionary change in various lineages and many extinctions, and there were seemingly a number of 'experimental lines' that did not survive beyond the end of the period. Some of these had certain marsupial attributes. Thus the Late Cretaceous *Deltatheridium* of Mongolia had marsupial-like cranial characters but otherwise was morphologically quite distinct from both marsupials and placentals (Butler and Kielan-Jaworowska, 1973; Lillegraven, 1974).

It now has been amply demonstrated that placentals and marsupials came from a common stock that retained a range of 'primitive therian' characters, including tribosphenic molars. Differentiation of marsupials and placentals may now be pinpointed as probably having occurred between Albian and Campanian time (mid-Cretaceous), about 100 – 106 m.y.B.P. (Slaughter, 1968; Lillegraven, 1974; Tedford, 1974). Thus dentitions characteristic of both groups are found in the Paluxy Formations of south-eastern Texas (Slaughter) and others approaching the eutherian type occur in Lower Cretaceous deposits in Manchuria (Barsbold, Voronin and Zhegallo, 1971). Late Cretaceous marsupials are recognisable on four characters (Tedford, 1974): (1) the upper molars have prominent labial cingular shelves studded with stylar cusps and a metacone labial equal to, and often larger than, the paracone; (2) the hypoconulid and entoconid of the lower molars are closely approximated; (3) the dental formula is I5 - I4, C1, P3, M4; and (4) only a single deciduous premolar is present. Such dental specialisations may, however, considerably have pos⁺-dated the marsupial-type reproductive system: again, Lillegraven (1974) stresses that such contemporary marsupial features as the lack of a developed corpus callosum and dominance of the yolk sac placenta are not exclusive marsupial characters but were probably shared by the primitive therian stocks that gave rise to both marsupials and placentals.

The Late Cretaceous North American marsupials, ranging in size from that of a mouse to that of a cat, had achieved considerable dental specialisations and regional differentiation in their distributions. This degree of diversity, although still somewhat modest, exceeded that of their cohabiting placental contemporaries. By the end of the Cretaceous, however, they had suffered a conspicuous decline relative to the placentals.

No fossil marsupials are known from Asia: the only Late Cretaceous mammals so far recorded were multituberculates and eutherians. There is, however, evidence that some of these Asian forms colonised North America in the Campanian and Maastrichtian (Kielan-Jaworowska, 1975). Notwithstanding some relationships in the Late Cretaceous dinosaurs of Asia and North America (Russell, 1970a, b), marked faunal differences show that a strong filtering effect was operative between the two continents at this time (Lillegraven, 1974). A theory allowing for separate continental origins of marsupials and placentals (e.g. in North America and Asia, respectively) has considerable appeal in that it would account for these two comparable groups originating and radiating contemporaneously. It will be noted, however, that in both North America and South America the two occurred together quite early.

Marsupials lingered on in North America and in Europe until the Miocene, but latterly were never more than a minor element. They recolonised North America again after the formation of the Panamanian isthmus at the end of the Pliocene.

Only in South America and Australia did the marsupials undergo a wide Tertiary radiation. This was already well established in the former in the Palaeocene (Simpson, 1950; Patterson and Pascual, 1968, 1972). The Australian fossil record

extends no further back than the Lower Miocene or Uppermost Oligocene
(Tedford *et al.*, 1975). By this time, nearly all major families and genera are
present, indicating a long previous history of evolution and differentiation. The
oldest and 'most primitive' Australian fossil is the Upper Oligocene *Wynyardia*, a
possum-like arboreal type with certain didelphoid characteristics (Ride, 1964).

That marsupials arose only once, Australian and South American marsupials
sharing a common origin, is indicated by their common anatomical characteristics
and a range of shared cytological attributes (Sharman, 1961; Hayman *et al.*,
1971). Further, the two groups represent a basic dichotomy within the marsupial
lineage – note the consistent differences in stylar cusp structure between the most
primitive Australian (dasyurid) stocks and South American ones, consistently more
complex auditory development in Australian groups (Tedford, 1974), differences
in morphology and development of spermatozoa (Biggers and De Lamater, 1965)
and in blood serology (Kirsch, 1968; Hayman *et al.*, 1971). The South American
marsupials are parasitised by lice which are restricted to South America, but
these are related at the family level to those of Australian marsupials (Vanzolini
and Guimaraes, 1955). Three families of Australian fleas also exhibit solid phylogenetic ties with South America (Traub, 1972), but this order is less host-specific
and is not limited to marsupials.

Figure 5.6, modified from Tedford (1974), summarises the evolutionary relationships and geographic history of the marsupials, subsequent to their differentiation
from a parental metatherian – eutherian stock in Albian - Cenomanian times.
Modifications to the original published figure include the incorporation of newer
data on the age and relationships of some of the Australian families, following
newer evidence from Tedford *et al.* (1975), and some of the newer unpublished
findings of this author, vide that syndactyly and polydactyly may represent a
basic division. The Necrolestidae are eliminated from the tree as they are apparently
not marsupials (Tedford, personal communication). The figure shows that the
North American marsupials differentiated into three families during the Cretaceous.
In Tertiary times the South American stocks radiated into six families and the
Australian ones into 14 families.

Critical areas in the early dispersal history of
the marsupials, and their geological history

If marsupials differentiated in Albian - Campanian times (i.e. about 100 m.y. B.P.),
their major dispersals must have occurred between that time and the Eocene, the
latest date they could have arrived in Australia. Possible dispersive pathways
during this time could have been, or were: Bering Strait, the North Atlantic,
North America to South America, South America to Africa to Australia, South
America to Antarctica to Australia.

(1)   *Bering Strait*. In the middle Mesozoic the Pacific Ocean covered two-
thirds of the Earth's surface. Extensive subduction around the periphery (Asia,
Alaska, Americas) contracted it to about half the original area (Larson and
Pitman, 1972; Larson and Chase, 1972), producing a progressive shortening of
the distance from western Alaska to eastern Siberia. This contraction began about
81 m.y. B.P., with Alaska reaching its present proximity to Siberia about 63 m.y.

B.P. (Pitman and Talwani, 1972). A variety of stratigraphic data supports these deduction, e.g. increased continentality in northern Alaska (Detterman, 1973), and the deposition of extensive terrigenous sediments and magmas in north-eastern Siberia (Semenovich, Gramberg and Nesterov, 1973; Vinogradov *et al.*, 1973). They are also supported by the fossil bivalve data. Prior to the mid-Cretaceous there were separate north-eastern and north-western Pacific faunas but by the Late Cretaceous they merge, presumably as the result of the develop ment of shallow-water connections (Kauffman, 1973).

The fossil floras around Beringia show the following picture. In the Early Cretaceous (Aptian through Albian) those of the north-eastern USSR were dominated by Ginkoales, conifers and deciduous forms with a luxuriant under-growth of ferns and possible cycadophytes: this represents a degree of luxuriance suggestive of a warm climate (Samylina, 1968). Albian - Cenomanian times (about 94 m.y. B.P.) in coastal Alaska are marked by a warm-adapted gymnosperm – fern flora, equivalent to forms growing at latitudes of 25 - 30 degrees today (Smiley, 1966): these were succeeded, in Turonian times, by a cool-adapted flora equivalent to one that would grow at latitudes of about 35 -45 degrees today. The Early Tertiary of the northern regions (e.g. Greenland, Spitzbergen, Alaska) was characterised by temperate floras (references in Frakes and Kemp, 1973).

A corollary of the above is that mammalian dispersion through Beringia should have become possible in the middle Late Cretaceous. Calculations by McKenna (1973), by contrast, show that the north rotational pole lay near Beringia in the Late Cretaceous; dispersion at such high latitudes must have been very difficult. The mammalian fossil evidence, moreover, indicates that a marked filter barrier was in effect – see next section.

(2)   *North Atlantic.* The fossil record indicates that didelphoid marsupials reached Europe in the Early Eocene. Szalay and McKenna (1971) and McKenna (1973), in listing similarities between the North American and European mammal faunas at this time, postulate a land connection through Spitzbergen and the Barents Shelf (the 'De Geer route'). This lay at a lower palaeolatitude than any Beringian connection at this time.

(3)   *North America – South America.* As Simpson (1950) and Patterson and Pascual (1968, 1972) have noted, the striking differences between the Late Cretaceous – Early Tertiary fossil faunas of the two Americas indicate a wide water-gap between them at this time. These differences extend also to the dinosaur faunas.

Stratigraphic studies fully confirm the presence of this extensive water-gap (Woodring, 1954; Harrington, 1962).

The plate tectonic history of the central American – Caribbean region is, however, very poorly understood. Workers have variously seen the Caribbean as a piece of 'proto-Atlantic' interposed between the Americas, Europe and Africa, as an eastern extension of the Tethys Sea (note similarities in the Early Cretaceous lamellibranchs of the Tethys and Caribbean; Kauffmann, 1973), or as an eastwards extension of the West Pacific Plate (Edgar, Ewing and Hennion, 1971; Malfait and Dinkelman, 1972). Phillips and Forsyth (1972) and Kesler (1973), however, regard it as oceanic crust formed *in situ*, and not as an extension of older Pacific crust.

There is now a considerable literature on the geology of the circum-Caribbean area (list in Edgar *et al.*, 1971). North America and South America converged from perhaps the mid-Cretaceous to the Early Eocene (Dietz and Holden, 1970; Malfait and Dinkelman, 1972): this was associated with the extensive tectonic and igneous activity in the area during these periods. Parts of Belize and Honduras were already land when the section from southern Mexico to Nicaragua was up-lifted in the Late Cretaceous (Mills *et al.*, 1967; Dengo 1969, 1973; Dengo and Bohnenberger, 1969). The central cordillera of Colombia (Anderson, 1972), the Cordillera de la Costa of Venezuela (Bell, 1971) and Trinidad (Barr and Saunders, 1971) are also of Late Cretaceous age. Figure 5.1, adapted from Dengo (1973), shows land distribution in the Caribbean area in the Late Cretaceous. A volcanic arc extends southwards from nuclear central America towards South America and undoubtedly improved the potential dispersive pathway between the Americas.

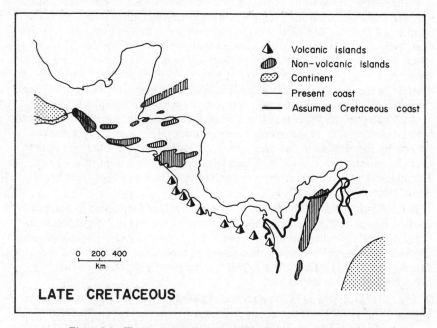

Figure 5.1   The water-barrier between North America and South America in the Late Cretaceous; postulated land distribution accord-ing to Dengo (1973). Any mammalian colonisation would have had to be along island chains. There is evidence that a volcanic island arc stretched between nuclear central America and South America.

This route would have been more effective and involved shorter water-gaps than one through the West Indies at this time.

A distance between the major 'islands' of nuclear central America (Nicaragua) and the newly emerged Andean-Venezuelan cordillera of about 1200 km in the Late Cretaceous is indicated.

(4)   *South America – West Africa.* Deep-sea drilling and sea-floor spreading data indicate that the initial rift between Africa and South America began in the South in the late Triassic to Mid-Jurassic, probably at about 125 - 135 m.y.B.P.

(Maxwell *et al.*, 1970; Le Pichon and Hayes, 1971; Larson and Ladd, 1973). The continents finally became separated in tropical latitudes about the middle Cretaceous, at 100 m.y. B.P. (Allard and Hurst, 1969; Grant, 1971; Reyment and Tait, 1972; Douglas, Moullade and Nairn, 1973). Raven and Axelrod (1974) have estimated that at the close of the Cretaceous about 800 km would have separated Africa and South America at their closest points, although some intermediate islands possibly were present along the mid-Atlantic ridge. They also estimate that South America would have been equally distant from Africa and North America at 50 m.y. B.P., inferring that chances of trans-oceanic colonisation from Africa would have been greater up to that time. This is not by any means the whole story, however, as the chances of a 'propogule' making a landfall is equally a function of the number, size and distribution of intermediate islands (if any), and the directions of prevailing winds and currents.

Latterly Hoffstetter (1970, 1972) and Hershkovitz (1972) have suggested that the caviomorph rodents and monkeys that first appear in the South American fossil record in the Late Eocene – Early Oligocene may have rafted across the South Atlantic rather than have entered from the north. This contrasts with the generally accepted view (Simpson, 1950; Patterson and Pascual, 1972) that they came from the north.

(5) *Africa – Antarctica – Australia.* Cox (1970) has advanced the novel hypothesis that marsupials may have entered Australia by way of Africa. Subsequently he came to accept the South America – Antarctica – Australia pathway (Cox, 1973). Recently, as noted, Raven and Axelrod (1974) have resurrected this theory.

Various dates have been advanced for the separation of Africa from Antarctica, i.e. the separation of East and West Gondwanaland. Dietz and Holden (1970) suggested a Triassic date but later workers have suggested a Late Jurassic or Early Cretaceous separation (150 – 100 m.y. B.P.); see Smith and Hallam (1970) and McElhinny (1970). The reconstruction of Smith *et al.* (1973) allows a continuing land link between Africa and the other southern continents in the area of Tanzania at 100 m.y. B.P. (Figure 5.2) subsequent to separation further south. Laughton *et al.* (1973) allow a similarly late separation. The history of this part of the world is, of course, closely tied up with the movements of India. Sclater and Fisher (1974) suggest that India started to separate from Antarctica - Africa – Australia near the Early Cretaceous – Late Cretaceous boundary, at about 100 m.y. B.P., initially moving north-east for about 20 million years, then north-west.

The date of this final separation is, of course, critical. The stratigraphic data of Dingle and Klinger (1971) and Dingle (1973), however, show the first incursion of the sea around the south-east tip of Africa as probably occurring in the latest Triassic or early Jurassic, i.e. about 180 m.y. B.P. This produced a shallow shelf in the area of the Agulhas Plateau and a marine connection from the Knysna region to Madagascar (which may then have been further south). Jones (1972) has challenged that these marine sediments necessarily correspond to the beginning of continental rift and that they indicate the conclusions of Veevers, Jones and Talent (1971) that rifting only began in the Cretaceous. In reply Dingle (1972) reiterated that the geology of the South African continental margin, as now known, is of Upper Jurassic (at the latest) age and that this evidence of the beginning of marine conditions there cannot be refuted. Scrutton (1973) and Dingle

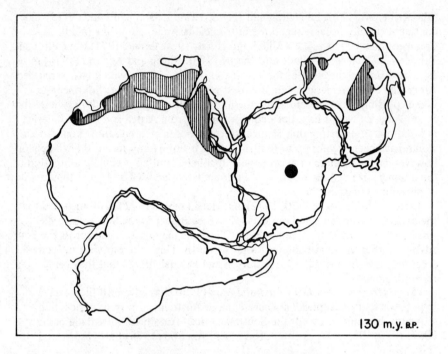

Figure 5.2   Late Jurassic fit of the southern continents prior to
break-up (after Smith and Hallam, 1970), and at about 130 m.y. B.P.
Distribution of the epicontinental seas is as per Tedford (1974), based
on Kummel's compilations. Subsequent to the opening of the South
Atlantic 125 – 130 m.y. B.P., direct colonisation between Africa,
India, Antarctica, Australia may have been possible until 100 – 110 m.y.
B.P. (Smith *et al.*, 1973). Raven and Axelrod (1974) suggest that,
because of the lower latitudes involved, this was a more logical entry
route to Australia than that via South America and Antarctica. The
evidence is, however, that it closed too early for this to have been
possible.

and Scrutton (1974), after considering the seismic, magnetic and bathymetric data,
give a mid-Jurassic date for the Africa - Antarctica separation. Figure 5.3 shows
Scrutton's postulated separation history for southern Africa.

Raven and Axelrod (1974) develop a range of biogeographic conclusions on
the basis of land continuity between Africa, India, Antarctica and Australia, up
to 100 m.y. B.P. (they, incidentally, use the term 'mid' Cretaceous for this date:
Smith *et al.*, 1973, refer to it as 'early' Cretaceous). Raven and Axelrod also
reproduce the reconstruction of Laughton *et al.* (1973) showing land distribution
in the southern Indian Ocean at 75 m.y. B.P.(their figure 3) This map shows a
wide separation of Africa from Antarctica at this time and India isolated well to
the north. These authors remark with respect to it, however, that 'more or less
direct migration between Africa and Australia via East Antarctica was possible' at
this time. The map does not support this statement, even if volcanic islands on

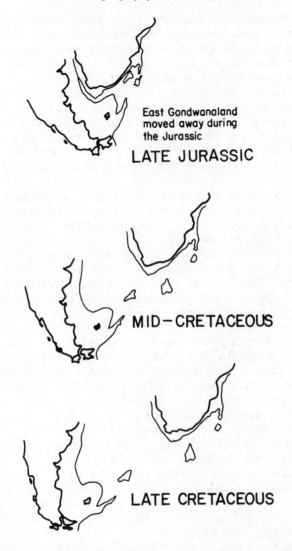

East Gondwanaland
moved away during
the Jurassic

LATE JURASSIC

MID–CRETACEOUS

LATE CRETACEOUS

Figure 5.3 Arguing for a Late Jurassic, rather than Cretaceous, separation of East Gondwana (Australia, Antarctica) are the stratigraphic data from eastern South Africa (coastal landforms and sediments); see reconstructions of Scrutton (1973) shown here. The fine line shows probably continental shelf limits. Note the far eastward extension of the Falkland Plateau.

mid-oceanic ridges were present. Presumably it is the 100 m.y. B.P. map that is meant.

In summary, the validity of the proposed Africa - Antarctica - Australia pathway for colonising marsupials is very dubious. For the advantage of gaining a more northern (warmer) pathway it goes against all the fossil evidence on mid-Cretaceous distributions and requires that the very earliest marsupial stocks dispersed across considerable distances, and also across a climatic and (it can reasonably be inferred) an oceanic barrier in the south as well as the north. The presence of island arcs would only slightly have alleviated this difficulty.

There is, of course, no denying that some marsupials are efficient 'rafters'. The presence of *Alphodon* in both North and South America in the Late Cretaceous, and the occurrence of *Phalanger* in Timor and the Celebes is ample evidence not only of the ability of some groups of marsupials to cross water-gaps but also of their hardiness and ecological versatility. The Albian - Campanian marsupials were, however, 'primitive' stem forms and it would seem unlikely that these attributes had yet been evolved.

The Africa - Antarctica - Australia dispersal pathway for marsupials can safely be rejected on three grounds: (1) the land connections were disrupted too early; (2) it is unnecessarily circuitous and complicated; and (3) fossil marsupials are unknown from Africa. Negative evidence is, of course, inconclusive.

(6)   *South America - Antarctica - Australia.* Mesozoic geological continuity between South America and the Antarctic Peninsula of West Antarctica has now been amply established (Hamilton, 1966, 1967; Barker, 1970; Dalziel and Elliot, 1971), the southern Andes and Antarctandes forming a continuous cordillera and sharing a common continental shelf until near the end of that period. However, a marginal basin opened and andesite island arcs had formed in the earliest Cretaceous in southern South America and there was sea-floor spreading in the mid-Cretaceous (Dalziel, De Wit and Palmer, 1974). Subsequently, disruption of the Andean cordillera in the form of a 90 degree bend led to formation of the Scotia Arc (Barker, 1970; Elliot, 1972). Parts of the Drake Passage are, however, only mid- to Late Tertiary in age (Barker, 1970; Griffiths and Barker, 1972).

The date of the actual break between South America and Antarctica is of the greatest biogeographic interest. Originally Dalziel and Elliot (1971) and Elliot (1972) suggested that junction may have persisted through to the Late Cretaceous or Early Tertiary; subsequently, however, Dalziel *et al.* (1973) have suggested that the initial break may have occurred during the Late Jurassic. Foster (1974), noting that the breaching of the Drake Passage by ocean would have drastically changed current patterns, has endeavoured to explain the sudden change from warm-water to cold-water echinoid groups in Eocene deposits in southern Australia (estimated as occurring at 36-41 m.y. B.P.) by this event.

Nevertheless, South America and Antarctica have remained in close proximity; see Figure 5.5 for positions of the southern continents at 70 m.y. B.P.

Pre-drift Australia lay against East Antarctica: its precise position is, however, uncertain since certain correlations cannot be made between geological elements in the two regions (Griffiths, 1974). The sea-floor spreading data indicate an Early Eocene date for the northward movement of Australia (Heirtzler, 1968; Le Pichon, 1968). A date of 50-60 m.y. B.P. is given by Hayes and Ringis (1973). The geology along the suture zone between Antarctica and Australia gives no indication of any significant distributional barrier to animal migration up to about 49 m.y. B.P.: by about 47 m.y. B.P. however, the stratigraphic record indicates the beginning of the 'full influence of marine and oceanic conditions' (McGowran,

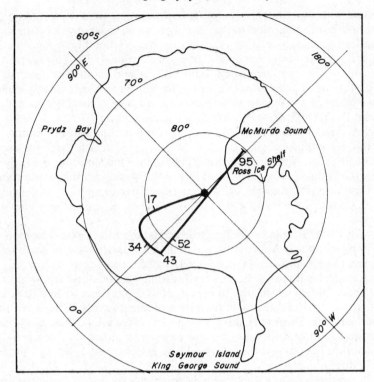

Figure 5.4 Postulated wander path of the South Pole from 95 m.y.
B.P. to present and based on data from Australian shield volcanos
(not Antarctic data), as per McElhinny and Wellman (1969). This
indicates that the pole did not deviate more than about 12 degrees
from its present position during this time and, hence, polar wander,
in itself, would not have led to a significant amelioration of the
Antarctic climate. World temperatures were, however, significantly
warmer in the Eocene. Fossil pollens of this age from Prydz Bay,
McMurdo Sound and the Ross Ice Shelf indicate the presence of
proteaceous plants, as well as *Nothofagus*, on the Antarctic mainland,
and megafossils, some of Eocene age, from Seymour Island, show
that a diversified flora occurred at far southern latitudes at this time.

1973; p. 760). As noted, Late Eocene deposits along the south coast of Australia
between Albany and Adelaide have a diverse echinoid fauna.

Deep-sea drilling indicates that the Tasman Rise between Antarctica and
Australia may be continental (Kennett *et al.*, 1972). A date of 30 m.y. B.P. is
suggested for the final separation of Antarctica from it and the beginning of the
circum-Antarctic current (Kennett *et al.*, 1974).

New Zealand's original position was probably adjacent to Marie Byrd Land
east of Australia. Northward drift of the Campbell Plateau – New Zealand plates
was initiated in the Late Cretaceous, about 80 m.y. B.P. (Le Pichon, 1968;
Griffiths and Varne, 1972). This would also be the date of the latest direct

connection with Australia. The central Tasman sea was created by spreading
about a now buried ridge that was active only from 80 to 60 m.y. B.P.: the south
Tasman sea formed sixty million years ago in direct relationship to Australia's
northward migration from Antarctica (Hayes and Ringis, 1973). This earlier
separation, and the more archipelagic nature of the land on the New Zealand
plate, plus recorded extinctions (e.g. of Eocene Proteaceae),undoubtedly explains
the basic faunistic and floristic differences between New Zealand and Australia
(Fleming, 1962, 1963; Couper, 1960; Raven and Axelrod, 1972).

West Antarctica, it must be noted, is itself largely archipelagic in structure
under the ice. Archipelagos also characterised the orogenic belt bordering the
Antarctic plate and Australian peninsula in both the Late Mesozoic and Early
Tertiary (Griffiths and Varne, 1972). Tedford (1974) suggests that marsupial dispersal between South America and Australia could have been partly through
archipelagos where as a result of the marine influence, the climate may have been
less severe.

To what latitudes would colonising marsupials have had to extend to move
from South America to Australia via Antarctica? Reconstructions of south polar
wanderings (McElhinny and Wellman, 1969), based on data derived from
Australian shield volcanos, indicate a north-easterly displacement of the south
pole of about 10 degrees (relative to present position) at about 95 m.y. B.P.,
and a south-westerly displacement thereafter, reaching a point 10 degrees south-
west of the present position at about 34 m.y. B.P. (Figure 5.4). These positional
changes would not be sufficiently great to make any significant difference in the
climate of the dispersal pathway. It would have entailed passing through the 60 - 70
degree latitudinal belts.

World temperatures as a whole were, however, markedly warmer both in the
middle Cretaceous and in the Eocene, as indicated by plant and invertebrate fossil
distribution patterns. Note fossil biotas described by Samylina (1968), Smiley
(1966), Frakes and Kemp (1973) and Hopkins (1974). Albian assemblages from
Melville Island, Canadian Arctic, for example, indicate a moist, warm-temperate
climate (Hopkins). In the Early Tertiary high latitudes were marked by warm-
adapted fossil floras, coral reefs and tropical weathering profiles (references in
Frakes and Kemp, 1973). Sixty per cent of the genera of megafossil remains in
the Early Eocene London Clay flora are known today from the Malay Peninsula,
India and Ceylon (Reid and Chandler, 1933). Greenland, Spitzbergen and Alaska
supported temperate-adapted floras at this time (Frakes and Kemp, 1973).

The Cretaceous -Early Tertiary fossil and climatic history of Antarctica is
discussed by Plumstead (1962, 1964), Adie (1972), Denton, Armstrong and
Stuiver (1971), and others, Cretaceous data from the mainland are scanty. In New
Zealand, however, Clayton and Stevens (1968) record an apparent rise in
temperature between the Albian and Turonian – Santonian and decline in the
Maestrichtian: they note that there is no evidence of an unduly cool marine
climate during the Late Cretaceous. Mid-Cretaceous belemnites, however, had a
mean growth temperature of about $15 - 16\,^{\circ}$C, and post-Aptian temperatures
as determined by the oxygen isotope method, are relatively lower than that of
other parts of the world at this time (Lowenstam and Epstein, 1954). New
Zealand Late Cretaceous marine invertbrate faunas are dominated by ammonites,
bivalves and crinoids: these and plesiosaurs and ichthyosaurs suggest a warm

climate (Brown, Campbell and Crook, 1968). Further south, in West Antarctica, there was a rich invertebrate fauna dominated by molluscs (Adie, 1972), suggesting that higher latitude waters also were not cold.

Eocene plant data from the Antarctic mainland are available from Kemp (1972), Wilson (1968), McIntyre and Wilson (1966) and Cranwell, Harrington and Speden (1960). Kemp's studies are of reworked palynomorphs contained in muds taken at depths of 320 - 1500 m from a wide shallow submarine platform in the West Ice Shelf (Prydz Bay) area. The material was eroded away from sedimentary rocks by the growing ice-caps of the later Tertiary and carried seawards. The material is of three ages, Permian, Early Cretaceous, and Uppermost Cretaceous - Lower Tertiary, to judge from the time chronology of these same genera in Australian, New Zealand and South American sequences. The presumed Early Cretaceous spores belong to 13 genera. The Late Cretaceous - Early Tertiary material is dominated by Fagaceae (the three species-groups of *Nothofagus*), Proteaceae (several genera with one, *Proteacidites*, having eight species), and the genera *Gambierina* and *Triorites,* and there is some Myrtaceae. On the other hand, no forms are present that make their first appearance in post-Eocene strata in Australia and New Zealand; Graminae, Compositae and Restionaceae are absent. The deposits also contain a variety of dinoflagellate cysts that have been identified with presumed Eocene strata elsewhere.

Widely scattered localities in the Ross Sea area have contributed Early Tertiary microspore, pollen grain and microplankton-bearing bottom sediments (Wilson, 1968). Pollen of *Nothofagus* and proteaceous plants has been described from McMurdo Sound by Cranwell *et al.* (1960) and McIntyre and Wilson (1966). Some authorities, however, have suggested that this pollen was actually wing-blown; its presence does not necessarily mean that these plants grew in the area (J. Schopf, personal communication; Raven and Axelrod, 1972), it being pointed out that *Nothofagus* pollen has been carried as far as Kerguelen and Juan Fernandez.

Kemp (1972), noting the striking diversity of Early Tertiary pollens from Prydz Bay, suggests that periods of considerable warmth are indicated. No tropical or sub-tropical forms (palms, *Cuponieidites, Malvaciopollis*) are, however, present. McIntyre and Wilson suggest that the *Nothofagus* pollen at McMurdo Sound indicates a temperate, cool to moderately warm climate there. Tedford (1974), noting that included in the *Nothofagus* pollen from Antarctica is that of the *N. brassi* group, confined today to the warm temperate, equable, uplands of New Guinea and New Caledonia, suggests that the climate was warmer than that favoured by the southern temperate beech — pine forests today. Mandra (1969), on the basis of silicoflagellate studies, considers that the Upper Eocene was a period of considerable warmth in Antarctica.

Unfortunately the only actual plant megafossils described from Antarctica are from the Antarctic Peninsula and offshore islands: Seymour Island (Dusén. 1908; Cranwell, 1963), and King George Island in the South Shetlands (Orlando, 1964; Barton, 1964). The former, which have been identified as Lower Miocene in age but are probably at least part Eocene (Cranwell, 1959), belong to some 70 species of plants, including gymnosperms, some *Nothofagus*, Cruciferae, Myrtaceae, Proteaceae, Loranthaceae and Onagreae (lists in Plumstead, 1962, 1964). At least at this latitude the climate must have been warm temperate when this flora flourished.

In seeming conflict with the evidence of extensive plant development on the Antarctic mainland during the Eocene is evidence of glaciation. Thus Margolis and Kennett (1971) argue that sand grains with etch-marks from deep-sea cores, of Lower Eocene, Upper Middle Eocene and Oligocene, indicate ice-rafting and suggest that reduced foraminiferal diversity at these times confirms low temperatures. Fitzpatrick and Summerson (1971), however, do not agree that such surface characteristics on sand grains indicate ice action. Glaciation would, of course, not necessarily be inconsistent with extensive plant development if the glaciers were montane ones. Thus in the south island of New Zealand today glaciers, *Nothofagus* forests and palm trees are to be found within a few miles of each other. By Oligocene times, and subsequently, there is plenty of evidence of increasing glaciation in Antarctica (e.g. Denton, Armstrong and Stuiver, 1971; Hayes *et al.*, 1973).

Eocene temperatures for New Zealand were warm. Thus Webb (1968) records a gradual increase in marine temperatures from the Cretaceous to the Palaeocene – Eocene, with the amelioration becoming more pronounced in the Eocene when numerous corals, warm-water bivalves and some plants with tropical - subtropical affinities appear (Brown *et al.*, 1968). In some areas tropical conditions were achieved by the Late Eocene (Fleming, 1962).

The total evidence indicates that the South America – Antarctica – Australia route could readily have been traversed by a marsupial propogule. Only short water-gaps were involved. Temperatures, though cool, were well within the range that mammals can handle (see later). Finally, the existence of a diversified flora over extensive areas of the Antarctic mainland is now indicated (even if, in the absence of plant megafossils, it has not been finally proved): this flora would have constituted adequate marsupial habitat.

(7)   *Antarctica - New Zealand*. In the absence of terrestrial, fossil-bearing strata of early age it cannot be proved that marsupials never reached New Zealand. The remarkably cool temperate adapted tuatara (Bogart, 1953 a and b), ratite birds and leiopelmid frogs must have entered New Zealand via Antarctica. The fact that these have survived the subsequent climatic vicissitudes would make it seem highly likely that there would be some record of marsupials there had they ever reached these islands.

New Zealand, which broke free from Antarctica in the Late Cretaceous, lacks marsupials; Australia, which remained attached until the Lower Eocene, obtained them. These facts suggest that marsupials dispersed, not in the Cretaceous but at the time of high Early Tertiary temperatures.

(8)   *South-east Asia*. Since Australia was far to the south in the Early Tertiary and marsupial entry via Indonesia can now effectively be ruled out, a consideration of the early geological history of this area is not now relevant. It became important as a mammalian dispersal pathway into Australia only when this continent approached its present position in the Late Tertiary. See discussion of the continuing faunal differences across Wallace's Line in Raven (1935), Mayr (1944), Simpson (1961), and Keast (1972b).

Several recent writers have reviewed the paleogeography of Malaysia relative to Gondwanaland. Whilst it is generally agreed that its sedimentary rocks indicate a former juxtaposition with a larger land mass there is disagreement as to whether this was India, Tibet, or even Africa (Ridd, 1971; Stauffer, 1974; Crawford, 1974).

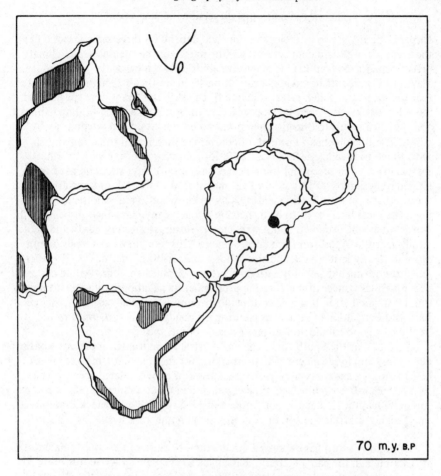

70 m.y. B.P

Figure 5.5   Position of the southern continents relative to Antarctica
in the Late Cretaceous, about 70 m.y. B.P., a disposition that was maintained (except for the continued northward movement of India, etc.)
until Australia left Antarctica in the Eocene, about 45 m.y. B.P.
Fine line shows continental shelves. Reconstructions as per Dietz
and Holden (1970) and Jardine and McKenzie (1972). Note continued
approximation of South America to Antarctica and Australia's
junction with the latter. Epicontinental seas are shown for the Late
Cretaceous, as per Tedford (1974), based on Kummel's compilations.

(9)   *Trans-Pacific origins.* This hypothetical route, suggested by Martin
(1970) as a dispersive pathway between North America and Australia, lacks
both geological and geophysical support. Later reconstructions of the tectonic
history of the Pacific (Larson and Chase, 1972; Larson and Pitman, 1972) do
not provide support for this postulated pathway. The theory is reviewed in the
light of these later data by Martin elsewhere in this book.

Theories of marsupial dispersion: an assessment

Dispersal of any group in geological time will depend on three main factors: (1) the availability of land connections: (2) the nature of the ecological and climatic belts to be crossed; and (3) the dispersive abilities of the particular group concerned. Mayr (1953), discussing trans-oceanic dispersal in south-west Pacific birds, noted that the successful water-crossers shared three attributes: they were species that moved in flocks; they were ecologically versatile (and of versatile feeding habits); and they had habits that made them prone to be caught up and swept away by winds. The flocking habit ensured, of course, the presence of other individuals with which to breed. (See Simpson 1952, for a discussion of the various other factors involved in successful trans-oceanic dispersal.) Mayr's criteria need to be altered only slightly to fit a non-volant mammal. An arboreal, or semi-arboreal, species that lives near beaches and is likely to be swept to sea on a tree or floating log is the most likely type to be dispersion-prone. Many marsupials (especially the North American *Didelphis* and Australian *Trichosurus*) are very hardy and ecologically versatile. Apart from primates, some rodents, etc., most mammals do not move in closely knit groups, and this is not a basic habit of marsupials. However, one or two female didelphids with 6 – 8 uterine or pouched young would certainly be a potentially successful colonising propogule: its adaptibility would certainly stand it in good stead in a new continent. By contrast, a quadrupedal herbivore such as the small Late Cretaceous Neotropical condylarth *Perutherium* would hardly be a good candidate for trans-oceanic or archipelago-type dispersal.

The fact that the South American and Australian marsupials represent a basic dichotomy and independent radiations argues for the trans-Antarctic coloniser being a single adaptive type or perhaps a cluster of closely related species. The cool temperatures, archipelago-type crossings and modest range of plants in the Antarctic habitat all argue for the route being a dispersive bottleneck that only a single highly versatile species, or at most a small number of species, was able to cross.

Darlington (1965) has expressed doubts that, because of the cool temperatures of Antarctica in the early Tertiary, marsupials were able to traverse this route. He notes that no marsupials penetrate south into the *Nothofagus* forests of Tierra del Fuego today. This argument is not, however, critical; a diversity of mammalian types extend into high latitudes in both hemispheres. Tierra del Fuego, for example, has an otter, a fox, a camel, a cavioid rodent and several small cricetine rodents. Arctic North América has a fox, lemmings and others. Other rodents (Squirrel, Porcupine), martens and others extend to the limits of the tree-line, where, incidentally, the tree diversity (80 – 90 per cent of individuals may be black spruce) is no greater than that of the homogeneous southern *Nothofagus* forests. There is no reason to believe that marsupials, despite their somewhat poorer temperature regulation, are physiologically less able to withstand cold than placentals. In fact, the North American opossum (*Didelphis*) successfully overwinters in northern New York, where the snow-cover lasts 4 months and deciduous trees are leafless for 7 months. Hibernation, and such behavioural attributes as group denning, are widespread in the mammals, from the most primitive forms upwards. Various Australian marsupials withstand cold temperate winter temperatures in Tasmania and the higher regions of the Australian Alps with snow cover of one or two months' duration. This includes phalangers such as *Trichosurus* (which as an introduction has proven highly successful in the southern beech forests of New Zealand) and the mouse-size *Burramys*.

Rather than cold temperatures limiting the distribution of contemporary marsupials, the distributional limits are probably dictated more by ecological factors such as habitat diversity and food availability.

To sum up, there can be no doubt that a reasonably advanced, hardy, ecologically versatile, generalised marsupial could have inhabited Antarctica in the Early Tertiary, even at latitudes of 60 – 70 degrees S. The presence of a flora of proteaceous plants, plus the three species groups of *Nothofagus*, would have provided a diversified habitat. Several Australian marsupial genera utilise thickets of Proteaceae as habitat and nest-sites (e.g. *Pseudocheirus, Pseudantechinus, Cercartetus*), while *Tarsipes* feeds on the nectar. Of the seven or so theories that have been advanced to explain marsupial evolutionary biogeography, only those allowing a North American and/or South American origin are reasonable, and the only pathway to Australia that fits all the requirements is the South America - Antarctica one.

Distributional pathways and patterns in the contemporaries of the early marsupials

Since, in both animals and plants, groups and species differ markedly in their ecological requirements and dispersive capacities, confirmation of marsupial colonising pathways should not necessarily be expected from associated biotas. Nevertheless land junctions should be reflected in the palaeobiogeography of more than one group.

In a previous paper (Keast, 1973) I noted the much closer affinities of the Late Cretaceous - Early Tertiary, and contemporary, cold temperate biotas of Australia and South America, and New Zealand and South America, compared with Africa/Australia and Africa/South America ones. Thus, *Nothofagus*, which has enjoyed a long and continuous history in the other southern land masses, is unknown from Africa. Of the contemporary forms, closer Australia - South American affinities are marked in the Proteaceae (Johnson and Briggs, 1963), chelydid turtles, parastacid crayfishes, stoneflies, peloridiid bugs, chironomids, trichopterans, and certain spiders, dragonflies, beetles, and so on (original authors listed in Keast, 1973). Such links, involving as they do forms with widely differing ecological tolerances and dispersive capacities, would certainly seem to confirm the much later (Eocene) connection between, or approximation of, these land masses. Several of these groups either do not cross water-gaps or do so only poorly and, hence, would have required direct land connections.

Against this, of course, the possibility that they formerly also occurred in southern Africa and have become extinct there must be recognised. This continent does not extend so far south as the other land masses and is deficient in cold, wet, cold termperate habitats. It is also drier and its southern parts have been subject to intermittent aridity (Moreau, 1966).

Africa - South America have, of course, various affinities in their tropical biotas.

The Tertiary radiation of marsupials

After their initial separation both the South American and the Australian marsupials initiated remarkable radiations. That of the former is documented by an excellent fossil record (Simpson, 1950; Patterson and Pascual, 1972). The early evolution of the Australian fauna is, however, unknown since the fossil record

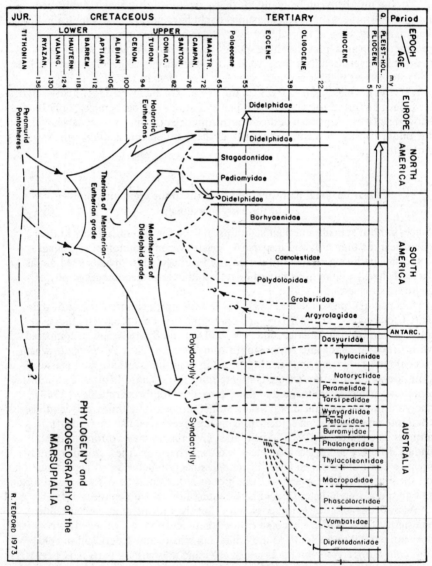

Figure 5.6 Evolutionary relationships of the marsupials, slightly modified from Tedford (1974). Reprinted with kind permission of Dr R. Tedford, the Society of Economic Paleontologists and the editor of the book *Paleogeographic Provinces and Provinciality*.

only begins in the Early Miocene. Two things are significant about these earliest Australian fossils. The 'most primitive' of them, the unique *Wynyardia*, has certain didelphid characteristics (Ride, 1964), and the rest recognisably belong to Pleistocene and recent families. The major evolutionary lines, in other words, had already differentiated by that time (Stirton, Tedford and Woodburne, 1968; Ride, 1970; Tedford, 1974). For general accounts of the range of body forms and ecological types in the Australian marsupials see Jones (1923, 1924), Troughton (1948), Ride (1970), Sharman (1970), and Keast (1972b).

An interesting comparative facet of Tertiary marsupial biogeography is the different patterns that evolution took in South America and Australia. The two areas differed in that at the beginning of the Tertiary the former had a mixed fauna of placental herbivores, edentates and marsupials, so that right from the beginning the last-named were channelled into carnivore and insectivore niches. This was to have profound effects on the adaptive zones that marsupials were to occupy subsequently. Australian marsupials were not subject to this restriction: their only mammalian competitors were monotremes. Nothing is known, of course, of any Late Cretaceous – Early Tertiary monotreme radiation and, hence, whether all but specialised types were replaced by marsupials.

The two continents independently produced moderate-sized to large 'doglike' carnivores (*Borhyaena* and *Thylacinus*), saltatorial types (Argyrolagidae and the Macropodidae). Generalised weasel- and shrew-like ecological types are represented by *Caenolestes* and the smaller Dasyurinae. South America alone produced aquatic types (*Chironectes*) and a sabre-toothed carnivore (*Thylacosmilus*). Australia alone produced a full range of herbivorous types from generalised small to medium-sized arboreal types (phalangers) to large saltatorial herbivores (macropodids) and large terrestrial quadrupeds (Phascolomidae, Diprotodontidae). Neither continent produced true flying types or long-necked treetop browsing giraffid or camelid counterparts. The smaller Australian macropodids fill the lagomorph adaptive zone.

The 'ant-eating' adaptive zone in Australia is shared by a monotreme (*Tachyglossus*) and a marsupial (*Myrmecobius*): in South America edentates fill it. The tree-climbing arboreal herbivore one in South America is filled by monkeys and sciurids, not marsupials.

These differences provide excellent demonstrations of the effects both of isolation and of associated fauna on patterns and directions taken by evolution. Nevertheless there are some surprising features — for example, the continued division of the herbivore and insectivore – carnivore roles between placentals and marsupials in South America.

Acknowledgement

I thank the National Research Council of Canada for support during the tenure of a grant from which the present manuscript was compiled, and Dr R. H. Tedford, the Society of Economic Paleontologists, and the editor of the book *Paleogeographic Provinces and Provinciality* for permission to reproduce Figure 5.6. Dr Tedford has kindly updated the diagram.

This paper was written while the author was Visiting Professor at Griffith University, Brisbane, and I should like to thank Professor Calvin Rose there for many courtesies.

References

Adie, R. J. (ed.) (1972). *Antarctic Geology and Geophysics,* Universitetsforlaget, Oslo.

Allard, G. O. and Hurst, V. J. (1969). Brazil-Gabon geologic link supports continental drift. *Science,* **163,** 528 - 32.

Anderson, T. A. (1972). Paleogene nonmarine Gualanday Group, Neiva Basin, Colombia, and regional development of the Colombian Andes. *Bull. Geol. Soc. Am.,* **83,** 2423 - 38.

Barsbold, R., Voronin, Yu. I. and Zhegallo, V. I. (1971). The work of the Soviet-Mongolian paleontological expedition in 1969 - 70. *Paleontol. J. (Paleontol. Zh.),* **5,** 272 - 6.

Barker, P. F. (1970). Plate tectonics of the Scotia Sea. *Nature,* **228,** 1293 - 6.

Barr, K. W. and Saunders, J. B. (1971). An outline of the geology of Trinidad, in *Field Guides to the Geology of Trinidad* (ed. Tomblin, J. F.) Internat. Field Institute; Guidebook to the Caribbean Island-Arc System; Am. Geol. Institute, pp. 2 - 12.

Barton, C. M. (1964). Significance of the Tertiary fossil floras of King George Island-South Shetland Islands, in *Antarctic Geology,* New York, Wiley, pp. 603 - 8.

Bell, J. (1971). Tectonic evolution of the central part of the Venezuelan Coast Ranges. *Mem. Geol. Soc. Am.,* **130,** 107 - 18.

Biggers, J. D. and De Lamater, E. D. (1965). *Nature,* **208,** 402.

Bogart, C. M. (1953a). Body temperatures of the tuatara under natural conditions. *Zoologica,* **38,** 63 - 4.

Bogart, C. M. (1953b). The tuatara. *Sci. Monthly,* **76,** 163 - 70.

Brown, D. A., Campbell, K. S. W. and Crook, K. A. W. (1968). *The Geographical Evolution of Australia and New Zealand,* Pergamon Press, New York.

Butler, P. M. and Kielan-Jaworowska, Z. (1973). Is *Deltatheridium* a marsupial? *Nature,* **245,** 105 - 6.

Clayton, R. N. and Stevens, G. R. (1968). Paleotemperatures of the New Zealand Jurassic and Cretaceous. *Tuatara,* **16,** 3 - 7.

Clemens, W. A. (1968). Origin and early evolution of the marsupials. *Evolution,* **22,** 1 - 18.

Clemens W. A. (1970). Mesozoic mammalian evolution. *Ann. Rev. Ecol. Systemat.,* **1,** 357 - 90.

Couper, R. A. (1960). New Zealand Mesozoic and Cainozoic plant microfossils. *N.Z. Geol. Surv. Palaeont. Bull.,* **32,** 1 - 87.

Cox, C. B. (1970). Migrating marsupials and drifting continents. *Nature,* **226,** 767 - 70.

Cox, C.B. (1973). Systematics and plate tectonics in the spread of marsupials, in *Organisms and Continents through Time* (ed. Hughes, N. F.), *Spec. Pap. Palaeontol.* no. 12, 175 - 87.

Cracraft, J. (1973). Continental drift, paleoclimatology, and evolution and biogeography of birds. *J. Zool. London,* **179,** 455 - 545.

Cracraft, J. (1974). Continental drift and vertebrate distribution. *Ann. Rev. Ecol. Systemat.,* **5,** 215 - 61.

Cranwell, L. C. (1963). *Nothofagus,* living and fossil, in *Pacific Basin Biogeography: a Symposium* (ed. Gressitt, L. J.), University of Hawaii Press, Honolulu, pp. 387 - 400.

Cranwell, L. C., Harrington, H. J. and Speden, I. G. (1960). Lower Tertiary microfossils from McMurdo Sound, Antarctica. *Nature,* 186, 700 – 2.

Cranwell, L. M. (1959). Fossil pollen from Seymour Island, Antarctica. *Nature,* 184, 1782 – 5.

Crawford, A. F. (1974). A greater Gondwanaland. *Science,* 184, 1179 – 81.

Dalziel, I. W. D. and Elliott, D. H. (1971). Evolution of the Scotia Arc. *Nature,* 233, 246 – 5.

Dalziel, I. W. D., Lowrie, W., Klingfield, R. and Opdyke, No. C. (1973). Paleo-magnetic data from the southernmost Andes and Arctandes, in *Implications of Continental Drift to the Earth Sciences* (ed. Tarling, D. H. and Runcorn, S. K.), vol. 1, Academic Press, London, pp. 87 - 101.

Dalziel, I. W. D., De Wit, M. J. and Palmer, K. F. (1974). Fossil marginal basin in the southern Andes. *Nature,* 250, 291 – 4.

Darlington, P. J., Jr. (1965). *Biogeography of the Southern End of the World,* Harvard University Press, Cambridge, Mass.

Dengo, G. (1969). Problems of tectonic relations between Central America and the Caribbean. *Trans. Gulf. Assoc. Geol.,* 19, 311 – 20.

Dengo, G. (1973). *Estructura Geologica, Historia Tectonica y Morfologia de America Central,* Ed. 2, Centro Regional de Ayuda Tecnica, A.I.D. Mexico.

Dengo, G. and Bohnenberger, O. (1969). Structural development of northern Central America. *Mem. Am. Assoc. Petrol. Geol.,* 11, 203 – 20.

Denton, G. H., Armstrong, R. L. and Stuiver, M. (1971). The late Cenozoic glacial history of Antarctica, in *The Late Cenozoic Glacial Ages* (ed. Turekian, K. K.), Yale University Press, New Haven, pp. 267 – 306.

Detterman, R. L. (1973). Mesozoic sequence in Arctic Alaska, in *Arctic Geology* (ed. Pitcher, M. G.), vol. 19, pp. 376 - 87, Tulsa, Oklahoma: Am. Assoc. Petrol. Geol. Mem.

Dietz, R. S. and Holden, J. C. (1970). Reconstruction of Pangaea: breakup and dispersion of continents, Permian to present. *J. Geophys. Res.,* 75, 4939 – 56.

Dingle, R. V. (1972). Reply to J. G. Jones (1972). *Nature Physical Science,* 235, 60.

Dingle, R. V. (1973). Mesozoic paleogeography of the southern Cape, South Africa. *Palaeogeogr. Palaeoclimatol. Palaeocecol.,* 13, 203 – 13.

Dingle, R. V. and Klinger, H. C. (1971). Significance of upper Jurassic sediments in the Knysna Outlier (Cape Province) for timing of the breakup of Gond-wanaland. *Nature,* 232, 37 – 8.

Dingle, R. V. and Scrutton, R. A. (1974). Continental breakup and the development of post-Paleozoic sedimentary basins around southern Africa. *Bull. Geol. Soc. Am.,* 85, 1467 – 74.

Douglas, R. G., Moullade, M. and Nairn, A. E. M. (1973). Causes and consequences of drift in the South Atlantic, in *Implications of Continental Drift to the Earth Sciences* (ed. Tarling, D. H. and Runcorn, S. K.), vol. 1, Academic Press, London, pp. 517 – 37.

Dusén, P. (1908). Uber die Tertiare flora der Seymour Insel. *Wiss. Erg. Schwed. Sudpolar Exped. 1901 – 03,* 3, 1 – 27.

Edgar, N. T., Ewing, J. T. and Hennion, J. (1971). Seismic refraction and reflection in Caribbean Sea. *Am. Assoc. Petrol. Geol. Bull.,* 55, 833 – 70.

Elliot, D. H. (1972). Aspects of Antarctic geology and drift reconstructions, in *Antarctic Geology and Geophysics* (ed. Adie, R. J.), Universitetsforlaget, Oslo, pp. 849 – 58.

Fitzpatrick, K. T. and Summerson, C. H. (1971). Some observations on electron micrographs of quartz sand grains. *Ohio J. Sci.,* 71, 106 – 19.

Fleming, C. A. (1962). New Zealand biogeography. A paleontologist's approach. *Tuatara,* 110, 53 - 108.

Fleming, C. A. (1963). Paleontology and southern biogeography, in *Pacific Basin Biogeography* (ed. Cressitt, J. L.), Bishop Museum Press, Honolulu, pp. 369- 85.

Fooden, J. (1972). Breakup of Pangaea and isolation of relict mammals in Australia, South America, and Madagascar. *Science,* 175, 894 – 8.

Foster, F. J. (1974). Eocene echinoids and the Drake Passage. *Nature,* 249, 751.

Frakes, L. A. and Kemp, E. M. (1973). Palaeogene continental positions and evolution of climate, in *Implications of Continental Drift to the Earth Sciences* (ed. Tarling, D. H. and Runcorn, S. K.), vol. 1, Academic Press, London, pp. 539 – 59.

Grant, N. K. (1971). South Atlantic, Benue Trough, and Gulf of Guinea Cretaceous triple junction. *Bull. Geol. Soc. Am.,* 82, 2295 - 8.

Griffiths, D. H. and Barker, P. F. (1972). Review of marine geophysical investigations in the Scotia Sea, in *Antarctic Geology and Geophysics* (ed. Adie, R. J.), Universitetsforlaget, Oslo, pp. 3 – 11.

Griffiths, J. R. (1974). Revised continental fit of Australia and Antarctica. *Nature,* 249, 336 -8.

Griffiths, J. R. and Varne, R. (1972). Evolution of the Tasman Sea, Macquarie Ridge, and Alpine Fault. *Nature Physical Science,* 235, 83 - 6.

Hamilton, W. (1966). Formation of the Scotia and Caribbean arcs. *Geol. Surv. Canada Pap.,* 66 - 15, 178 - 87.

Hamilton, W. (1967). Tectonics of Antarctica. *Tectonophysics,* 4, 455 – 68.

Harrington, H. J. (1962). Paleogeographic development of South America. *Bull. Am. Assoc. Petrol. Geol.,* 46, 1773 -814.

Harrison, L. (1924). The migration route of the Australian marsupial fauna. *Austr. Zool.,* 3, 247 – 63.

Hayes, H. E. and Ringis, J. (1973). Seafloor spreading in the Tasman Sea. *Nature,* 243, 454 -8.

Hayes, H. E. *et al.* (1973). Leg 28 deep -sea drilling in the southern ocean. *Geotimes* 18 (6), 19 -24.

Hayman, D. L., Kirsch, J. A. W., Martin, P. G. and Waller, P. F. (1971). Chromosomal and serological studies of the Caenolestidae and their implications for marsupial evolution. *Nature,* 231, 194 - 5.

Heirtzler, J. R. (1968). Sea -floor spreading. *Sci. Am.,* 219 (6), 60 - 70.

Hershkovitz, P. (1972). The recent mammals of the Neotropical Region: a zoogeographic and ecological review, in *Evolution, Mammals, and Southern Continents,* (ed. Keast, A., Erk, F. C. and Glass, B.), State University of New York Press, Albany, pp. 311 - 432.

Hoffstetter, R. (1970). Radiation initiale des mammiferes placentaires et biogeographie. *C. R. Acad. Sci. Paris, Ser. D,* 270, 3027 - 30.

Hoffstetter, R. (1972). Relationships, origins, and history of the ceboid monkeys and caviomorph rodents: a modern reinterpretation. *Evol. Biol.,* 6, 323 - 47.

Hopkins, W. S. (1974). Some spores and pollen from the Christopher Formation (Albian) of Ellef and Amund Ringnes Island, and northwestern Melville Island, Canadian Arctic Archipelago. *Geol. Surv. Canada Pap.,* **73** - 112.

Howden, H. F. (1974). Problems in interpreting dispersal of terrestrial organisms as related to continental drift. *Biotropica,* **6**, 1 - 6.

Jardine, N. and McKenzie, D. (1972). Continental drift and the dispersal and evolution of organisms. *Nature,* **235**, 20 - 4.

Johnson, L. A. S. and Briggs, B. G. (1963). Evolution in the Proteaceae. *Aust. J. Bot.,* **11**, 21 - 61.

Jones, F. Wood (1923). *The Mammals of South Australia.* Part I: *The Monotremes and the Carnivorous Marsupials,* Government Printer, Adelaide.

Jones, F. Wood (1924). *The Mammals of South Australia.* Part II: *The Bandicoots and the Herbivorous Marsupials,* Government Printer, Adelaide,

Jones, J. G. (1972). Significance of Upper Jurassic sediments in the Knysna Outlier (Cape Province). *Nature Physical Science,* **235**, 59 - 60.

Kauffman, E. G. (1973). Cretaceous bivalvia, in *Atlas of Palaeobiogeography* (ed. Hallam, A.), Elsevier, New York, pp. 353 - 83.

Keast, A. (1971). Continental drift and the evolution of the biota on southern continents. *Quart. Rev. Biol.,* **46**, 335 - 78.

Keast, A. (1972a). Reprint of the above in *Evolution, Mammals, and Southern Continents* (ed. Keast, A., Erk, F. C. and Glass, B.), State University of New York Press, Albany, pp. 23 - 88.

Keast, A. (1972b). Australian mammals: zoogeography and evolution, in *Evolution, Mammals, and Southern Continents* (ed. Keast, A., Erk, F. C. and Glass, B.), State University of New York Press, Albany, pp. 195 - 246.

Keast, A. (1973). Contemporary biotas and the separation sequence of the southern continents, in *Implications of Continental Drift to the Earth Sciences,* vol. 1 (ed. Tarling, D. H. and Runcorn, S. K.), Academic Press, London, pp. 309 - 43.

Kemp, E. M. (1972). Reworked palynomorphs from the West Ice Shelf area, East Antarctica, and their possible geological and palaeoclimatological significance. *Marine Geol.,* **13**, 145 - 57.

Kennett, J. P. *et al.* (1972). Australian-Antarctic continental drift, palaeocirculation changes and Oligocene deep-sea erosion. *Nature Physical Science.* **239**, 51 - 5.

Kennett, J. P. *et al.* (1974). Development of the Circum-Antarctic current. *Science,* **186**, 144 - 7.

Kesler, S. E. (1973). Basement rock structural trends in southern Mexico. *Bull. Geol. Soc. Am.,* **84**, 1059 - 64.

Kielan-Jaworowska, Z. (1975). Migrations of the Multituberculata and the Late Cretaceous connections between Asia and North America. *Ann. S. Afr. Mus.,* **64** (in press).

Kirsch, J. A. W. (1968). Prodromus of the comparative serology of Marsupialia. *Nature,* **217**, 418 - 20.

Larson, R. L. and Chase, C. G. (1972). Late Mesozoic evolution of the western Pacific Ocean. *Bull. Geol. Soc. Am.,* **83**, 3627 - 44.

Larson, R. L. and Ladd, J. W. (1973). Evidence for the opening of the South Atlantic in the Early Cretaceous. *Nature,* **246**, 209 - 12.

Larson, R. L. and Pitman III, W. C. (1972). World-wide correlation of Mesozoic magnetic anomalies, and its implications. *Bull. Geol. Soc. Am.,* **83**, 3645 - 62.

Laughton, A. S., McKenzie, D. P. and Sclater, J. G. (1973). The structure and evolution of the Indian Ocean, in *Implication of Continental Drift to the Earth Sciences* (ed. Tarling, D. H. and Runcorn, S. K.), Academic Press, London, pp. 203 - 12.

Le Pichon, X. (1968). Sea floor spreading and continental drift. *J. Geophys. Res.,* **73**, 3661 -97.

Le Pichon, X. and Hayes, H. E. (1971). Marginal offsets, fracture zones, and the early opening of the North Atlantic. *J. Geophys. Res.,* **76**, 6283 - 93.

Lillegraven, J. A. (1969). Latest Cretaceous mammals of upper part of Edmonton Formation of Alberta, Canada, and review of marsupial-placental dichotomy in mammalian evolution. *Paleontol. Contrib. Univ. Kansas,* Art. 50 (Vertebrate 12).

Lillegraven, J. A. (1974). Biogeographical considerations of the marsupial-placental dichotomy. *Ann. Rev. Ecol. System.,* **5**, 263 - 83.

Lowenstam, H. A. and Epstein, S. (1954). Paleotemperatures of the Post-Aptian Cretaceous as determined by the oxygen isotope method. *J. Geol.,* **62**, 207 -48.

McElhinny, M. W. (1970). Formation of the Indian Ocean. *Nature,* **228**, 977 - 9.

McElhinny, M. W. and Wellman, P. (1969). Polar wandering and sea-floor spreading in the southern Indian Ocean. *Earth Planet. Sci. Lett.* **6**, 198 - 204.

McGowran, B. (1973). Rifting and drift of Australia and the migration of mammals. *Science,* **180**, 759 - 61.

McIntyre, D. J. and Wilson, G. J. (1966). Preliminary palynology of some Antarctic Tertiary erratics. *N. Z. J. Bot,* **4**, 315 - 21.

McKenna, M. C. (1972). Was Europe connected directly to North America prior to the Middle Eocene? *Evol. Biol.,* **6**, 179-89.

McKenna, M. C. (1973). Sweepstakes, filters, corridors, Noah's Arks, and beached Viking funeral ships in paleogeography, in *Implication of Continental Drift to the Earth Sciences* (ed. Tarling, D. H. and Runcorn, S. K.), Academic Press, London, pp. 291 - 304.

Malfait, B. T. and Dinkelman, M. G. (1972). Circum-Caribbean tectonic and igneous activity and the evolution of the Caribbean plate. *Bull. Geol. Soc. Am.,* **83**, 251 -72.

Mandra, Y. T. (1969). Silicoflagellates: a new tool for the study of Antarctic Tertiary climates. *Antarct. J. U.S.,* **4**, 172 -4.

Margolis, S. V. and Kennett, J. P. (1971). Cenozoic Paleoglacial history of Antarctica recorded in Subantarctic deepsea cores. *Am. J. Sci.,* **271**, 1 - 36.

Martin, P. G. (1970). The Darwin Rise hypothesis of the biogeographic dispersal of marsupials. *Nature,* **225**, 197 - 8.

Maxwell, A. E. *et al.* (1970). Deep sea drilling in the South Atlantic. *Science,* **168**, 1047 - 59.

Mayr, E. (1944). Wallace's Line in the light of recent zoogeographic studies. *Quart. Rev. Biol.,* **19**, 1 - 14.

Mayr, E. (1953). *Animal Species and Evolution,* Harvard University Press, Cambridge, Mass.

Mills, R. A., Hugh, K. E. Feray, D. E. and Swolfs, H. C. (1967). Mesozoic stratigraphy of Honduras. *Bull. Am. Assoc. Petrol. Geol.,* **51**, 1711 –86.

Moreau, R. E. (1966). *The Bird Faunas of Africa and its Islands,* Academic Press,
London.

Orlando, H. A. (1964). The fossil flora of the surroundings of Ardley Peninsula
(Ardley Island), 25 de Mayo Island (King George Island), South Shetland
Islands, in *Antarctic Geology,* (ed. Adie, R. J.), Wiley, New York,
pp. 629–36.

Patterson, B. and Pascual, R. (1968). The fossil mammal fauna of South America.
Quart. Rev. Biol., **43**, 409 –451.

Patterson, B. and Pascual, R. (1972). As above, republished in *Evolution, Mammals,
and Southern Continents* (ed. Keast, A., Erk, F. C. and Glass, B.), State
University of New York Press, pp. 247 – 310.

Phillips, J. D. and Forsyth, D. (1972). Plate tectonics, paleomagnetism, and the
opening of the Atlantic. *Bull. Geol. Soc. Am.,* **83**, 1579 – 600.

Pitman, W. C. III and Talwani, M. (1972). Sea-floor spreading in the North
Atlantic. *Bull. Geol. Soc. Am.,* **83**, 619 – 46.

Plumstead, E. P. (1962). Fossil floras of Antarctica. *Sci. Reps. Trans-Antarctic
Exped.,* **9**, 1 – 154.

Plumstead, E. P. (1964). Paleobotany of Antarctica, in *Antarctic Geology* (ed.
Adie, R. J.), North -Holland, Amsterdam, pp. 637 – 54.

Raven, H. C. (1935). Wallace's Line and the distribution of Indo-Australian
mammals. *Bull. Am. Mus. Nat. Hist.,* **68**, 179 – 293.

Raven, P. H. and Axelrod, D. I. (1972). Plate tectonics and Australasian paleobiogeography. *Science,* **176**, 1379 - 86.

Raven, P. H. and Axelrod, D. I. (1974). Angiosperm biogeography and past continental movements. *Ann. Missouri Bot. Garden,* **61**, 539 – 673.

Reid, E. M. and Chandler, M. E. J. (1933). *The London Clay Flora,* British Museum
(Nat. History), London.

Reyment, R. A. and Tait, E. A. (1972). Biostratigraphical data of the early history
of the South Atlantic Ocean. *Trans. Roy, Soc. London,* **264**, 55 - 95.

Rich, P. V. (1975). Antarctic dispersal routes, wandering continents, and the origin
of Australia's non-passerine avifauna. *Mem. Nat. Mus. Victoria,* No. 36,
63 – 125.

Ridd, M. F. (1971). South-East Asia as a part of Gondwanaland. *Nature,* **234**,
531 – 3.

Ride, W. D. L. (1964). A review of the Australian fossil marsupials. *J. Roy. Soc.
Western Aust.,* **47**, 97 - 131.

Ride, W. D. L. (1970). *A Guide to the Native Mammals of Australia,* Oxford
University Press, Melbourne.

Russell, D. A. (1970a). The dinosaurs of central Asia. *Can. Geogr. J.,* **81**, 208 - 15.

Russell, D. A. (1970b). Tyrannosaurs from the Late Cretaccous of western Canada.
Publ. Palaeontol., Nat. Mus. Nat. Sci. Can., No. 1.

Samylina, V. A. (1968). Early Cretaceous angiosperms of the Soviet Union based
on leaf and fruit remains. *J. Linn. Soc. (Bot.),* **61**, 207 - 16.

Sclater, J. G. and Fisher, R. L. (1974). Evolution of the East Central Indian Ocean,
with emphasis on the tectonic setting of the Ninetyeast Ridge. *Bull. Geol.
Soc. Am.,* **85**, 683 - 702.

Scrutton, R. A. (1973). Structure and evolution of the sea floor south of South
    Africa. *Earth Planet. Sci. Lett.,* 19, 250 - 6.
Semenovich, V. N., Gramberg, I. S. and Nesterov, I. I. (1973). Oil and gas possibil-
    ities in the Soviet Arctic, in *Arctic Geology* (ed. Pitcher, M. G.), Am. Assoc.
    Petrol. Geol. Mem., Tulsa, Oklahoma, pp. 194 - 203.
Sharman, G. B. (1961). The mitotic chromosomes of marsupials and their bearing
    on taxonomy and phylogeny. *Aust. J. Zool.,* 9, 38-60.
Sharman, G. B. (1970). Reproductive physiology of marsupials. *Science,* 167,
    1221 - 8.
Sigé, B. (1972). La faunule de Mammiferes du Cretace superieur de Laguna Umayi
    (Andes Peruvienres). *Bull. Mus. Nat. Hist. Naturelle, Sci. Terre,* 19, 375 - 409.
Simpson, G. G. (1940). Antarctica as a faunal migration route. *Proc. 6 Pacific Sci.
    Congr.,* 755 - 68.
Simpson, G. G. (1950). History of the fauna of Latin America. *Am. Sci,* 38 (3),
    361 - 89.
Simpson G. G. (1952). Probabilities of dispersal in geologic time, in *The Problem
    of Land Connections across the South Atlantic, with Special Reference to
    the Mesozoic* (ed. Mayr, E.). *Bull. Am. Mus. Nat. Hist.,* 99, 163 - 76.
Simpson G. G. (1961). Historical zoogeography of Australian mammals. *Evolution,*
    15, 431 - 46.
Slaughter, B. H. (1968). Earliest known marsupials. *Science,* 162, 254 - 5.
Smiley, C. J. (1966). Cretaceous floras from Kuk River area, Alaska: stratigraphic
    and climatic interpretations. *Geol. Soc. Am. Bull,* 77, 1 - 14.
Smith, A. G. and Hallam, A. (1970). The fit of the southern continents. *Nature,*
    225, 139 - 44.
Smith, A. G., Briden, J. C. and Drewry, G. E. (1973). Phanerozoic world maps, in
    Organisms and Continents through Time (ed. Hughes, N. F.), Palaeontol.
    Assoc. London, Spec. Paper Palaeontol. 12, 1 - 43.
Smith, J. M. B. (1972). Southern biogeography on the basis of continental drift: a
    review. *Aust. J. Mammalogy,* 1, 213 - 29.
Stauffer, P. H. (1974). Malaya and southeast Asia in the pattern of continental
    drift. *Geol. Soc. Malaysia Bull.,* 7, 89 - 138.
Stirton, R. A., Tedford, R. H. and Woodburne, M. O. (1968). Tertiary deposits
    in the Australian region containing terrestrial mammals. *Univ. Calif. Publs.
    Geol. Sci.,* 7, 1 - 30.
Szalay, F. S. and McKenna, M. C. (1971). Beginning of the age of mammals in Asia:
    the late Paleocene Gashato fauna, Mongolia. *Bull. Am. Mus. Nat. Hist.,* 144,
    271 - 317.
Tedford, R. H. (1974). Marsupials and the new paleogeography, in *Paleogeographic
    Provinces and Provinciality* (ed. Ross, C. A.), Soc. Econ. Paleontol. Mineral.
    Spec. Pub. no. 21, pp. 109 - 26.
Tedford, R. H., Banks, M. R., Kemp, N. R., McDougall, I. and Sutherland, F. L.
    (1975). Recognition of the oldest known fossil marsupials from Australia.
    Nature, 255, 141 - 2.
Traub, R. (1972). The zoogeography of fleas (Siphonaptera) as supporting the
    theory of continental drift. *J. Med. Ent.,* 9, 584 - 9.
Troughton, E. (1948). *Furred Animals of Australia,* Angus and Robertson, Sydney.
Vanzolini, P. E. and Guimaraes, L. M. (1955). South American land mammals and

their lice. *Evolution*, **9**, 345.

Veevers, J. J., Jones, J. G. and Talent, J. A. (1971). Indo-Australian stratigraphy and the configuration and dispersal of Gondwanaland. *Nature*, **229**, 383 - 8.

Vinogradov, V. A. *et. al.* (1973). Main features of geologic structure and history of north-central Siberia, in *Arctic Geology* (ed. Pitcher, M. G.), Am. Assoc. Petrol. Geol. Mem., Tulsa, Oklahoma, pp. 181 - 8.

Webb, P. (1968). Comments on Late Cretaceous marine climates in New Zealand and adjacent areas. *Tuatara*, **16**, 8 - 10.

Wilson, G. J. (1968). On the occurrence of fossil microspores, pollen grains, and microplankton in bottom sediments of the Ross Sea, Antarctica. *N.Z. J. Mar. Freshwater Res.*, **2**, 381 - 9.

Woodring, W. P. (1954). Caribbean land and sea through the ages. *Bull. Geol. Soc. Am.*, **65**, 719 - 32.

6 Marsupial biogeography and plate tectonics

P. G. Martin

Introduction

It would not be difficult to append to this paper a very long list of questions, each reflecting an uncertainty, the resolution of which is probably relevant to the problem of marsupial biogeography. Some of these questions would be biological, some palaeontological, and a large group would be geophysical. While reconstructions of former continental positions, such as those of Dietz and Holden (1970) or Briden, Drewry and Smith (1974), may be tempting points of reference for the biologist, they are continually under challenge. Doubt exists both on obvious points such as the positions of India and Madagascar and on less obvious ones like those raised by Burrett (1974), who suggests that Asia comprised nine blocks that only fused well into the Mesozoic. Similar doubts about the nature of Antarctica will be discussed later and might also be raised in relation to Australia (McElhinny and Embleton, 1974).

While I have no doubt that marsupial dispersal has been very strongly influenced by continental drift, it seems unlikely that, in the face of these uncertainties, a definitive explanation can yet be given. Nevertheless there is one broad hypothesis to which I believe more students of the subject will subscribe at this time than to any other; for example, a similar one has been adopted in a recent review of marsupial phylogeny (Sharman, 1974). After describing this relatively orthodox viewpoint and briefly discussing its weaknesses and rivals, I propose concentrating on two topics which seem to hold the promise of important new clues, viz. Mesozoic plate tectonics of the Pacific ocean and the distribution of primitive angiosperms. Although the particular reason for discussing early angiosperms will be made clear later, at this stage it is worth drawing attention to the hypothesis of Clemens (1971) that the early radiation of mammals was related to the origin and diversification of angiosperm-dominated floras and their associated invertebrate faunas.

*Except for three periods of study leave in England or the USA, Peter Martin's career as a scientist has been spent in the University of Adelaide. He obtained his Ph.D. in genetics, was then appointed lecturer in biology and has spent 10 years in each of the Zoology and Botany Departments. His research interests have been mostly concerned with chromosomes, both plant and animal, and he has published a book on marsupial cytogenetics with his colleague, David Hayman, also a contributor to this volume. His interest in continental drift was aroused by a lecture by S.K. Runcorn and for 14 years he has followed the literature closely.

An orthodox account of marsupial biogeography

The following has been stated many times, either in part or completely, and no attempt will be made to trace its history. The Early Cretaceous therian ancestors of the marsupials are all from Europe, Asia and North America, and this is also true of the later therians of eutherian - metatherian grade (Clemens, 1971; Olson, 1971). Of the latter, one currently thought likely to be a marsupial is *Deltatheridium* from the Coniacian or Santonian (88 - 77 m.y. B.P.) of Mongolia (Butler and Kielan-Jaworowska, 1973; Van Valen, 1974). It has been reported that the Mongolian Cretaceous faunas were largely distinct from those of North America, although there was limited dispersal, apparently one-way from Asia, near the end of the Cretaceous (Kielan-Jaworowska, 1974).

The earliest fossils which are unquestionably marsupial occur in the Campanian (76 - 73 m.y. B.P.) of western North America (Fox, 1968), and this will be taken as the site of their first significant radiation. During the Santonian - Campanian, North America was divided by a seaway from the Arctic ocean to the Gulf of Mexico (Figure 6.1A). By the end of the Cretaceous (66 m.y. B.P.) this seaway had become dry land, which, no doubt, accounts for the appearance of marsupials in eastern North America at that time (Clemens, 1971). One North American genus, *Peratherium*, migrated to Europe in the Eocene (54 - 39 m.y. B.P.). The faunal connections between North America and Europe in the Eocene have been discussed

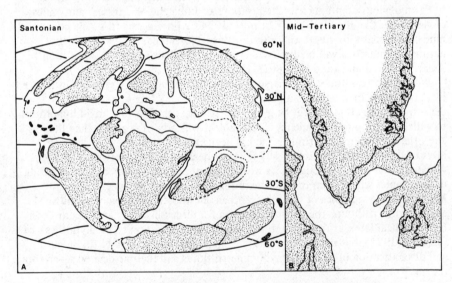

Figure 6.1 Maps showing coastlines at times important in marsupial history. A is redrawn from Gordon (1973) and shows Santonian (82 - 77 m.y. B.P.) land (shaded) and sea with edges of continents in fine outline. The earliest certain marsupial fossils are found at this time in western North America. B is redrawn from Vogt (1972) and shows that a proto-Iceland could have formed a continuous land bridge from Greenland to the Faeroes between 60 and 40 m.y. B.P. when *Peratherium* occurred in western Europe; shaded at or above sea level.

by McKenna (1972) and the disjunction of these two continents has been described by Vogt and Avery (1974); it seems highly probable that overland migration via Greenland occurred. The most revealing map is that of Vogt (1972), which indicates that there could have been a land bridge from Greenland to the Faeroes between 60 and 40 m.y. B.P. (Figure 6.1B). By the early Miocene, marsupials were extinct in North America and Europe.

Several orders of placentals were present with didelphoid marsupials in North America at the end of the Cretaceous and, since three of these appear with didelphoids in South America in the Palaeocene (65 - 55 m.y. B.P.), it is thought that they all migrated at similar times and by similar routes from North America (Patterson and Pascual, 1968). How this migration might have been made possible will be discussed later but, because not all North American orders accomplished it, it is suggested that an element of chance, as in 'island-hopping', was involved. The discovery (Grambast *et al.*, 1967; Sigé, 1968, 1971) of didelphoid and placental teeth in Upper Cretaceous strata of Peru does not appear to change this interpretation beyond setting the migration at an earlier time. The important point is that there is no evidence to suggest that marsupials migrated to South American before placentals. With the joining of North and South America in the Pliocene (5.5 -1.8 m.y. B.P.), didelphids migrated back to North America where they had become extinct 20 million years earlier.

The last two paragraphs would, I believe, be relatively non-controversial if it were not for the fact that, in trying to explain the additional problem of dispersal to Australia, some authors have been forced to adopt alternative hypotheses. In the last 3 years there has been an increasing consensus that migration took place from South America across the Scotia Arc (straightened as

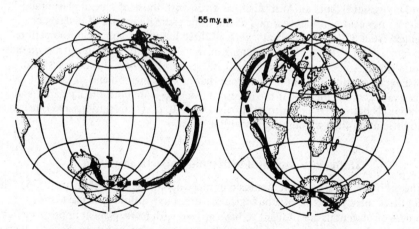

55 m.y. B.P.

Figure 6.2 The heavy black arrows indicate the dispersal of marsupials according to an orthodox hypothesis (see text). Dashes in these arrows suggest migration across chains of islands. The map is redrawn from Jurdy and Van der Voo (1974) and shows relative positions of continents at 55 m.y. B.P. but does not purport to show coastlines at that time. It should be noted that the reconstruction is based in part on McKenzie and Sclater (1971) and is subject to the same possible criticisms that are discussed in the text.

in Hawkes, 1962) and Antarctic Peninsula ultimately to East Antarctica and
Australia - New Guinea. This migration, if it occurred, must have been accomplished
before Australia and East Antarctica parted, the best estimate of the time of this
separation being 49 m.y. B.P. (McGowran, 1973). Dalziel *et al.* (1973) have suggested that the initial break between the Andes and the Antarctandes occurred some
time before the Late Cretaceous, and if this is correct, the migration from South
America to Antarctica must have been by 'island hopping'.

The time of marsupial — placental divergences

One basis for possible criticism of the last section is the assumption that the first
site of significant marsupial radiation was in western North America while it was
isolated in the Coniacian - Santonian (88 - 77 m.y. B.P.). Slaughter (1968) believed
that some mammals present in Texas in the Albian (106 - 100 m.y. B.P.) were
already differentiated; thus *Pappotherium* was a eutherian and *Holoclemensia*
was a didelphoid marsupial. It should be noted that a seaway did not divide North
America in the Early Cretaceous and there was probably a connection to Western
Europe (Gordon, 1973). Clemens (1971) agreed that it was probable that
differentiation of marsupials and placentals occurred prior to the Albian.

Since these palaeontological opinions were expressed, evidence has come
forward from a relatively independent source, the amino acid sequencing of
proteins. Although the logical basis of this method of dating is not firmly
established, the empirical evidence is largely, though not completely, self-
consistent; while this evidence should not be accepted uncritically, neither should
it be ignored. Air *et al.* (1971) compared three proteins (α- and β-haemoglobins
and myoglobin) from an Australian kangaroo with those of several placentals.
The 95 per cent confidence limits of the three estimated times of divergence
ranged from 85 to 170 m.y. B.P., with all three limits overlapping in the period
136 to 123 m.y. B.P., i.e. the beginning of the Cretaceous. Now the first sequence
for a North American opossum is available (Stenzel, 1974). This diverges from
placentals considerably more than does that of the kangaroo and, if taken at its
face value, would mean that the opossum diverged from both the kangaroo and
placentals before the Cretaceous. It is too early to draw conclusions and one can
do little more than watch progress with interest.

Difficulties associated with trans-Antarctic migration

These difficulties have been discussed in more detail elsewhere (Martin, 1975). There
are three main problems of a biological nature. First, although at least three
orders of placentals are thought to have arrived with marsupials in South America,
for them the migration route to Australia was apparently not open. Second, marsupials must have arrived in Australia before 49 m.y. B.P., yet their first recorded
occurrence in the fossil record is about 25 m.y. B.P. (McGowran, 1973). The corresponding period after the postulated invasion of South America is exceedingly rich in fossils
(Patterson and Pascual, 1968) and it is not at all apparent why Australia should be
so different. Third, if a migration bridge like that postulated had existed, it would

surely be the physical explanation for the existence of Hooker's 'south temperate flora' and its associated invertebrate fauna. This has been reviewed elsewhere and an alternative explanation offered (Martin, 1975). Briefly, the present objection must be that marsupials would surely be components of these ecosystems, yet they do not seem to be. The epitome of this flora is *Nothofagus*, the genus of southern beeches, but even in South America, where the genus has existed since the Cretaceous, beech forests are devoid of marsupials (Hershkovitz, 1972), as indeed is the southern 1500 km of the continent. In an otherwise excellent review, Keast (1973) implies similar biogeographical histories for *Nothofagus* and marsupials but does not comment on the fact that, although the opportunity would have existed for 60 million years, there has been no obvious radiation of marsupials into *Nothofagus* forests. It is not difficult to agree with Darlington (1965) that marsupials are not a south-temperate group.

The geophysical difficulties require more explanation. The hypothesis of trans-Antarctic dispersal has been stated most authoritatively by Jardine and McKenzie (1972); the second author having been one of the pioneers of plate tectonics theory, it is not easy for a biologist to criticise with confidence. Nevertheless I feel this should be done lest biogeographers who rely on their paper accept its authority uncritically. The geophysical work on which their paper was based was described by McKenzie and Sclater (1971) and concerned the Indian Ocean, although they also considered other plates. At one point in that paper 'it was arbitrarily decided to reduce the motion between South America and Antarctica to as little as possible'. This *assumption* in a geophysical paper has become the established fact in the biogeographical paper of Jardine and McKenzie (1972), although the objective testing of the assumption does not appear to have been very rigorous and depended on a particular selection of palaeomagnetic poles, especially for India and Antarctica. The effect of this assumption is, of course, to minimise the distance between South America and Australia, a situation which maximises the ability to visualise migration routes.

In the reconstructions of McKenzie and Sclater (1971) the relative positions of Australia, Antarctica and South America are the same at both 75 m.y. and 45 m.y. B.P. (Figure 6.3A). Either South America and Africa are required to move about 14 degrees of latitude towards the south pole from 75 m.y. to 45 m.y. B.P., and then back again, or Australia is placed significantly further south than a reliable pole position indicates (Wellman, McElhinny and McDougall, 1969; McElhinny and Embleton, 1974). If Australia is parted from East Antarctica at about the present position of the Southeast Indian ridge, and if South America did not execute the '14 degree bob' that McKenzie and Sclater (1971) show, then it seems rather less likely that a bridge could have existed from South America to East Antarctica *before East Antarctica parted from Australia*; after that event it would be too late for marsupial migration into Australia. This is illustrated in Figure 6.3B. As a further basis for comparing Figures 6.3A and 6.3B, the distance in Figure 6.3B from Cape Horn to the intersection of its meridian with the junction of Australia – Antarctica is about 75 degrees of latitude. The equivalent in Figure 6.3A is about 55 degrees of latitude, i.e. there is a difference of about 20 degrees of latitude or 2200 km. The most accurate pole position for Australia (Wellman *et al.*, 1969; McElhinny and Embleton, 1974) would reduce the distance in Figure 6.3B by 5 degrees of latitude (95 per cent confidence limits 0 – 10 degrees) and increase the distance in Figure 6.3A by 15 degrees (95 per cent confidence limits 10 – 20 degrees), assuming in both cases that South America is fixed. Thus, on the basis of this pole, Figure 6.3B may be closer to the true position than Figure 6.3A.

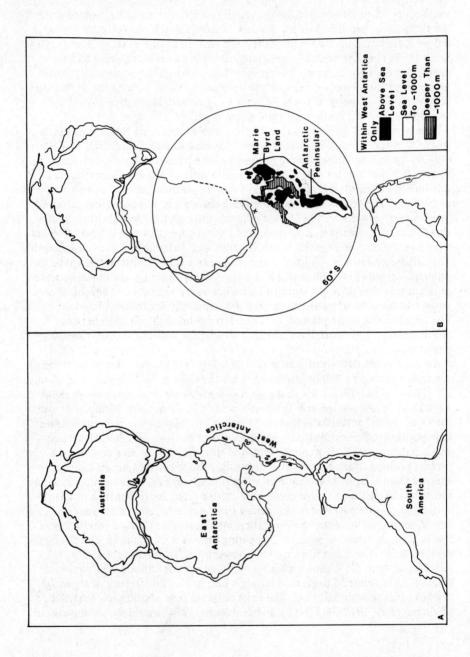

In reconstructions Hayes and Ringis (1973) encountered difficulties that could best be solved by moving East Antarctica and West Antarctica relative to each other. The importance of the Pacific Ocean plates, which, because the information about them was unpublished, were not considered by McKenzie and Sclater (1971), will be made clear in the next section. Until new models which consider all plates around the Antarctic plate have been devised and tested. I believe some scepticism by biogeographers is justified.

A different sort of criticism of Jardine and McKenzie (1972) is more easily justified. Even after removal of the ice cap, West Antarctica would now be an archipelago (Figure 6.3B) with sea depths down to 2500 m *within* the usual continental outline (Anderson, 1965; Hamilton, 1967; Shopf, 1970). For the purposes of a geophysical paper it was probably not important that only a continental outline was shown, as in Figure 6.3A. For the biogeographical paper of Jardine and McKenzie (1972) it seems improper to ignore the true present geography. This is not to suggest that the Early Tertiary geography was the same as at present.

Having stated these difficulties, of both a biological and geophysical nature, nevertheless the hypothesis of trans-Antarctic dispersal probably presents fewer difficulties today than other hypotheses. However, some readers may wish to keep open minds in the hope that better hypotheses will come with new facts and ideas.

Other hypotheses about marsupial biogeography

In this section only four markedly different hypotheses will be outlined. The 'garland' hypothesis of Hoffstetter (1970) does not differ substantially from the 'orthodox' hypothesis and so will not be discussed.

Cox (1970) favoured an origin of marsupials in Africa with entry from there to Australia via East Antarctica. Cox (1973) has since favoured a hypothesis like the 'orthodox' one but now a hypothesis similar to his original one has been supported by Raven and Axelrod (1974). However, there is no evidence that marsupials have ever existed either in Africa or in New Zealand, to which, on this hypothesis, they should have gained access at the time proposed. Moreover, recent evidence that the sea-floor in the Indian Ocean off Western Australia is of late Jurassic age (Heirtzler *et al.*, 1973; Markl, 1974) suggests that Eastern Gondwanaland (Australia plus East Antarctica) probably separated from the remainder at

---

Figure 6.3 (opposite) Drawings which form a basis for discussing the relative positions of Australia, Antarctica and South America at 50 m.y. B.P., after which Australia and East Antarctica parted. A shows present coastlines and continental shelves and is redrawn from the reconstruction of McKenzie and Sclater (1971) for 45 m.y. B.P. The relative positions being the same at 75 m.y. B.P., they are assumed to be correct from 50 m.y. B.P. B is not a reconstruction but has been drawn to illustrate two points. First, that if the junction of Australia – East Antarctica is placed along the present position of the south-east Indian ridge (about 50°S) and if South America is placed close to its present latitude (though further east towards the mid-Atlantic ridge), it is doubtful whether West Antarctica could form a bridge from South America to East Antarctica. In B West Antarctica has been moved to its present position relative to South America (i.e. further away than in A) but this is not to suggest its whereabouts at 50 m.y. B.P. Second, some detail is drawn within West Antarctica to show that the coastline in A is misleading. Within West Antarctica there is water up to 2500 m deep (Anderson, 1965), and, even after isostatic adjustment following ice removal, it would still be an archipelago (Hamilton, 1967). For further discussion see text.

about this time, leaving a marsupial-less gap of more than 110 million years in the Australian fossil record. Dingle and Scrutton (1974) date the separation even earlier, at about 180 m.y. B.P.

Fooden (1972) favoured the concept that, with the break-up of 'Pangaea', samples of its mammalian fauna were isolated successively in Australia – New Guinea (prototherians and metatherians), South America (metatherians and eutherians) and Madagascar (eutherians only). The weakness of his hypothesis is its lack of support from the fossil record, the arguments being similar to those in the last paragraph but perhaps stronger because they must be extended to another group (prototherians) and another continent (South America). Thus prototherian fossils have never been found in South America and marsupial fossils not until 70 million years after Fooden predicts them.

Tyndale-Biscoe (1973) suggested that in the Early Cretaceous all continents had prototherians, metatherians and eutherians but on different continents different groups have prevailed. This hypothesis again lacks support from the fossil record but Tyndale-Biscoe does make a valuable contribution in arguing convincingly that marsupials should not be thought of as necessarily being adaptively inferior to placentals; they would not inevitably have lost ground to eutherian competitors.

Clemens (1968) favoured dispersal of didelphids from North America through the Pacific coastal area of Asia to Australia; he suggested that there was distinct differentiation between faunas on the coast and those in the interior of Asia, thus accounting for the postulated absence of marsupials from Mongolian fossil faunas. With the discovery that Australia had been adjacent to East Antarctica until the mid-Eocene, Clemens (1971) weakened his support for this hypothesis. Several authors have rejected the possibility that marsupials entered Australia when it collided with Indonesia because of the absence of living or fossil marsupials in Asia or Indonesia and because the Tertiary placentals of Asia did not enter Australia. However, in view of the information to be discussed in the next section, it may be desirable to keep in mind Clemens's original hypothesis, or some modification of it.

The Mesozoic Pacific

Concentration on the hypotheses about Pangaea and its break-up may have left the impression with some readers that the rest of the Mesozoic world, about half of it, was a vast Pacific Ocean, tectonically quiet, devoid of land and so of little interest to biogeographers of land plants and animals. That the Pacific region was tectonically very active in the Cretaceous is now clear from the papers of Larson

---

Figure 6.4 A. Map redrawn from Larson and Pitman (1972) showing the Pacific, Kula, Farallon of North America so that the palaeco-equator appears as a curved line. B. Map redrawn from the Atlantic continents at that time and the plot construction is based on the present location of North America so that the palaeo-equator appears as a curved line. B: Map redrawn from Menard (1973) and Larson and Chase (1972) showing the Pacific ocean as it is today. The isochrons connect sea-floors of equal ages and are those suggested by Larson and Chase (1972). Note that only isochrons on the original Pacific plate are shown; sea-floors of these ages on the Kula, Farallon and Phoenix plates have been subducted. In the South-east Pacific are shown the original positions of the ridge system at 150 m.y. and 110 m.y. B.P., as suggested by Menard (1973). An idea of movement can be gained both by comparing the original positions with the isochrons and also by noting in Figure 6.5 where the northern triple junction was subducted.

and Pitman (1972) and Larson and Chase (1972). A ridge system, including two triple junctions (Figure 6.4A), divided the Pacific into four plates; the Farallon plate has now been subducted beneath America with 7000 km of lithosphere, the Kula plate beneath Alaska and Asia also with 7000 km, and the Phoenix plate

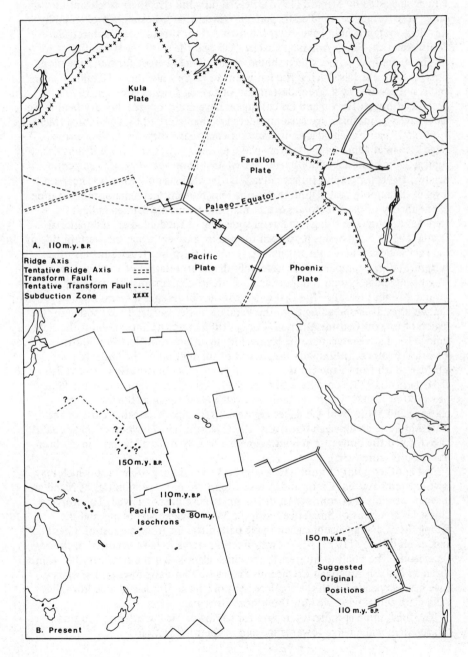

Kula
Plate

Farallon
Plate

Palaeo-Equator

A. 110 m.y. B.P.

Ridge Axis	═══
Tentative Ridge Axis	════
Transform Fault	───
Tentative Transform Fault	----
Subduction Zone	xxxx

Pacific
Plate

Phoenix
Plate

? · ·, ?
? · ·, ?
150 m.y. B.P.

110 m.y. B.P.

Pacific Plate 80 m.y.
Isochrons

150 m.y. B.P.

Suggested
Original
Positions

110 m.y. B.P.

B. Present

has mostly disappeared beneath Chile and West Antarctica with 5000 km of lithosphere. The Pacific plate remains and the probable isochrons for 150, 110 and 80 m.y. B.P. (parts of which are still detectable as magnetic lineations) are indicated in Figure 6.4B. Figure 6.4B also shows the position of the ridge at 110 m.y. B.P., as suggested by Menard (1973). Bearing in mind that these isochrons are only those formed on the Pacific plate, it is clear that lithosphere broken up by this ridge system would have dispersed by sea-floor spreading to Antarctica, South America, North America, Eastern Asia and the south-west Pacific.

There are three points which should be mentioned before further discussion. First, the present East Pacific rise has been active for only about 10 million years and is not apparently a direct descendant of Larson's ridges (Herron, 1972). Second, although Larson and his colleagues have not proposed any dry land movements away from this ridge system, ridges are characterised by up-arching (Menard, 1969, 1973) and volcanic islands, and there may have been dry land over the ridge system at some stage. As lithosphere moves away from a ridge it subsides, a typical figure being about 3000m in the first 45 million years (see McKenzie and Sclater, 1971, for graph); it may not always be obvious now what was near sea level in the remote past. Third, it must be remembered that a major modification of the theory of plate tectonics is that marginal basins are formed behind trenches through the rising and lateral spreading of subducted crustal material (Karig, 1971). Such basins have been formed extensively along the western side of the Pacific ocean, no doubt largely as a result of spreading from Larson's ridge system, and this complicates considerably the interpretation of what we see now.

Larson's ridge system is important to students of marsupial biogeography for several different reasons. The first and most tangible is that it seems highly probable that, as a result of spreading from the Farallon plate, an island arc formed in the region of present Central America (Coney, 1971) so providing, at about the right time in the Late Cretaceous, the bridge for migration from North to South America (Figure 6.5). Similarly, as a result of subduction of the Phoenix plate, a bridge south from Cape Horn (i.e. a straightened Scotia arc: Hawkes, 1962; Dalziel *et al.*, 1973) could have been formed. Again, the extension of the Phoenix rise at 80 m.y. B.P. almost certainly separated New Zealand (the Campbell plateau) and Marie Byrd Land (see Figure 6.3B) and perhaps other parts of the West Antarctic archipelago (Griffiths, 1971; Martin, 1975). Whether these could have formed the bridge from South America to East Antarctica has already been discussed (Figure 6.3B).

Yet another point of importance concerns the bridging of any gap which may have existed between Siberia and Alaska. Apart from the possibilities of island arcs it seems possible that continental drifting may have been involved. The Kolyma block, i.e. that part of Siberia between the Verkhoyansk mountains and the Bering Strait, has a palaeomagnetic pole path different from the rest of Siberia and, as McElhinny (1973) has shown, junction seems to have occurred in the Cretaceous. The point of interest to mammalogists is that the similarity between Asian and North American Cretaceous faunas (Kielan-Jaworowska, 1974) may have been associated with the emplacement of land and/or island arcs from the Kula plate across what we now think of as Beringia.

The final, more speculative, reason for paying particular attention to this ridge system will emerge towards the end of the next section.

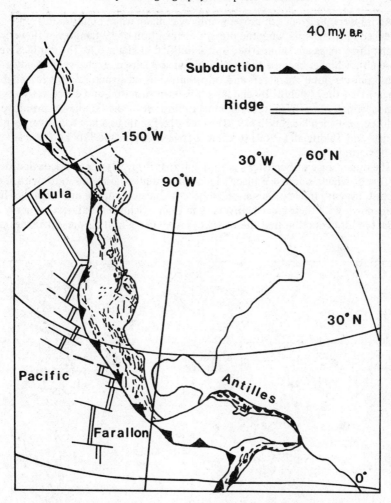

Figure 6.5 Map redrawn from Coney (1971) and showing the development of the Cordillera at 40 m.y. B.P. The northern triple junction
of Figure 6.4A is about to be subducted in the vicinity of present
Vancouver Island. The island arc of the Antilles and Venezuelan
coast range may have been the land-bridge which, near the
Cretaceous – Tertiary boundary (65 m.y. B.P.), allowed migration
from North to South America.

Wallace's line and the biogeography of primitive angiosperms

Since it become known that, after separating from East Antarctica at 49 m.y. B.P.
(Le Pichon, 1968; McGowran, 1973), Australia moved north to approximately its
present position, it has been clear that Wallace's line is the zoogeographic boundary
that marks the collision zone with Indonesia. Since the collision, comparatively
few placentals have migrated east into New Guinea and even fewer marsupials west
into Indonesia. (For a discussion, see Keast, 1972.)

An important advance in biogeography was made when Schuster (1972)
produced new maps showing the present distribution of 19 families of flowering
plants; these maps are summarised and simplified in Figure 6.6. The families were
chosen by Schuster because they are 'truly unspecialised, exclusively woody groups
with primitive floral characters and/or primitive wood structure'; it is reasonable
to agree that they 'include by and large the most primitive and conservative extant
Angiospermae'. They represent the most important of the families of primitive
angiosperms which led Smith (1970) to see 'the Pacific as a key to flowering plant
history' and Takhtajan (1969) to name 'between Assam and Fiji' as 'the cradle of
the flowering plants'.

The importance of Schuster's grouping is that it reveals major discontinuities,
the first of which is across Wallace's line. Nine families lie to the south with their
greatest diversity in New Guinea and adjacent Queensland; ten other families lie
to the north with their greatest diversity in China. Schuster is almost certainly
correct in his deduction that these families were living in their present ranges when

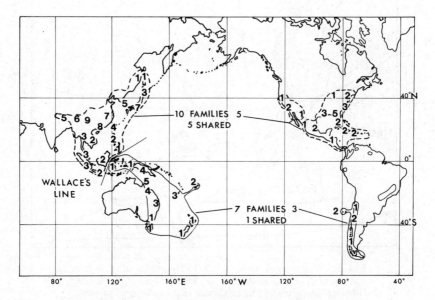

Figure 6.6 The distribution of 19 families of primitive angiosperms.
A number denotes how many families occur at that point. Note the
concentration in China. The map is adapted from Schuster (1972 :
Figures 29, 30 and 31) but simplified by omitting a handful of species
from each of two families, viz.: from the Magnoliaceae (210 species,
northern hemisphere), one or two species in South India and Ceylon
(probably introduced cultivars), two species in northern South
America and a very few which occur south-east of Wallace's line
(only two extend as far east as Papua – New Guinea); from the
Winteraceae (90 species, southern hemisphere), one species from
Madagascar and one which occurs north-west of Wallace's line. The
unnumbered isolates in South America are two species of *Drimys*
(Winteraceae), one of which is also found in Central America (not
shown).

Australia - New Guinea collided with Indonesia and that there has been little migration by these families across Wallace's line since. A notable feature is that for the first time phytogeography and zoogeography are telling similar stories about the area.

It seems prudent to go a little beyond Schuster (1972) and briefly mention other primitive angiosperm families. In his chapter on 'the cradle of the flowering plants' Takhtajan (1969) discusses all the 19 families mapped by Schuster (1972) but also mentions about the same number again, most of which fall into two groups. Families in the first and larger group have their primitive members on the borders of the Pacific Ocean but maps of their distributions would be distorted by the wide dispersal of only a few species; their omissions from Schuster's maps are, in my opinion, justified. However, there is another small group which is more significant in that they are confined to the ranges of the first 19 families but in both northern and southern hemispheres. Thus the Lardizabalaceae has six genera between Assam and Japan and then, without any in between, two more genera in Chile. Another group which is primitive (Takhtajan, 1969; Sporne, 1971) and has a disjunct distribution in both hemispheres is the subfamily Fagoideae, which comprises two genera. *Fagus* species, the northern beeches, have their centre of diversity in China - Japan (Takhtajan, 1969). (The fact that one species, closely related to an oriental one, spreads to Europe is probably not significant in the present context but the occurrences of upper Cretaceous fossils in south-east USA probably are.) *Nothofagus* species, the southern beeches, are found in New Guinea, New Caledonia, New Zealand, south-east Australia, Tasmania and Chile. Thus the Fagoideae have a distribution similar to that of Schuster's families. Two more points should be noted about *Nothofagus;* it is not dispersed for long distances over sea (Preest, 1963; Van Steenis, 1971), and it has an Upper Cretaceous fossil record in New Zealand, Australia, Chile and the Antarctic peninsula (Van Steeenis, 1971).

The main biogeographical problem which Schuster tackled was the disjunction across Wallace's line; he advanced no hypotheses to account for three additional problems posed by his maps. The first of these is the disjunction between eastern Asia and North America shown by five of his ten northern families. It seems probable that this may have an explanation similar to that for mammals (Kielan-Jaworowska, 1974), discussed in the last section.

The second problem is more difficult. Leaving aside the possibility that the angiosperms are polyphyletic, if the northern and southern groups came near each other at Wallace's line in the mid-Tertiary, their ancestors must have been separated much earlier. Can Larson's ridge system offer an explanation for this? Larson and Chase (1972) speculated that the Japanese ridge (i.e. between the Kula and Pacific plates in Figure 6.4A) 'extended west to join the north Atlantic system from the other side of the world, or at least occupied the Tethyan seaway present at that time' ('that time' was probably late Jurassic; Larson and Pitman, 1972, suggested that the Japanese ridge had been active for 30 or 40 million years prior to 110 m.y. B.P.). They went on to point out how biophysical evidence has long since disappeared down subduction zones. Can the disjunction of primitive angiosperms between China and New Guinea - Queensland be equivalent botanical evidence that has survived? Such a hypothesis requires that we examine the structure of New Guinea.

While the southern part of New Guinea (the Fly River Platform) is undoubtedly
part of continental Australia, much of the rest has been accreted during
Australia's northerly drift in the Tertiary. There is considerable agreement that an
island arc lay north of central New Guinea in the late Cretaceous (Johnson and
Molnar, 1972; Audley-Charles, Carter and Milsom, 1972) and that these islands
have been accreted into the Highlands (Johnson and Molnar, 1972). The Kubor
range, which has been a land mass since the Cretaceous (St. John, 1970) (possibly
since the upper Triassic: Bain, 1973; Doutch, 1972) and presumably carried
flora and fauna, is separated from the Fly River Platform by the Southern
Highlands Basin containing 7000 m of Mesozoic sediments (St. John, 1970) (see
Figure 6.7). Could this putative island arc have had its origin on Larson's ridge
system and be the carrier of primitive angiosperms to New Guinea – Australia?
If so, it does not seem unreasonable to take the further step and ask if it could
also have carried marsupials.

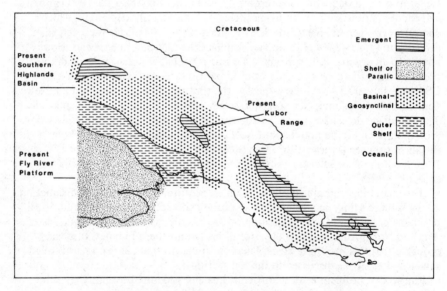

Figure 6.7 Map redrawn from St. John (1970) showing the
conditions during the Cretaceous of different tectonic elements of
eastern New Guinea. The present Kubor range has been land since
the Cretaceous. The present Southern Highlands Basin contains
7000 m of Mesozoic sediments.

Briefly, such a hypothesis would fit in well with several points. Isolation in
islands in the Early Cretaceous is probably compatible with the time of marsupial
origin. Accretion of the islands into the mainland at the end of Australia's
northerly drift is compatible with the first appearance of marsupials in the
Australian fossil record in the Early Miocene (McGowran, 1973). As Darlington
(1965) has emphasised, marsupials are a group of tropical origin. It would explain

the exclusion of placentals and, perhaps, the biogeography of monotremes (Martin, 1975). This last point would be particularly plausible if, as has been suggested (Kielan-Jaworowska, 1970; Kermack and Kielan-Jaworowska, 1971), monotremes are closely related to multituberculates which were prevalent in the Mesozoic northern continents but fossils of which have never been found in a remnant of Gondwanaland.

The third problem posed by Schuster's maps is the inclusion of some of his southern families in New Caledonia (3), New Zealand (1) and Chile (3). Several botanists (Schuster, 1972; Melville, 1969; Good, 1964; Aubreville, 1973; Wardle, 1973) have emphasised the strong floral connections between the highlands of New Guinea, these islands of this 'inner Melanesian arc' and Chile. A former closer connection, and its break-up by part of Larson's ridge system, is suggested. It does not seem profitable to expand on this aspect now since marsupials, like dinosaurs (Fleming, 1962), did not reach either New Caledonia or New Zealand; moreover, as has been pointed out already, the extensive marginal seas make interpretations in terms of plate tectonics very difficult.

Speculation could be endless and I hope I have already made clear why I like to keep a part of my mind open about marsupial biogeography. If, as I have intimated, Larson's ridge system holds the key to an understanding of the biogeography of primitive angiosperms and perhaps of monotremes, might it not also hold the key to marsupial biogeography?

Conclusion

At this time, the safest hypothesis about marsupial biogeography is that the first significant radiation was in western North America at about 75 m.y. B.P. and that marsupials dispersed from there near the end of the Cretaceous to South America and to eastern North America; later one genus reached Europe. From South America they dispersed, probably by island hopping, across the Scotia arc and the archipelago of West Antarctica to East Antarctica and so to Australia, arriving no later than 49 m.y. B.P. Although this may be the safest hypothesis, the difficulties that have been discussed are sufficiently serious to delay its acceptance. Immense tectonic events occurred in the Mesozoic in the Pacific region and, while it is too early to assess the biogeographical effects of these, it seems possible that they hold the key to the understanding of the biogeography of primitive angiosperms and, perhaps, of monotremes and marsupials.

References

Air, G. M., Thompson, E. O. P., Richardson, B. J. and Sharman, G. B. (1971). Amino-acid sequences of kangaroo myoglobin and haemoglobin and the date of marsupial-eutherian divergence. *Nature, 229,* 391 - 4.

Anderson, J. J. (1965). Bedrock geology of Antarctica: a summary of exploration 1831 - 1962. *Am. Geophys. Union, Antarctic Res. Ser., 6,* 1 - 70.

Aubreville, A. (1973). Distribution des Coniferes dans la Pangee. Essais. *Adansonia,* ser. 2, **13,** 125 - 33.

Audley-Charles, M. G., Carter, D. J. and Milsom, J. S. (1972). Tectonic develop-
    ment of eastern Indonesia in relation to Gondwanaland dispersal. *Nature
    Physical Science,* 239, 35 - 9.
Bain, J. H. C. (1973). A summary of the main structural elements of Papua New
    Guinea, in *The Western Pacific; Island Arcs, Marginal Seas, Geochemistry* (ed.
    Coleman, P. J.), University of Western Australia Press, Nedlands.
Briden, J. C., Drewry, G. E. and Smith, G. A. (1974). Phanerozoic equal-area world
    maps. *J. Geol.,* 82, 555 - 74.
Burrett, C. F. (1974). Plate tectonics and the fusion of Asia. *Earth Planet. Sci.
    Lett.,* 21, 181 - 9.
Butler, P. M. and Kielan-Jaworowska, Z. (1973). Is *Deltatheridium* a marsupial?
    Nature, 245, 105 - 6.
Clemens, W. A. (1968). Origin and early evolution of marsupials. *Evolution,* 22,
    1 - 18.
Clemens, W. A. (1971). Mammalian evolution in the Cretaceous, in *Early Mammals*
    (ed. Kermack, D. M. and K. A.), Suppl. No. 1 to *J. Linn. Soc. (Ecol.*), 5,
    Academic Press, London.
Coney, P. J. (1971). Cordilleran tectonic transitions and motion of the North
    American plate. *Nature,* 233, 462 - 5.
Cox, C. B. (1970). Migrating marsupials and drifting continents. *Nature,* 226,
    767 - 70.
Cox, C. B. (1973). Systematics and plate tectonics in the spread of marsupials.
    Spec. Pap. Palaeontol. 12, 113 - 9.
Dalziel, I. W. D., Klingfield, R., Lowrie, W. and Opdyke, N. C. (1973). Palaeo-
    magnetic data from the southernmost Andes and the Antarctandes, in
    Implications of Continental Drift to the Earth Sciences, vol. I (ed. Tarling,
    D. H. and Runcorn, S. K.), Academic Press, London, pp. 87 - 101.
Darlington, P. J. (1965). *Biogeography of the Southern End of the World,*
    Harvard University Press, Cambridge, Mass.
Dietz, R. S. and Holden, J. C. (1970). Reconstruction of Pangaea: breakup and
    dispersion of continents, Permian to present. *J. Geophys. Res.,* 75,
    4939 - 56.
Dingle, R. V. and Scrutton, R. A. (1974). Continental breakup and the develop-
    ment of post-Palaeozoic sedimentary basins around Southern Africa.
    Geol. Soc. Am. Bull., 85, 1467 - 74.
Doutch, H. F. (1972). The paleogeography of Northern Australia and New Guinea
    and its relevance to the Torres Strait Area, in *Bridge and Barrier: the Natural
    and Cultural History of Torres Strait* (ed. Walker, D.), Australian National
    University Publication BG/3, Canberra, pp. 1 - 10.
Fleming, C. A. (1962). New Zealand biogeography – a palaeontologist's approach.
    Tuatara, 10, 53 - 108.
Fooden, J. (1972). Breakup of Pangaea and isolation of relict mammals in
    Australia, South America and Madagascar. *Science,* 175, 894 - 8.
Fox, R. C. (1968). Early Campanian (Late Cretaceous) mammals from Alberta,
    Canada. *Nature,* 220, 1046.
Good, R. (1964). *The Geography of Flowering Plants,* 3rd ed., Longmans, London.
Gordon, W. A. (1973). Marine life and ocean surface currents in the Cretaceous.
    J. Geol., 81, 269 - 84.

Grambast, L., Martinez, M., Mattauer, M. and Thaler, L. (1967). *Peratherium altiplanense* Nov. Gen., nov. sp., premier mammifere Mesozoique d'Amerique du sud. *C. R. Acad. Sci. Paris,* 264, 707 - 10.

Griffiths, J. R. (1971). Reconstruction of the south-west Pacific margin of Gondwanaland. *Nature,* 234, 203 - 7.

Hamilton, W. (1967). Tectonics of Antarctica. *Tectonophysics,* 4, 555 - 68.

Hawkes, D. D. (1962). The structure of the Scotia Arc. *Geol. Mag.,* 99, 85 - 91.

Hayes, H. E. and Ringis, J. (1973). Seafloor spreading in the Tasman Sea. *Nature,* 243, 454 - 8.

Heirtzler, J. R., Veevers, J. V., Bolli, H. M., Carter, A. N., Cook, P. J., Krasheninnikov, V. A., McKnight, B. K., Proto-Decima, F., Renz, G. W., Robinson, P. T., Rocker, K. and Thayer, P. A. (1973). Age of the floor of the eastern Indian ocean. *Science,* 180, 952 - 4.

Herron, E. M. (1972). Seafloor spreading and the Cenozoic history of the east-central Pacific. *Geol. Soc. Am. Bull.,* 83, 1671 - 92.

Hershkovitz, P. (1972). The recent mammals of the neotropical region: a zoogeographic and ecological review, in *Evolution, Mammals, and Southern Continents* (ed. Keast, A., Erk, F. C. and Glass, B.), State University of New York Press, Albany, pp. 311 - 431.

Hoffstetter, M. R. (1970). L'histoire biogeographiques des marsupiaux et la dichotomie marsupiaux-placentaires. *C. R. Acad. Sci. Paris,* 271, 388 - 91.

Jardine, N. and McKenzie, D. (1972). Continental drift and the dispersal and evolution of organisms. *Nature,* 235, 20 - 4.

Johnson, T. and Molnar, P. (1972). Focal mechanisms and plate tectonics of the southwest Pacific. *J. Geophys. Res.,* 77, 5000 - 32.

Jurdy, D. M. and Van der Voo, R. (1974). A method for the separation of true polar wander and continental drift, including results for the last 55 m.y. *J. Geophys. Res.,* 79, 2945 - 52.

Karig, D. E. (1971). Origin and development of marginal basins in the western Pacific. *J. Geophys. Res.,* 76, 2542 - 61.

Keast, A. (1972). Australian mammals: zeoogeography and evolution, in *Evolution, Mammals and Southern Continents* (ed. Keast, A., Erk, F. C. and Glass, B.), State University of New York Press, Albany, pp. 195 - 246.

Keast, A. (1973). Contemporary biotas and the separation sequence of the southern continents, in *Implications of Continental Drift to the Earth Sciences,* vol. 1 (ed. Tarling, D. H. and Runcorn, S. K.), Academic Press, London, pp. 309 - 43.

Kermack, K. A. and Kielan-Jaworowska, Z. (1971). Therian and non-therian mammals, in *Early Mammals* (ed. Kermack, D. M. and K. A.), Suppl. No. 1 to *J. Linn, Soc. (Ecol.),* 5, Academic Press, London.

Kielan-Jaworowska, Z. (1970). Unknown structures in multituberculate skull. *Nature,* 226, 974 - 6.

Kielan-Jaworowska, Z. (1974). Migrations of the Multituberculata and the late Cretaceous connections between Asia and North America. *Ann. S. Afr. Mus.,* 64, 231 - 43.

Larson, R. L. and Chase, C. G. (1972). Late Mesozoic evolution of the western Pacific. *Geol. Soc. Am. Bull.,* 83, 3627 - 44.

Larson, R. L. and Pitman, W. C. (1972). World-wide correlation of Mesozoic magnetic anomalies and its implications. *Geol. Soc. Am. Bull.*, 83, 3645 - 62.

Le Pichon, X. (1968). Sea floor spreading and continental drift. *J. Geophys. Res.*, 73, 3661 - 97.

McElhinny, M. W. (1973). *Palaeomagnetism and Plate Tectonics*, Cambridge University Press, London.

McElhinny, M. W. and Embleton, B. J. J. (1974). Australian palaeomagnetism and the Phanerozoic plate tectonics of Eastern Gondwanaland. *Tectonophysics*, 22, 1 - 29.

McGowran, B. (1973). Rifting and drifting of Australia and the migration of mammals. *Science*, 180, 759 - 61.

McKenna, M. C. (1972). Was Europe connected directly to North America prior to the Middle Eocene? *Evol. Biol.*, 6, 179 - 89.

McKenzie, D. and Sclater, J. G. (1971). The evolution of the Indian Ocean since the late Cretaceous. *Geophys. J. R. Astr. Soc.*, 25, 437 - 528.

Markl. R. G. (1974). Evidence for the breakup of eastern Gondwanaland by the early Cretaceous. *Nature*, 251, 196 - 200.

Martin, P. G. (1975). Marsupial biogeography in relation to continental drift. *Mem. Mus. Nat. Hist. Nat. N. S.* (A), 88, 216 - 37.

Melville, R. (1969) (no title). *Phil. Trans. Roy. Soc. B*, 225, 617 - 9.

Menard, H. W. (1969). Growth of drifting volcanoes. *J. Geophys. Res.*, 74, 4827 - 37.

Menard, H. W. (1973). Does Mesozoic mantle convection still persist? *Earth Planet. Sci. Lett.*, 20, 237 - 41.

Olson, E. C. (1971). *Vertebrate Paleozoology*, Wiley, New York.

Patterson, B. and Pascual, R. (1968). The fossil mammal fauna of South America. *Quart. Rev. Biol.*, 43, 409 - 51.

Preest, D. S. (1963). A note on the dispersal characteristics of the seed of the New Zealand podocarps and beeches and their biogeographical significance, in *Pacific Basin Biogeography* (ed. Gressitt, J. L.), Bishop Museum Press, Honolulu, pp. 415 - 24.

Raven, P. H. and Axelrod, D. I. (1974). Angiosperm biogeography and past continental movements. *Ann. Missouri Bot. Garden*, 61 (2), 539 - 673.

Shopf, J. M. (1970). Ellsworth mountains: position in West Antarctica due to sea -floor spreading. *Science*, 164, 63 - 5.

Schuster, R. M. (1972). Continental movements, 'Wallace's line' and Indomalayan — Australasian dispersal of land plants: some eclectic concepts. *Bot. Rev.*, 38, 3 - 86.

Sharman, G. B. (1974). Marsupial taxonomy and phylogeny. *J. Aust. Mammal Soc.*, 1, 137 - 54.

Sigé, M. B. (1968). Dents de micromammiferes et fragments de coquilles d'oeufs de dinosauriens dans la fauna de vertebres du cretace superieur de laguna Umayo (Andes Peruviennes). *C. R. Acad. Sci. Paris*, 267, 1495 - 8.

Sigé, M. B. (1971). Les Didelphoidea de Laguna Umayo (formation Vilquechico, Cretace superieur, Perou) et le peuplement marsupial d'Amerique du Sud. *C. R. Acad. Sci. Paris*, 273, 2479 - 81

Slaughter, B. H. (1968). Earliest known marsupials. *Science*, 162, 254 - 5.

Smith, A. G. (1970). The Pacific as a key to flowering plant history. University of Hawaii, Harold L. Lyon Arboretum lecture number 1.

Sporne, K. R. (1971). *The Mysterious Origin of Flowering Plants,* Oxford University Press, London.

Stenzel, P. (1974). Opossum Hb chain sequence and neutral mutation theory. *Nature,* **252,** 62 - 3.

St. John, V. P. (1970). The gravity field and structure of Papua and New Guinea. *Aust. Petrol. Explor. Assoc. J.,* **10** (2), 41 - 55.

Takhtajan, A. (1969). *Flowering Plants. Origin and Dispersal,* Oliver and Boyd, Edinburgh.

Tyndale-Biscoe, H. (1973). *Life of Marsupials,* Edward Arnold, London.

Van Steenis, C. G. G. J. (1971). *Nothofagus,* key genus of plant geography, in time and space, living and fossil, ecology and phylogeny. *Blumea,* **19,** 69 - 98.

Van Valen, L. (1974). *Deltatheridium* and marsupials. *Nature,* **248,** 165 - 6.

Vogt, P. R. (1972). The Faeroe-Iceland-Greenland aseismic ridge and the western boundary undercurrent. *Nature,* **239,** 79 - 81.

Vogt, P. F. and Avery, O. E. (1974). Detailed magnetic surveys in the Northeast Atlantic and Labrador sea. *J. Geophys. Res.,* **79,** 363 - 89.

Wardle, P. (1973). New Guinea, our tropical counterpart. *Tuatara,* **20,** 113 - 24.

Wellman, P., McElhinny, M. W. and McDougall, I. (1969). On the polar-wandering path for Australia during the Cenozoic. *Geophys. J. R. Astr. Soc.,* **18,** 371 - 95.

7 Evolution of New Guinea's marsupial fauna in response to a forested environment

Alan C. Ziegler*

Introduction

Four orders of Recent land mammals reached the island of New Guinea before the arrival of prehistoric man: the Monotremata, Marsupialia, Chiroptera and Rodentia. In the Marsupialia the four families now represented in New Guinea are all found also among the six extant families occurring in Australia. These correspond to the 6 New Guinea and 13 Australian families in the check-list of Kirsch and Calaby in this volume. While 14 of the 22 New Guinea genera are found also in Australia, only 11 of New Guinea's 53 species are shared with Australia. Six of these 11 shared species, along with most of the 42 New Guinea endemic species, are, as described in this paper, primarily forest-adapted forms. This is in marked contrast to the situation in Australia, where the bulk of the marsupial fauna is adapted either to open-land living or at least to habitats other than closed-canopy forests.

Various portions of the present island of New Guinea and of the Australian mainland, however, formed parts of the same land mass for most of the Age of Mammals, a circumstance which would, on first thought, suggest that the present New Guinea and northern Australian mammal faunas should be more closely related at the specific and generic levels than is actually so. The major portion of this report consists of an attempt to reconstruct the series of zoogeographic events which led both to the high degree of endemism now evident in New Guinea's marsupial fauna and to the equally high degree of forest-adaptation characteristic of this fauna. Although entirely reasonable on the basis of information at present available, this account is obviously only one of numerous possible reconstructions of the actual zoogeography and evolutionary diversification of Recent New Guinea Marsupialia.

Present geography of New Guinea

To provide some background for the ecological and other accounts to follow, a generalised and necessarily oversimplified statement of the topography and

*Dr Alan Conrad Ziegler (Division of Vertebrate Zoology, Bishop Museum, Honolulu, Hawaii, USA) has been Head of his Division since 1967 and an Affiliate of the Graduate Faculty in Zoology of the University of Hawaii since 1969. His main research interests are in the systematics and evolution of New Guinea mammals, and the identification and analysis of archaeological faunal remains.

vegetation of present-day New Guinea might be useful: more specific information
may be found in various pertinent articles in Ryan (1972) and Walker (1972). The
term 'forest' (alone or in combination) is used here in the sense of Ride (1970:
p. 4): 'crowns of trees more or less touching or overhanging to give a continuous
canopy'. As Ride notes, those habitats often knows as 'rain forest', 'wet sclerophyll
forest' and 'dry sclerophyll forest' are included. My term 'open-land' rather
loosely encompasses all other types of habitat such as Ride's 'woodland' (and
'savannah woodland'), 'heath', 'grassland', 'steppe', and so on.

Mainland New Guinea, next to Australia the world's largest island, is
dominated by a substantial and geologically young central mountain chain,
extending essentially the entire 2400 k east-west length of the country. Several
other forested mountain ranges, mostly below 1400 m in elevation, occur along
much of the north coast of New Guinea, more or less isolated by lowlands from
this main cordillera. As shown in Figure 7.1, the highest elevations of almost
4900 m in the Central Range bear fields of ice or snow. Below this, down to an
average elevation of perhaps 3200 m, follow, successively, talus or rockfields,
alpine grasslands and alpine scrub forests. Wet, and usually mossy, forests then
extend down to about 1500 m, and from this level rain forests continue down to
500 m or less in most regions.

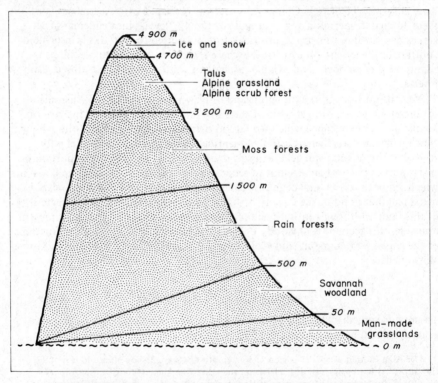

Figure 7.1 Schematised diagram of altitudinal distribution of major
vegetation types on the island of New Guinea. Elevations are
approximate.

Rain forest may extend essentially unbroken to the mangrove swamps at sea level in some areas, while in others such as the extensive south-coast portions of New Guinea, broad plains with savannah woodland and grassland maintained largely by periodic native burning intervene between the forests and shore. Rain forest, as such, covers well over half of the land area of New Guinea (perhaps as much as 70 per cent; see van Deusen, 1972: p. 692) and even in the open-land south-coast region it may extend from the foothills for scores of kilometres along watercourses in the form of gallery forest. Man-maintained grasslands occur on lowland coastal plains and are also frequently encountered at middle and high elevations of the rest of New Guinea, in areas which were once densely forested.

For convenience, reference is occasionally made in this report to political and other sub-divisions of New Guinea. The eastern half of the island is currently termed Papua New Guinea; and the western half, Irian Jaya. The broad-based Huon Peninsula is located midway along the northeast-facing coast of Papua New Guinea while the Vogelkop is a large peninsula connected by an extremely narrow strip of land to far northwestern Irian Jaya (see Figure 7.2d).

Geographic history of region

For probably a major portion of the period from Late Cretaceous or Early *Palaeocene*, about 65 m.y. B.P., to the Early Pleistocene (2 m.y. B.P.) the land making up the southern lowlands of the present island of New Guinea formed the northern edge of the original Australian continent (Figure 7.2 a, b). (For clarity and conciseness, this combined lowland New Guinea - North Australia area will here be termed the 'North Australian land mass'.) As outlined by Doutch (1972), in the Oligocene (35 m.y. B.P.) portions of what is now New Guinea's Central Range began emerging from the sea as an east-west string of islands north of the North Australian land mass, and this mid-Tertiary island formation continued — and spread — until the Early Miocene (20 m.y. B.P.; Figure 7.2b). Then, for the next 15 million years or so, this major east-west string of still-uprising insular mountains and the North Australian land mass approached each other, and finally merged (Figure 7.2c). A second, minor, east-west island chain also formed during this period, as shown in Figure 7.2c, but did not become permanently united with the future Central Range until sometime during the later Pliocene (after 5 m.y. B.P.). (The New Guinea Central Range and all the lands to the north of it, at any stage in their history, will here be collectively termed 'Northern New Guinea'.)

However, because of fluctuations in both sea and land levels, in the 30 or so million years of emergence and uplifting of these major and minor island chains there were periods of tens — at times, undoubtedly hundreds — of thousands of years during which these predominantly highland areas were variously united to each other and to the lower-elevation North Australian land mass. Incursions of the sea onto different areas of the North Australian land mass itself also occurred (as in Figure 7.2c), and divided this extensive region into generally northern and southern portions on a number of occasions. The present Torres Strait area, for example, became covered by water for the most recent time about 8000 years ago, and would again revert to a land corridor about 100 km wide between northern and southern portions of the North Australian land mass if the present level of the ocean dropped as little as 20 m (see Figure 7.2d).

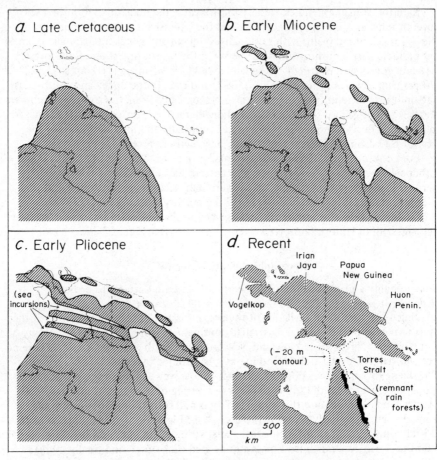

Figure 7.2 Palaeogeography of the North Australian land mass and
Northern New Guinea at four selected geologic periods. Adapted
from Doutch (1972: Figure 1.1) and Keast (1968: Figure 1).

In addition to the foregoing land interrelationships, it should be realised that
the entire group of land masses in this area had been steadily drifting north (the
latitude of the Torres Strait area probably changing from about 47° S to its
present 10° S in the last 100 million years), although the magnitude of the
probable climatic and possible resultant vegetational changes due to this continental drifting are still undetermined (see discussion in Jennings, 1972: p. 94).

The predominant vegetation type of the North Australian land mass from
Miocene to Pleistocene obviously had a profound effect on the geographic movements and course of speciation of New Guinea forest-adapted marsupials. Even
though direct evidence on this matter of habitat is lacking, some postulate based
on indirect evidence must be provided here for discussion purposes. I would suggest
that this lowland area was primarily open woodland (as opposed to dense closed-
canopy forest on the one hand or arid desert scrub on the other) throughout most

− but not quite all − times of this period. If there had been a heavy forest over this entire region during the Miocene and all of Pliocene into Pleistocene times, it seems that not only would numerous forest-adapted genera have been produced in this area but also, more pertinently, these many genera would also have remained common to both the present forests of Northern New Guinea and the restricted rain forests (with maximum elevations of below 1500 m; see Figure 7.2d); found in northeastern Australia today. For instance, if the well-differentiated New Guinea forest-adapted peramelid genus *Peroryctes,* or the distinctive New Guinea arboreal phalangerid *Distoechurus,* had evolved in rain forests of the North Australian land mass, one or both might be expected to appear also in the present north-eastern Australian rain forests. Instead, today in these remnant closed-canopy forests are found mostly endemic *Australian* forest-adapted taxa such as the peramelid *Perameles nasuta* and the arboreal phalangerid *Hemibelideus.* Apparently, essentially continuous rain forest did appear over the North Australian land mass during one or, at most, a few geologically brief periods, because present species or subspecies of the phalangerid *Pseudocheirus* or the tree kangaroo *Dendrolagus* in northeastern Australian rain forests were almost surely derived from Northern New Guinea forest-adapted ancestral forms of the same respective genera which somehow succeeded in travelling south across the lowland North Australian land mass in the latter part of the Tertiary.

In reality, the predominant vegetation type of Northern New Guinea during this earlier period is likewise unknown, but the mountains of this area have been steadily rising (to present elevations of 4500 m or more) since their first appearance above the sea in the Oligocene or Early Miocene. It thus seems reasonable to assume that whatever vegetation type predominated on the North Australian land mass at any given period, the Northern New Guinea uplands must always have been more densely forested than the lowland Australian area.

Present and past zoogeography of region

The earliest Australian Region marsupials presumably arrived on the main Australian land mass in the Middle or Late Cretaceous. Darlington (1957: pp. 335, 338) and Simpson (1961: p. 444) are correct in stating that, thereafter, Australia (rather than New Guinea) was the major evolutionary centre of the marsupials of this south-west Pacific area, at least if this fauna is considered at the family level. These early marsupials had probably differentiated into most or all of the modern families of the region by the beginning of the Miocene (Simpson, *loc. cit.*) − that is, by the time parts of Northern New Guinea had fully emerged from the sea.

Even before the initial physical contact between these scattered islands (which would eventually coalesce into much of present-day New Guinea) and the large North Australian land mass, at least primitive members of possibly two of the four living New Guinea marsupial families could have 'rafted' north to populate − and begin speciation on − one or more of the serveral islands. Certainly at the time when actual merger of the islands and the North Australian land mass began about 15 million years ago, there obviously was no longer any question of a physical barrier such as a seaway to the entrance of the heretofore uniquely Australian mammals into Northern New Guinea.

The fossil record, at least below the family level, gives little help in definitely assigning the geographic origin of species and genera to either Australia or Northern New Guinea because few pertinent pre-Pleistocene mammalian faunas are yet known from Australia, and the only significant Tertiary land fauna reported from all of New Guinea is a limited Pliocene one. In view of this incompleteness of the fossil record, for purposes of this zoogeographical discussion, I generally use the assumptions that (1) if an animal genus now occurs only on the island of New Guinea or only in Australia, the genus evolved on that particular island, and (2) that the greatest number of Recent species of a particular genus occur on the island of origin of the genus. Obviously, neither of these is a completely safe premise but both must serve now until good evidence to the contrary is forthcoming.

Table 7.1 lists all living species of New Guinea marsupials, along with altitudinal and geographic ranges, and selected additional known or speculative information. Because I have had essentially no first-hand experience with endemic Australian marsupials, information on such species has been freely drawn from Marlow (1962) and, especially, Ride (1970). In the very brief species' descriptions preceding each historical zoogeographic account of the family members, external characteristics are used almost exclusively, primarily because they are most easily observed in the live or freshly collected animal. Cranial and, especially, dental characters and measurements usually prove more reliable in species identification as well as more significant in determining relationships within families or genera.

As used here, the term 'endemic' means that a genus or species occurs either only in New Guinea or only in Australia, as the context indicates, while 'shared' means occurring in both places.

Table 7.1 Altitudinal, geographic and other selected information for all living New Guinea marsupials. In the two columns relating to area of origin of genera and species, 'Austr.' indicates any portion of the Australian continent, including the North Australian land mass, while 'New G.' means only Northern New Guinea as defined in the text. Nomenclatural deviations from the check-list of Kirsch and Calaby (this volume) are also explained in text.

Family	Probable area of origin of genus	Number of additional Austr. spp. in genus	Genus occurs also in Australia	New Guinea taxa Genus	New Guinea taxa Species	Species occurs also in Australia	Altitudinal range in New Guinea m	Probable area of origin of species
Dasyuridae	New G.			*Neophascogale*	*lorentzi*		1500–3400	New G.
	New G.			*Phascolosorex*	*dorsalis*		1500–3400	New G.
					doriae		1000–1800	New G.
	Austr.	9	X	*Antechinus*	*wilhelmina*		1700–3400	New G.
					naso		1000–2800	New G.
					melanurus		0–2800	New G.
	Austr.	4	X	*Dasyurus*	*albopunctatus*		0–3500	New G.
	New G.			*Murexia*	*rothschildi*		1000–2000	New G.
					longicaudata		0–1900	New G.
	New G.			*Myoictis*	*melas*		0–1600	New G.
	Austr.	3	X	*Planigale*	*novaeguineae*		0–300	New G.?
	Austr.	10	X	*Sminthopsis*	*virginiae*	X	0–250?	Austr.

Table 7.1 (*continued*)

Family	Probable area of origin of genus	Number of additional Austr. spp. in genus	Genus occurs also in Australia	New Guinea taxa Genus	Species	Species occurs also in Australia	Altitudinal range in New Guinea m	Probable area of origin of species
Peramelidae	New G.			*Microperoryctes*	*murina*		1900–2500	New G.
	New G.			*Peroryctes*	*longicauda*		1000–4500	New G.
					papuensis		1200–1500	New G.
					raffrayanus		60–3900	New G.
					broadbenti		0–2700?	New G.
	New G.	0	X	*Echymipera*	*kalubu*		0–1600	New G.
					clara		0–1200	New G.
					rufescens	X	0–800	Austr.?
	Austr.	2	X	*Isoodon*	*macrourus*	X	0–300	Austr.
Phalangeridae	Austr.	3	X	*Cercartetus*	*caudatus*	X	700–4000	New G.?
	New G.	3	X	*Pseudocheirus*	*cupreus*		1300–4000	New G.
					mayeri		1500–3600	New G.
					forbesi		500–2800	New G.
					corinnae		900–2700	New G.
					albertisii		600–2000	New G.
					schlegeli		< 2000?	New G.
					caroli		30–2000	New G.
					canescens		100–1500	New G.
	New G.	0	X	*Phalanger*	*vestitus*		1500–3800	New G.
					gymnotis		0–2700	New G.
					orientalis	X	0–1600	New G.
					maculatus	X	0–1000	New G.
					atrimaculatus		0–100?	New G.
	Austr.	2	X	*Petaurus*	*breviceps*	X	0–3000	Austr.
					(un-named sp.)		1200±?	New G.
	New G.	0	X	*Dactylopsila*	*palpator*		1200–2800	New G.
					trivirgata	X	0–2300	New G.
	New G.			*Distoechurus*	*pennatus*		0–1900	New G.
Macropodidae	Austr.	2	X	*Thylogale*	*brunii*		0–4200	New G.
					stigmatica	X	0–100?	Austr.
	New G.	1	X	*Dendrolagus*	*matschiei*		1000–3300	New G.
					goodfellowi		1000–3300	New G.
					dorianus	X	1000–3300	New G.
					ursinus		50–2000?	New G.
					inustus		50–2000?	New G.
	New G.			*Dorcopsis*	*vanheurni*		800–3100	New G.
					atrata		900–1800	New G.
					macleayi		1200–1300	New G.
					veterum		0–500	New G.
					hageni		0–500	New G.
	Austr.	12	X	*Macropus*	*agilis*	X	0–200?	Austr.

Family Dasyuridae

Anatomically little-specialised marsupials; active predators, which are roughly
the ecological counterparts of non-fossorial insectivores and smaller terrestrial or
semi-arboreal carnivores of other parts of world. Most of the 12 New Guinea
species are found, by day or night, on floors of various kinds of forest types, but
three or four species often forage high in trees. Distinguished externally by
having all five toes of the hind foot completely separated, as opposed to all other
New Guinea marsupials, which have second and third hind toes united almost or
quite to claw bases. Among New Guinea dasyurids, one species the size of
domestic cat; three species in old age as large as mature laboratory rats; remainder
mostly mouse-sized.

The New Guinea Marsupial cat (*Dasyurus albopunctatus*) has a unique body
pattern of numerous small white spots on blackish- or reddish-brown background,
and short fur. Endemic, primarily terrestrial; it occurs in most forests and open
woodlands, sometimes grassland and is rare below 1000 m but fairly common
otherwise.

Three genera of medium-sized New Guinea dasyurids are all endemic. The
Long-clawed marsupial mouse (*Neophascogale lorentzi*) has rather long fur,
reddish-brown with a slight greenish tint dorsally; many longer hairs of back and
sides, and back of ears and tip of tail are tipped whitish; body colour strongly
resembles that of certain tree shrews (order Primates) and many diurnal tree
squirrels (order Rodentia) of continental areas. Rarer than *Dasyurus*; partly
diurnal and probably largely arboreal, it is found in moss forests of the middle
half of the Central Range. The Three-striped marsupial mouse (*Myoictis melas*)
is occasionally partly or wholly melanistic, but in normal coloration shows a
conspicuous pattern of three dark dorsal stripes along a somewhat variegated
silvery-brown to chestnut body; thus it is similar to diurnal, largely terrestrial,
open-forest continental chipmunks and other sciurids (order Rodentia),
suggesting similar habitats and parallel activity patterns. Relatively uncommon;
it occurs in most rain forests of lowlands and mid-mountains. The Short-haired
marsupial mouse (*Murexia longicaudata*) is unpatterned, light brown overall with
fur distinctively short and close-set; old males in northern New Guinea attain
relatively very large size (head-and-body to 285 mm), southern males and all
females seldom reach half this length. Never abundant but consistently encountered
in all lowland and mid-mountain forests. The Broad-striped marsupial mouse
(*Murexia rothschildi*) has a single broad black back stripe, over 10 mm wide, on
background of greyish-brown. Only eight specimens are known from forests of a
limited area in extreme south-eastern Papua New Guinea. This and all remaining
species of New Guinea dasyurids are the size of a large laboratory mouse, although
occasional very old individuals may grow a little larger.

Two more endemic New Guinea dasyurids, both with striped back pattern, are
moderately abundant and primarily terrestrial. The Narrow-striped marsupial
mouse (*Phascolosorex dorsalis*) and slightly larger but otherwise extremely similar
Red-bellied marsupial mouse (*Phascolosorex doriae*) are both very dark chestnut-
brown above, with a single narrow (less than 10 mm wide) black mid-dorsal stripe
and bright chestnut below. The deep saturate tone of upper body suggest that their
usual habitat is poorly illuminated. *Phascolosorex dorsalis* occurs in wet forests of

the Central Range, including Vogelkop. *P. doriae,* for undetermined reasons, is restricted to the western half of New Guinea; it is generally a low-altitude replacement of *P. dorsalis,* occurring primarily in upper rain forest, but overlaps with highland *P. dorsalis* in lower moss forest.

Five remaining small dasyurids, all with unstriped backs, are mostly without distinctive patterning. The three *Antechinus* species are all endemic. The Lesser marsupial mouse (*Antechinus wilheimina*) occurs throughout most of the Central Range, excluding Vogelkop and south-eastern Papua New Guinea. In most of the eastern half of the range it is unpatterned and light- to medium-brown overall; thus, except for longer fur, it resembles small *Murexia longicaudata;* in western specimens a darker chestnut tint is found on the back, and especially on the rump. This is the most abundant dasyurid in moss forest, found usually on the ground but also on logs and fallen trees. The Long-nosed marsupial mouse (*Antechinus naso,* here including *A. mayeri* of Kirsch and Calaby's list) is also similar to *Murexia longicaudata* but the tail tip is almost invariably white. More widely distributed than *A. wilhelmina,* it is generally found throughout all mountain forests, but most typical of rain forest. The Black-tailed marsupial mouse (*Antechinus melanurus*) is distinguished from the foregoing small species by a solid-blackish (rather than brownish) tail and, in the eastern half of New Guinea, a chestnut-coloured patch of fur at the posterior base of the ear. General body colour is greyish-brown above, with a dull greenish tinge overall, and lower back and hind legs washed rufous; these are colours typical of partly diurnal semi-arboreal mammals. Never abundant, but probably widespread through New Guinea forests, usually on the ground, but occasionally as high as 12 m in trees. The Papuan Marsupial mouse (*Planigale novaeguineae*). is plain dorsally and generally the greyest of New Guinea dasyurids, lacking the warmer colours found in most other species; the hind foot is very short (13 - 16 mm), and the head somewhat flattened for life among rocks and in cracks of dried soil. This species is limited to central and eastern lowlands of New Guinea's south coast, but relatively common there; its preferred habitats are rather open grasslands and savannahs. The Red-cheeked marsupial mouse (*Sminthopsis virginiae*) has a dorsal coloration of greyish-brown, with silvery buff tips to many hairs, giving a finely streaked appearance; the head is chestnut with three black stripes, one from nose to crown, and one from nose to each eye. The hind foot is relatively narrow and long (19 - 25 mm), suggesting semi-hopping locomotion. A relatively abundant species with geographic distribution, altitudinal range and primary habitat similar to those of *Planigale novaeguineae.*

Four of the eight genera, and 11 of the 12 species, of New Guinea Dasyuridae are endemic. The ancestral stocks of the forest-adapted endemic genera *Murexia* and *Phascolosorex* (two species each) and *Myoictis* and *Neophascogale* (one species each) were very early Australian immigrants to Northern New Guinea, possibly even 'rafting' in before merger of the North Australian land mass and portions of Northern New Guinea. None of the four genera can definitely be assigned close relationship to any extant Australian genus. On the evidence of certain dental characters, at least two evolutionary lines are involved: one yielding *Murexia* and the other producing the three remaining genera.

In this and other cases concerning two or more related endemic New Guinea genera or species, although the *ultimate* ancestral stock had to be Australian, it would not necessarily have to represent the immediate ancestor of each endemic genus or species of the group in question. This immediate ancestor could, of course, have been a presently unknown taxon which had already attained a certain degree of differentiation after immigrating into an uplifting area of Northern New Guinea and which, in turn, then provided emigrant stock to colonise and adaptively radiate on other relatively isolated, forested regions or islands of Northern New Guinea.

Of the two *Murexia* species, the wide-spread *M. longicaudata* is presumed more primitive, and the restricted striped *M. rothschildi* more specialised. The latter evolved from a *M. longicaudata* population of south-eastern Papua New Guinea mid-mountains cut off by rising seas in, perhaps, the Early Pleistocene. Simultaneously, the two present species of *Phascolosorex* evolved from a ubiquitous ancestral *Phascolosorex* species in isolated or semi-isolated eastern and western portions, respectively, of the Central Range. Subsequent rejoining of these two areas allowed the eastern *P. dorsalis* to successfully invade the range of the western *P. doriae*, becoming increasingly adapted to higher elevations as it accomplished this sympatry.

The three endemic, forest-adapted, species of *Antechinus* are not strikingly differentiated morphologically from their respective closest relatives among the nine living congeners in Australia. Evolution of the New Guinea species resulted from a relatively recent (i.e. Late Pliocene or later) isolation of Australian members of the genus in Northern New Guinea during periodic east – west sea incursions across the North Australian land mass. Two such instances of isolation may be involved, the first yielding *Antechinus melanurus,* and the second *A. wilhelmina* and *A. naso.* The endemic *Dasyurus albopunctatus* evolved from an Australian open-land emigrant of this typically Australian genus at about this same time and in the same manner. Possibly, the endemic but little differentiated *Planigale novaeguineae* and, certainly, the shared *Sminthopsis virginiae* – both open-land forms of typically Australian genera – represent former North Australian land mass taxa relatively very recently isolated in lowland areas of southern New Guinea by Late Pleistocene Torres Strait region inundations.

Family Peramelidae

These are terrestrial and nocturnal animals with a scampering or semi-hopping locomotion reminiscent of less specialised kinds of continental rabbits (order Lagomorpha), primarily insectivorous and frugivorous. The snout is long and pointed, ears are rather short, foreclaws strong and elongated: these are adaptations for rooting or digging out food and for making burrows in forest litter or under logs. The hind feet are rather elongated, with the innermost digit ('big toe') often rudimentary, never set off at right angle to remaining digits, and not opposable. The second and third hind digits are united to claw bases. The tail is non-prehensile, short in most species and easily damaged or broken off. The pouch opens to the rear, rather than anteriorly or ventrally as in other New Guinea marsupials. Adults of nine mainland New Guinea species range from small rat- to large rabbit-size.

The Brindled bandicoot (*Isoodon macrourus*) has a head-and-body length up to 450 mm, tail to 170 mm. The fur is harsh but not spiny, with overall colour light brown, very finely streaked with black. A shared species of grassland, sometimes found in underbrush of open woodlands of south and south-east Papua New Guinea coast, avoiding forests.

Three species of *Echymipera* are externally all very similar, and best identified on dental characters; head-and-body to 400 mm, tail seldom more than 90 mm; overall colour variegated medium- to darkish-brown, resulting from mixture of yellowish-tipped hairs and blackish, flattened, flexible spines. In many areas two or possibly all three species occur together, the ecological differences allowing such a high degree of sympatry are unknown. The Spiny bandicoot (*Echymipera kalubu*) is endemic; generally the most widespread and abundant species of the genus it occurs throughout most rain forests, also readily venturing into man-made grasslands. The Rufescent bandicoot (*Echymipera rufescens*) is shared, co-existing widely with *E. kalubu* below 800 m. The Long-fanged bandicoot (*Echymipera clara*) is an endemic species, restricted to rain forests north of the Central Range in the north-western quadrant of Papua New Guinea and northern half of Irian Jaya, excluding Vogelkop.

Four species of *Peroryctes* are all endemic. All are relatively long-tailed, distinguished from *Echymipera* by hispid or bristly (rather than spiny) pelage, and by an adult tail of 100-220 mm. Raffray's bandicoot (*Peroryctes raffrayanus*) is unpatterned, vaguely similar to *Echymipera* except that the limbs are longer and slimmer, with body colour more uniform medium-to-dark brown. Primarily found in dense dark forests throughout New Guinea, it usually avoids woodland and grasslands, and is rare below 500 m; on high mountains of Huon Peninsula, in the absence of highland species *Peroryctes longicauda*, it ranges to over 3900 m. The Giant bandicoot (*Peroryctes broadbenti*) is sometimes considered conspecific with *P. raffrayanus* but seems to occur with this closely related species above 60 m in south-eastern Papua New Guinea rain forests. Few have been collected. Males reach relatively enormous size, with head-and-body more than 500 mm, weight over 4.7 kg; the body grades from dark reddish-brown dorsally, through light rufous-brown laterally, to light rufous-buff ventrally. The Striped bandicoot (*Peroryctes longicauda*) occupies the entire Central Range, occasionally occurring as low as 1000 m in rain forest but most abundant from lower moss forest through alpine grasslands. In the eastern third of New Guinea its general colour is light-brown above with a conspicuous pattern of a dark stripe down mid-back, a shorter one to either side of the rump, and another over each eye. The pattern progressively diminishes in intensity, and general colour darkens, from east to west; Vogelkop specimens have no dorsal striping. The Papuan bandicoot (*Peroryctes papuensis*) is limited to extreme south-eastern Papua New Guinea rain forest, where it is apparently sympatric with closely related *P. longicauda*. In colour and dorsal pattern it resembles the eastern form of *P. longicauda* but is only two-thirds the size; both species usually have a white-tipped tail.

The Mouse bandicoot (*Microperoryctes murina*) lives in moss forest: very small (head-and-body 150-175 mm, tail 105-110 mm). It is a poorly known species. Three specimens from west-central Irian Jaya show an unpatterned back, and fur very dark grey, short and soft, suggesting semi-fossorial existence. A fourth specimen, tentatively referred, from eastern Vogelkop differs in having light greyish-buff fur, not especially soft; the back has a dark grey mid-dorsal stripe, with tail tip white.

Two of the four New Guinea peramelid genera, and seven of nine species, are endemic. Morphological adaptations of this family make it unlikely that any member 'rafted' to insular portions of Northern New Guinea. However, soon after the first island contact with the North Australian land mass the ancestral stock of the endemic forest-adapted *Peroryctes* entered the highland forests and began differentiation. This same Australian stock probably also provided the immediate ancestor of the endemic Australian *Perameles*, which, like *Peroryctes*, is morphologically a relatively primitive peramelid group. This hypothesised Australian ancestral stock — or a continuation of it — is possibly represented by fossil material resembling *Peroryctes* and *Echymipera* from a Late Pliocene or Early Pleistocene fauna of south-eastern Australia (see Keast, 1968: p.396).

The higher-elevation *Peroryctes longicauda - papuensis* complex is externally more differentiated from a presumed common ancestor than is the lower-elevation *P. raffrayanus - broadbenti* complex, indicating that the former resulted from the earlier of two evolutionary radiations. An early, temporarily isolated, western highland population of *Peroryctes longicauda* eventually evolved into the enigmatic and distinctive endemic genus of dwarf peramelid *Microperoryctes*.

The shared genus *Echymipera*, also fairly well differentiated from any living endemic Australian genus, presumably evolved only slightly later than *Peroryctes*, but in lower elevations of the Northern New Guinea forests. The endemic *Echymipera clara* evolved from primitive *E. kalubu* stock stranded on one of the north coast ranges during a sea incursion in, perhaps, Late Pliocene, The shared, relatively low-elevation *E. rufescens* evolved in transitory rain forest of the low-lying North Australian land mass at about this same time. With retreat of this rain forest towards the Northern New Guinea highlands, the younger species *E. rufescens* was brought into contact with its ancestor, *E. kalubu*, and proved well enough differentiated to co-exist with it. The open-land *Isoodon macrourus*, ubiquitous to the North Australian land mass, was divided into New Guinea and Australian populations by Late Pleistocene flooding of a former Torres Strait area land connection.

Family Phalangeridae*

Various members, rather analogous to tree sloths (order Edentata), smaller arboreal primates, or 'flying' squirrels (order Rodentia) of other parts of the world. Most of the larger species are almost completely arboreal, nocturnal and herbivorous; medium and smaller species are mostly found nearer ground in forest understory, foraging at night; generally insectivorous but they are also partly herbivorous and frugivorous. The family has unique opposable first hind toes, large, thumb-like and claw-less, extending at right angles to the rest of the toes; second and third hind toes are united only near their bases. Often called 'possums' but only distantly related to 'opossums' of the New World marsupial family Didelphidae.

There are five New Guinea species of *Phalanger*, large (head-and-body length to 550, or even 650, mm), slow-moving, and highly specialised for arboreal existence; the tail is relatively long, with distal half hairless and downwardly prehensile. The fore claws are strongly curved and sharply pointed for climbing. Where not overhunted, most species are fairly common; natural geographic spread determined primarily by forest limits. The Spotted cuscus (*Phalanger maculatus*)

*Corresponding to the three families Phalangeridae, Burramyidae and Petauridae of Kirsch and Calaby's list is sometimes all white or buffy white with rufous extremities, more often with grey, brown or reddish splotches on lighter background; colour varying with geography, often also with sex and age; the pelage is close and woolly, almost concealing the ears. A shared species, occurring throughout most of the New Guinea lowlands. The Black-spotted cuscus (*Phalanger atrimaculatus*) is closely related to *P. maculatus*, if not conspecific. A little-known species, identified from two males with body golden-brown, posterior back closely spotted jet-black, and one referred female, dark rusty red anteriorly, posterior body-half solid jet-black. This form is endemic, and restricted to northern New Guinea lowlands. The Common cuscus (*Phalanger orientalis*) is a shared species, abundant throughout New Guinea rain forests. Adult males vary from white, through greyish-white, to almost clear medium- or dark-grey; adult females are reddish-brown to brownish-grey. This species is unique in that all the young are reddish-brown. Specimens of all ages have a darker mid-dorsal line of variable intensity. In the Grey cuscus (*Phalanger gymnotis*) adults and young of both sexes are rich grey above, with a sharply marked wavy narrow black mid-dorsal line. Endemic, and moderately abundant throughout the forests of most of New Guinea, they often forage on the ground, sometimes nesting in rock outcrops or under tree roots: they may be partly carnivorous. The Silky cuscus (*Phalanger vestitus*) is an endemic species, very common in the deep undisturbed moss forest of the Central Range, Vogelkop and Huon Peninsula, sometimes extending into alpine scrub forest. Both young and adults are chocolate-brown above and creamy white below; the pelage is usually long and lax, with a dense underfur, providing protection in a constantly wet habitat.

There are eight New Guinea species of *Pseudocheirus*, all endemic, with habits and external appearance fairly similar to those of *Phalanger*; they too, have a downwardly prehensile tail (thus 'Ring-tail') but with the distal half naked only along the median ventral surface. This species has the most highly specialised dentition of all New Guinea phalangerids for a diet of leaves. Larger species equal *Phalanger* in size; medium species have a head-and-body length of about 300 – 350 mm, with the smallest as low as 200 – 210 mm. Most *Pseudocheirus* are relatively abundant in undisturbed forests, the three or more species sometimes replacing each other altitudinally on high mountains. The Silvery ring-tail (*Pseudocheirus corinnae*) is a large animal, somewhat glistening golden-brown above, duller brown below, often with an all-white chest patch. It has an indistinct dark mid-dorsal stripe, usually outlined by a distinctive light silvery brown area, often with an additional short dark stripe on either side. Relatively common in rain forest and moss forest of the eastern half of the Central Range, it extends sparingly to central Irian Jaya. The Coppery ring-tail (*Pseudocheirus cupreus*) is also large; dark saturate reddish-brown to coppery brown above, it has a narrow dark mid-dorsal stripe, and usually an all-white patch on upper chest and/or lower abdomen, sometimes extending between. Its geographic range is similar to that of *P. corinnae* but generally at higher altitude; the preferred habitat is dense moss forest, and also alpine scrub forest; this is the only species of *Pseudocheirus* which occasionally ventures out into alpine grasslands. The Northern ring-tail (*Pseudocheirus albertisii*) is quite similar to *P. cupreus* but slightly smaller; and hair on the distal half of the tail is dense and rather bushy. It occurs in mountain forests of Vogelkop and north-coastal Irian Jaya, extending

east into Papua New Guinea's north-west coast ranges, replacing closely related
P. cupreus in far-western and northern New Guinea. Of the Arfak Mountains
ring-tail (*Pseudocheirus schlegeli*) only a single specimen is recorded; it was
collected in the late 1800s at an unrecorded elevation in the Arfak Mountains of
Vogelkop. Small-to-medium, and dorsally brownish-grey, it has no facial or back
stripes; the underparts are buffy. The Weyland Mountains ring-tail (*Pseudocheirus
caroli*) is an uncommon species, occurring in a few lowland and mid-mountain
forests in west-central Irian Jaya. Of medium size, it is plain grey or brownish-
grey above, white to buffy below with the terminal third to half of the tail white.
The Moss-forest ring-tail (*Pseudocheirus forbesi*) is a medium-sized animal, grey-
to medium-brown, occasionally with an obscure dark mid-dorsal stripe, and
tail darker than the body. In the western half of New Guinea this species has
black mid-crown and lower-cheek stripes, a large rufous area around the eye,
and a conspicuous whitish patch below the ear; intensity of the facial pattern
diminishes to the east, only a suggestion of it remaining in far-eastern Papua
New Guinea. The hair on the distal half of the tail is appressed. The Lowland
ring-rail (*Pseudocheirus canescens*) is of small-to-medium size and similar to
closely related *P. forbesi*, but the body colour is often greyer, and hair of the distal
half of the tail is relatively bushy. Its facial pattern is intermediate between those
of far western and far eastern forms of *P. forbesi*. The two species replace each
other altitudinally over similar extensive geographic ranges, with *P. forbesi* is most
abundant above 1500 m, *P. canescens* below. Both are found in forests of the
Central Range, including Vogelkop and Huon Peninsula, only *P. canescens* occurs
in other north coast ranges. The Pygmy ring-tail (*Pseudocheirus mayeri*) is a small
unpatterned animal, medium-brown above, lighter brown below; often it has a
slightly lighter tan or greyish-tan patch just below the ear; occasionally the distal
fifth or less of the tail is white. It occurs in moss forest and alpine scrub forest
of Irian Jaya Central Range (excluding Vogelkop) east to central Papua New
Guinea; very abundant in the western two-thirds of range, it is inexplicably
scarce in the eastern third.

The New Guinea species of *Petaurus* are essentially duplicates of Holarctic
flying squirrels, in appearance and habits, with gliding membranes along the
sides of their limbs and body from wrist to ankle. Their entire pelage is soft and
fine: the tail is long and well-furred throughout, and only slightly prehensile.
Completely nocturnal and arboreal, they eat invertebrates, fruit, flowers and
tree sap, gliding from high in one tree to the base of another, foraging as they
ascend the trunk or climb among the foliage. Sugar gliders (*Petaurus breviceps*)
are small animals (head-and-body to 105 mm in lowlands, 25 mm less in
highlands) with medium- to brownish-grey back varying with locality and a
darker mid-dorsal stripe; ventrally they are white to dark-buff. A shared species,
they occur in any area with trees, including coastal coconut groves, very common
below 2000 m, but often much rarer above. A yet un-named species, similar but
much larger, was recently collected in the thick upper rain forest of the isolated
north coast range of Papua New Guinea. It is apparently not closely related to the
equally large endemic Australian *P. australis*; it is sympatric with *P. breviceps* but
its ecology is yet to be determined!

Two species of New Guinea *Dactylopsila* have a unique longitudinally striped
black-and-white body pattern, and rank (but non-ejectile) anal-gland secretion –

characters which indicate convergence with continental skunks and polecats (order Carnivora). The tail is long, often white-tipped, slightly prehensile, and densely haired throughout except for a small naked ventral area at the extreme tip. Arboreal, nocturnal and largely insectivorous, they have anterior-most incisors elongated and strengthened for tearing into dead wood, and elongated fourth fore-digits which they use for extracting invertebrates from holes. Reported to tap on bark with fore-digits to locate or drive out insects, in these and other characteristics they are strikingly convergent with the primitive primate Aye-aye (*Daubentonia madagascariensis*). The Striped possum (*Dactylopsila trivirgata*, here including *D. tatei* and *D. megalura* of Kirsch and Calaby's list) is larger and longer-tailed (head-and-body to 330 mm, tail always longer); the tail is occasionally very bushy (hair length to 80 mm); the fourth fore-digit moderately lengthened, with a normally sized claw (cf. below). The Long-fingered possum (*Dactylopsila palpator*) is similar but smaller (head-and-body to 260 mm, tail usually shorter); the fourth fore-digit is greatly elongated and very thin, with claw only half the size of others. Both species relatively abundant, to some degree replacing each other altitudinally. *D. trivirgata*, a shared form, inhabits coastal woodlands up to the middle moss forest zone throughout New Guinea; endemic *D. palpator* is commonest in moss forests of the Central Range (excluding Vogelkop) and Huon Peninsula.

The two remaining New Guinea phalangerids are very small (head-and-body 100–130 mm), with tail noticeably longer than the head-and-body and weakly prehensile near the tip; both are nocturnal and arboreal, living primarily in smaller trees or forest undergrowth. Their diet is largely insects and other invertebrates, possibly also fruit. The Feather-tailed possum (*Distoechurus pennatus*) has a distinctive head pattern of white face, with a broad black band from the side of the snout through each eye to the crown, and a smaller blackish mark behind the ear base. The rest of the body is plain light-brown; hairs in a single line on either side of the tail are stiffened and elongated, but their function is unknown. This is an endemic species of rain forest and lower moss forest throughout New Guinea, abundant in some areas, scarce in others. The Long-tailed pygmy possum (*Cercartetus caudatus*) is somewhat similar to *Distoechurus* but lacks the white face; it is slightly smaller-bodied, and has a short-haired tail without elongated lateral hairs. Lowland forms have a light-brown body, highland forms are dark brown; there is a small black patch from the side of the snout back through each eye. A shared species, fairly abundant in all New Guinea moss forest and alpine scrub forest, it appears also in one surprisingly low 700 m locality in the rain forest of south-western Papua New Guinea.

Among New Guinea Phalangeridae, only one of the six genera is endemic, while 14 out of 19 species are so restricted. At least some early North Australian land mass phalangerids were presumably well morphologically adapted for 'rafting' to — and among — insular portions of Northern New Guinea. In the case of some Recent genera it seems certain that this form of dispersal did take place. For example, representatives of both *maculatus*-like *Phalanger* (i.e. *P. ursinus*) and *orientalis*-like animals (i.e. *P. celebensis*) occur on the islands of Celebes, 900 km west of New Guinea, as well as on other islands which likewise were never connected by land to either Australia or New Guinea during the time period involved here. These particular insular cuscus populations are too well differentiated to have been introduced by relatively recent prehistoric man, although some modern species of *Phalanger* and other marsupials undoubtedly have been transported to islands by native peoples within the last few thousand years.

The genus *Phalanger* originated at a very early date in the lower-elevation forests of uplifting Northern New Guinea. The ancestral species could well have been quite similar in morphology and ecology to the living, relatively generalised, shared species *Phalanger orientalis*. However, an early representative of this initial *Phalanger* species 'rafted' either to insular regions of other parts of New Guinea's future Central Range or to islands of the incipient north coast ranges, and produced the line leading to the large-toothed species complex exemplified by *P.maculatus*. After final land connection of the north coast and central mountains with each other and with the North Australian land mass (in Late Pliocene?), *P. maculatus* spread from its northern area of origin throughout the forested lowlands of all Northern New Guinea, and also onto the North Australian land mass during a period of rain forest development in this lowland area, part of the population being isolated in extreme northern present Australia by Pleistocene formation of the Torres Strait. *Phalanger atrimaculatus*, if indeed a valid species, evolved only very recently — probably within the Pleistocene — from a *maculatus* population isolated in an area north of the Central Range, to which the range of this nominate species is still restricted.

A few million years after providing the ancestor of the *P. maculatus* complex, another portion of the still-evolving *P. orientalis* stock living in the rising Early or mid-Pliocene mountains of Northern New Guinea differentiated into the endemic and highly forest-adapted *P. vestitus*, while somewhere in this general northern New Guinea region a later isolated group of *orientalis* produced the endemic and possibly partly carnivorous *P. gymnotis*. The present-day representatives of *P. orientalis* in northern Australia reached this remnant forest area in the same manner as described above for *P. maculatus* (present populations on various Solomon Islands east of New Guinea were possibly transported there by early human voyagers).

The forest-adapted shared genus *Pseudocheirus* appears to have originated in Northern New Guinea at least as early as *Phalanger* but, beyond this point, little can be said with certainty about its subsequent course of distribution and speciation. The geographical distribution of living species and the remarkable present degree of sympatry of the different species in the forests of, especially, New Guinea suggest an amazingly complex history of invasion and re-invasion of a number of areas by variously speciating populations during formation of the eight endemic species of New Guinea and the three of Australia.

A conjectural account of these events is as follows. The entire genus may be considered to consist of two groups: a 'primitive' one of seven smaller species with cranial and dental characters presumed little evolved from those of the ancestral *Pseudocheirus*; and an 'advanced' group of four larger species with more specialised cranial and dental characters. The initial member referable to the genus evolved in the lower-elevation forests then covering most of Northern New Guinea and, morphologically and ecologically, resembled one of the present 'primitive' species: *Pseudocheirus canescens*. Subsequent isolation of one population of this original species in rapidly uplifting areas ultimately yielded the small and very distinctive high-elevation *P. mayeri*, which in turn provided stock to the then-insular New Guinea western mountains for differentiation into the species *P. caroli* and *P. schlegeli.* One of the later reconnections of Northern New Guinea with the North Australian land mass, concurrent with a time of rain forest development over this lowland area, allowed the ancestral *canescens*-like animal to reach the north-east coast upland areas of present Australia and eventually differentiate into the two living 'primitive' endemic Australian species of the genus. At a relatively much more recent time, other geographic isolates of fully differentiated *P. canescens* in Northern New Guinea speciated into a high-elevation altitudinal replacement: the morphologically very similar *P. forbesi.*

Concurrent with this spread and speciation of the 'primitive' members of *Pseudocheirus,* two 'advanced' lines were developing in Northern New Guinea, probably also from the same original early *canescens*-like stock. One of these lines resulted in the development of today's large high-elevation *P. cupreus,* which, much later, itself furnished stock to New Guinea's north coast ranges for the formation of its very close relative *P. albertisii.* The other resulted in the mid-to-high-elevation *P. corinnae.* And then, unless the similarity in coloration of the endemic Australian rain forest species *P. archeri* to New Guinea's *P. corinnae* is due to convergent evolution rather than to close genetic relationship, representatives of *P. corinnae* must have been able to travel south through a former continuous — and extremely well developed — rain forest on the North Australian land mass to yield this single 'advanced' species now endemic to Australia's north-east coast remnant rain forest.

In New Guinea the genera *Dactylopsila* and *Petaurus* each contain one shared and one endemic species. Both genera could have initially differentiated in the Late Miocene from a mutual (unknown) common ancestor which once occurred widely over the North Australian land mass: the non-gliding *Dactylopsila* in the thick forests of a recently attached region of Northern New Guinea, and the gliding *Petaurus* (plus the closely related non-gliding endemic Australian *Gymnobelideus*) in woodlands of northern Australia. Subsequently in *Dactylopsila,* an isolate of the more-generalised, low-elevation, shared *D. trivirgata* yielded the more-specialised, high-elevation, endemic New Guinea *D. palpator.* A geologically recent tree cover with a more-or-less open canopy — possibly a degenerating rain forest — on the North Australian land mass allowed *D. trivirgata* to reach present-day north-eastern coastal Australia, where isolated populations remain in rain forest and certain open woodlands. The Australian-derived *Petaurus* (of the small *breviceps* type rather than of the larger *australis* form) utilised this same kind of former North Australian land mass forest to reach at least the Central Range of Northern New Guinea, over which it promptly spread widely. It also eventually occupied most of the future north coast ranges of New Guinea, but one stock of this form of *Petaurus* had reached a still-insular portion of this north coast early enough to fully develop into the recently discovered, very large, endemic species by the time this newer species' range became connected with that of surviving *P. breviceps* of the remainder of Northern New Guinea.

Lack of morphological similarity to any extinct or living Australian genus indicates that the ancestor of the endemic forest-adapted *Distoechurus pennatus* arrived and differentiated very early in Northern New Guinea. This ancestral stock is unknown, but certain dental and other characteristics of the monotypic *Distoechurus* suggest closer ultimate relationship to one or more of the three

The biology of marsupials

living Australian-derived pygmy possum genera (*Cercartetus*, *Acrobates* and *Burramys*) than to any group of larger phalangerids. The shared *Cercartetus caudatus*, a member of a presumed primitive section of the typically Australian genus, could have reached New Guinea's mountainous areas (and successfully colonised habitats throughout an extremely extensive altitudinal range) from the south during a Late Pliocene or Early Pleistocene period of continuous rain forest cover on the North Australian land mass. But, because its Australian range as presently known is restricted to a remnant rain forest area on the north-east coast, it seems more plausible that this particular species developed in Northern New Guinea — although from an Australian emigrant of the same genus — and only then spread back through transitory rain forest on the North Australian land mass to reach its present Australian range.

Family Macropodidae

This includes New Guinea's medium-to-large mammalian herbivores, which are to varying degrees ecological counterparts of continental hoofed mammals and other large terrestrial plant-eaters. The group includes one good-sized open-land species and seven small-to-medium species (head-and-body 400 – 700 mm) with habitats ranging from grasslands to dark wet forests; a further five medium-sized arboreal species inhabit the deep forest. None is uncommon except where overhunted; all are primarily nocturnal but also forage near dawn and dusk. All New Guinea macropodids have non-prehensile tails; the first hind-digit is missing, and the small second and third digits are united along most of their length; in terrestrial species the fourth is greatly enlarged and lengthened.

The Agile wallaby (*Macropus agilis*) is a large species (head-and-body to 1500 mm), with body shape and dental characters like those of the large open-land grazing kangaroos of Australia; the body is light sandy brown. This is a shared species of savannah woodlands and adjacent open grasslands, found in lowlands of southern New Guinea and also on the north coast of the south-east tip of the island and possibly in other northern localities; it was perhaps transported to these areas north of Central Range by man (Schodde and Calaby, 1972: p. 280; van Deusen, 1972: p. 692).

The two New Guinea species of *Thylogale* are medium-to-small animals, somewhat less specialised than *M. agilis* for open-land existence. The Red-legged wallaby (*Thylogale stigmatica*) has a light brownish-grey body with limbs distally rufous. This is a shared species, usually found in underbrush of gallery rain forests, but also in savannah woodland and occasionally in grasslands of the south-central New Guinea lowlands. The Dusky wallaby (*Thylogale brunii*) is dark brownish-grey above, whitish below; lowland forms often have dark and light marks on the face and a whitish hip-stripe. An endemic species, occurring in the eastern half of New Guinea, it inhabits man-made and alpine grasslands and areas of secondary growth, seldom straying far into thick forest.

Four species of the endemic genus *Dorcopsis* (here including *Dorcopsulus*) occur on mainland New Guinea, with a fifth on Goodenough Island off the south-east coast: superficially resembling *Thylogale*, they are even less advanced in open land specialisation; the ears are short, the muzzle is not especially narrowed or elongated, and the tail is neither greatly thickened basally nor significantly lengthened. Primarily animals of dense forests, they occasionally inhabit second-growth brush and man-made grasslands. The Common forest wallaby (*Dorcopsis veterum*) is a medium-sized animal with plain greyish medium-brown body, found in lowland rain forest along the entire southern half of New Guinea, also on Vogelkop. The Greater forest wallaby (*Dorcopsis hageni*) is similar to *D. veterum* but with a narrow white mid-dorsal back stripe making it unique among New Guinea macropodids; it occurs in lowland rain forest but only along the north-central New Guinea coast. The Black forest wallaby (*Dorcopsis atrata*) is slightly smaller and blackish overall, inhabiting thick forests of Goodenough Island between, at least, 900 and 1800 m. The Lesser forest wallaby (*Dorcopsis vanheurni*) is smaller than the preceding species with long, soft grey- to reddish-brown fur; it is a higher-altitude replacement of *D. veterum* and *D. hageni*, living in upper rain forest and moss forest throughout the Central Range (excluding Vogelkop), Huon Peninsula and other north coast ranges. The Papuan Forest wallaby (*Dorcopsis macleayi*) is similar to *D. vanheurni* but with shorter fur, and restricted to rain forest near Port Moresby on the south coast of south-eastern Papua New Guinea.

There are five currently recognised New Guinea species of *Dendrolagus,* all highly specialised for arboreal existence, with fore- and hind-limbs approximately of equal size; the hind foot is short and broad with curved (instead of the usual straight) claws; the tail is long, non-tapering and fully haired, and the head is broad and unlengthened, with very short ears. Feeding on foliage and fruit in trees and on the ground, in or near thick forests, these are proficient climbers; normally resting and sleeping in trees, they are capable of long leaps between trees or from tree to ground. One New Guinea species is probably represented also in Australia by a subspecies; the remaining four are endemic. They display three general body-colour patterns: (1) similarly coloured above and below; (2) reddish-brown above, yellowish below; and (3) dark-grey to blackish above, obviously lighter coloured below. The three groups are very difficult to distinguish dentally or cranially. The Unicolored tree kangaroo (*Dendrolagus dorianus*) is generally some shade of brown or grey above (including tail), and essentially the same colour below; sometimes there is an ill-defined darker mid-dorsal stripe and a lighter fur patch at the upper base of the tail. Most widely distributed of all New Guinea species of *Dendrolagus*, it inhabits rain forest and moss forest throughout the Central Range (excluding Vogelkop); *D. 'bennettianus'* of the Australian rain forest is presumably a race of this species. Matschie's tree kangaroo (*Dendrolagus matschiei*) and the Ornate tree kangaroo (*Dendrolagus goodfellowi*) have not yet been found to live sympatrically, and could possibly be conspecific; both have the body deep reddish-brown above, usually with an obvious darker mid-dorsal stripe, and yellowish underparts. In *D. matschiei* the tail is about the same yellowish shade as the body underparts, and there are no markings anywhere along its length. This species occurs in the upper rain forest and moss forest of Huon Peninsula, and in a limited area (elevation unrecorded) on south slopes of the Central Range in central Papua New Guinea. *D. goodfellowi* has the tail generally of the same darkish colour as the back, the terminal half may be banded or splotched with light colour of body underparts, and there is a light-yellowish stripe to either side of the dark mid-dorsal line on the lower back and rump. This species occurs in the upper rain forest and moss forest throughout the eastern two-thirds of the Central Range; an isolated population has also been recorded at an undetermined elevation in south-central Papua New Guinea foothills. The Vogel-kop tree kangaroo (*Dendrolagus ursinus*) and Grizzled tree kangaroo (*Dendrolagus inustus*) are closely related but possibly not conspecific, because they are apparent-ly sympatric in the Vogelkop area; both are generally dark-coloured, on the back, with underparts lighter. *D. ursinus* shows considerable colour variation, not yet fully documented, including a variable amount white 'frosting' on the tips of the fur and usually a dark-brown to blackish back and tail; white or whitish-tan underparts are usually sharply delimited from the dark back, and sometimes there is a distinct white cheek patch as well as white tail tip. The hair around the ears is long. This species occurs in rain forest near sea level, rising to lower moss forest, but restricted to Vogelkop and its vicinity. In *D. inustus* the back is often as in *D. ursinus,* sometimes noticeably lighter; underparts are usually light-brown, merging gradually into the darker back. Hair around the ears is short. Altitudinal distribution is similar to that of *D. ursinus*; this species lives on Vogelkop and its vicinity, extending east along the north coast of Irian Jaya and north-west Papua New Guinea.

Only one genus of living macropodid is restricted to New Guinea, but 10 of the 13 species there are endemic. Almost certainly no North Australian member of the family 'rafted' to Northern New Guinea. The shared *Macropus agilis* was ubiquitous to non-forested portions of the Northern Australian land mass throughout the Pleistocene, and the present population of the south coast of New Guinea was isolated for the most recent time by the last interglacial for-mation of Torres Strait. The typically Australian genus *Thylogale* provided immigrant stock to the Northern New Guinea highlands, probably in the Late Pliocene; this developed into the New Guinea endemic *T. brunii.* The shared *T. stigmatica* evolved in forested areas of northern and eastern Australia and, in the very late Pleistocene, entered New Guinea across a temporary land corridor in the Torres Strait area.

The endemic forest wallaby genus *Dorcopsis* evolved within Northern New Guinea but from an Australian ancestor. This presumed ancestor, *Dorcopsoides,* occurred in central and northern Australia at least as long ago as Early Miocene (see Keast, 1968: p. 395), and entered the rising Northern New Guinea forests upon its merger with the southern land mass. As increasingly higher elevations were attained, it differentiated into the endemic *Dorcopsis vanheurni.* Subse-quent isolation of a population of *vanheurni* in a mountain range on the south-east New Guinea coast yielded the little-differentiated endemic *D. macleayi. Dorcopsoides* subsequently became extinct in Australia but it survived in the isolated lower-elevation forests of Northern New Guinea, there evolving into *Dorcopsis veterum* on the lower southern slopes of the Central Range. Indi-viduals of either late-stage *Dorcopsoides* or early *Dorcopsis veterum* crossed a low point of the incipient Central Range and reached a semi-isolated area on the future north coast mountains, where *Dorcopsis hageni* was produced. Indi-viduals of the later *D. veterum* line reached the present islands off south-eastern Papua New Guinea by land during a Pleistocene glacial low sea-stand, and there speciated into *D. atrata,* which survives at least on Goodenough Island today.

The tree kangaroo genus *Dendrolagus* originated within Northern New

Guinea when a North Australian land mass macropodid — quite possibly a rock-inhabiting form similar to Recent *Petrogale* — occupied the uplifting area soon after former insular Northern New Guinea merged with this southern land mass. Three species groups eventually developed in New Guinea, but it is difficult to determine which is most primitive. Here the widespread *Dendrolagus dorianus* is tentatively assumed to be the species least modified from the ancestral stock (although the skull of the living form is remarkably broad and massive compared with that of rock-wallabies). Representatives of this early species then reached the Northern New Guinea north coast ranges and differentiated into the *Dendrolagus ursinus – inustus* complex, which still remains essentially restricted to areas north of the Central Range. Another portion of the original *dorianus*-like stock living in the rising Central Range yielded the brightly coloured *goodfellowi – matschiei* complex. During the mid-Pliocene or Late Pliocene, and again in the Pleistocene, *D. dorianus* was able to expand its range south through transitory rain forest cover on the North Australian land mass to the area of present-day north-eastern Australia. Survivors of the earlier invasion of primitive *D. dorianus* differentiated into the present Australian endemic species *D. lumholtzi*; and survivors of the much later second invasion are today represented by the little-changed taxon *D. 'bennettianus'*.

Concluding observations

The foregoing accounts have concentrated on the description and evolution of forest adaptations in New Guinea marsupials. Some perspective regarding the strikingly high degree of specialisation in this one evolutionary course attained by the New Guinea fauna may be gained by comparing the particular type of habitat adaptation of endemic Australian marsupials with that of endemic New Guinea members of the same families.

In the Dasyuridae four out of five Australian endemic genera are wholly or predominantly adapted for life on open land (*Dasyuroides, Dasycercus, Antechinomys, Myrmecobius*), while none of New Guinea's four endemic genera is so adapted. Likewise, in the Peramelidae the three Australian endemic genera (*Macrotis, Chaeropus* and, for the most part, *Perameles*) represent open-land animals, but neither of the two native New Guinea genera could be termed anything but almost entirely forest-adapted.

Although all of Australia's eight or so endemic phalangerid genera are dependent for their continued existence on trees *per se*, at least half of these genera (*Wyulda, Trichosurus, Acrobates, Petropseudes*) occur primarily in open woodlands rather than in closed-canopy forests. New Guinea's sole endemic phalangerid genus seems unquestionably forest-adapted. Incidentally, it is interesting to speculate that the independent evolution of a gliding membrane in three separate lines of Australian-origin phalangerids (*Acrobates, Petaurus, Schoinobates*) may not have been an adaptation to *increasing* forestation but, conversely, a mechanism selected to permit safe and rapid locomotion between trees becoming increasingly further separated from one another in areas which were changing from forest to open woodland. Compare, for instance, the apparent primary habitat preferences of the two very closely related Australian phalangerids: the non-gliding *Gymnobelideus* and the gliding *Petaurus* (species *breviceps* or *norfolcensis*). The former genus is restricted to dense forests, while the latter is widely spread – and eminently successful – in areas of almost any degree of tree density.

Of the four families considered here, the Macropodidae show the greatest degree of difference between respective major habitat adaptations of Australian and New Guinea genera. While New Guinea's sole endemic genus is almost completely forest-adapted, 11 out of 12 Australian endemic genera are primarily adapted to some habitat type other than pure closed-canopy forest. *Hypsiprymnodon* of the north-east Australia coastal rain forest is the single obvious exception.

Although a detailed account is beyond the scope of this paper, the same phenomenon of evolution of forest-adapted infrafamilial taxa in New Guinea, as opposed to Australia, is encountered in the only two placental orders which have undergone any significant degree of adaptive radiation in the Australian Region. These are the orders Chiroptera (considering only the fruit- and blossom-eating suborder Megachiroptera: 9 species – shared or otherwise – occurring in Australia; compared with 22 in New Guinea) and Rodentia (considering only known or presumed forest-dwelling *arboreal* members of the Family Muridae: 4 or 5 endemic species in Australia; over 15 endemic species in New Guinea).

References

Darlington, P. J., Jr. (1957). *Zoogeography,* Wiley, New York.

Doutch, H. F. (1972). The paleogeography of northern Australia and New Guinea and its relevance to the Torres Strait area, in *Bridge and Barrier: the Natural and Cultural History of Torres Strait* (ed. Walker, D.), Australian National University, Canberra, pp. 1 - 10.

Jennings, J. N. (1972). Discussion on the physical environment around Torres Strait and its history, in *Bridge and Barrier: the Natural and Cultural History of Torres Strait* (ed. Walker D.), Australian National University, Canberra, pp. 93 - 108.

Keast, A. (1968). Australian mammals: zoogeography and evolution. *Quart. Rev. Biol.,* **43,** 373 - 408.

Marlow, B. J. (1962). *Marsupials of Australia,* Jacaranda, Brisbane.

Ride, W. D. L. (1970). *A Guide to the Native Mammals of Australia,* Oxford University Press, Melbourne.

Ryan, P. (ed.) (1972). *Encyclopaedia of Papua and New Guinea* (2 vols.), Melbourne University Press, Melbourne.

Schodde, R. and Calaby, J. H. (1972). The biogeography of the Australo-Papuan bird and mammal faunas in relation to Torres Strait, in *Bridge and Barrier: the Natural and Cultural History of Torres Strait* (ed. Walker, D.), Australian National University, Canberra, pp. 257 - 300.

Simpson, G. G. (1961). Historical zoogeography of Australian mammals. *Evolution,* 15, 431 - 46.

van Deusen, H. M. (1972). Mammals, in *Encyclopaedia of Papua and New Guinea* (2 vols.) (ed. Ryan, P.), Melbourne University Press, Melbourne, pp. 688 - 94.

Walker, D. (Ed.) (1972). *Bridge and Barrier: the Natural and Cultural History of Torres Strait,* Australian National University, Canberra.

Section 3   Population, species and behaviour studies of marsupials

The seven chapters in this section deal with various aspects of the life of marsupials, based mainly on field or laboratory studies by Australian biologists. Arundel, Barker and Beveridge (Chapter 8) review some aspects of the diseases of marsupials both in the wild and in captivity, including some reference to the role of endemic disease in mortality among wild populations. Lee, Bradley and Braithwaite (Chapter 13) focus attention on the curious mortality pattern found in the dasyurid genus *Antechinus*, in some species of which practically all males die shortly after a single mating season. Their account, based mainly on studies of the Brown antechinus, *A. stuartii*, is extended to consider semelparity in small mammals generally. Dr. Alan Newsome's contribution (Chapter 14) arises from long-term population studies of that archetypal marsupial, the Red kangaroo *Macropus rufus*. It shows the effects of episodic rainfall and heavy hunting pressure on a population of large herbivores, stressing the superb adaptation of these animals to a difficult environment and their struggle to maintain numbers against odds lengthened by man. Buchmann and Guiler's study (Chapter 9) of the Tasmanian devil *Sarcophilus harrisii* is similarly part of a long-term study. The subject in this case is a carnivore, well adapted for its natural role of forest scavenger but demonstrably – almost pathetically – inept in the role of sheep-stealing predator which man wishes upon it, and for which he has persecuted it remorselessly for a century and more.

It was for long the tradition to regard marsupials as 'second-class' mammals, successful on their own ground but destined to disappear when faced with competition from 'higher' mammals. In Chapter 10 Dr. Desmond Gilmore underlines the success which many species of marsupials have achieved as introduced species, in New Zealand, Hawaii, Germany and Britain, often in apparent competition with native species and usually under severe hunting pressure from man. He draws attention to the continuing spread of some species – notably the arboreal Brush-tailed possum *Trichosurus vulpecula* and Virginia opossum *Didelphis virginiana* –in areas which they have recently colonised. Mrs. Eleanor Stodart's account of bandicoots (Chapter 11) reviews recent field and enclosure studies of behaviour, ecology and breeding of a group of relatively common Australian species, and may serve as a reminder that life history studies of very few marsupials are yet available in the general literature. Dr. Kirkby's contribution (Chapter 12) on learning and problem-solving behaviour in marsupials identifies another neglected field of study. Despite the intrinsic interest of marsupials as mammals-with-a-difference, he has found very few papers which allow critical ethological comparisons to be made either between different species of marsupials or between marsupials and eutherians. We hope that his review may stimulate further behaviour studies of a wide range of marsupials.

8   Diseases of marsupials

J. H. Arundel,* I. K. Barker† and I. Beveridge‡

This review is not intended to cover the subject completely, but rather to reflect the interests of the authors over the last few years. Readers wishing to consult a more detailed review are referred to Barker, Calaby and Sharman (1963).

Macropodidae

Coccidia

Coccidia have been described only from the Red kangaroo, *Macropus rufus*; from population of Grey kangaroos, *Macropus giganteus* and *Macropus fuliginosus* (Mykytowycz, 1964); and from Bennett's wallabies, *Macropus rufogriseus* (Triffett, 1926; Wenyon and Scott, 1925). Outbreaks of coccidiosis typically occur among young *M. giganteus* held under crowded conditions, or artificially concentrated in the wild, and usually in association with flooding (Winter, 1959; Barker *et al.*, 1963; Barker, Harrigan and Dempster, 1972). Death results from acute haemorrhagic enteritis, associated with the presence of schizonts in the lamina propria of the small intestine, and the course may be so short that blood does not appear in the faeces. Prophylactic measures are available to protect animals in captivity (Finnie, 1974). In the wild, coccidiosis may be of some local significance in populations subject to flooding, crowding and food shortage, although in the Barmah Forest outbreak (Barker *et al.*, 1972) starvation was probably the most important component of mortalities during the Murray flood, especially in mature animals (our unpublished observations).

In some cases of coccidiosis in *M. giganteus* the only sexual stages present were those of *Eimeria kogoni* (our unpublished observations), and it is possible that

*J. H. Arundel is Chairman of the Department of Paraclinical Sciences, School of Veterinary Science, University of Melbourne. He is a parasitologist whose main interests have been control of parasitic disease in domestic animals. Some wildlife work has been done in his Department for five or six years and this was formalised by the establishment of a Wildlife Pathology Laboratory three years ago.

†I. K. Barker is a veterinary graduate from Guelph, Canada, who completed his Ph. D. in the Melbourne Veterinary School on the pathogenesis of *Trichostrongylus colubriformis* in sheep. He has been employed in the Wildlife Pathology Laboratory since its commencement and has been particularly interested in protozoan diseases.

‡I. Beveridge is a graduate of the Melbourne Veterinary School. His work on the taxonomy of the genus *Taenia* gained him his Ph. D. He has been employed in the Wildlife Laboratory since its commencement and has completed a revision of the cestodes of marsupials and contributed greatly to the taxonomy of parasites of native animals.

this may be the species whose schizonts are pathogenic.

Large schizonts of an unidentified coccidium have been associated with haemorrhages found in the pyloric antrum of *M. fuliginosus*, but not in other species (unpublished data).

Filarioids

The filarioid nematode *Dirofilaria roemeri* is a common parasite of several species of large macropod (*Macropus rufogriseus;* the Euro, *Macropus robustus; M. rufus; M. giganteus*), usually being found subcutaneously or in fascial tissues, frequently in the vicinity of the knee joints (Spratt and Varughese, 1975). Although generally unencapsulated and provoking little pathological change in *M. robustus*, nematodes in the tissues of *M. giganteus* are often encapsulated and provoke an inflammatory lesion composed of eosinophiles, plasma cells, macrophages, fibroblasts and giant cells (Spratt, 1972). Intermediate hosts in Southern Queensland include 12 species of tabanid flies of the genera *Dasybasis, Tabanus* and *Mesomyia* (Spratt, 1974). Another filarioid species, yet un-named, is found in the thoracic cavity of Tammars, *M. eugenii*, on Kangaroo Island, S.A., and is capable of causing a severe pleuritis and pericarditis with fibrous adhesions (our unpublished observations). *Dipetalonema mundayi* also causes a significant pericarditis in *M. rufogriseus* in Tasmania (Munday, personal communication, 1974).

Trichonematids

Species of the large strongylate nematodes *Labiostrongylus* and *Parazoniolaimus* are ubiquitous parasites of the stomachs of macropods (Mackerras, 1958). During development of the larvae of *L. eugenii* in the stomach of *M. eugenii*, small pedunculated papillae, usually only one per stomach, and composed of granulation tissue, are produced at the junction of the squamous and mucous gastric epithelia (L. Smales, personal communication, 1974). Similar nodules occur in Swamp wallabies, *Wallabia bicolor*, due to *P. collaris*, and in *M. giganteus* with mixed infections of *L. kungi* and *L. bipapillosus* (our unpublished observations), while comparable lesions in the stomach of Pademelons, *Thylogale billardierii*, containing unidentified *Labiostrongylus* sp. larvae have been described associated with a squamous cell carcinoma (Munday, 1971).

Large numbers of *Rugopharynx, Pharyngostrongylus* and *Cloacina* species were found in the stomach of juvenile *M. giganteus* in a confined population near Melbourne, building up to heavy burdens by the end of the first winter in which the young reached independence (see below).

Trichostrongylids

Deaths due to *Globocephaloides trifidospicularis* infection of the Eastern Grey kangaroo, *Macropus giganteus,* were investigated in a water catchment area close to Melbourne in 1971 - 2 (Arundel, Dempster and Black, unpublished data). One of use (J.K.D.), investigating the population dynamics of this confined group for a number of years, observed that mortality with anaemia occurred each year in July and August.

Autopsy of eight dead or moribund animals in July 1971 confirmed that anaemia was the cause of death, a haemoglobin value of 3.5 gm/100 ml and a PCV of 10 being the lowest recorded (see Table 8.1). All animals were heavily infected with gastro-intestinal parasites, but the blood-sucking nematode *G. trifidospicularis* appeared to be the only serious pathogen. Other lesions were widespread small areas of hyperaemia associated with a *Strongyloides* sp. and a severely eroded area in the mucosa of the saccular portion of the stomach adjacent to the margo plicatus. Histological examination showed that this was due to heavy larval penetration.

Table 8.1 Number of *G. trifidospicularis* and haematological parameters in a group of 8 juvenile *M. giganteus*

Kangaroo No.	*G. trifidospicularis*	Total plasma protein	PCV	Hb	Condition
W6929	1300	–	–	–	dead
W6930	800	3.8	10.0	3.5	moribund
W6931	1200	–	–	–	dead
W6932	1000	3.5	21.0	7.0	very weak
W6933	1500	2.8	18.5	6.0	very weak
W6971	500	–	–	–	dead
W6972	1200	–	–	–	dead
W6973	400	4.2	21.0	7.3	very weak

The following year some animals were taken every 6 weeks. Total differential worm counts were made and haematological parameters were measured. On each occasion an effort was made to obtain some young and some mature animals so that an understanding of the epidemiology of these parasites could be obtained.

Juvenile kangaroos left the pouch in December – January but significant numbers of parasites were not picked up until May. Numbers then built up quickly until a peak in August, after which they declined. This epidemiological pattern is similar to that seen with sheep nematodes in this state (Arundel, 1971; Anderson, 1972). The haemoglobin concentration, PCV values and total plasma protein fell as the numbers of *G. trifidospicularis* increased, the lowest value recorded being Hb 2.9 g per cent, PCV 7.0 and total plasma protein of 2.3 per cent (Table 8.2).

Significant numbers of nematode eggs were passed by the juveniles from June to October, and as few adult worms and very low numbers of eggs were seen in the adult Grey kangaroos killed at the same time (Table 8.3), it appears that the eggs passed by the juvenile kangaroos in the late winter and spring survive over the hot dry summer and infect the juveniles of the following year. This pattern, in which the young of one year infect the young of the next, is similar to that seen with *Nematodirus battus* infections in lambs (Gibson, 1963).

Seasonal anaemia has been noted in the Euro (Ealey and Main, 1967) and it is possible that at least part of this may be due to nematodes. Seasonal anaemia in

The biology of marsupials

Table 8.2 Number of *G. trifidospicularis* and mean haematological
parameters in juvenile *M. giganteus*

Date	No. killed	G. trifidospicularis			e.p.g.*	Hb	PCV	Total plasma protein
		mature	immature	total				
14.2.72	3	3	0	3	50	13.9	42.0	4.9
28.3.72	2	26	0	26	50	14.3	40.5	4.9
9.5.72	4	24	9	33	50	14.2	42.0	5.2
15.6.72	3	164	88	252	500	12.7	37.0	4.5
1.8.72	4	597	31	628	650	6.0	16.5	2.7
12.9.72	4	146	10	156	300	9.4	25.8	4.2
23.10.72	5	157	1	158	140	10.2	32.8	4.0
4.12.72	3	27	0	27	50	12.8	36.7	4.9

*e.p.g. = eggs per gram of faeces, 50 e.p.g. being the lowest number measured.

the Quokka, *Setonix brachyurus*, was ascribed to nutrition (Barker, 1960),
although the contribution made by nematodes was not adequately examined.
Later work has suggested that nematodes do not affect this species (Barker,
personal communication, 1973). Mortality in Red kangaroos in South-eastern
Queensland has been associated with burdens of hookworms, presumably *Hypo-
dontus* (Copeman, personal communication, 1973), and anaemia would be an
important symptom.

The trichostrongyles *Filarinema flagrifer* and *F. moennigi* produce small
nodules in the stomach and intestines of *M. rufus*, although unencapsulated
worms are found in the pyloric region of the stomach (Mykytowycz, 1964).
Further observations on other *Filarinema* spp. have failed to associate it with any
pathology (unpublished data).

Table 8.3 Mean number of *G. trifidospicularis* and mean haematological
parameters in adult *M. giganteus*

Date	No. killed	G. trifidospicularis			e.p.g.*	Hb	PCV	Total plasma protein
		mature	immature	total				
14.2.72	1	0	0	0	0	15.2	47.0	6.8
28.3.72	2	18	0	18	50	15.2	46.0	6.4
9.5.72	1	0	0	0	0	14.4	41.0	6.0
15.6.72	2	6	1	7	0	14.9	44.0	6.3
1.8.72	2	6	1	7	0	13.0	36.0	6.3
12.9.72	2	31	8	39	0	13.3	38.5	6.1
23.10.72	1	0	0	0	0	13.7	38.0	4.8
4.12.72	2	41	0	41	0	14.1	40.0	6.0

*e.p.g. = eggs per gram of faeces; 0 signifies less than 25 e.p.g.

Strongyloidids

Gastritis, associated with the presence of an incompletely described species of *Strongyloides*, has been implicated as a cause of death in *M. giganteus* (Winter, 1958), *M. robustus* and *M. rufus* (our unpublished data) held in captivity. The nematodes burrow in the epithelium of the cardiac mucosa which lines the saccular portion of the stomach, laying ova in tunnels, and causing severe inflammation. The gastric mucosa appears reddened and is cast in irregular folds or nodules up to 5 mm in diameter. Treatment has not been attempted, since the diagnosis has only been made at necropsy. To our knowledge this condition has not been seen in animals from the field, although widespread small areas of hyperaemia associated with a *Strongyloides* sp. were seen in an outbreak of mortality of *M. giganteus* due to *G. trifidospicularis* (see above).

Toxoplasma

Toxoplasma antibody was not detected in a sample of 49 large macropods from Queensland (Cook and Pope, 1959). However, Munday (1972c) has reported serologic evidence of toxoplasma infection among small macropods in Tasmania, and Gibb *et al.* (1966) reported that antibody to *Toxoplasma gondii* or tissue cysts were present in a significant number of Quokkas on Rottnest Island. Variations in local prevalence may be related to the distribution of feral cats, and the environmental conditions to which oocysts are subject. There appear to be no reports of clinical toxoplasmosis in wild macropods.

However, clinical disease has been recorded in a number of species held in captivity or recently captured or handled, and it is apparent that under these conditions possibly latent infections may become fulminant, resulting in acute death with few premonitory signs and gross lesions. Histologic examination reveals *Toxoplasma*.

White Muscle Disease

A degenerative myopathy has been described from a number of macropodid species, principally from the Quokka. Animals of this species, when kept in captivity, commonly develop a syndrome of weight loss, progressive wasting, particularly of hind limb muscles, paralysis of the hind limbs and death (Kakulas, 1961, 1966). At autopsy pallor and atrophy of the pelvic and femoral muscles is marked and the lesion is characterised histologically by hyaline degeneration of muscle fibres with pycnosis of muscle nuclei and infiltrates of lymphocytes, plasma cells and polymorphs (Kakulas, 1961, 1966). The myopathy in the Quokka is reversible if treated with α-tocopherol but not selenium (Kakulas, 1961, 1963a), and the incidence of myopathy is apparently influenced by the size of the enclosure in which the animals are kept (Kakulas, 1963b). The development of myopathy in animals placed in small enclosures has been noted in *M. eugenii* (Smales, personal communication). Myonecrosis has been found in recently captured macropods (*M. rufogriseus* and *Thylogale billardieri*) in Tasmania (Munday, 1972b), although the aetiology in this case is uncertain. Apart from the possibility of it being a vitamin E responsive disorder, Munday (1972b) suggested that in this instance the overexertion related to capture should be considered as a possible precipitating agent.

A similar clinical syndrome to that in the Quokka also occurs in *W. bicolor* and *M. giganteus* under captive conditions, often in young animals receiving milk as well as having access to pasture (our unpublished observations). The syndrome in *W. bicolor* is further characterised by erythema of the skin of the inside of the thigh and by severe hyperkeratosis and desquamation of flakes of cornified epithelium all over the body.

Although myopathies have not been reported from marsupial groups other than the Macropodidae, a syndrome of posterior paralysis has been seen in wild and captive gliders, *Petaurus breviceps*, and Leadbeater's possums, *Gymnobelideus leadbeateri*, with the typical histological picture seen in macropods (our unpublished observations).

'Lumpy jaw'

'Lumpy jaw' has long been recognised as a scourge of macropods held under captive conditions (Fox, 1923; Foss, 1947), although the aetiology of the disease is uncertain. The condition usually begins in the periodontal region but may extend to a local osteomyelitis, or involve the other mucous membranes of the oral cavity and the tongue, the eye or nostril. Similar necrotising lesions may also be found in lungs, stomach or liver, indicating a septicaemic spread to these sites. Clinical signs in individual animals are related to sites of localisation (see Barker *et al.*, 1963). The condition is frequently not noticed until an advanced stage is reached, and is evident by severe wasting (as the animal is unable to feed) or by a swelling on the side of the jaw. The infection may respond to antibiotic treatment if detected sufficiently early (Watts and McLean, 1956).

The histological appearance of such jaw lesions has been described by Beveridge (1934). In acute lesions there is a central core of caseous material surrounded by tissue undergoing necrosis. There is a significant neutrophil infiltrate and, surrounding it, a zone of active fibroplasia. In lesions described as peracute the reaction is more diffuse; there are extensive regions of cellular necrosis but fewer neutrophils and no fibroplasia. Beveridge (1934) found large numbers of *Sphaerophorus necrophorus* (syn. *Bacillus necrophorus*) in infected areas, the filaments of bacteria frequently extending into areas of normal tissue.

Considerable confusion exists as to the aetiology of the disease. Fox (1923) isolated *Nocardia macropodidarum* from lesions and considered this to be the cause of the disease. Subsequently the disease has consistently been referred to as nocardiosis (Le Souef and Seddon 1929; Tucker and Millar, 1953; Barker *et al.*, 1963), although the actual identity of the original organism is by no means certain. Jensen (1913), Mouquet (1923), Beveridge (1934) Tomlinson and Gooding (1954) isolated the anaerobe *S. necrophorus* (syn. *B. necrophorus*) from lesions in kangaroo jaws, and in the case of Beveridge (1934), at least, it was identified in detail and its pathogenicity compared with strains from bovine liver abscesses. As already noted, Beveridge (1934) detected the organism histologically in advancing lesions with filaments extending into areas of normal tissue. Furthermore, he inoculated isolates of the organism into the jaw of a wallaby (species not stated) producing typical lesions of 'lumpy jaw' which led to the death of the animal.

Isolates from macropods made by Tucker and Millar (1953) were identified only as *Nocardia* spp., although they noted the difficulty of further identifications due to taxonomic difficulties and also the problem of the possible role of 'streptothrix' (*S. necrophorus*) also found in lesions as a pathogen. Watts and McLean (1956) isolated a species of *Bacteroides* from lesions and considered this to be the causative agent.

Our observations suggest that a number of anaerobes can be detected in smears from lesions, and these include *S. necrophorus, Bacteroides* spp. and other saprophytic bacteria.

The confused taxonomic tangle of purported causative agents of 'lumpy jaw' is beyond the scope of this review, but it seems possible that many of the isolates reported are referable to *S. necrophorus* or other normal saprophytes of the gastro-intestinal tracts of herbivora. In domestic animals *S. necrophorus* is the cause of necrobacillosis, usually invading tissues otherwise compromised. Were the periodontal tissues of macropods to become damaged or compromised in some way, then invasion by *S. necrophorus* and other saprophytes is a likely consequence. The experimental transmission of 'lumpy jaw' by Beveridge (1934) lends support to this view in that, once the integrity of the periodontal mucosa has been breached and organisms introduced, the genesis of 'lumpy jaw' lesions is automatic.

The factors which might compromise the integrity of the periodontal membranes of macropods are less clear. Vitamin A deficiency is well known to predispose mucous membranes to bacterial invasion (Jubb and Kennedy, 1970), but there is no evidence yet that this occurs in macropods. The suggestion has also been made that tough or fibrous food may predispose to infection (Le Souef and Seddon, 1929; Barker *et al.*, 1963), but again there is no supporting evidence.

The observation that the origin of the lesions is the periodontal region suggests that tooth loss or some other normal function which breaks the integrity of the membranes might be involved. In macropods molars and premolars erupt posteriorly in the jaw and migrate anteriorly before being lost adjacent to the diastema (see Tyndale-Biscoe, 1973). In cases of 'lumpy jaw' seen by us and Finney (personal communication) there has invariably been an abnormality in the dentition, usually in the form of a retained tooth, leading to the suggestion that an abnormality of this type initiates the disease.

While it is obvious that 'lumpy jaw' occurs commonly in captive animals, it is also seen in wild macropods. Tomlinson and Gooding (1954) recorded an outbreak in *M. rufus* in Western Australia, and the disease has been seen in occasional wild *W. bicolor* (our unpublished observations) and *M. eugenii* (Smales, personal communication).

Phascolarctos cinereus – Koala

In the early general literature on the Koala reference is made to epizootics of an extensive or local nature which seriously reduced wild populations (Le Souef, 1923; Fleay, 1937; Troughton, 1941); however, there is little information on their aetiology. Fleay (1937) drew attention to the fastidious and capricious appetite of the Koala, and suggested that a condition characterised by diarrhoea and loss of condition culminating in death, among animals in Melbourne Zoo, and possibly in the wild, was associated with the poor quality of feed available, particularly during the winter. This syndrome continues to be a potential problem among koalas in sanctuaries and zoos, and can only be prevented by maintaining an adequate supply of fresh fodder on a daily basis (A. D. Wood, personal communication, 1974).

Recently Cockram and Jackson (1974) isolated *Chlamydia psittaci* from the conjunctival sacs of koalas with conjunctivitis and keratitis. The condition is frequently bilateral and is characterised by conjunctivitis, a mucopurulent ocular discharge, and corneal opacity. It may persist for months, and the authors concluded that this disease, which appears to have a sporadic incidence in most koala populations, may cause local reductions in members of this species. A similar condition was described by Troughton (1941) as being responsible for significant mortality as early as 1887.

Koalas, frequently from the wild, were diagnosed as having pneumonia, possibly of bacterial origin, by Backhouse and Bolliger (1961), and Rahman (1957) refers to the isolation of *Corynebacterium* from the pneumonic lungs of a koala. Cryptococcosis has also been reported as a cause of death in 5 of 39 koalas examined from around Sydney, N.S.W. (Backhouse and Bolliger, 1961; Bolliger and Finkh, 1962) and in animals from a fauna reserve in West Australia (Gardiner and Nairn, 1963). Syndromes represented included necrotising rhinitis and sinusitis, pneumonia, meningitis and encephalitis, in some instances with disseminated infection of many organs. The West Australian cases were associated with large pigeon populations, and infection may have originated from their droppings.

Spontaneous neoplasms, particularly cystadenoma of the ovaries and lymphocytic leukaemia, have been recorded, in some instances from wild or recently captured animals (Backhouse and Bolliger, 1961; Heuschele and Hayes, 1961; Finkh and Bolliger, 1964). The latter authors recorded cystadenomas in the ovaries of 12 of 18 mature koalas, and frequently both ovaries were involved, undoubtedly rendering the animal infertile. The presence of two cases of lymphocytic leukaemia among the low number of koala necropsies reported in the literature (about 50) suggests that this condition may have a relatively high prevalence, and the possibility of a viral aetiology must be considered. Other blood 'dyscrasias' of unknown aetiology, manifest initially as moderate anaemia unresponsive to haematinic therapy, and ultimately fatal, have also been described (Bolliger and Backhouse, 1961).

Internal parasitism is of no consequence in the Koala, the cestode *Bertiella obesa* being the single species encountered. The only records of disease due to external parasites are references to mange in individual captive animals, associated with the presence of *Notoedres cati* (Seddon, 1968) and *Sarcoptes scabiei* (Barker, 1974), probably contracted accidentally from other host species.

Vombatidae

The major infectious disease known in wombats is sarcoptic mange due to the mite *Sarcoptes scabiei* which is enzootic in many areas, and may erupt and possibly cause significant mortality among *Vombatus ursinus*, the Common wombat, although evidence of its importance has been based largely on hearsay (Gray, 1937). Sweatman (1971) illustrated a severe case with typical encrustations on the face, muzzle and anterior extremities. Scabies, presumably acquired by contact, is a common problem in people who handle wombats frequently, and individuals rearing these animals frequently use a malathion wash to control infection on the wombat. *Acaroptes vombatus* is also found on *V. ursinus*, through not associated with disease.

Toxoplasmosis, with organisms disseminated throughout many organs and frequently associated with encephalitis, has been seen in wombats in captivity (Hamerton, 1933; our unpublished observations) but has not to our knowledge been observed in animals from the wild.

Coccidia have been described from the Wombat, and our observations indicate that the large protozoon cysts described by Gilruth and Bull (1912) in the lamina propria of the small intestine of a Hairy-nosed wombat (*Lasiorhinus latifrons*) were probably gametogenous stages of a species of *Eimeria*. Similar organisms of a different species occur in *V. ursinus* (I. K. Barker and B. L. Munday, unpublished), but their pathogenicity is doubtful.

Barker *et al.* (1963) discussed at some length the genera *Ileocystis, Lymphocystis* and *Globidium* in reference to Gilruth and Bull's observations (1912). These were referred to the genus *Eimeria* (syn. *Globidium*) by Triffett (1926).

The strongylate nematodes *Oesophagostomoides* spp. (syn. *Phascolostrongylus* Cavanan, 1931: see Chabaud, 1965) inhabiting the large intestine are as yet unassociated with disease. An undescribed lungworm has been encountered (Harrigan, Dunsmore and Spratt, personal communication), again unassociated with systemic disease, although heavy infections may be potentially pathogenic.

Although Boray (1969) failed in attempts to infect wombats experimentally with *Fasciola hepatica*, the parasite has been found occurring naturally in the bile ducts of this host (Spratt, personal communication; our unpublished observations), in one instance causing severe hepatic fibrosis and biliary hyperplasia, but it is probably only found in animals grazing areas in which *F. hepatica* is enzootic in local ovine or bovine population.

Antibody titres to *Leptospira pomona* were detected in *V. ursinus* in Tasmania (Munday, 1972a; Munday and Corbould, 1973). Interstitial nephritis, with leptospires in the renal tubules, was detected in six seropositive animals, and two experimentally infected wombats succumbed to leptospirosis within 2 weeks. However, the significance of leptospirosis as an enzootic disease in wombat populations is questionable, since all seropositive animals came from properties which carried infected cattle, possibly infecting or sharing a common source of infection with wombats.

Dasyurids

Brown antechinus, Antechinus stuartii

The aetiology of the synchronous total mortality of males which occurs during the post-breeding period over the entire range of this species (Woolley, 1966; Wood, 1970) has been the subject of considerable investigation. Males after breeding show physiologic changes consistent with a response to stress (Woollard, 1971; Barnett, 1973, 1974). Parasitism by a range of internal and external parasites, with a higher prevalence of infection with some nematodes in males during the breeding and post-breeding period, has been observed in a Victorian population (our unpublished observations). However, it was not possible to implicate para-sitism as a significant contribution to mortality of male *A. stuartii*. Similarly, although a haematologic study revealed that a proportion of males and females became moderately anaemic during the period of the field mortality, it was not possible to identify the cause of the anaemia or implicate it as a cause of death.

Recent work, done over two breeding seasons, has suggested a possible mechanism for the synchronous decline in numbers of males (unpublished data). In the first year splenic follicles in male *A. stuartii* trapped and killed during the die-off were found to be significantly smaller than those in females killed at the same time. In the subsequent year an attempt was made to observe the mortality in the laboratory. A total of 17 males and 4 females were trapped during the month prior to the expected onset of mortality and held indefinitely. Eleven males were trapped after breeding, during the period 17 – 22 August, and begin-ning 19 August regular observations were begun at 2 hourly intervals in attempts to detect moribund animals. All males trapped after breeding died between 21 and 24 August when mortality in the field occurred, while the 6 trapped prior to breeding had all died by the beginning of September. Two of 4 females survived well beyond the period of male mortality.

Deaths among males were acute, with only 6 animals actually observed mori-bund. Moderate anaemia was observed in 3 of 4 moribund animals from which adequate blood was obtained. Anisocytosis, reticulocytosis and high parasitaemias with a pleomorphic *Babesia* were seen in most blood smears obtained from mori-bund or recently dead animals. Terminal plasma corticosteroid levels were elevated in moribund animals.

Post mortem examination revealed a number of pathologic processes, including severely involuted splenic lymphoid follicles, gastro-intestinal ulceration and haemorrhage, and haemoglobinuria. In 9 out of 17 males, and 1 female, foci of hepatic necrosis containing Gram-positive rods were found, consistent with infection by *Listeria monocytogenes*, which was grown from the 4 affected livers cultured. In addition a number of sporadic lesions were observed.

These observations suggest that males, reacting to stressful behaviour during the mating season, suffer involution of a substantial mass of lymphoid tissue, probably as a result of elevated plasma corticosteroid levels. Impairment of the immune system may ensue, permitting recrudescence or *de novo* infections of *Listeria* and *Babesia*. The significance of babesial infection is yet to be determined, but if it does incite an haemolytic anaemia, it may contribute to the development of listerial infection, whose growth is enhanced by the availability of free iron, and the probable elevation of plasma glucose levels expected to ensue in animals with high circulating corticosteroid levels.

Since *L. monocytogenes* is an organism which may be present in soil or water, and may establish latent infection within the body, there is the potential for a high proportion of *A. stuartii* to be susceptible to development of listeriosis should resistance be impaired. Similarly, the potential for babesiosis to be wide-spread exists, since infestation with a possible vector, *Ixodes antechinii* is common in populations we have studied in Victoria, and a related species *I. fecialis* is found in northern populations of *A. stuartii* (Roberts, 1970). The third major condition which may contribute to mortality, gastrointestinal haemorrhage, is associated in other species with physiologic response to stress (Selye, 1950). The local synchrony of the mortality of males is probably related to the simultaneous induction of stressful behaviour among males associated with the concentrated breeding season.

Other dasyurids

There is virtually no information on infectious disease in other dasyurid marsupials.

Toxoplasma gondii infection is probably enzootic among wild Kowari, *Dasyuroides byrnei* (Attwood and Woolley, 1972), and this organism may cause significant mortality among laboratory stock (ibid.). However, there is no indication of its significance in the wild.

Neoplasia was also a frequent observation among small dasyurids dying in captivity (Attwood and Woolley, 1973), tumours observed including fibrosarcomas, medulloblastoma, pulmonary adenomatosis, lymphosarcoma and lymphocytic leukaemia. Other reports of neoplasia in dasyurids (reviewed by Barker *et al.*, 1963, and Attwood and Woolley, 1973) usually emanate from zoological gardens or other captive colonies.

References

Anderson, N. (1972). Trichostrongylid infections of sheep in a winter rainfall region. i. Epizootiological studies in the Western District of Victoria 1966-7. *Aust. J. Ag. Res.*, **23**, 1113-29.

Arundel, J. H. (1971). Control of sheep nematodes in the winter rainfall area. Proceeding No. 11 Post-Graduate Committee in Veterinary Science, University of Sydney.

Attwood, H. D. and Woolley, P. A. (1972). Toxoplasmosis in dasyurid marsupials. *Pathology*, **7**, 50 [Abstract].

Attwood, H. D. and Woolley, P. A. (1973). Spontaneous malignant neoplasms in dasyurid marsupials. *J. Comp. Pathol.*, **83**, 569-81.

Backhouse, T. C. and Bolliger, A. (1961). Morbidity and mortality in the koala (*Phascolarctos cinereus*). *Aust. J. Zool.*, **9**, 24-37.

Barker, I. K. (1974). *Sarcoptes scabei* infestation of a koala (*Phascolarctos cinereus*) with probable human involvement. *Aust. Vet. J.*, **50**, 528.

Barker, I. K., Harrigan, K. E. and Dempster, J. K. (1972). Coccidiosis in wild grey kangaroos. *Int. J. Parasitol.*, **2**, 187-92.

Barker, S. (1960). The role of trace elements in the biology of the quokka (*Setonix brachyurus* - Quoy and Gaimard). PhD Thesis, University of Western Australia.

Barker, S., Calaby, J. H. and Sharman, G. B. (1963). Diseases of Australian laboratory marsupials. *Vet. Bull.*, **33**, 539-44.

Barnett, J. L. (1973). A stress response in *Antechinus stuartii* (Macleay). *Aust. J. Zool.*, **21**, 501-13.

Barnett, J. L. (1974). Changes in the hydroxyproline concentration of the skin of *Antechinus stuartii* with age and hormonal treatment. *Aust. J. Zool.*, **22**, 311-18.

Beveridge, W. I. B. (1934). A study of twelve strains of *Bacillus necrophorus,* with observations on the oxygen tolerance of the organism. *J. Pathol. Bacteriol.*, 38, 467 – 91.

Bolliger, A. and Backhouse, T. C. (1961). The blood of the koala (*Phascolarctos cinereus*). *Aust. J. Zool.*, 8, 363 – 70.

Bolliger, A. and Finkh, E. S. (1962). The prevalence of cryptococcosis in the koala (*Phascolarctos cinereus*). *Med. J. Aust.*, 1, 545 – 7.

Boray, J. C. (1969). Experimental fascioliasis in Australia. *Advan. Parasitol.*, 7, 96 – 204.

Chabaud, A. G. (1965). Ordre des Strongylida, *Traite de Zoologie*, 4, No. 3 (ed. P. P. Grasse), Masson and Co., Paris, 883.

Cockram, F. A. and Jackson, A. R. B. (1974). Isolation of a chlamydia from cases of kerato-conjunctivitis in koalas. *Aust. Vet. J.*, 50, 82 – 3.

Cook, I. and Pope, J. H. (1959). Toxoplasma in Queensland. III. A preliminary survey of animal hosts. *Aust. J. Exp. Biol. Med. Sci.*, 37, 253 – 62.

Ealey, E. H. M. and Main, A. R. (1967). Ecology of the Euro, *Macropus robustus* (Gould), in north-western Australia. III. Seasonal changes in nutrition. *C.S.I.R.O. Wildlife Res.*, 12, 53 – 65.

Finkh, E. S. and Bolliger, A. (1964). Serous cystadenoma of the ovary in the koala. *J. Pathol. Bacteriol.*, 85, 526 – 8.

Finnie, E. P. (1974). Prophylaxis against coccidiosis in kangaroos. *Aust. Vet. J.*, 50, 276.

Fleay, D. (1937). Observations on the koala in captivity. Successful breeding in Melbourne Zoo. *Aust. Zool.*, 9, 68 – 80.

Foss, D. (1947). Autopsies reveal causes of death among zoo animals. *Field Mus. Nat. Hist. Bull.*, 18, 4.

Fox, H. (1923). *Disease in Captive Wild Mammals and Birds*, Lippincott, Philadelphia.

Gardiner, M. R. and Nairn, M. E. (1963). Cryptococcosis in koalas in Western Australia. *Aust. Vet. J.*, 40, 62 – 3.

Gibb, D. G. A., Kakulas, B. A., Perret, D. H. and Jenkyn, D. G. (1966). Toxo-plasmosis in the Rottnest Quokka (*Setonix brachurus*). *Aust. J. Exp. Biol. Med. Sci.*, 44, 665 – 71.

Gibson, T. E. (1963). Experiments on the epidemiology of nematodiriasis. *Res. Vet. Sci.*, 4, 258 – 68.

Gilruth, J. A. and Bull, L. B. (1912). Enteritis associated with infection of the intestinal wall by cyst-forming protozoa (Neosporidia) occurring in certain native animals. *Proc. Roy. Soc. Victoria*, 24, 432 – 50.

Gray, D. F. (1937). Sarcoptic mange affecting wild fauna in New South Wales. *Aust. Vet. J.*, 13, 154 – 5.

Hamerton, A. E. (1933). Report on deaths occurring in the Society's gardens during the year 1932. *Proc. Zool. Soc. London*, 451 – 82.

Heuschele, W. P. and Hayes, J. R. (1961). Acute leukaemia in a New South Wales koala (*Phascolarctos cinereus*). *Cancer Res.*, 21, 1394 – 5.

Jensen, C. D. (1913). In: *Handbuch der Pathogenen Mikro-organismem* (ed. W. Kolle and A. von Wassermann), Fischer, Jena, 2 Aufl., Bd. vi., p. 235.

Jubb, K. V. F. and Kennedy, P. C. (1970). *Pathology of Domestic Animals,* 2nd edn., Academic Press, New York and London.

Kakulas, B. A. (1961). Myopathy affecting the Rottnest Quokka (*Setonix brachyurus*) reversed by α-tocopherol. *Nature*, **191**, 402.

Kakulas, B. A. (1963a). Trace quantities of selenium ineffective for the prevention of nutritional myopathy in the Rottnest Quokka (*Setonix brachyurus*). *Aust. J. Sci.*, **25**, 313.

Kakulas, B. A. (1963b). Influence of the size of the enclosure on the development of myopathy in the captive Rottnest Quokka. *Nature*, **198**, 673.

Kakulas, B. A. (1966). Regeneration of skeletal muscle in the Rottnest Quokka. *Aust. J. Exp. Biol. Med. Sci.*, **44**, 673 - 88.

Le Souef, A. S. (1923). The Australian native animals. How they stand today and the cause of the scarcity of certain species. *Aust. Zool.*, **3**, 108 - 11.

Le Souef, A. S. and Seddon, H. R. (1929). Streptothrix disease in kangaroos. *Aust. Vet. J.*, **5**, 79.

Mackerras, M. J. (1958). Catalogue of Australian mammals and their recorded internal parasites. Part I. Monotremes and marsupials. *Proc. Linnean Soc. N.S.Wales*, **53**, 101 - 25.

Mouquet, A. (1923). Maladie de Schmorl chez un kangourou. *Bull. Soc. Cent. Med. Vet.*, **76**, 419.

Munday, B. L. (1971). Gastric squamous tumor in the stomach of a pademelon (*Thylogale billardierii*) associated with an infestation with *Labiostrongylus* sp. larvae. *J. Wildlife Dis.*, **7**, 125.

Munday, B. L. (1972a). A serological study of some infectious diseases of Tasmanian wildlife. *J. Wildlife Dis.*, **8**, 169 - 75.

Munday, B. L. (1972b). Myonecrosis in free-living and recently-captured macropods. *J. Wildlife Dis.*, **8**, 191 - 2.

Munday, B. L. and Corbould, A. (1973). *Leptospira pomona* infection in wombats. *J. Wildlife Dis.*, **9**, 72 - 3.

Mykytowycz, R. (1964). A survey of the endoparasites of the red kangaroo, *Megaleia rufa* (Desmarest). *Parasitology*, **54**, 677 - 93.

Rahman, A. (1957). The sensitivity of various bacteria to chemotherapeutic agents. *Brit. Vet. J.*, **113**, 175 - 8.

Roberts, F. H. S. (1970). *Australian Ticks.* C.S.I.R.O., Melbourne.

Seddon, H. R. (1968). *Arthropod Infestations.* Commonwealth of Australia Department of Health Service Publication Number 7 (rev. H. E. Albiston), Commonwealth Government Printer, Canberra.

Selye, H. (1950). *The Physiology and Pathology of Exposure to Stress*, Acta Inc. Montreal.

Spratt, D. M. (1972). Aspects of the life history of *Dirofilaria roemeri* in naturally and experimentally infected kangaroos, wallaroos and wallabies. *Int. J. Parasitol.*, **2**, 139 - 56.

Spratt, D. M. (1974). Comparative epidemiology of *Dirofilaria roemeri* infection in two regions of Queensland. *Int. J. Parasitol.*, **4**, 481 - 8.

Spratt, D. M. and Varughese, G. (1975). A taxonomic revision of filaroid nematodes from Australian marsupials. *Aust. J. Zool.*, supplement number 35.

Sweatman, G. K. (1971). Mites and pentastomes, in *Parasitic Diseases of Wild Mammals* (ed. J. W. Davis and R. C. Anderson), Iowa State Press, pp. 3 - 64.

Tomlinson, A. R. and Gooding, C. G. (1954). A kangaroo disease. Investigations into "lumpy jaw" on the Murchison, 1954. *J. Agric. W. Australia, 3*, 715 - 8.

Triffett, M. (1926). Some sporozoan parasites found in the intestinal wall of Bennett's wallaby (*Macropus bennetti*). *Protozoology*, 2, 31 – 46.

Troughton, E. (1941). *Furred Animals of Australia*, Angus and Robertson, Sydney.

Tucker, R. and Millar, R. (1953). Outbreak of nocardiosis in marsupials in the Brisbane botanical gardens. *J. Comp. Pathol.*, 63, 143 – 6.

Tyndale-Biscoe, H. (1973). *Life of Marsupials*, Arnold, London.

Watts, P. S. and McLean, S. J. (1956). Bacteroides infection in kangaroos. *J. Comp. Pathol.*, 66, 159 – 62.

Wenyon, C. M. and Scott, H. H. (1925). *Transaction of the Royal Society of Tropical Medicine and Hygiene*, 19, 7 [Exhibit] .

Winter, H. (1958). Gastric strongyloidosis in kangaroos. *Aust. Vet. J.*, 34, 118 – 20.

Winter, H. (1959). Coccidiosis in kangaroos. *Aust. Vet. J.*, 35, 301 – 3.

Wood, D. H. (1970). An ecological study of *Antechinus stuartii* (Marsupialia) in a south-east Queensland rain forest. *Aust. J. Zool.*, 18, 185 – 207.

Woollard, P. (1971). Differential mortality of *Antechinus stuartii* (Macleay): nitrogen balance and somatic changes. *Aust. J. Zool.*, 19, 347 – 53.

Woolley, P. (1966). Reproduction in *Antechinus* spp and other dasyurid marsupials in: *Comparative Biology of Reproduction in Mammals* (ed. I. W. Rowlands), *Symp. Zool. Soc. London*, 15, 281 – 94.

9 Behaviour and ecology of the Tasmanian devil, Sarcophilus harrisii

Othmar L. K. Buchmann* and Eric R. Guiler†

The Tasmanian devil, *Sarcophilus harrisii* (Boitard), is the largest of living dasyurid marsupials, except for the doubtfully extant thylacine or Tasmanian tiger, *Thylacinus cynocephalus*. Recent ecological studies have investigated its numbers, diet and economic importance (Green, 1967; Guiler 1970a, b and c), and some aspects of its behaviour have also been reported. Early accounts (for example, Le Souef and Burrell, 1926) gave some details of agonistic behaviour, vocalisations and maintenance activities, which have been augmented in more recent reviews of marsupial biology (Walker, 1964; Troughton, 1965; Ride, 1970). Fleay (1935, 1952) described methods of fighting, grooming and comfort behaviour, resting, feeding, drinking, and changes in behaviour associated with individual development, and Guiler (1964, 1971) and Green (1967) gave further details of maintenance activities and social behaviour. Other studies by Ewer (1969) and Moeller (1972a, b, c; 1974) focused exclusively on prey-killing and feeding behaviour of the species. The Tasmanian devil is of considerable interest as a carnivorous marsupial successfully adapting to a changing environment. Field studies are difficult because of their mainly nocturnal activities, dark colour, cryptic habits and timidity towards man. The present study is based mainly on captive devils observed over several years at the University of Tasmania, and forms part of a series of continuing projects on selected problems of dasyurid behaviour. Our conclusions on prey-killing and feeding, agonistic interactions and social communication, and cloacal dragging, are drawn mainly from quantitative records; full reports will be published elsewhere.

Prey-killing and feeding

Our first observations on prey-killing concerned a large and exceptionally pugnacious male devil presented with a live laboratory rat. The devil blundered about the cage, approaching the rat and occasionally biting it. The rat remained uninjured for several minutes, biting the devil on the snout and causing it to bleed. Eventually the devil killed the rat, first snapping at the lumbar area of its back and partly paralysing it, and subsequently crushing the skull. Further studies

*O. L. K. Buchmann is a graduate of the University of Aberdeen, now a lecturer in zoology at the University of Tasmania. He is especially interested in behaviour of marsupials, population regulation, learning, stress and experimentally induced neurosis.

†Dr E. R. Guiler is a graduate of Queen's College, Belfast, and of the University of Tasmania, where he is at present a reader in zoology. He has worked on Tasmanian devils since 1966, and is interested in field ecology and its relation to whole-animal biology.

involving other animals confirmed a general ineptitude among devils for killing, which contrasted with the animal's reputation as an accomplished killer of vertebrate prey, including domestic livestock. Of 19 devils tested by presentation with mature male Wistar rats, 8 failed altogether to kill the prey; they either ignored the rats, or responded to them with hesitancy, conflict and even fear. Some snapped (apparently *in vacuo*) and made tentative attacks, but seldom completed them. Attempts to bite the prey were usually ill-oriented, clumsy and unsuccessful, and retaliatory bites caused the devils to growl, shake their heads and withdraw. The rats often evaded capture altogether, sometimes establishing themselves in the devils' cages, moving about freely and confidently, and even taking over part of the bedding. Devils which failed to kill the rats usually became less active, typically huddling in a corner of the cage, sometimes yawning repeatedly when the rat approached. Slow, spasmodic yawning actions are associated with tension or conflict in many mammals (Ewer, 1969). One rat even crossed safely between the yawning jaws of an adult devil.

Devils which succeeded in killing rats also showed elements of conflict or hesitancy in their behaviour. They were, however, more responsive to the prey, turning towards them, approaching, investigating and attempting to catch them more frequently. Prey-oriented behaviour increased in frequency and intensity with repeated exposure to rats, and with previous success. A few individuals improved their killing technique with practice, but most did not. Capture involved stabbing and grasping with one forepaw, and biting. These actions were performed separately or in combination, in succession or simultaneously, and improvement in their performance resulted from their coordination becoming more efficient. Nevertheless, the impression produced by all devils was one of clumsiness and ineptitude: even after repeated confrontations the bites remained inaccurate, and prey were seldom despatched immediately or decisively. The time taken to kill prey usually decreased over successive trials, apparently owing more to increased motivation and persistence in attacking than to refinement in the methods of killing themselves.

The ontogeny of prey-oriented responses was investigated in a litter of four young devils. Initially these showed little interest in rats placed in their cage, and made no attempts at capture. The development of their killing response was apparently aided by a chance occurrence — the accidental drowning of one rat in a water trough. All four devils shared in eating the rat. Following this episode, all types of specifically prey-directed behaviour previously noted in adult subjects developed rapidly. One of the young devils, the smallest of the litter, made no attempt to capture rats, but all of the others did so, albeit with varying degrees of success. One individual killed a disproportionately large number of rats. This subject was the highest-ranking animal in an incipient dominance hierarchy and its more efficient performance was perhaps related to its relatively greater activity.

Young devils are conspicuously endowed with manual dexterity and can be readily induced to grip objects such as rolls of paper or cloth in their highly prehensile forepaws. These skills are clearly useful in capturing prey, and early attempts at doing so appeared to be somewhat more skilful than those of the adult devils previously tested. Paradoxically, there was a subsequent decline in efficiency, associated with a general increase in conflictual and hesitancy elements, apparently as a result of the subjects' painful discovery of the retaliatory abilities of the rats. The course of development of prey-related responses resembled, in most respects, the behavioural changes exhibited by adult subjects. Orientation, approaches and attacks directed at prey, as well as investigative elements and following, increased dramatically with successive exposures to rats (Figure 9.1

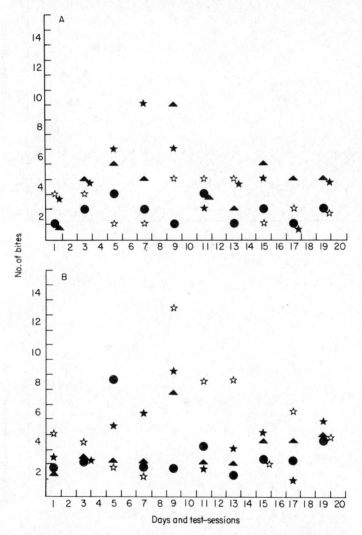

Figure 9.1 A. Numbers of biting attempts at rats by 4 adult *S. harrisii* in tests conducted over a 20 day period. Prey were presented during alternate test sessions. B. Corrected frequencies (number per 10 min period of observation) of biting attempts by the same subjects. Three individuals were males (solid symbols) and one was female (open symbols). Subjects were housed individually and a single rat was provided in each observation session. Note that the prey was seldom despatched at the first attempt and occasionally as many as 9 bites were required to kill a single rat.

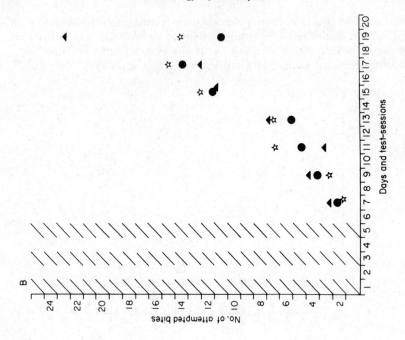

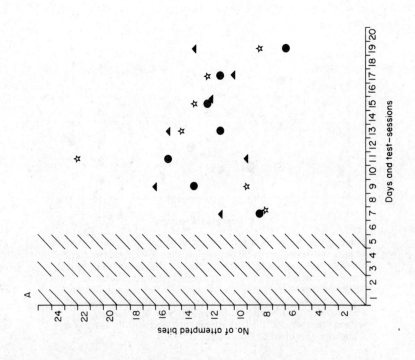

and 9.2). The time taken to kill prey also decreased rapidly over a series of tests; as in adult devils, this appeared to result more from changes in motivation than from improvement in killing technique. The mother of this litter, housed with her young, did not attempt to kill rats and was generally inactive, remaining motionless in the corner of the cage. She usually fed on the rats killed by her offspring, joining them a few minutes after they had begun feeding.

A second litter of four young devils, with no previous experience of live prey, killed an adult male Southern Short-nosed bandicoot, *Isoodon obesulus*, at their first attempt. The bandicoot escaped from its holding cage and ran into the room which held the devils. One snapped at it, damaging its spine; after initial caution, the four approached the injured animal, which was struggling on the floor but unable to escape, and formed a semicircle around it, watching, gaping frequently and making treading movements. Eventually one edged forward and bit the bandicoot, then jumped back; this was repeated by all members of the litter over a dozen times until the prey stopped moving, when they ate it communally.

Moving prey invoked stronger responses from both adult and young devils; prey running away along a line $90-120°$ to the body axis of the devil were attacked most frequently. Retreating prey were often followed; motionless ones were commonly prodded or nudged with the snout. Visual and olfactory cues were used in hunting. The devils frequently sniffed at their prey.

Feeding behaviour

The general inefficiency of killing contrasted strangely with the efficient feeding which followed. After the killing, devils occasionally assumed alert postures and sniffed the air, but more usually began to feed immediately. Prey were sometimes eaten at the site of the killing, but more commonly carried to a secluded part of the cage; corners of the pen were often selected as preferred sites and used regularly by individuals. Small food items like rats were picked up and carried, larger ones dragged. Moeller (1974) filmed and described a devil which galloped at remarkable speed with food in its jaws. Our own animals were more restricted, but those housed in groups carried their food vigorously under harassment from cage mates. Small birds or mammals were consumed completely in a single session of five to ten minutes, only the large feathers of some of the birds remaining. Chunks of larger carcases were detached and bolted, large bones crushed in the powerful jaws and swallowed. Viscera, including entrails, were also consumed, the forepaws being used to cram lengths of intestine into the mouth like spaghetti. Blood spilt during the meal was licked up meticulously. Only the most robust bones – for example, the pelvic bones of wallabies and other large prey – remained when feeding ended.

Figure 9.2. A. Number of biting attempts at rats by 3 young *S. harrisii* (litter mates) in tests conducted over a 20 day period. Prey were presented during alternate test sessions. B. Corrected frequencies (number per 10 min period of observation) of biting attempts by the subjects. Two male subjects (solid symbols) and one female (open symbols) were observed. The subjects were housed communally and 6 rats were introduced simultaneously in each session. On days 1, 3 and 5 (shaded) rats were present but no attempts at biting were recorded.

We were unable to confirm Ewer's observation (1969), based on the behaviour
of a captive animal and supported by experiments using models of skin filled with
flesh, that feeding usually starts at the head end, orientation being determined by
the feeding animal rubbing its muzzle repeatedly over the carcase to determine the
lie of the fur. Our own devils tended to begin feeding at the end of the prey
(usually a rat) which happened to lie nearest after killing or carrying. This was
generally the head, because the final, definitive bite was invariably directed at the
skull. We saw no indication of the fur-testing activity described by Ewer. Our own
experiments suggested that devils could be induced to begin feeding at almost any
part of a carcase by anointing it with fresh blood; although the possible importance
of the texture of the fur cannot be disregarded, the role of olfactory orientation is
by no means negligible. Skinning, also reported by Ewer, did not occur among
our animals, though kneading and massaging movements – probably related to
restraining and positioning the prey during killing – were noted. The Quoll,
Dasyurus quoll, does skin its prey, using distinctive lateral scrabbling movements
in peeling the skin from the flesh (our observations).

Agonistic behaviour and social communication

Captive devils are generally belligerent towards their own kind. Most agonistic
encounters, which are often frequent and prolonged, are associated with feeding,
and fighting over food is notably intense. Dyadic interactions typically result in
the establishment of stable dominance – subordination relationships, followed by
a progressive decline in frequency of aggressive incidents, apparently because of the
increasing reluctance of the lower-ranking animal to challenge its superior cage
mate (Figure 9.3). The agonistic behavioural repertoire of *S. harrisii* consists of an
impressive number of distinctive elements, including postures, acts, vocal signals
and characteristic configurations of minor components of the body, organised
into elaborate display sequences. Displays generally begin with unilateral or
mutual orientation (of the head alone or involving the whole body), followed by
approach by one or both contestants. Approaches may be direct or ambivalent,
and frequently result in close (1 m or less) proximity of the antagonist. A series of
postural changes usually follows accompanied by vocalisations and other behavioural changes. The succession of typical elements, investigated by transaction
flow analysis, has disclosed consistent sequential relationships and the existence
of a considerable degree of predictability with respect to motivational changes
and the potential signalling functions of various behavioural events. In general, the
aggressive motivation of both participants appears to escalate gradually until one
individual is prepared to attack. Until the terminal phase of the interaction, the
antagonists usually show a tendency for gradual transition from a crouched to an
erect, quadrupedal posture and also exhibit changes in the types of vocalisations
emitted. Initially these usually consist of a low, even growl (termed *monotone*),
but the latter is eventually superseded by a second type of sound (*vibrato*), rising
and falling in pitch and increasing in volume. This may, in some circumstances,
be succeeded by a loud, high-pitched screech (*crescendo*). Vocalisations are
interspersed with mutual gaping. At the beginning of displays the jaws are
generally opened only slightly; graded changes in intensity occur subsequently,
and wide gaping by both contestants is a characteristic precursor of attacks. These

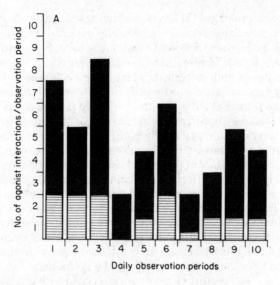

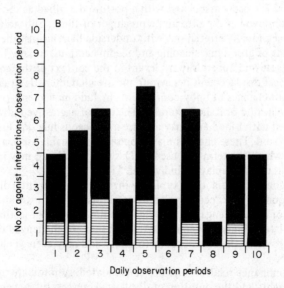

Figure 9.3 Numbers of agonistic displays recorded in two series of dyadic interactions (A, B) between adult *S. harrisii* (cage mates), based on individual observation periods of 5 h, over 10 successive days. Solid blocks represent episodes associated with feeding; barred blocks represent interactions not associated with feeding activity.

interactions may be prolonged for several minutes and normally do not involve
physical contact. Attack behaviour may be briefly preceded by snapping actions in
the direction of the opponent followed occasionally by stabbing with the forefeet,
but its most characteristic feature is jaw-wrestling. This occurs as a result of a
violent lunge by one or both individuals, occasioning a collision of their heads, and
the teeth, partly exposed, are clashed together usually with an audible crunching
sound. The antagonists then struggle with interlocked jaws, and may rear up
briefly on their hind-legs, apparently in an attempt to gain supremacy. Eventually
they disengage and one individual (rarely both) effects a retreat, generally by
backing away. It may be pursued for short distances but more often the winner
of the encounter merely glares at its withdrawing opponent, occasionally vocalising.
The signal most consistently effective in eliciting attack by a conspecific appears
to be *sneezing*, a vocalisation generally produced at the onset of the terminal phase
of the display. The sound is a sharp, exhalant noise with a somewhat querulous
note. It is usually first given by the eventual winner of the encounter and may
induce immediate retreat, but if the antagonist does not withdraw, or if it returns
the same vocalisation, it is almost invariably attacked without further delay.
Sneezing appears to be construed as a challenge by individuals of an appropriate
motivational state. Several other elements are associated with displays. Treading
movements of the feet and alternate extension and shortening of the body are
more frequently shown by the eventual losers of interactions and appear to be
intention locomotory movements, indicating hesitancy. Conflict and fear are
associated with yawning (a wide opening of the jaws with a spasmodic action)
and this element is closely associated with a posture described as squatting. The
latter is often observed as the proximate precursor of flight. Subordinate animals
under attack occasionally stretch out with underside hard against the ground.

Various kinds of grooming, sleeping and feeding actions, usually brief, perfunc-
tory or incomplete and apparently irrelevant in the context of the ongoing
behaviour, suggest displacement behaviour and are characteristic of conflictual
situations. Minor changes of body components include movements of the tail,
which may be elevated or lashed laterally. The ears of aggressive individuals
become suffused with blood and may acquire a rosaceous hue. Piloerection is
usually also marked. These changes may represent serviceable associated habits
in the sense in which the term was used by Darwin (1872) and may increase the
conspicuousness of the displaying individual.

Although agonistic interactions typically involve mutual frontal displays,
fighting over food may be accompanied by attempts to gain possession of dead
prey or pieces of flesh by engaging in a vigorous tug-of-war, or jostling for
position using lateral body movements. Animals that are challenged over food
often attempt to carry their possessions to safety and may even shield the food
by lying over it.

As stable dominance relationships are consolidated, subordinate individuals
initiate fewer fights and the duration of displays decreases, the lower-ranking
animal often retreating before physical contact is made. Subordinates may be
dispossessed of food. When devils are kept in captivity, it is advisable to distribute
food sources as widely as possible so that the lower-ranking animals can feed with
a minimum of disturbance.

The development of dominance relationships in young individuals generally parallels corresponding changes in adults. A litter of four young devils established a linear relative rank-order, associated with different priorities of access to food, within three weeks of full weaning. All of the behavioural elements noted in the displays of adult subjects were exhibited by the young animals at the end of this period and their agonistic behaviour was also comparable with that of adults in other respects. Aggressive encounters appeared, in general, to be earnest and there were few instances of interactions that could be construed as play-fighting.

Cloacal dragging

This is a form of behaviour commonly shown by captive animals of either sex. The rump is lowered so that the perianal area is pressed to the ground. The hind legs are fully bent under the body and project forward, slightly splayed. The forequarters remain slightly elevated and the forelegs are extended. The snout is held horizontally or somewhat raised, and the mouth may be partly opened in a fixed gape. The eyes are typically open, but occasionally narrowed or completely closed. The animal pulls itself forward with its forelimbs, using them together or some-what asynchronously. These somewhat stereotyped actions are usually performed repeatedly, resulting in a slow, jerky progression. The path of the drag is usually irregular but may follow a circular or semicircular track, sometimes with a zig-zag component.

Cloacal dragging is closely associated with agonistic behaviour and intraspecific dominance; animals performing it are often challenged by others. When two devils are confined together, both drag vigorously at first, but dragging by the lower-ranking animal declines progressively as its subordinate status is established. Threat displays are often preceded, interrupted or followed by cloacal dragging, and dominant individuals often start to fight immediately after dragging. High-ranking devils almost invariably drag after successful encounters when their antagonist has withdrawn, but losers may also drag after fights. Dragging often accompanies aggression over food; sometimes an animal drags in a circle or semi-circle around or over the food, and one carrying food may drag either incompletely or fully without dropping it. Dragging does not appear to be associated with defaecation or micturition, although the possibility of glandular marking secretions cannot be overruled. It first appears among young devils at age 210-214 days, and may developmentally be related to social dominance interactions which also appear at about this age. Adults and young appear to use this behaviour pattern in similar contexts.

Relevance to behaviour in the wild

Extrapolation from studies of captive animals to field situations must always be made with caution, especially when field data are fragmentary, as in this case. However, our observations are relevant to public attitudes and policy towards devils in the wild, and we hope that they will stimulate further field investigations to see whether our conclusions, at present largely conjectural, are justified.

The Tasmanian devil is generally considered by farmers and some naturalists to be an efficient predator, able to kill quite large vertebrates, including sheep and other livestock. This belief has engendered bitter and intensive persecution

campaigns against devils, especially in farming districts. The behaviour of laboratory animals, described above, suggests that this reputation is unmerited, and that
time, money and effort are wasted in trying to control an animal which can only
be regarded as a dilettante among predators. Our conclusion is in agreement with
Moeller (1974), who noted that an adult devil which failed to kill a rabbit in a
confined space was far more successful in catching insects and small vertebrates
(fish, frogs, birds and small mammals), occasionally pursuing them actively but
requiring several attempted bites to kill them. Ewer's observations (1969), like
ours, reflect the ineptitude of devils when faced even with rat-sized prey. Native
cats or quolls tested in similar conditions kill rats with ease, often in a few
seconds (Guiler and Buchmann, unpublished). Although devils are capable of
moving at considerable speed, as Moeller's observations have shown (see above),
it is doubtful whether they can pursue prey actively in their characteristic habitats
of forest and heavy undergrowth. Nor are they well constructed for dragging down
large prey; despite heavy build and powerful jaws, the hindquarters are relatively
weak.

Field observations on food and feeding are sparse, but generally support the
conclusion that Tasmanian devils do not hunt large, active prey. Guiler (1970a)
found that large vertebrates such as sheep, wallabies, pademelons and wombats
were well represented in the diet, but the remains of these organisms generally
contained maggots — indicating that they were eaten when the animal was already
dead. Injured or moribund sheep may well be taken; there is no doubt that devils
are able to take quick and opportunistic advantage of natural mortality, and of
cast lambs and offal. They are also known to eat colonial ascidians — for example,
Boltenia (Guiler, 1970a) — and perhaps small birds are also captured while
roosting — young devils especially are adept climbers. They can, in fact, use an
extraordinarily broad spectrum of diet, and given the opportunity may even
ingest such indigestible materials as cartridge wadding and pieces of plastics and
rubber boots with the indiscrimination of sharks. Thus they appear to be versatile
scavengers rather than hunters. Le Souef and Burrell's suggestion (1926) that they
capture animals by lying in wait and pouncing requires confirmation. There were
no indications of this behaviour in the laboratory subjects, and their strong
characteristic scent, readily detected even by man, would limit prospective prey
to creatures with virtually no sense of smell.

Very little is known of social organisation of Tasmanian devils in the field,
although sporadic observations suggest that they feed communally over large
carcases and squabble freely over food. There is no evidence of conventional
territoriality, but devils may maintain 'portable' territories comparable with
those of some birds (McBride, 1964). Individuals which come into regular contact
may establish truce relationships within feeding groups. There is some evidence
from the field that such groups are at least partly closed to outsiders, and that
strangers seeking entry are attacked vigorously and concertedly by established
members. Where food is restricted, size of local populations may be controlled by
the exclusion of low-ranking outsiders from feeding sites. Young unweaned devils
are sometimes threatened by marauding, cannibalistic adults; after weaning, their
intraspecific agonistic behaviour develops rapidly, and aggressive encounters with
both litter mates and adults may assist in inducing their dispersal. In the laboratory, competition often arose between mothers and their young. Mothers often

appropriated food from the young and were threatened by them, retaliating by pushing them away and occasionally tossing them violently by butting with the head.

Population structure

Examination of the size structure of two devil populations, one at Cape Portland in eastern Tasmania (1966, 1968 – 9) and the other at Granville Harbour (1966 – 74), showed that in many years populations were dominated by large adult animals. The Cape Portland population in 1966 had 21 per cent of the devils weighing over 9 kg, while only 13 per cent weighed less than 5 kg and were considered to be juvenile. Juvenile and immature animals comprised only 15.7 per cent of the total population at Cape Portland in 1966. Similar dominance of the population by the larger adult groups was found in a nine year survey at Granville, juvenile and immature animals accounting for 16.5 - 43 per cent with a mean of 26 per cent (Figure 9.4). Very small animals of less than 2 kg (or three months post weaning) are but rarely encountered in the wild, only three such being captured at Granville in a total of 276 animals. This may imply a behavioural or feeding response towards the trap which differs from that of adults. Laboratory observations indicate that young devils eat the same types of food as adults, and it would seem that the apparent scarcity of very young animals is real, and the survival of the young is low. A survival curve constructed for the Granville population showed a form illustrating

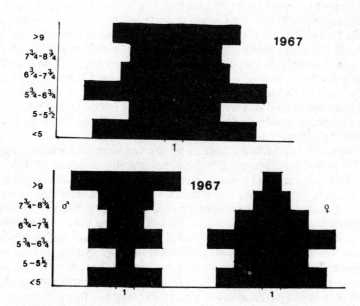

Figure 9.4 Weight and sex structure of a western Tasmanian (Granville Harbour) population of *Sarcophilus harrisii.* Animals of less than 5 kg are immature; those greater are mature. Weights in kg.

a very high first-year mortality with subsequently a higher survival rate over the next five years. In view of the existence of laboratory and field evidence indicating that agonistic interactions between devils occur principally at food sources, and that the size of the animal plays a large part in its hierarchial position (including competitive social interactions between a female and her litter) it is tempting to relate their agonistic behaviour to a natural situation and speculate on the chances of young devils of up to 12 months of age, weighing less than 5 kg, in encounters around carrion sources. Reference has been made earlier to dead or ailing sheep and other pastoral by-products as food sources for devils. These must play an important part in the maintenance of the very large populations found in some areas. A north-east Tasmanian property was estimated to carry 136 devils in 15.5 km^2, a very dense population for a medium-sized carnivore. The Granville area has fewer devils over a much greater area (80 over 26 km^2) but the food supply is less, sheep being absent in the vicinity. The influence of man, his garbage dumping and other practices, has been considerable on devils.

Generally, the agonistic behaviour of *S. harrisii* appears to be highly ritualised (*sensu* Huxley, 1914), consisting of elaborately organised display sequences of considerable motivational predictability and reliable information content. The impressive motor equipment of powerful jaws and well-developed teeth are employed during intimidatory displays, and the threat is not always an idle one, as the formidable injuries borne by the skulls of most devils, as well as broken teeth, dislodged while fighting, bear eloquent testimony. However, not all the lacerations commonly visible on the muzzles of devils are the result of intraspecific aggression. Devils appear to be relatively insensitive to pain and frequently sustain cuts from jagged pieces of bone while feeding. The wounds acquired in fighting are usually more serious, but considering the extent of the damage that devils could inflict on adversaries if they attempted to bite without restraint, they are negligible. Defeated opponents are usually permitted to withdraw, although if they turn around while retreating, they may be bitten on the rump. In general bluff and intimidatory actions are substituted for prolonged and injurious fighting. Vanquished animals subsequently initiate fewer interactions and tend to avoid offering challenge to their socially superior peers. The importance of threat, rather than fighting, is indicated by observations on captive devils housed singly in adjacent cages. These frequently establish a difference in dominance rank, one animal consistently withdrawing from its threatening neighbour, although physical contact between them is prevented and fighting cannot occur. Most of the sensory modalities are probably used in social communication, although the importance of some of these, notably olfaction, is difficult to assess. Cloacal dragging may be employed to reinforce the effectiveness of visual and vocal displays and perhaps in attempting to maintain a monopolistic integrity of some food sources. It is also possible that it may reassure the individual performing it or increase its aggressive motivation. Visual display elements include posturing and dynamic actions as well as gaping, tail movements and other minor changes. The apparent usefulness and large number of the visual signals produced by devils is somewhat unexpected in view of their nocturnal habits. However, when feeding or fighting in exposed situations, the animals are generally conspicuous even to human observers. The muzzle and ears are sparsely furred and pale, drawing attention to the gape, when the teeth are prominently displayed, and possibly also to slight movements of the ears. Apart from these features, devils lack the range of facial expressions exhibited by many social eutherian carnivores (Fox, 1970). The tail is largely devoid of hair, particularly in adults, and not unlike a necrotic turnip. Its movements are also easily seen even at a distance. The white patches marking the body exhibit a high degree of individual variability and may be useful in social recognition of conspecifics.

Devils are also extremely vocal on occasion and their loud and rather eerie noises may enable them to effect efficient long-range communication. The sneezing challenge call is of particular interest; laboratory experiments have indicated that devils are highly responsive to this sound when it is played back to them on tape or even imitated by humans. Similar methods have already proved useful as census techniques in sampling the populations of some eutherian carnivores (Pimlott, 1960). Preliminary observations on the effects of broadcast calls on devil populations have indicated that these elusive animals readily respond to them by emerging from cover and vocalising in turn.

Hodos and Campbell (1969) noted that the concept of the *scala naturae* has proved to be an extraordinarily persistent and tenacious belief in the behavioural sciences. This is often seen in approaches to behaviour studies in marsupials, a group which many workers implicitly consider to be behaviourally homogeneous, and representing a transitional stage between reptiles and eutherian mammals. Our own observations indicate that the Tasmanian devil, far from being anachronistic or possessing 'behavioural fossils' (Ewer, 1969), is a highly adaptable and successful mammal coping well with the problems of an environment in which man and his activities are becoming increasingly prominent. We hope that, as field investigations develop further, more will be learned of this remarkable species, and the currently speculative aspects of its behaviour will gradually be elucidated.

References

Darwin, C. (1872). *The Expression of Emotions in Man and Animals*, Appleton, New York.

Ewer, R. F. (1969). Some observations on the killing and eating of prey by two dasyurid mammals, the mulgara, *Dasycercus cristicauda* and the Tasmanian devil, *Sarcophilus harrisii. Z. Tierpsychol.*, **26**, 23 - 38.

Fleay, D. (1935). Notes on the breeding of Tasmanian devils. *Vict. Nat.*, **52**, 100 - 105.

Fleay, D. (1952). The Tasmanian or marsupial devil. Its habits and family life. *Aust. Mus. Mag.*, **10**, 275 - 80.

Fox, M. W. (1970). A comparative study of the development of facial expressions in canids; wolf, coyote and foxes. *Behaviour*, **35**, 259 - 72.

Green, R. H. (1967). Notes on the devil (*Sarcophilus harrisii*) and the quoll (*Dasyurus viverrinus*) in north eastern Tasmania. *Rec. Q. Vict. Mus.*, **27**, 1 - 13.

Guiler, E. R. (1964). Tasmanian devils. *Aust. Nat. Hist.*, **14**, 360 - 2.

Guiler, E. R. (1970a). Observations on the Tasmanian devil, *Sarcophilus harrisii* (Marsupialia: Dasyuridae) I. Numbers, home range, movements and food in two populations. *Aust. J. Zool.*, **18**, 49 - 62.

Guiler, E. R. (1970b). Observations on the Tasmanian devil, *Sarcophilus harrisii* (Marsupialia: Dasyuridae) II. Reproduction, breeding and growth of young. *Aust. J. Zool.*, **18**, 63 – 70.

Guiler, E. R. (1970c). Tasmanian devils and agriculture. *Tasm. J. Agric.*, **42**, 134 – 7.

Guiler, E. R. (1971). The Tasmanian devil, *Sarcophilus harrisii*, in captivity. *Int. Zoo. Ybk.*, **11**, 32 – 3.

Hodos, W. and Campbell, C. B. G. (1969). *Scala naturae*: why there is no theory in comparative psychology. *Psychol. Rev.*, **76**, 337 – 50.

Huxley, T. S. (1914). The courtship habits of the great crested grebe (*Podiceps cristatus*). *Proc. Zool. Soc. London*, 491 – 562.

Le Souef, A. S. and Burrell, H. (1926). *The Wild Animals of Australasia*, Harrap, London.

McBride, G. (1964). A general theory of social organization and behaviour *Univ. of Queensland Papers, Faculty of Veterinary Science*, **1** (2), 73 – 110.

Moeller, H. (1972a). *Sarcophilus harrisii* (Dasyuridae). Gebrauch der Vorderbeine bei Beuteerwerb und Fressen. *Film F 1834 des Inst. Wiss. Film*, Gottingen.

Moeller, H. (1972b). *Sarcophilus harrisii* (Dasyuridae). Beutefang und Fressen. *Film E 1835 des Inst. Wiss. Film*, Gottingen.

Moeller, H. (1972c). *Sarcophilus harrisii* (Dasyuridae). Fressen von Eiern. *Film E 1836 des Inst. Wiss. Film*, Gottingen.

Moeller, H. (1974). Bewegungsweisen, Beutefang und Fressen beim Beutelteufel, *Sarcophilus harrisii. Film C 1105 des Inst. Wiss Film*, Gottingen.

Pimlott, D. H. (1960). The use of tape-recorded wolf howls to locate timber wolves. *22nd Midwest. Fish Wildlife Conf.*, 1 – 7.

Ride, W. D. L. (1970). *A Guide to the Native Animals of Australia*, Oxford University Press, Melbourne.

Troughton, E. le G. (1965). *Furred Animals of Australia*, Angus and Robertson, Melbourne.

Walker, E. P. (1964). *Mammals of the World*, Johns Hopkins Press, Baltimore.

10 The success of marsupials as introduced species

D. P. Gilmore*

Marsupials are frequently regarded as inferior and archaic in comparison with their placental counterparts. However, certain marsupials have proved themselves to be capable of not only holding their own, but also rapidly extending their range when introduced into a new environment. This is clearly shown by the successful establishment and rapid spread of the Brush-tailed possum, *Trichosurus vulpecula*, in New Zealand and the colonisation of certain localities there by six species of wallaby (Figure 10.1) a species of wallaby, *Petrogale penicillata*, is also firmly entrenched on the island of Oahu in the Hawaiian group. In Britain wallabies are successfully established both in the Peak District and in the vicinity of Ashdown Forest, Sussex, and there are large free-ranging groups within the perimeter fences of Whipsnade Zoo, and at Prince Reuss's Estate in the Black Forest, Germany. Natural spread of numerous marsupials, especially the phalangerids, has also occurred from Australia to New Guinea and on as far as the Solomons to the north-east, Talaut to the north and the Celebes to the west. In North America the opossum, *Didelphis virginiana*, has extended its range northwards for hundreds of kilometres in historic times. This opossum adapts well to suburban living, too, and has been successfully introduced to some areas of the western United States.

Before organised European settlement of New Zealand began early in the nineteenth century, almost all the country below 900-1000m and with a rainfall of 100-115 cm per annum (that is, about two-thirds of the total land area) was forested. Apart from forest, tussock vegetation was the main cover. With the arrival of the European much of the forest was cleared and the soils suitable for agriculture occupied. Today 5.7 million ha of forest remain (covering about 22 per cent of New Zealand), most of which is in the heavier rainfall areas of the mountain ranges, the narrow plain of the South Island's west coast and on the soils in the central North Island derived from pumice showers (Poole and Adams, 1963). In pre-European times there was a very rich bird fauna in New Zealand, but the only native mammals were seals and bats. Also present were a small dog and a rat (*Rattus exulans*) introduced by the Polynesians. During the latter half of the eighteenth and throughout the nineteenth century New Zealand settlers introduced a large influx of exotic animals, from which 34 species of mammals and 31 species of birds became established (Wodzicki, 1950). Among the mammals

*Dr D. P. Gilmore is a graduate of the University of Canterbury, New Zealand, where he completed a doctorate on the biology of the Brush-tailed possum. After post-doctoral study at the Royal Veterinary College in London and the Worcester Foundation in Massachusetts, he now holds a Lectureship in Physiology at the University of Glasgow.

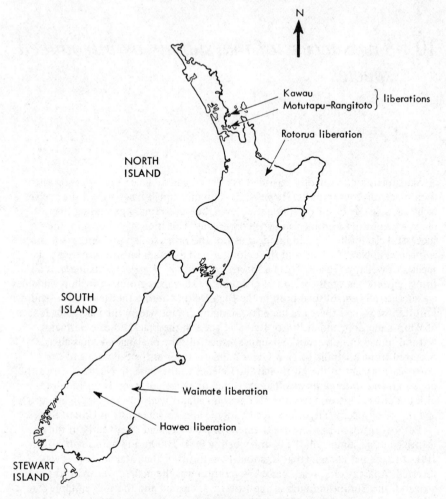

Figure 10.1   New Zealand: localities referred to in text

brought to New Zealand were about 12 species of marsupials, including the Brush-tailed possum (*Trichosurus vulpecula*) and several macropods (Gilmore, 1965, 1968).

The Brush-tailed possum was originally brought to New Zealand because of its value as a fur-bearing animal. Both Tasmanian and mainland Australian possums were introduced, and although the first liberations were made as early as around 1840, the peak introductory period was not until 1890–8, which also coincided with the main phase of releasing New Zealand-bred stock by the Acclimatization Societies (Pracy, 1962). Possums were liberated on the three main islands and on many offshore ones. Initial increase and spread was slow because of the small numbers originally set free, but after the many secondary liberations late in the

nineteenth century, population increase and spread became much more rapid.
Pracy pointed out that the establishment and colonisation by the possum was
successful to the point of surpassing all expectations, and the few failures in
Fiordland and South Westland could be attributed to three main factors: insufficient numbers released, lack of suitable dry nesting at the point of liberation,
and the advent of extremely wet and cold conditions before full establishment.
Today the possum is probably the most numerous mammal in New Zealand,
surviving wherever there is sufficient cover and a suitable and varied food supply,
and thriving in the heavy rainfall areas as well as in the drier districts. Only parts
of the higher mountains are unsuitable for permanent occupation by the animal
because of the presence of seasonal or persistent snow. The two main colour
phases, 'black' and 'grey', occur in varying proportions throughout the country
and freely interbreed in many districts.

Damage to orchards and gardens caused by possums had been noticed in some
areas by the turn of the century, but it was concluded that the damage caused to
the native forests by these animals was negligible. However, during the 1930s and
early 1940s increasing evidence of damage by possums to both exotic and indigenous tree species became apparent. This eventually led to the introduction of a
bounty system whereby a sum of money was paid for each possum destroyed. The
bounty, introduced in 1951, lasted for eight years and resulted in the destruction of
eight million possums. Even so, the number destroyed each year would probably
have been below the annual increase in population. Following the dropping of
the bounty, a vigorous extermination policy was pursued. Rabbits Boards, first
formed late in the nineteenth century to reduce losses in grazing capacity caused
by the rabbit, undertook to destroy possums in farming localities. In the main
forest and high country regions the Forest Service organised campaigns for possum
destruction. Measures involved the dropping of poisoned baits from aircraft, and the
use of dogs, traps and gun as well as the laying of poisoned bait by hand (Gilmore,
1967a). However, these widespread and extremely costly blanket attempts at
eradication had little success, and today control is generally taken at a more
realistic level and restricted to regions where obvious damage is apparent. An
exception, however, is in the Buller district of the South Island's west coast. Here
the possum has been implicated in the spread of bovine tuberculosis, and Buller
has been declared a special area with regard to possum control. Enormous sums of
money are at present being spent on aerial drops of poisoned baits. In addition,
throughout the country much private trapping is still undertaken and the amount
of money realised each year from the sale of possum skins is very substantial, as
indicated in Table 10.1.

Early surveys, including that by Kirk (1920), failed to note any damage to the
indigenous vegetation by the possum, but the increase and spread of deer in the
years following opened up the lower tier of vegetation, helping possum numbers
to grow more rapidly. The interaction of deer and other browsing animals with
the possum has undoubtedly caused damage to the forest canopy in certain areas,
and young conifers can be ruined for future timber production by possums.
However, in many localities, where the population has stabilised, it would
appear that the present vegetation is now little affected.

In Australia the Brush-tailed possum has a varied diet including the leaves of

Table 10.1 Number and value of possum skins exported July 1962-June 1974

July – June	Number exported	Value of exports, (NZ) dollars
1962 – 1963	848 875	1 198 000
1963 – 1964	916 589	795 000
1964 – 1965	418 056	362 000
1965 – 1966	512 993	419 000
1966 – 1967	710 086	668 000
1967 – 1968	757 872	592 000
1968 – 1969	1 292 475	1 863 000
1969 – 1970	1 605 000	2 052 000
1970 – 1971	346 000	262 000
1971 – 1972	517 000	595 000
1972 – 1973	1 240 000	2 021 000
1973 – 1974*	1 572 000	3 869 000

*Provisional figures.

several species of eucalypt and a narrow range of other indigenous and introduced trees and shrubs (Owen and Thomson, 1965). In New Zealand foods eaten by possums living in indigenous forest and settled farmland have been investigated by Mason (1958) and Gilmore (1967b). These studies have shown that in most circumstances possums feed on plants growing in their immediate vicinity, but occasionally travel considerable distances to feed on a particular food such as the catkins of pine. In summer and autumn a much greater choice of foods is available, leading to a more extensive variety of plants being eaten. Flowers of many species are taken throughout the spring and summer, and in the autumn fruits are especially favoured. Although some plants, such as the tree fuschia or kotukutuku (*Fuschia excorticata*), rata (*Metrosideros robusta*), kamahi (*Weinmannia racemosa*) and fivefinger (*Nothopanax arboreum*), are favoured foods throughout New Zealand, their preference may be superseded by other palatable species in some areas. Where forest borders farmland, red and white clover (*Trifolium pratense* and *T. repens*) generally form a large proportion of the diet. Kean and Pracy (1949) postulated that the reason for the seasonal and geographical preferences for various species probably lies in the balancing of different food combinations, the building up of adequate proteins and the blending of semi-poisonous alkaloids. Certainly the spread of the possum in New Zealand does not appear to have been hindered by the lack of suitable food, although lack of a wide range of palatable plant species undoubtedly controls the population density in some localities.

Several studies, including one by Dunnet (1956), have indicated that in Australia the Brush-tailed possum has two breeding periods: a main one in March – April and a minor one in September – October, although in certain areas the spring breeding season appears to be suppressed. In New Zealand the pattern of breeding seems to be basically the same (Gilmore, 1969). As pointed out by Kean (1959), favourable conditions, probably low population pressures and abundant food, enable many females to commence breeding during their second year and to raise two young annually. Sexual maturity in the male is rarely achieved before the animal is two years old. Breeding continues to the tenth year or later.

Unlike the Brush-tailed possum, now widespread throughout New Zealand, the wallabies established there have a very restricted distribution. In 1870 Sir George Grey, a former Governor, brought a number of species of wallaby to New Zealand. These were released on Kawau Island in the Hauraki Gulf near Auckland. In the same year wallabies were also liberated on Motutapu Island in the Hauraki Gulf and from there spread to the adjoining Rangitoto Island, where they soon became abundant. In 1912 wallabies taken from Kawau Island were set free near Lake Kareka in the Rotorua district. In the South Island none of the kangaroos liberated near Christchurch, Dunedin and Bluff survived. However, wallabies liberated close to Waimate in South Canterbury successfully established themselves in the nearby Hunter Hills. These have now spread north and west from the original site of liberation. A few wallabies are also located near Lakes Wanaka and Hawea in Central Otago, well south of the main colony, having become established from escaped pets. The six species of wallaby now in New Zealand are listed in Table 10.2 (p. 177).

Five species of wallaby are found on Kawau Island, an area of just under 3000 ha of infertile, leached yellow-brown soil that was once farmed but has now gradually reverted to scrub consisting mainly of teatree, *Leptospermum ericoides*. This is felled in blocks for firewood on a continuous yield basis, with the result that stands of different ages are interspersed. Only remnants of indigenous forest remain and the ground is bare. This habitat is particularly suitable for the larger Black-tailed and Black-striped wallabies, which feed on marginal grass, occasional open pasture land and sedges growing along water courses and swamps. The Tammar occurs over the whole island. Scattered numbers are present throughout the dominant scrublands, and the species is especially numerous on the old remnant grasslands near the eastern coast, where full scrub has not yet developed, and on the southern section of the island. The Parma is also widespread on Kawau Island, although only discovered there in 1966. Wodzicki and Flux (1971) report that, of 837 wallabies trapped on Kawau Island and exported to various zoos between 1960 and 1970, 384 were Parmas. This species appears to be the second most numerous on the island, being particularly common within the eastern grasslands. The Black-tailed wallaby occurs in much lower numbers and in more localised areas, preferring the scrubland – grassland margins of the developed country in the valley floors. The Black-striped wallaby has only been sighted on ridges within the scrubland, but the Brush-tailed rock wallaby occurs in light to scattered numbers on the steep shoreline faces on the east and south of the island. Elsewhere they occur on localised shoreline faces and appear to have declined in numbers where scrubland and fern have invaded grassland. These animals appear to supplement a diet of grasses by eating shoots put out from the lower trunks of pohutukawa (*Metrosideros*) trees and are able to climb to a height of about 3 m.

Motutapu Island (1200 ha) is joined to Rangitoto Island (1600 ha) at low tide and is now farmed extensively. Once wallabies were found over the whole island, but as the forest and scrub were cleared to make way for pasture, the animals became restricted and reduced to low numbers by control operations. They are now confined to about 20 ha of bluffs on the east coast, where only scanty food is available. Rangitoto Island is a scenic reserve with a scattering of weekend cottages and a few houses on it. The island is a quiescent volcano and parts are still devoid of vegetation. The dominant cover is the pohutukawa tree with a ground layer of ferns. Here the

Brush-tailed rock wallaby occurs in light to scattered numbers in localised areas, being most numerous below the top of the central volcanic cone. The animals have been kept in check by periodic government control operations and by persistent trapping by successive government caretakers.

Little attention was paid to the wallabies in the Rotorua area for many years and a survey in 1949 estimated that they inhabited only about 500 ha. However, in 1968 the area had increased to 1100 and in 1973 to more than 2500 ha. Today the Tammar is scattered thinly throughout the area around the Rotorua lakes except where scrub or forest borders farmland and where moderate to high numbers occur. During the past 25 years progressive milling of indigenous timber in the vicinity of the lakes has maintained a considerable area of very favourable habitat. Spread of the species from the point where it was liberated has been somewhat restricted by lakes and by the city of Rotorua, but the wallabies are now well established in large areas north and west of Lake Tarawera where there is a considerable amount of forest mixed with scrub. Over the last few years it would appear that there has been a migratory drift east with a slight southern tendency.

The country which the Brush wallaby colonised in the South Island of New Zealand contrasts markedly with that inhabited by wallabies in the North Island. The density of the species is greatest in the Hunter Hills, which extend for about 50 km and rise to between 1000 and 1500 m. The warmer eastern faces carry tussock grassland and tongues of native forest climb up the valley floors. The cooler western faces carry tussock and snowgrass on the shady slopes and the hilltops are frequently snow-covered during the winter months. To the north the Hunter Hills join the Two Thumb range; to the west are the Southern Alps; and to the south, the Waitaki river forming a natural boundary to wallaby spread. Cattle runs and sheep stations occupy much of the area, while many crops are grown on the lower ground. Wallabies occur in light to scattered numbers over a very wide area, being most numerous where dense cover prevails and in the forest remnants along the eastern flank of the Hunter Hills. Wodzicki (1950) reported that various grasses and sedges, bracken (*Pteridium esculentum*) and the leaves of trees such as *Fuchsia* sp. and *Leptospermum* sp. form the bulk of the diet of the wallabies in the Waimate area. Serious damage is also sometimes caused to such crops as turnips, chou moellier and oats as well as to the produce of market gardens.

Extensive control operations have been carried out against all the wallaby species in New Zealand. Hunting and poisoning have been undertaken on the islands of the Hauraki Gulf for many years, but the nature of the vegetation and the terrain affords the animals a high degree of protection. A similar situation exists with regard to the wallabies near Rotorua. Because they have little contact with farmland, the local incentive for extermination is less than in South Canterbury, where the wallabies are restricted to private property. However, poisoning was commenced in 1961 and has continued intermittently along with some private hunting. In 1966, 150 wallabies were shot in the Rotorua area.

The wallabies in the Waimate area, undoubtedly the most common in the country, have been hunted since early this century. Wodzicki and Flux estimated that they increased to 750 000 before intensive control operations were undertaken. In the 10 years up to 1956, 69 000 were destroyed, and in 1960 large-scale poisoning operations commenced. This has continued along with organised hunting in an area of 750 000 ha over which the species is spread. Sodium monofluoro-acetate (1080) is the main poison used, and those responsible for control operations maintain that the populations in both the Rotorua and Waimate areas are being kept relatively static.

In Britain the wallabies successfully established in both the Peak District and Ashdown Forest also belong to the species *Macropus rufogriseus*. Those in the Peak District, originating from a group of five that escaped from a private zoo around the beginning of the World War II, have been the subject of an extensive study by Yalden and Hosey (1971).

Although from time to time scattered wanderers are found over a wide area, the main population remains located close to the original site of introduction and lives on a tract of predominantly heather, moorland, grasses and bilberry. Some pine and birch also make up a scrubby woodland in the locality. It appears that the population rose as high as 40 – 50 at the beginning of the 1960s, but the severe winter of 1962 – 3 caused considerable mortality, and Yalden and Hosey suggested it unlikely that the 1970 population exceeded 12 animals. Little is known about the breeding of these wallabies. Heather (*Calluna vulgaris*) provides 70 per cent of the diet, being especially important in the winter. Grasses such as *Festuca ovina* and *Molinia caerulae,* bracken (*Pteridium aquilinum*) and pieces of Scots pine (*Pinus sylvestris*) are also eaten. Yalden and Hosey maintain that because the wallabies survived the very severe winter of 1962 – 3, they should be able to cope if such a situation arose again, and therefore the species should be included on the roster of British mammals.

Taylor-Page (1970) reported that the wallabies in Sussex are found in the woods enclosed by the triangle East Grinstead – Crawley – Cuckfield. They appear to have arisen from a colony kept at Leonardslee, Lower Beeding, 8 km south-east of Horsham. Fitter (1959) mentioned that escapes from here are probably the origin of a number of wallabies at large in the neighbourhood, although the animals have been kept at three different places in Sussex since World War II. Little is know about the present population, but occasional sightings confirm that the species is surviving.

The wallabies on the Hawaiian island of Oahu belong to the same species (*Petrogale penicillata*) as found on Kawau, Mototapu and Rangitoto islands in the Hauraki Gulf of New Zealand. Tomich (1969) described how the colony arose from two adults which escaped from the Alewa Heights area of Honolulu shortly after being brought to Hawaii in 1916. They soon gave rise to a small population extending for a distance of approximately 7 km across a series of ridges and valleys of the lower Koolau Range, and distributed on rocky slopes between 90 and 425 m where there are many caves and recesses to afford them protection and shelter. Tomich estimated the entire Oahu population to probably consist of only a few dozen individuals concentrated in the cliffs on the north-west side of the Kalihi valley, but scattered over a wider area including the nearby Nuuanu valley. The wallabies apparently prefer the drier and less heavily vegetated slopes, carefully avoiding the wetter caves and not penetrating far into the forest.

A more detailed account of the wallabies on Oahu is given by Kramer (1971). He considered that although the wallaby population probably approached 100 animals in the early 1940s, increasing urbanisation and other factors combined to force the expanding population back to its original stronghold on the Kalihi cliffs. Kramer observed that young wallabies were born throughout the year, and from an examination of faecal pellets was able to determine that the main foods were introduced grasses and seeds of sulei (*Osteomeles anthyllidifolia*), a woody shrub covering the cliff areas. Kramer has suggested that the remnant colony continues to exist merely because the primary breeding and resting grounds are in excess of 60° slopes and cannot be traversed by dogs or any but the most determined humans.

The Didelphid marsupials of the American Cretaceous had entirely disappeared from North America by the Eocene, and *Didelphis virginiana*, the opossum now so widespread throughout the temperate zones of the United States and western Mexico, is a comparatively recent immigrant from South America. It invaded North America only after the Pliocene union of the two continents by a land bridge. The species was absent from the west coast of the United States until the late nineteenth century. The great plains of southern Canada, along with the semi-desert country of southern New Mexico and north-west Mexico, constituted a natural barrier to the opossum's spread from the east. Its movement up into Canada appears to be controlled by a climatic factor, for its northern limit at 40°N coincides with the $-7°C$ January isotherm. Tyndale-Biscoe (1972) has pointed out that the species is unable to maintain body temperature below $-7°C$; since it cannot conserve energy by hibernation, its survival is limited by the proportion of winter nights on which foraging is possible.

Immigrants entering California from the eastern United States liberated opossums in various counties from about 1870 through to 1915. Grinnell, Dixon and Linsdale (1937) summarised these early introductions and described how most of the opossums were brought to California from Missouri and Tennessee. Spread of the animal throughout the State, both by natural means and with repeated human assistance, was at an amazing rate. Apparently the species found in California extremely favourable conditions for survival and increase, becoming most numerous along the lower warmer levels west of the main mountain divides, in the vicinity of streams where there is a dense cover of deciduous trees and low vegetation. Much of the opossum's success is due to its feeding on a wide range of small animals, both invertebrates and vertebrates, as well as on fruits and grasses.

Although *Didelphis* grows to a size comparable with that of *Trichosurus*, it is not long-lived. However, it is a prolific breeder, so that population turnover is rapid. In most parts of the United States two litters, totalling an average of 12 young, are reared annually. The animals are not territorial and they suffer little predation apart from man. Hall and Kelson (1959) have reported that *Didelphis* has now spread along the North American Pacific coast from southern Canada to northern Mexico, being found everywhere to an altitude of 1500 m. Besides being well established in the states of Washington, Oregon and California, the opossum has also been introduced into the vicinity of Grand Junction, Colorado, and into parts of Arizona.

In Australia the Brush-tailed possum is only one of a number of arboreal phalangerids all competing for similar food and for nesting sites. It is preyed upon by snakes, owls, foxes and dingoes. In contrast, there is in New Zealand no competition with the possum by other arboreal browsers for highly palatable plant species, no shortage of suitable nesting sites exists, and predators, other than man, are virtually absent. These factors have undoubtedly contributed to the animal's markedly successful establishment in this country.

Wallabies on the islands of the Hauraki Gulf are severely restricted by a limited food supply and suitable shelter as well as being subject to heavy predation by man. On the other hand, those in the Rotorua and Waimate Districts seem to be well able to make use of the food and habitat available and their continued spread is only slowed by hunting and poisoning campaigns. As with the possum, there is little competition for food and the Australian predators, dingoes, foxes, eagles, pythons and goannas are all absent from New Zealand.

In the two areas of England where wallabies have managed to establish themselves, the climate does not differ markedly from certain areas in Australia where the species exists, although particularly severe winters tend to induce a high mortality and so limit the rate of expansion. Grzimek (1967) has described how colonies of wallabies established in the Rhineland, Silesia and the Channel Islands during the late nineteenth century survived for varying periods before being hunted out, and Yalden (personal communication) believes that although wallabies get little chance to survive in Europe, they would probably do well if allowed to. Certainly there seems to be no lack of suitable habitat, and availability of food is not a limiting factor. Few predators exist, and it appears that both in Europe and in New Zealand it is only hunting pressure from man that has restricted the establishment and severely limited the spread of marsupials.

Table 10.2 Distribution of Wallabies in New Zealand

Species	Locality Established
Tammar or Silver-grey wallaby (*Macropus eugenii*)	Kawau Island: Rotorua district
Swamp or Black-tailed wallaby (*Wallabia bicolor*)	Kawau Island
Black-striped wallaby (*M. dorsalis*)	Kawau Island
Parma or White-throated wallaby (*M. parma*)	Kawau Island
Brush-tailed rock wallaby (*Petrogale penicillata*)	Kawau, Rangitoto and Motutapu Islands
Brush wallaby (*M. rufogriseus*)	Waimate and Hawea districts

Acknowledgements

Appreciation is expressed to Mr K. H. Miers of the New Zealand Forest Service and to Mr C. C. Taylor for the New Zealand Ministry of Agriculture and Fisheries for the unpublished information they so generously supplied. Sincere gratitude is also due to Dr D. W. Yalden of Manchester University and to Dr A. C. Ziegler of the Bishop Museum, Hawaii, for consultation during the preparation of this paper.

References

Dunnet, G. M. (1956). A live trapping study of the brush-tailed possum. *Trichosurus vulpecula* Kerr. *C.S.I.R.O. Wildlife Res.*, 1, 1 - 18.

Fitter, R. S. R (1959). *The Ark in our Midst*, Collins, London.

Gilmore, D. P. (1965). New Zealand's possum pest. *Animals*, 7 (18), 498 - 501.

Gilmore, D. P. (1967a). Present ways of opossum control held to be wasteful
    failure. *New Zealand J. Agr.*, 115 (4), 50 - 5.

Gilmore, D. P. (1967b). Foods of the Australian opossum (*Trichosurus vulpecula*
    Kerr) on Banks Peninsula Canterbury, and a comparison with other selected
    areas. *New Zealand J. Sci.*, 10 (1), 235 - 79.

Gilmore, D. P. (1968). Wallabies in New Zealand. *Animals*, 11 (2), 62 - 6.

Gilmore, D. P. (1969). Seasonal reproductive periodicity in the male Australian
    brush-tailed possum (*Trichosurus vulpecula* Kerr). *J. Zool., London*, 157
    (1), 75 - 98.

Grinnell, J., Dixon, J. S. and Linsdale, J. M. (1937). *Fur-bearing Mammals of
    California*, University of California Press, Berkeley.

Grzimek, B. (1967). *Four-legged Australians*, Collins, London.

Hall, E. R. and Kelson, K. R (1959). *The Mammals of North America*, Ronald
    Press, New York.

Kean, R. I. (1959). Bionomics of the brush-tailed opossum *Trichosurus vulpecula*
    in New Zealand. *Nature*, 184, 1388 - 9.

Kean, R. I. and Pracy, L. T. (1949). Effects of the Australian opossum (*Trichosuru:
    vulpecula* Kerr) on indigenous vegetation in New Zealand. *Proc. 7th Pacif.
    Sci. Cong. IV Zool.*, 696 - 705.

Kirk, H. B. (1920). Report on Australian opossums in New Zealand. Appendices,
    House of Reps., 1920 H28.

Kramer, R. J. (1971). *Hawaiian Land Mammals*, Tuttle, Honolulu.

Mason, R. (1958). Foods of the Australian opossum (*Trichosurus vulpecula* Kerr)
    in New Zealand indigenous forest in the Orongorongo Valley, Wellington.
    New Zealand J. Sci., 1, 590 - 613.

Owen, W. J. and Thomson, J. A. (1965). Notes on the comparative ecology of the
    common brushtail and mountain possums in Eastern Australia. *Victoria
    Nat.*, 82, 214 - 7.

Poole, A. L. and Adams, N. M. (1963). *Trees and Shrubs of New Zealand*,
    R. E. Owen, Govt. Printer, Wellington.

Pracy, L. T. (1962). Introduction and liberation of the opossum (*Trichosurus
    vulpecula*) into New Zealand. *New Zealand Forest Serv. Inf. Ser.* 45.

Taylor-Page, F. J. (1970). The Sussex Mammal Report 1969, Sussex Natur-
    alists Trust, Henfield.

Tomich, P. Q. (1969). *Mammals in Hawaii*, Bishop Museum Press, Honolulu.

Tyndale-Biscoe, H. (1973). *Life of Marsupials*, Edward Arnold, London.

Wodzicki, K. A. (1950). *Introduced Mammals of New Zealand*, New Zealand
    D.S.I.R. Bull. 98.

Wodzicki, K. A. and Flux, J.E.C. (1971). The parma wallaby and its future.
    Oryx, 11, 40 - 47.

Yalden, D. W. and Hosey, G. R (1971). Feral wallabies in the Peak District.
    J. Zool., London, 165 (4), 513 - 20.

11 *Breeding and behaviour of Australian bandicoots*

Eleanor Stodart*

Introduction

Eleven or twelve species of bandicoots in five genera (Ride, 1970) previously occurred over the whole of Australia. Since the advent of the white man, his domestic animals and the rabbit, those inhabiting the drier regions have been greatly reduced in number and distribution. Some have become extinct, others almost so. Bandicoots inhabiting the coastal regions of eastern Australia are by contrast still common, and can be trapped even in the suburbs of some cities. Despite their wide range of habitats, bandicoots form a fairly uniform group. They are small animals, weighing 1 - 2 kg when mature, and are mainly nocturnal and insectivorous, hiding in a nest or burrow by day. Taxonomically they hold an interesting position among marsupials, having polyprotodont dentition but syndactylous toes; see Kirsch and Calaby (this volume).

Interest in bandicoots has lagged behind that in other marsupials; early attempts to breed them often resulted in fights to the death or destruction of pouch young by mothers (Wood Jones, 1924; Mackerras and Smith, 1960). The last decade, however, has brought more success. I have bred *Perameles nasuta* in enclosures (Stodart, 1966) and Lyne (1974) has established breeding colonies of *P. nasuta* and *Isoodon macrourus* in small cages and outdoor enclosures. Field populations of *P. gunnii* and *Isoodon obesulus* have been studied by Heinsohn (1966) in Tasmania, and Gordon (1971, 1974) has investigated *I. macrourus* in Queensland and New South Wales. Other studies are currently in progress. The first part of this paper, comparing the biology of the four species of bandicoot listed above, is based mainly on my own observations and those of Heinsohn and Gordon. The second part reviews observations of other species of bandicoots on the Australian mainland.

Bandicoots of sub-tropical and temperate coastal regions of eastern Australia

Enclosure studies of P. nasuta

Perameles nasuta, the Long-nosed bandicoot, lives in open grassland and scrub regions of eastern Australia. The animals used in this study were held in two

*Eleanor Stodart began her study of marsupials as part of an honours degree at the University of Sydney. Later she worked on rabbit parasites in the C.S.I.R.O. Division of Wildlife Research, but maintained her interest in marsupials with part-time studies of the behaviour of bettongs and bandicoots. She now works at home with her young family, spending free time writing nature study books for young children.

179

enclosures at Canberra, outside the natural range of the species but in a habitat suitable for them. Constructed originally for rabbit studies, the enclosures were designed to exclude cats and foxes. The smaller one was 0.05 ha in area. Of four males and four females released into it at intervals, only one male and two females survived for more than two weeks; the most held at one time was two males and three females. The larger enclosure, of 0.10 ha, contained two males and four females. Rabbit burrows, with plugged observation holes, were left for the bandicoots to use. Boxes and low shelters of wire netting filled with straw provided alternative cover. Most of the grass was mown for observation, but clumps were left to provide cover. Fresh minced meat, a mixture of milk and eggs and bread, and dried meat crumbs were supplied in summer, when the centre of the enclosure was watered to keep the ground moist. Behavioural observations were made with the aid of a spotlight and binoculars on 104 nights between sunset and 10 p.m.; later on six further nights. The bandicoots were identified by different markings of black dye. Handling was restricted to the times around which births were expected, except for occasional checks on weight and state of health.

Field and enclosure studies of other species

Heinsohn (1966) studied wild populations of the Barred bandicoot *P. gunnii* and *I. obesulus* over two years on farmland on the north coast of Tasmania; his work involved shooting, trapping over 16 months, observations at night with the aid of a spotlight and reflective tags, and observations on animals in enclosures. The two species were sympatric, but *P. gunnii* was more evident in the open areas, *I. obesulus* in areas where cover was more readily available. More data were obtained from *P. gunnii* than from *I. obesulus*.

Gordon (1974) studied the Short-nosed bandicoot, *I. macrourus*, in pastureland and open forest in south-eastern Queensland, by trapping, tracking toe-clipped animals and radiotelemetry. This species used areas where there was cover, but not dense cover. He also (1971) studies *I. macrourus* on the central coast of New South Wales, in open forest on three islands in a lake, and on one mainland site. He trapped the islands and mainland site monthly or bi-monthly, usually observing the animals under laboratory conditions before releasing them.

Nesting, feeding and grooming

All four species made nests of hay, usually simple heaps placed in depressions in the ground, often in tall vegetation, with a neat hollow in the centre about the size of the bandicoot. The nests were extremely well camouflaged and usually lay in tall ground cover, with no visible entrance. Entrance and exit appeared to remain constant, although disturbed bandicoots broke out in any direction when an inter-loper was almost on top of them. Nests of *I. macrourus* found in wet weather had a layer of soil on top, and were quite waterproof. *P. nasuta* and *P. gunnii* and *I. obesulus* used rabbit burrows for shelter when abailable. Captive *P. nasuta* made grass nests in the burrows similar to those on the surface. They did not use the artificial shelters provided. Two female *P. nasuta*, seen nest-making in the evening after feeding, scraped several heaps of hay backwards with the forelegs to form the nest.

P. nasuta, P. gunnii, and *I. obesulus* emerged after sunset but before dark, although *I. obesulus* was occasionally seen in daytime. *P. nasuta* was most active in the first two hours of evening, with short intermittent periods of activity throughout the night. In contrast, wild *I. macrourus* in Queensland, without an artificial supply of food, were active most of the night. *P. gunnii* and *I. obesulus* spent most of their time feeding when out of the nest. All four species fed on grubs or worms, digging in the dirt with their forepaws to make small conical holes just big enough for the snout. Scent is apparently the guide to food.

Analysis of the stomach contents of 26 *P. gunnii* and 11 *I. obesulus* showed that earthworms and adult and larval beetles and other insects were the main food. Blackberries were eaten by *P. gunnii* in late summer when animal food was scarce, and box-thorn berries were eaten by *I. obesulus.* A sample of eight stomachs of *P. nasuta* in Queensland showed that insects were the main food, but grass leaves and a lizard were also ingested, and a sample of 22 stomachs of *I. macrourus* showed both insect and plant material (raspberries, seeds, roots, and sugar cane) as the main foodstuffs (Harrison, 1963).

Although bandicoots normally carry numerous ectoparasites, *P. nasuta* groomed only briefly and infrequently during their periods of activity. *P. gunnii* groomed in spells of up to seven minutes; caged animals groomed for a considerable part of the day, which free animals normally spend hidden in their nests.

None of the four species is gregarious. *P. nasuta* appeared to disregard each other for most of the time, using the whole of the enclosure more or less equally without territorial defence. Where two males were together, one was clearly dominant over the other. In the small enclosure only one male survived more than a fortnight. In the large enclosure the two males lived side by side, apparently ignoring each other, though only one, which was dominant, paid attention to the females when in oestrus. When the subordinate male injured a foot, the dominant one showed aggression, chasing him several times: no fighting was seen, although fur was missing from the rump of one. Two males released into the company of an established male died; they may have been prevented from feeding by continual harassment. The dominant male of the large enclosure and the male in the small enclosure threatened each other through the fence. The threats were accompanied by curling of the lips at the corners of the mouth, occasional baring of the teeth and restlessness. Females appeared to ignore or avoid each other most of the time, although they sometimes nested together. Only once was a female seen to threaten another female. Responses to males were similar, and once a female in the large enclosure chased a subordinate male away from her nest. Males sometimes nested with females.

P. gunnii and *I. obesulus* moved about and fed independently at night, although their distribution tended to be clumped, probably around food supply.

Heinsohn did not see encounters between the two species in the wild; when placed in enclosures together, animals of different species threatened but then ignored one another, whereas aggression within a species continued for longer.

Wood Jones (1924) described in detail the fights of captive *I. obesulus*: 'The aggressor will tirelessly follow his victim until he wears it down. Each time they come to close quarters a curious series of little sounds, half-way between a grunt and a squeak, is emitted . . . . When one animal overtakes the other . . . the assault is made by a jump and an endeavour to strike with the claws of the hind feet. Each stroke carried home, removes some hair from the victim's back and scratches the thin skin . . . as the less aggressive animal tires, the stronger one will attack with a rapid scrambling motion of its fore feet . . . . They seem never to fight face to face.'

Both radiotelemetry and tracking on smoked boards revealed that there was very little contact between individuals of *I. macrourus* in Queensland. Only once were two adult animals — a male and a female — recorded as moving together. The female gave birth 12 days later, making it probable that the association was one of sexual attraction at oestrus. Although eight litters reached late pouch life, only one juvenile was tracked with its mother.

Sexual behaviour and reproduction

Sexual attraction of *P. nasuta* was mostly limited to a few nights close to oestrus, reaching a peak on one night only. Mating was observed eight times, only at night; the full sequence of following, mating and waning of attraction was observed on two nights. The male followed the female persistently for several hours before attempting to mount her. While following he appeared restless and his lips curled as when threatening. As the attraction grew more intense, the male kept closer to the female and often sat behind her, his nose pointed up to her rump. He made no attempt to smell the pouch or urogenital opening. The male then started to place his forelegs on the female's hindquarters. On two occasions the female approached the male, and once a female tried to mount the male. The male mounted the female from behind, pawing at her tail with his front legs and appearing to hold her still by this means. Then the male stood up with his forelegs free and pressed his pelvic region towards the female. The female raised her hind-quarters slightly and then moved forward. Intromission lasted 2 – 4 s. The male was slow and deliberate in his movements.

Although the males mounted the females from July onwards, the first litter appeared in August; the first litter to survive appeared in October. Lyne and Bradford (1974) reported that *P. nasuta* breeds throughout the year in Sydney, with a well-defined peak in late winter, spring and early summer.

Mating of *P. gunnii* was essentially similar to that of *P. nasuta*. Once mating ceased, the animals paid little attention to one another and did not groom their genitals.

Gestation, birth and care of young

Twelve litters of *P. nasuta*, containing a total of 35 bandicoots, were born in the enclosures. Six were still in the pouch at the end of the observation period; of the rest only 12 survived to weaning. At least three litters were ejected from the pouch when the mother was caught, and failed to reattach to the teats. Lyne (personal communication) has avoided ejection of young by using traps to catch animals in enclosures, and a small shaped black bag for handling.

The birth of three litters of *P. nasuta* was timed to within a few hours. Each female was observed to attract males 12 nights before giving birth in daytime, indicating a gestation period of 12½ days. The most accurately measured gestation

period was 12 days 14 hours; mating was observed to finish at 1 : 25 a.m. and
birth took place 12 days later at about 3 : 30 p.m. in a bag in the laboratory.
These figures agree with others previously published (Hughes, 1962; see below).
When this female was examined less than ten minutes after the birth of her litter,
the young were already firmly attached to the teats and the allantoic stalk still
anchored the young to the mother so that their hindquarters were pulled towards
the urogenital opening (Figure 11.1). Thus the birth of the three young, their
passage to the pouch and attachment to the teats must have taken less than ten
minutes. The female made no attempt to clean the young.

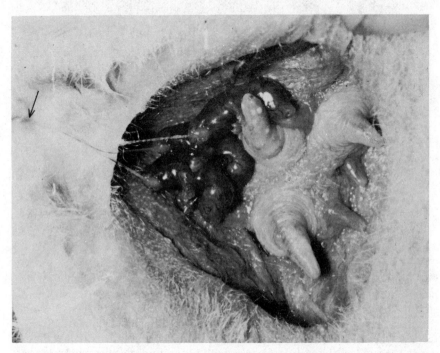

Figure 11.1 Newborn young (40 min old) of *Perameles nasuta* with allantoic stalks anchoring
them to the urogenital opening (arrow). The young are all attached to small teats. Two spare
teats of the same size, and three large ones suckled by the previous litter are visible. Reproduced from Stodart, 1966.

Lyne (1974) observed the birth of *I. macrourus* (Figures 11.2 - 11.4). The
female urinated 23 min before the birth but was otherwise inactive. Immediately
before giving birth she lay on her side with one hind-leg raised, steadying herself
with her front feet, and with her head twisted so that she was facing her rear end.
Lyne reported that vigorous licking of the urogenital opening preceded the sudden
emergence of two young, and the first was born about two minutes before the
second. They appeared to be born free of membranes but they had intact allantoic stalks. The mother did not lick a path for the young but licked around them
as they crawled towards the pouch. In this instance the gestation period was
between 12 days 8 hours and 12 days 11 hours.

Figure 11.2 *Isoodon macrourus* in birth position. The expectant mother is on her left side with right hind-leg raised. The pouch, seen below the muzzle, opens backwards. Photograph taken immediately before birth of two young. Reproduced from Lyne (1974).

Pouch young of *P. nasuta* were carried for 50 – 54 days, when the bulge of the pouch could easily be seen and the animal moved more slowly than usual. Thereafter the pouch was often empty at night, the young remaining in the nest. At this stage they were covered in short, yellow-brown fur, and ran aimlessly from the nest when disturbed. After 62 – 63 days, when the next litter was due, the young appeared on the surface, foraging with the mother, gleaning food from her diggings and digging for themselves. They ran straight for cover when disturbed. Their pelage by this time was similar to that of the adult. Other adults in the enclosures normally avoided newly emerged litters. Two litters born a few days apart are believed to have changed pouches before they were 52 – 60 days old; these young followed both females indiscriminately.

Breeding records of five female *P. nasuta* (summarised diagrammatically in Stodart, 1966) show that the time from one oestrus, through pregnancy and lactation to the next oestrus, was 62 – 63 days. The females mated when the previous litter was 49 – 50 days old and still in the pouch. Records of oestrus without pregnancy or lactation were obtained in one female where injections of pregnant mare serum and human chorionic gonadotrophin induced the first oestrus. Successive nights of attraction were 28, 13, 14, and 15 days apart, and after the last one the female became pregnant. These figures suggest that the oestrous cycle may be a little shorter than the 17 – 34 days of Hughes (1962). One

Figure 11.3   Two young *I. macrourus* with intact allantoic stalks leading back into the urogenital opening (arrow). Reproduced from Lyne (1974).

Figure 11.4   Female *I. macrourus* disturbed while giving birth, showing how small the young is relative to the mother, and the importance during disturbance of anchoring by the allantoic stalk. Reproduced from Lyne (1974).

other record of successive oestrus without pregnancy or lactation was 35 - 40 days. When pouch young of two litters were removed at 28 - 31 days, the females became attractive 6 and 7 days later, 11 - 15 days earlier than if lactation had lasted the full period.

Males of both *P. gunnii* and *I. obesulus* were capable of breeding throughout the year, but females had a non-breeding season in autumn; this appeared to be longer in *I. obesulus*, although fewer were caught. Both species had litters of one to four young; both had eight teats, allowing one litter to succeed another immediately without waiting for the used, enlarged teats to revert to normal size.

Gordon's studies indicate that *I. macrourus* breeds throughout the year in southern Queensland, with a mean litter size of 3.4, and up to seven young in large litters. In New South Wales females begin breeding between mid-July and early September, and stop lactating from late February to late April. In the four years of Gordon's study breeding appeared to be related to food supply, occurring each year just after the females increased in weight. In two out of three years when the end of the season was studied, females were found to have lost weight; the third year was a year of drought, and the breeding season was short.

Early litters ranged from 2.6 to 2.9 (mean of two seasons), mid-season ones between 3.6 and 3.8, and late litters between 2.1 and 2.6. Survival of young was greatest for the earliest litters, possibly because these young were weaned at the height of the breeding season when the supply of food was presumably at its best, and because the earliest young were first to reach maturity. Young which established a home range in the area did so adjacent to or apart from the mother's home range; few overlapped in range with the mother.

Heinsohn made no attempt to estimate the length of gestation and lactation of *P. gunnii*, but his figures for live-trapped animals reveal that the most common interval between births was about 60 days. Young began to leave the pouch at 48 - 53 days, emerged from the nest at 58 - 59 days, and followed their mother until 71 - 73 days old. Suckling stopped at 59 - 61 days, about the time that the next litter was due. Thus reproduction appears to be very similar to that of *P. nasuta*, except that the young are independent a little earlier and the succession of litters can be a little quicker.

In southern Queensland six litters of *I. macrourus* left the pouch at 50 ± 7 days. One female produced six litters in 13 months, averaging a litter every 65 days. Data from other females showed a similar rate of turnover. Of 26 young known to survive pouch life, only three reached maturity within the study area; others were either killed or moved to new areas. In New South Wales the succession of litters was slightly quicker; the most frequent interval was 58 days, with most taking from 56 to 62 days.

Female *P. gunnii* became sexually mature from three months onward, and males were sexually mature at four to six months old. As far as could be determined, the figures for *I. obesulus* were similar. Female *P. nasuta* can become sexually mature at four months and males at five months (Lyne, 1964). Female *I. macrourus* in New South Wales start to breed when about 100 days old.

Home ranges

In Heinsohn's study 11 females and 14 males were caught more than ten times in a period of 13 months. He estimated the size of home ranges by plotting the locations where each animal was trapped, joining the outermost points to form a convex polygon, and calculating the area of the polygon. Ranges for bandicoots captured more than ten times are plotted in Figure 11.5. Females appeared to have a limited range of 1.01 – 4.05 ha (with one exception of 11.34 ha), while males ranged over areas of up to 40 ha. Home ranges, particularly those of the males, overlapped extensively. Estimated home range of females did not increase much with more than five captures, but those for males did, suggesting that figures for males are probably low. Only five specimens of *I. obesulus* were caught more than six times. Four males had home ranges of 4.05 – 6.48 ha and one female had a home range of 2.31 ha.

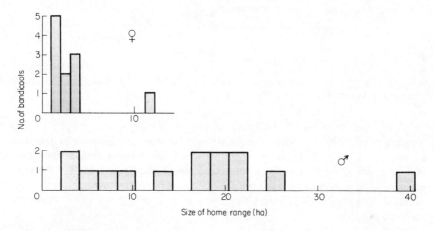

Figure 11.5   Size of home ranges of *Perameles gunnii* caught 10 or more times in Tasmania (from data by Heinsohn, 1966).

By tracking with radiotelemetry Gordon found that seven females of *I. macrourus* in Queensland had home ranges between 0.9 and 2.1 ha (a mean of 1.9 ha) and four males ranged over 1.7 – 5.2 ha (mean of 2.8 ha). Each bandicoot had more than one nest and a core area of the home range which moved gradually with time. Males concentrated their activity in one area but toured most of their range, usually in the early part of the night. Thus range-patrolling seemed to account for their larger home ranges. The home ranges of neighbouring animals overlapped, but their core areas were discrete.

Male *P. nasuta* may also have a large home range, making an enclosure as large as 0.05 ha too small for more than one animal. In the suburbs of Sydney *P. nasuta* tends to feed in one garden for a few nights and then disappear for a while, which suggests that they are wide - ranging.

Survival

Survival curves of young *I. macrourus* born in three succeeding seasons in New South Wales (Figure 11.6) indicate that survival was as low in the pouch as at any other time of the life cycle. No direct information was obtained at the time of

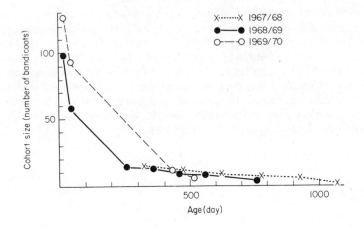

Figure 11.6   Survival curves of *Isoodon macrourus* in New South Wales from three annual cohorts of young. Reproduced from Gordon, 1971.

weaning or immediately afterwards. Although the shape of the curve is uncertain for this critical period, death rate appears to have fallen considerably after 250 – 300 days. All of the first season's young had died by 1070 days and the second season's by 855 days of age. The graphs exclude animals lost owing to trapping or handling. Lyne (1964) found no evidence of increasing mortality of *P. nasuta* up to 50 days.

Conclusions

These studies, based on four different species and emphasising different aspects of ecology, indicate many common features in the lives of bandicoots of sub-tropical and temperate coastal regions. They appear to be mainly solitary animals, males ranging widely over areas inhabited by several females, but paying no attention to them except during their short periods of oestrus. The young develop quickly, foraging with the mother for only a few nights after weaning before assuming their solitary life-pattern. Few survive to maturity, and the life expectancy of those which do (in *I. macrourus* at least) is 2½ – 3 years.

The main differences to emerge are that both species of *Perameles* prefer open areas, *Isoodon* keeping to covered ground where growth is not too dense for easy movement; *Perameles*, particularly male, have larger home ranges than *Isoodon*, perhaps arising from their preference for open terrain. Another difference is that, in response to heat stress, *Perameles* pants deeply, while *Isoodon* depends more on licking to increase heat loss (Hulbert and Dawson, 1974).

The ecology of other species of bandicoot

Very little is known about the Spiny-haired bandicoot, *Echymipera rufescens* (Peters and Doria), which lives in rain-forest areas of eastern Australia. Recently rediscovered by Hulbert, Gordon and Dawson (1971), this species appears to be reasonably common in a limited area round the Nesbit River and McIlraith Range in northern Queensland. Numerous conical diggings, similar to those of other bandicoots, were seen in its habitat, especially near creeks. The hair of this species, though much spinier and less dense than that of other bandicoots, has the same insulative properties as the summer hair of temperate zone species (Hulbert and Dawson, 1974).

Rather more is known of bandicoots in the arid zones of central and western Australia. Removal of saltbush and grass tussocks by grazing cattle has deprived native species of shelter and nest sites (Newsome, 1971). The Desert bandicoot, *Perameles eremiana* Spencer, is now extinct where cattle graze in Western Australia, but still known to aborigines in the unstocked desert. Wood Jones (1924) recorded that its habits were similar to those of *P. myosura* (*P. bougainville* Quoy and Gaimard), a Barred bandicoot, in that it nests in hollows scooped out in sand and produces two young at a time. *P. bougainville* now seems to be common only on Bernier and Dorre Islands of Western Australia (Ride and Tyndale-Biscoe, 1962), where they were reported to be active at night but occasionally seen in daytime if disturbed from their nests. It breeds in May and June, and is very aggressive in captivity.

The Pig-footed bandicoot, *Chaeropus ecaudatus* (Ogilby), is probably extinct today and little is known of its behaviour. Its nest was similar to that of other bandicoots previously discussed, and the species was not entirely nocturnal (Wood Jones, 1924).

Best-known of the arid zone species is the Rabbit-eared bandicoot or Bilby, *Macrotis lagotis* (Reid). This species avoids high temperatures and water loss by living in deep burrows, which Wood Jones (1924) describes as 'descending with a fairly steep, but ever opening, spiral to a depth of five feet [1.5 m] or more'. Smyth and Philpott (1968) depict two burrows in central eastern Western Australia, each over 1.5 m deep, with one or more offshoots branching off in the main burrow. Newsome (personal communication) found that the burrows about 30 miles north of Alice Springs, Northern Territory, were not so deep. Again they had one entrance, but the blind end rose to within 15 cm of the surface, so that the bandicoot could easily dig its way out. Several small holes branched off the main burrow. Of all bandicoots tested under heat stress, this was the only species unable to maintain a stable body temperature at 40°C (Hulbert and Dawson, 1974), although no other species of the arid zone, which mainly nest at the surface, were tested for comparison.

Faeces of *M. lagotis* analysed by Smyth and Philpott (1968) contained more than 50 per cent by volume of soil and grit. The hard heads and mandibles of soldier and worker termites of two species, *Hamitermes* (*Drepanotermes*) *rubriceps* (Froggat) and *Eutermes tumuli* Froggat, and of workers of two species of ant, *Camponotus* sp. and *Iridomyrmex* sp., were present in abundance. The termites apparently formed the main diet. Bilbies seem not to break into termite mounds, but their characteristic scratchings in the soil often intersect the underground passages of the termites. They occur in the mulga (*Acacia aneuria*) and spinifex (*Triodia* sp.) ecosystem. In Central Australia they are sparse; Newsome (1962) found only four localities where bilbies occurred in an area 80 km long and 50 km wide. Smyth and Philpott (1968) also record them as living in small isolated groups over a large area near Warburton Mission, Western Australia.

Newsome (personal communication) observed the movements of one bilby in an area 50 km north of Alice Springs, Northern Territory. Where rabbits burrowed only in sandy soil, the bilby burrowed in soil of clayey sand where there was a dense cover of mulga; it had dug five deep burrows which it used for sleeping quarters, and 12 short burrows, presumably for escape. It also made occasional use of the rabbit burrows. Seventy-five per cent of its feeding holes were dug in the mulga zone; 85 per cent were within 30 m of a burrow or warren. Smyth and Philpott (1968) recorded that all soil scratchings were within 180 m of the burrows.

In December 1960 the home range occupied about 10 ha. Following two months of high temperatures and no rain, the size of the home range was suddenly extended to about 14 ha, possibly in response to food shortage. As the home range grew, parts of it were little used and may eventually have been abandoned.

M. lagotis may have a lower reproductive rate than other bandicoots, breeding only in March or May (Wood Jones, 1924; Hulbert, 1972) or opportunely with rains and food supply, with litters of only one or two young. In growth and development of pouch young it is similar to *P. nasuta* and *I. macrourus* (Hulbert, 1972). Litters were first seen out of the pouch at 67 – 68 days and the young left the pouch altogether 4 – 5 days later, as in other bandicoots.

Numbers and distribution of bilbies have declined during this century. According to Newsome (1971), cattle grazing on mulga communities at first created open areas which favoured termites, and their colonies increased enormously; together with the stock, they kept the land denuded. As the large colonies of termites died, out, food for bilbies was reduced, so that they are now very rare in cattle country but quite common in unstocked desert.

Acknowledgements

I thank Dr A. G. Lyne for his photographs and Dr A. E. Newsome who gave additional data; both also read the manuscript. Dr G. Gordon kindly allowed me to reproduce a figure from his Ph.D. thesis, and I thank Miss B. Staples for help in the library of the Division of Wildlife Research, C.S.I.R.O.

References

Gordon, G. (1971). A study of island populations of the short nosed bandicoot, *Isoodon macrourus* Gould. Thesis submitted for the degree of Doctor of Philosophy, University of New South Wales.

Gordon, G. (1974). Movements and activity of the shortnosed bandicoot *Isoodon macrourus* Gould (Marsupialia). *Mammalia*, 38, 405 – 31.

Harrison, J. L. (1963). The food of some Innisfail mammals. *Proc. R. Soc. Queensland*, 73, 37 – 43.

Heinsohn, G. E. (1966). Ecology and reproduction of the Tasmanian bandicoots (*Perameles gunnii* and *Isoodon obesulus*).*Univ. Calif. Publ. Zool.*, 80.

Hughes, R. L. (1962). Role of the corpus luteum in marsupial reproduction.*Nature*, 194, 890 – 1.

Hulbert, A. J. (1972). Growth and development of pouch young in the rabbit-eared bandicoot, *Macrotis lagotis* (Peramelidae). *Aust. Mammal.* 1, 38 - 9.

Hulbert, A. J. and Dawson, T. J. (1974). Thermoregulation in Perameloid marsupials from different environments. *Comp. Biochem. Physiol.*, 47A, 591 - 616.

Hulbert, A. J., Gordon, G. and Dawson, T. J. (1971). Rediscovery of the Marsupial *Echymipera rufescens* in Australia. *Nature*, 231, 330 - 1.

Lyne, A. G. (1964). Observations on the breeding and growth of the marsupial *Perameles nasuta* Geoffroy, with notes on other bandicoots. *Aust. J. Zool.*, 12, 322 - 39.

Lyne, A. G. (1974). Gestation period and birth in the marsupial *Isoodon macrourus. Aust. J. Zool.*, 22, 303 - 9.

Lyne, A. G. and Bradford, H. M. (1974). Some observations on reproduction in bandicoots. *Aust. Mammal.*, 1, 293 - 4.

Mackerras, M. J. and Smith, R. H. (1960). Breeding the short-nosed bandicoot, *Isoodon macrourus* (Gould), in captivity. *Aust. J. Zool.*, 8, 371 - 82.

Newsome, A. E. (1962). Rabbit-eared bandicoots or bilbies. *Aust. Nat. Hist.*, 14, 97 - 8.

Newsome, A. E. (1971). Competition between wildlife and domestic livestock. *Aust. Vet. J.*, 47, 577 - 86.

Ride, W. D. L. (1970). *A Guide to the Native Mammals of Australia*, Oxford University Press, Melbourne.

Ride, W. D. L. and Tyndale-Biscoe, C. H. (1962). Mammals. In: The results of an Expedition to Bernier and Dorre Islands, Shark Bay, Western Australia in July 1959 (ed. A. J. Fraser), *Faun. Bull. Fish. Dep. West. Aust.*, 2, 54 - 97.

Smyth, D. R. and Philpott, C. M. (1968). Field notes on rabbit bandicoots, *Macrotis lagotis* Reid (Marsupialia), from central western Australia. *Trans. Roy. Soc. S. Australia*, 92, 3 - 14.

Stodart, E. (1966). Management and behaviour of breeding groups of the marsupial *Perameles nasuta* Geoffroy in captivity. *Aust. J. Zool.*, 14, 611 - 23.

Wood Jones, F. (1924). *The Mammals of South Australia*, Part II: *The Bandicoots and Herbivorous Marsupials*, Govt. Printer, Adelaide.

12 *Learning and problem - solving behaviour in marsupials*

Robert J. Kirkby*

Marsupials have been studied extensively by biologists (see, for example, Waring, Moir and Tyndale-Biscoe, 1966), offering unique characteristics as research subjects (Burns, 1956; Block, 1960; Martin, 1965; Mizell, 1968; Twyner and Allison, 1970). Despite equivalent advantages for behavioural research (Wiedorn, 1954; Herring *et al.*, 1966; Kirkby, 1969a) and an obvious potential for study, behavioural scientists have made little use of marsupials as research subjects. This has been particularly obvious in Australia (O'Neil, 1966), the country best provided with subject material for research into marsupial behaviour. It seems inappropriate that most experimental studies of marsupial behaviour have taken place in North America, the home of only one marsupial, *Didelphis virginiana* (Kirkby, 1969a). Of the relatively little work published on the psychological processes of marsupials, most is concerned with learning and problem-solving. This contribution reviews 19 research papers in the field of marsupial behaviour, grouping them according to the study techniques used.

Discrimination learning

The technique most consistently used to study learning in marsupialia has been discrimination learning, involving either spatial or non-spatial cues.

In a spatial discrimination task the subject is presented with two clearly separated alternatives, such as a pair of levers or, more usually, the opposite arms of a T-shaped maze. One of these alternatives, either the right or the left, is selected as the correct goal, and responses by the subject to the appropriate lever or maze alley are rewarded. In short, the positive goal is always in the same spatial position. The non-spatial discrimination task may also require the subject to respond to a lever or enter the arm of a maze. However, in this case the animal must respond not to a constant position but to where a specific stimulus is situated from trial to trial. This stimulus may be the presence or absence of light, a particular object or some other perceptual cue. Between successive trials, the stimulus may be moved randomly from lever to lever, or alley to alley. Animals generally learn spatial problems more easily than non-spatial problems.

*Dr Kirkby is Head of the Department of Behavioural Sciences, Lincoln Institute, Carlton, Victoria, Australia. He studied in the Universities of Oregon and New South Wales, and has published work on animal behaviour, psychology and physiology.

Spatial discrimination

Spatial discrimination studies with marsupials have been associated with 'learning set' experiments. The investigation of learning set through the spatial discrimination technique typically involves two discriminanda. Responses to one alternative are reinforced until a specific criterion of learning is achieved: for example, 18 correct responses in 20 trials. After this, responses to the original stimulus are no longer rewarded and the alternative is reinforced. Typically several such reversals are carried out. A significant decrease in errors over the successive reversal is taken to reflect the subject's ability to 'learn to learn'. Learning set has been used extensively to compare the learning abilities of different species (Warren, 1965a; Gossette, Kraus and Speiss, 1968).

The earliest investigations of learning set in marsupials were carried out by Friedman and Marshall (1965), who investigated the learning of successive spatial discriminations by the opossum *Didelphis virginiana*. In the first part of the experiment five adult (two males and three females) opossums were trained to enter either the right or left goal box of a two-compartment apparatus. Each animal was given four trials per day, and entry into the correct compartment was rewarded with 12 g of horsemeat. A non-correction technique was used: if the subject entered the incorrect box, the door was closed and the animal was not able to retrace its steps to enter the correct compartment. In this way the subject could not receive a reward until the next trial. When a criterion of 11 correct out of 12 successive trials was achieved, the problem was reversed – that is, the opossums originally rewarded to go right were now rewarded to go left, and vice versa. Under this procedure, which was carried out from 0 to 15 reversals, the opossums showed no consistent improvement over successive reversals in errors. Other species of mammals have consistently shown a decrease in errors over successive reversals. In the second part of the experiment, reversals 16 - 25, each daily session was run with as many trials as necessary for the subject to achieve criterion. Under this procedure the error rate dropped sharply, which indicated the acquisition of a 'learning set'. This suggests that to establish a learning set opossums must be given 'massed' rather than 'spaced' trials. The overall results indicated that the opossums learned spatial discrimination and reversal at a rate slower than that previously reported for rats. This difference between marsupials and placental mammals was not supported by later studies by Buchmann and Grecian (1974) and Williams and Kirkby (1976) with Australian marsupials.

Buchmann and Grecian (1974) tested the Brown bandicoot (*Isoodon obesulus*) on a spatial discrimination task. Three subjects (sex not stated) were trained initially to depress a pair of treadles to gain a reward of mealworms (larvae of *Tenebrio molitor*). The quantity of the reinforcement was not given. All subjects learned to operate the treadle within 60 trials spaced over three test sessions. Testing took place during the evening in a darkened room illuminated with red light. A non-correction procedure was used and the subjects were trained to a criterion of eight or fewer errors on two consecutive blocks (40 trials); however, even when criterion was reached within a single session, a further 100 trials were given.

The three subjects were trained to 8, 13 and 14 reversals, respectively. All subjects showed consistent 'one-trial' learning, defined as an error on the first trial of the initial session followed by 9 correct responses. The authors reported that the performance of the bandicoots was comparable with that of laboratory rats (*Rattus norvegicus*) tested under similar conditions. It is noteworthy that at the time of testing the bandicoots were 4 – 7 months old. *I. obesulus* are not sexually mature until about 4 – 6 months old. Experimental findings have indicated that mammals appear to achieve peak learning ability sometime after peak sexual maturity. This suggests that, in tests carried out at a later time of development, learning ability of the subjects may have been improved.

Williams and Kirkby (1976) carried out a similar study. They tested the Australian Brush-tailed marsupial possum (*Trichosurus vulpecula*) on discrimination learning. The subjects, 10 mature possums (three females and seven males), were habituated to the laboratory setting and were adapted to a reversed light cycle. In the part of the experiment concerned with spatial learning four possums (three females, one male) were trained on a right versus left spatial discrimination. The apparatus was a box-like enclosure with a pair of food-wells at one end. The food-wells were covered with small blocks. Musk solution, made by crushing musk-flavoured candy in water, was used as a reward.

In the first part of the experiment the subjects were trained to push away the blocks and drink the musk solution from the food-wells. Within 3 – 5 days the possums had learned to dislodge the blocks and the experiment proper was commenced. The subjects were given 20 trials daily using a non-correction procedure with a criterion set at 18 correct trials from any daily block of 20. Over a series of 19 reversals all subjects showed rapid reduction in errors and trials to criterion. 'One-trial' learning – that is, achievement of criterion after one or no error – was demonstrated consistently by three of the subjects. The performance of the marsupials was compared with that of several eutherians previously tested on the same type of spatial reversal task. It was found that the possums learned the problem about as rapidly as the carnivores Skunk (*Mephitis mephitis*) and Coatimundi (*Nasua narica*), and the primate Squirrel monkey.

To summarise: Three investigations of spatial discrimination learning in marsupials have involved learning set performance. It appears that *D. virginiana* develops a learning set only under specific conditions. *I. obesulus* and *T. vulpecula* rapidly acquire a learning set; *Trichosurus* at least appear to be as competent as some eutherian mammals.

Visual discrimination

There appear to have been seven published investigations of visual discrimination in marsupials. Although the range of marsupials represented in these studies was limited (most involving only *D. virginiana*), a wide variety of discriminanda was used, including light versus dark, colours, vertically and horizontally oriented lines, two-dimensional shapes and three-dimensional objects.

In the earliest of these investigations James (1960) tested one opossum (*D. virginiana*) on three visual discrimination problems. The test apparatus had two reward chambers which could be entered by doors on which were fixed 28 × 18 cm stimulus cards. Meat was placed in both reward chambers but only the door with the correct stimulus card could be opened. Thus a correction procedure was used. The correct card was randomly alternated from door to door. The subject was tested in the evening. James followed the unusual practice of giving the subject only one trial per night and allowing it to remain in the goal box. It is assumed that, apart from the meat (an unspecified amount) left in the goal box, the subject was not fed. Successes were tallied on the basis of blocks of 10 trials. The first problem required the opossum to discriminate between a black and white vertically striped card (+) versus a white card (−). A criterion of 5 successive correct responses out of a 10 trial block was reached within 60 trials. On the second problem, discriminating a black and white horizontally striped card (+) from a black card (−), the criterion was reached within 20 trials. The third problem, black card with small white triangle (+) versus white card with small black triangle (−), was learned in 30 trials. This report is liable to several criticisms. Only one subject was tested, and James failed to provide details of the sex, age and experimental experience of the subject. However, the study does indicate that *Didelphis* is able to learn visual discriminations with relative ease. This is particularly interesting in the light of reports by Buchmann and Grecian (1974) and Williams and Kirkby (1976) indicating the difficulties some Australasian marsupials have with this type of problem.

James and McFarland (1966) studied the learning of form discrimination in three opossums, *D. virginiana*. The ages of the subjects, two females and one male, were not given, although one of the females was described in the text as 'mature'. The subjects were tested in an apparatus with two levers. The stimuli were three-dimensional shapes which could be attached to the base of the levers. Nosing or otherwise moving the correct stimulus or lever would cause meat (quantity not stated) to drop into a pan situated between the levers. The position of the correct stimulus was moved randomly from side to side.

The authors state that testing took place at feeding time, but do not specify how long the animal was deprived of food prior to the test period; nor is it clearly indicated whether a correction or non-correction technique was used. The mature female opossum was tested first. After a day of pretraining, the experiment proper was commenced with the subject receiving 10 trials daily for 6 days, on a black versus white square. This problem was learned to a criterion of 9 out of 10 correct responses on the second day of training. The subject was trained on a further 6 problems to a criterion of 6 out of 10 correct responses. On one discrimination (white square versus white triangle) the subject failed to reach criterion within 80 trials; the other discriminations were acquired within 70 or fewer trials, the most rapidly learned being large black disc versus small black disc (20 trials). The other two opossums were trained on the black versus white squares. The female required 50 trials to learn this discrimination; the male required 170. Three other problems using modified forms of the black – white stimuli were presented. On all problems the female opossum required fewer trials to achieve criterion than the male. A similar sex difference in learning a discrimination problem was reported for the Australian possum, *Trichosurus vulpecula*, by Williams and Kirkby (1976). However, in their experiments it was the males who achieved the better performance.

Friedman (1967) trained two adult opossums (*D. virginiana*) to distinguish between red, blue, green and yellow hues, equated for brightness. The subjects, both females, had been raised and tamed in the laboratory. Although the author reported that the opossums were rewarded with a pellet of dog food for choosing the correct one of a pair of colours, it is not stated how, or for what length of time, the animals were deprived of food prior to the testing sessions. The reinforced colour was changed from side to side randomly. If the subject made an error, the trial was repeated with the stimuli in the same place until the correct response was made. Thus each opossum received 20 rewarded responses and a variable number of unrewarded responses per day. Both subjects learned to discriminate every colour from each of the others. Using a criterion of 35 out of 40 correct responses in two successive days of 20 trials, the opossums learned each of the 6 possible combinations in a mean of 11.2 days of training with a range of 3 – 15 days (first opossum) and 2 – 65 days (second opossum). A transfer of training effect was clearly demonstrated by the second opossum, which, after learning the blue (+) versus red (−) problem showed 18 out of 20 correct responses on the first day of testing of the blue (+) versus yellow (−) problem. Evidence for colour vision has also been reported for another marsupial, the Red kangaroo, *Macropus rufus* (Neumann, 1961).

Only two studies of learning behaviour in macropods appear to have been published, although there are some additional experiments with these species that have not reached publication (Jeeves, reported in Munn, 1964; Livesey and Di Lollo, 1969). In the first of these studies Neumann (1961) tested an adult female Red kangaroo on the multiple discrimination tasks developed by Rensch and Altevogt. The kangaroo was trained to push open the lids of boxes to gain access to tiger paw or cabbage leaves placed inside. Stimuli were painted on the lids of boxes and these were presented to the kangaroos in pairs. Eight such pairs of stimuli were presented to the kangaroos. The first problem, yellow versus blue, was learned to a criterion of 78 per cent correct choices (from at least 30 trials on three successive days) within 109 trials. The fifth problem took 252 trials. The remaining problems were learned within these limits. When all the problems were presented successively, the kangaroo could identify seven of the eight correct stimuli. A comparison with the results of other studies suggest that this individual, as measured by this task, was superior in learning ability to the domestic fowl and the domestic mouse (six pairs of stimuli) but inferior to the domestic rat (eight pairs) and the African elephant, *Loxodonta africana* (20 pairs).

Neumann also tested an adult male opossum (*D. virginiana*). Training was conducted between 10 p.m. and midnight. Three pairs of stimuli were presented. The first problem, black versus white, was learned by the opossum in 128 trials (as compared with the kangaroo in 179 trials, and the African elephant in 40 trials). The opossum was able to learn the second task, a triangle versus horizontal screen (223 trials), but not the third discrimination, white spot versus chessboard problem. Although the opossum learned to discriminate between two pairs of stimuli when each pair was presented quite separately, it failed when the two pairs were presented successively. Thus it appears that in learning multiple discriminations the ability of the opossum was well below that of the kangaroo. As the opossum died shortly after completion of the series, it was perhaps ill at the time of testing and not performing at its usual level. Other findings of this investigation were that at a distance of 30 cm the visual field angle of the kangaroo was 22° 54′, and the kangaroo appeared to discriminate colours better than the elephant.

Neumann reported that testing of the kangaroo took place during daylight hours in the open, and apparently in the presence of zoo visitors, who would surely be a serious distraction to the learning process. Munn (1964), under strict experimental conditions, studied the ability of three kangaroos, two males and one female, to learn to discriminate between vertically and horizontally oriented stimuli. The species was not stated but the animals were probably *Macropus giganteus* or *M. rufus*. Particular care was taken to calm the kangaroos and adapt them to the apparatus before training. The subjects were trained to push open large doors to gain access to a reward of bread and vegetables. To the outside of the doors were attached large black and white vertical or horizontal stripes. Initially the kangaroos were rewarded for entering the feeding compartment through the door with the horizontal stripes. A correction procedure was used, so that if the subject approached the incorrect stimulus and found that door locked, it could move over to the other door and so gain entrance to the food chamber. The initial discrimination and the first reversal were acquired at about the rate typical for rats. However, over successive reversals (up to ten for one of the subjects) little improvement occurred. These findings suggested that, unlike the monkey, cat or rat, the kangaroo does not develop a learning set, at least in the visual discrimination situation. At the start of the experiment, which took six months to complete, the female kangaroo was approximately eight months old, and the males were approximately one year old. Neither *M. giganteus* nor *M. rufus* is sexually mature at 14 – 18 months, and usually not before 24 months of age (Sharman, personal communication, 1969). Animals typically achieve maximum intellectual ability sometime after sexual maturation is reached. Thus the use of adult animals in this study might have resulted in quite different findings. Although the female was the youngest of the kangaroos, she performed consistently better in the learning situation, averaging only 63 errors per reversal; male subjects scored a mean of 88.9 errors per reversal.

In Buchmann and Grecian's (1974) investigation of the Brown bandicoot (*Isoodon obesulus*) mentioned earlier, three bandicoots (sex not stated) were trained to operate treadles to which were attached black or white cards. They were rewarded for choosing either stimulus, the cards being alternated randomly from treadle to treadle between trials. The performance of the animals tested on the visual reversal learning was markedly inferior to those tested on the spatial reversal learning. On the successive spatial discriminations all three subjects had shown improvement indicative of a learning set. In the visual discriminations only one subject, tested over 19 reversals, showed a gradual improvement. The other two animals showed no improvement, although tested for up to 30 reversals. Similar differences in spatial and visual discrimination learning abilities were found by Williams and Kirkby (1976).

The testing procedure used by these authors to investigate spatial learning has been described earlier. In the second part of their study they trained four Brush-tailed possums (*Trichosurus vulpecula*), one male and three females, on a light versus dark visual discrimination and reversal task. Only two subjects learned the original discrimination on the schedule of 20 trials per day. One of the remaining two subjects learned the original discrimination after the schedule was followed for 500 trials, then an additional 140 trials were given at the rate of 40 per day. The remaining possum did not achieve criterion after 500 trials at 20 per day and an additional 200 at 40 per day, when training ended. Similarly, for the two possums started on the first reversal of the visual discrimination, training was terminated after 200 and 500 trials, respectively, when neither animal had reached criterion. Later two other Brush-tailed possums, one male and one female, were trained on the same visual discrimination task. In this part of the study a correction procedure was used; after an incorrect response the subject was allowed to correct its error, approach the positive goal and receive a reward. Training for one animal had to be terminated after a week of testing (140 trials). The other subject achieved criterion on the original problem by the nineteenth day of training (360 trials). A calculation of the correct responses in the first 140 trials indicated that the two animals on the correction procedure performed significantly better than the four subjects tested under conditions of non-correction. Although it appeared that visual discrimination learning might be facilitated by the use of correction techniques, the performance of the possums on visual discrimination was clearly inferior to their performance on problems requiring spatial learning. Williams and Kirkby (1976) suggested that a reason for this difference could lie in the 'preparedness' of these animals to learn visual versus spatial problems (Seligman, 1970). The possum is primarily a nocturnal herbivore which probably does not rely on keen eyesight, its food, mostly vegetable, is gathered at night, does not have to be stalked or hunted, and is probably detected by spatial and olfactory cues rather than vision.

To summarise: Visual discrimination experiments have shown that the North American opossum (*D. virginiana*) can learn to discriminate black versus white (Neumann, 1961), different hues (Friedman, 1967), patterns (James, 1960; Neumann, 1961) and geometric forms (James and McFarland, 1966). Among Australasian marsupials it appears that *T. vulpecula* and *I. obesulus* can learn to discriminate between black and white. Additionally, Neumann (1961) reported that a kangaroo (*M. rufus*) was able to learn several pattern and hue discriminations, and Munn (1964) found that kangaroos (species not given) learned to discriminate between vertical- and horizontal-oriented visual stimuli.

Olfactory discriminations

Most investigations of discrimination learning in marsupials have involved spatial or visual stimuli. A different approach was taken by Tilley, Doolittle and Mason (1966), who investigated the ability of four adult opossums (*D. virginiana*) to discriminate between the presence or absence of an odour. A T-maze was used. In each maze alley were placed a glass dish (close to the choice point) and a glass plate (at the end of the alley). Two drops of oil of wintergreen were placed in one of the dishes and one drop on the plate in the same alley. The dish and plate containing the stimulus were randomly moved from alley to alley. The maze was fanned out between trials and the experimental room was continuously ventilated. The subjects were trained to enter the alley with the olfactory stimulus. Entry to the correct room was rewarded with candy placed on the glass plate. A non-correction procedure was used so that opossums entering the wrong alley could not retrace their steps, and so had to wait until the next trial to gain access to the reward. The subjects were deprived of food for most of the day and allowed only 30 minutes access to pellets of dog food at the end of each day's testing session. Although the authors state that the opossums were tested in darkness, they do not indicate whether these nocturnal animals were tested during day or night hours. Initially two subjects, both females, were tested. They were given a single block of

10 trials daily with the criterion set as 9 out of 10 trials correct. These subjects showed no improvement over six days, so on the seventh day each was given massed training of consecutive 10 trial blocks until the criterion was reached. In these circumstances each animal achieved the criterion in a further 40 trials. The remaining subjects, one female and one male, were also given a single day of massed trials. These animals reached the criterion in 30 and 60 trials, respectively.

   The authors point out that these results, taken with those of Friedman and Marshall (1965), indicate that opossums apparently require the massing of trials to learn certain problems and that this suggests a process of memory storage different from that of eutherian mammals.

Maze learning

The ability to solve maze problems has been used as a measure of learning since the last century. Three studies have investigated the performance of marsupials on maze learning tasks. In the first of these, five adult laboratory-reared opossums, *D. virginiana* (sex not stated), were tested in the Fink arrow maze by James (1959). The maze apparatus consisted of a choice compartment connected to a foodbox by four parallel alleys. The animals were trained to enter the correct alley to the foodbox, where they found a meat reward. The quantity of the reward was not specified. Testing took place immediately before feeding time. Each subject was allowed only one trial per day. After the opossum had learned to locate the correct alley on 10 successive days, the food was shifted to another alley. This procedure was repeated for all four alleys. Compared with the nine species investigated by Fink, the opossum scored second only to man and higher than other animals, including pig, dog, goat, chick, rat, cat, rabbit and turtle. It is noteworthy that Fink's findings indicated that the chick performed as competently as the rat and that both of these scored better than the cat (James, 1959: p. 201). This suggests some doubt as to the validity of the arrow maze as a test for comparative investigation of learning ability.

   James and Turner (1963) studied maze learning in seven young (60 days from the pouch) opossums (*D. virginiana*) and seven mature (90 days old) laboratory rats (*Rattus norvegicus albinus*). The apparatus was a Lashley II maze consisting of a central alley with two openings down each side, giving access to three culs-de-sac and one reward chamber. The subjects were given five trials daily. Correct responses were rewarded with food. In achieving the learning criterion of successive errorless trials the opossums (mean = 16.9) made significantly fewer errors than the rats (mean = 30.9). The experimenters suggested that a 'possible explanation for the relative superiority of the opossums was that these animals are less distractable and less curious in their exploratory behaviour than white rats and this tendency was conducive to faster learning in the present experimental situation' (James and Turner, 1963: p. 922). It is interesting to note that James (1960) suggested that the relatively superior performance of *D. virginiana* to most animals of another learning task, the Guthrie – Horton puzzle box, was also due to undistractability (see below). Because the authors failed to report many important experimental variables, including the sex of the subjects, the type of food used as a reward, the quantity of reward, the extent of food deprivation, the time of the daily testing sessions and the general testing procedure, this would be a difficult study to replicate.

The most recent investigation of maze learning in marsupials was carried out by Pollard and Lysons (1967). Six Brush-tailed possums (*Trichosurus vulpecula*) were tested on six maze problems adapted from the Hebb – Williams series (1946). The subjects, sex not given, were 9 – 18 months old at the commencement of the study. The training procedure was described in an earlier paper (Pollard, 1961); however neither the earlier nor the present paper specified how the possums were rewarded for finding the goal box. The six maze problems were presented as two pairs of three. Three problems allowed direct visual solution and three did not. In fact each pair had the same correct path, but for the non-visual problems additional barriers were added to obscure blind alleys. Following an adaptation period on a series of practice problems, the possums were given eight trials on each maze problem, a different problem being presented each day. The performance of the possums was compared with that of cats and rats tested in previous experiments (Pollard, Lysons and Hughes, 1965). The findings indicated that the possums were less efficient than rats or cats in learning both visual and non-visual closed field maze problems. It should be noted that Hebb – Williams problems were designed originally for the laboratory rat; it has been found that some species which are usually superior to the rat on most other measures of learning perform poorly on these problems (see Warren, 1965b: pp. 225 – 56).

To summarise: Two studies with *D. virginiana* indicated that mature animals were superior to most species, other than man, on one type of maze learning task, and that immature animals were superior to mature laboratory rats on another type of maze learning task. A third study reported that *T. vulpecula* was not as efficient in learning maze problems as laboratory rats and cats. It is noteworthy that there have been some recent criticisms of maze learning as a measure for species comparisons (Bitterman, 1960; Warren, 1965b).

Operant learning

In operant or instrumental learning the subject is required to make a specific response such as pressing a lever or pulling a ring. Correct responses are typically rewarded by food or water, or by escape from a painful stimulus.

In an investigation by James (1955) one opossum (*D. virginiana*) was tested in a puzzle box. The puzzle box technique was developed in the last century by Thorndike (1898) to study 'trial and error' learning. In Thorndike's experiments the subject, usually a young cat, was placed in a box-like apparatus. The door from the box could be opened by manipulation of a part of the apparatus – for example, turning a door handle, pressing a lever or pulling a string attached to an outside bolt. In James's study the door was operated by moving a vertical pole located in the centre of the box. During the first four days of the experiment the opossum was placed in the box with the escape door to the feeding chamber open. From the fifth day onwards the door was closed and could be opened only by manipulation of the pole. Pressure on the pole also activated a camera which photographed the subject moving the pole and the time, recorded on a clock-face in the background. It appears that within 23 nights the opossum learned to operate the puzzle box. Apart from the obvious fact that one example of *Didelphis* was capable of learning the appropriate operant response, little can be gleaned from this investigation. The age and sex of the subject were not given, and there was no attempt to judge performance objectively relative to that of other species. However in a later study the author reported that the opossum '. . . due to its undistractibility, performed better than most animals on this problem' (James, 1960: p. 127).

Cone and Cone (1970) studied operant learning in four opossums (*D. virginiana*). The subjects, two males and a female from one litter, were raised in captivity and tamed from time of weaning. At the start of the experiment the opossums were approximately 19 weeks out of the pouch. For most of the experiment they were deprived of water for 23 hours with only one hour of unrestricted drinking per day; presumedly this followed immediately after the testing session. The time of day that testing took place was not stated. The subjects were trained to press a lever, each press delivering 0.1 ml of water, increasing to 0.75 ml as the experiment progressed. Within 10 daily training sessions (each of 30 min) all opossums had learned to perform consistently on a FR 10 schedule (one reward for every 10 responses). Various experimental manipulations were carried out, and it was found that the opossums could respond in excess of 75 lever presses per minute over a sustained session of 2 h. Changing levels of deprivation from 12 to 48 h produced only minimal changes in response rates. When two opossums were placed on an FI 60 s schedule (one reward every minute regardless of the number of responses), they soon learned to time their responses so that all responding was delayed until the last 20 s before the delivery of the reinforcement. The findings indicated that the opossums adapted well to the lever pressing task, and the authors suggested that this technique should be utilised in further investigations with these animals.

Powell and Doolittle (1971) used an operant bar pressing task to investigate flexibility and learning in opossums (*D. virginiana*). They adapted a technique developed by Voronin (1962) involving the repeated acquisition and extinction of an operant response. In this task the animal was rewarded for pressing a bar (acquisition phase), then the rewards were terminated and the number of responses made during the non-reward period were recorded (extinction phase). These acquisition and extinction procedures were repeated several times. A reduction in responses made over consecutive extinction phases was taken to indicate the adaptability of the subject. Powell and Doolittle suggested that this task could be likened to a successive discrimination – reversal learning problem. The experiment involved a training and a testing component. During the training period Powell and Doolittle gave the subjects a daily 30 min session of training to press a bar for a water reward until a criterion of 15 min continuous responding was reached. The authors did not state their definition of continuous responding. During this pretraining period the access to water for the rats was restricted to 10 min per day and the opossums were allowed approximately 18 ml of water per kg of body weight daily. After the criterion had been reached, the subjects were given a further 25 barpresses per day for three days. For the experiment proper, carried out over the next 10 days, each subject was allowed 15 reinforced barpresses, followed by a 10 min extinction period where any responses made by the subjects were not reinforced. Both groups of animals showed a significant decrease in responses made during the extinction periods over the 10 days of testing, and this was interpreted as indicative of a progressive improvement in reversal learning performance.

Unfortunately, the investigators did not make any statistical comparisons between the rats and the opossums.

To summarise: Although several experimenters have used operant techniques in investigations of other learning phenomena in marsupials (see, for example, James and McFarland, 1966; Buchmann and Grecian, 1974) there appear to be only three published papers concerned with operant learning *per se*. All three were concerned with *D. virginiana.* The first paper indicated that *Didelphis* could learn a simple operant and the second showed that they were able to adapt their operant performance according to the system of reward. The third paper showed that *D. virginiana* could learn to extinguish or give up a non-rewarded behaviour, and that the process of learning to give up improved with practice.

Other problem-solving and learning tasks

Avoidance and escape learning

In an avoidance task the subject is presented with a signal, such as the onset of a sound or a light, which is followed after some time by an aversive stimulus, typically an electric shock. If between the onset of the signal and the negative reinforcement the subject carries out the appropriate response, such as leaving the compartment, pressing a bar or moving a limb, it can avoid the aversive stimulus. Escape learning differs from avoidance learning in that the signal and the negative reinforcement are presented simultaneously, and while the subject cannot avoid at least some punishment, the continuation of the aversive stimulation can be halted by making the appropriate response.

Two papers have reported the results of avoidance learning experiments with marsupials (James, 1937, 1958). Both papers were concerned with the opossum, *D. virginiana,* and unfortunately both were vague in regard to several experimental variables, including the sex and age of the subjects and intensity of negative stimulus. In the earlier study three experiments were carried out. In the first of these an attempt was made to condition an opossum to lift a foreleg in response to a bell sounded 3 s before the onset of electric shock. After more than 300 trials there were no indications of a conditioned leg reaction. The author stated that '. . . it may be definitely concluded that the foreleg reaction of the opossum cannot be conditioned' (James, 1937: p. 96). The second experiment investigated the response of an opossum in an escape learning situation using a two-compartment apparatus. After the animal was placed in one compartment, an electric shock was passed through the floor, and the animal was allowed to escape into the other compartment. The opossum never learned to escape by moving out of the compartment. Instead it responded to the onset of shock by curling up and lying on its side. In this way the subject was insulated from the source of shock by its fur. Such curling up or 'playing possum' behaviour is a usual response of *Didelphis* faced with danger. In the third experiment the two opossums tested in the previous experiments were trained to respond with defensive behaviour towards a stick presented as a threat after a buzzer was sounded. The response, head turned towards threat, mouth opened, was conditioned with only one pairing of the stick and buzzer. The investigator apparently overlooked the possibility that the attack response could be elicited by the sound of the buzzer alone. The interpretation of these studies is further clouded by a later report (James, 1958) indicating that the opossums used in the 1937 study were not reared in the laboratory and were relatively untamed.

The James (1958) paper involved two experiments. In the first experiment a single subject was administered 550 pairings of a buzzer and shock to the left hind-leg. Although the opossum failed to condition a leg lift response, hyperventilation was conditioned to the stimulus and near the end of the investigation the animal showed evidence of a total body lift in response to the stimulus. In the second experiment, three opossums were trained to avoid a shock by moving from one compartment to another at the sound of a buzzer. The opossums learned to react to the signal with 10 successive correct responses in an average of 32.7 presentations. The conditioned responses were extinguished very rapidly. James (1958) does not clearly indicate whether the second experiment involved an 'active avoidance' or 'active escape' task.

Probability learning

Probability learning tasks, similar to discrimination problems, usually involve two discriminanda. Responses to one or other of these stimuli are rewarded randomly, but to a specific percentage. For example, in a T-maze situation the reward would be placed for some set ratio (say 70 per cent of trials) in the left arm and for the remainder (30 per cent of trials) in the right. The changes in the position of the goal arm would be made in a random sequence. Placental mammals such as the laboratory rat tend to 'maximise' on such a task — that is, they tend to enter the most-rewarded arm in the majority of trials.

Probability learning of a spatial problem was investigated in five Sprague - Dawley rats (*R. norvegicus*) and five opossums (*D. virginiana*) by Doolittle and Weimer (1968). The rats (all male, 4 months old) were rewarded with an 0.1 g food-pellet per correct response, and the opossums (three males, two females, 6 months old) with half an 'M & M' candy. The subjects were given 20 trials per day in a single massed block, for 30 days. One arm of the T-maze was randomly rewarded for 70 per cent of trials and the other for 30 per cent. A correction technique was used so that if the subject entered the wrong arm of the maze and failed to find a reward, it was allowed to retrace its path and enter the correct arm. In this way each trial terminated with a positive reinforcement. The learning curves for the rats and the opossums were not significantly different. Both groups tended to maximise, and by the end of the 30 day test period were responding on 90 per cent of trials to the more rewarded arm.

Exploration-habituation

It has been suggested that the responsiveness of an animal to a novel environment reflects its level of intellectual development (see, for example, Wünschman, 1963). Furthermore, work with eutherian laboratory animals has indicated that the ability to learn and remember may be reflected in the intensity of an animal's exploratory responses to novelty and the habituation, or rate at which these exploratory responses diminish (Thompson, 1953; Thompson and Khan, 1955).

Kirkby and Preston (1972) investigated exploratory responses in terms of locomotion about a novel environment in eight male Marsupial mice (*Sminthopsis crassicaudata*) and groups of similar sized male Eutherian mice (*Mus musculus*). At the approximate time of sexual maturity, the subjects were tested in a T-shaped apparatus containing various objects made of different materials and of different shapes and colours. The apparatus and the objects were completely novel to the animals. For a 10 min period the exploratory responses of the subjects, in terms of movement about the novel environment, were recorded. The findings suggested that the *Sminthopsis* were more reactive to the novel environment and habituated to the novelty at a faster rate. These results were viewed as evidence of a superior learning ability of the Marsupial mice to the laboratory Eutherian mice.

To summarise: Two studies of avoidance and escape behaviour in *D. virginiana* have been published. As several variables were not controlled in these experiments, their interpretation is difficult: that *Didelphis* will learn a relatively simple response involving a total body lift to avoid receiving a shock to a hind-leg appears to be the only clear finding from these studies. An investigation of probability learning in *D. virginiana* indicated that this species performed similarly to the laboratory rat and other mammals. Finally, in an exploration-habituation experiment, *S. crassicaudata* appeared to show more rapid learning and extinction than *M. musculus*.

Conclusions

It is obvious from this survey that, compared with the vast literature on learning and problem-solving in eutherian and submammalian animals (Bitterman, 1960, 1965; Warren, 1965a), little study has been made of marsupials in this field. Only five species have been investigated, and 14 of the 19 papers have been concerned specifically with *D. virginiana*. As Kirkby (1969b) has pointed out, generalisations about marsupial behaviour based upon the performance of *Didelphis* are akin to equating eutherian behaviour with the ability of a single species, such as the White rat.

Unfortunately most of the published studies of marsupial learning are open to serious criticism, of errors in experimental design or procedure, or of the way in which the experiment was reported. Many studies used small groups of animals. In some cases the experimental samples were as low as one or two subjects (James, 1937, 1955, 1958, 1960; Neumann, 1961; Friedman, 1967). Several authors failed to give the sex, age or even the species of the subjects (James, 1937, 1955, 1958, 1959, 1960; James and Turner, 1963; Munn, 1964; James and McFarland, 1966; Buchmann and Grecian, 1974). Others did not report motivational data concerned with levels of deprivation (James and McFarland, 1966; Friedman, 1967) or details of reinforcement (James, 1937, 1958; James and Turner, 1963; Pollard and Lysons, 1967; Buchmann and Grecian, 1974). Some investigations used measures of performance that seem to be of little value for interspecies comparisons – for example, such esoteric techniques as the Guthrie - Horton puzzle box (James, 1955), Fink's arrow maze (James, 1959) and Rensch and Altevogt's multiple discrimination

(Neumann, 1961). In several experiments the investigators appear to have over-looked one or more of the experimental factors, thus weakening the value of their results. For example, it seems that intellectually immature subjects were tested in some experiments (Munn, 1964; James and McFarland, 1966; Buchmann and Grecian, 1974; Williams and Kirkby, 1976).

Both male and female subjects were tested in several experiments. Although the authors generally did not report learning performance according to sex, it was possible in three experiments to extract this information. The data from two studies suggested that females were superior to males in learning (Munn, 1964; James and McFarland, 1966) and in another it appeared that the males were superior (Williams and Kirkby, 1976). In the only study which allowed a direct comparison of learning abilities between marsupials, Neumann (1961) reported that a kangaroo was superior to an opossum in visual discrimination learning. By interpolating the results of several papers it is possible to make some comparisons between similar sized Australasian and North American marsupials. It appears that *D. virginiana* is possibly superior to *I. obesulus* and *T. vulpecula* in learning visual discrimination problems. Both the Australasian marsupials are clearly superior to *Didelphis* in learning spatial discrimination problems.

In conclusion, one point is worthy of particular emphasis: the results of recent studies of learning and problem-solving processes in marsupials (particularly Australasian marsupials) are strongly inconsistent with the often-reported view of marsupials as 'primitive' mammals intellectually inferior to eutherians (Neumann, 1961; Glickman and Sroges, 1966; Ewer, 1968).

References

Bitterman, M. E. (1960). Toward a comparative psychology of learning. *Am. Psychol.*, **15**, 704 – 12.

Bitterman, M. E. (1965). Phyletic differences in learning. *Am. Psychol.*, **20**, 396 – 410.

Block, M. (1960). Wound healing in the newborn opossum (*Didelphis virginiana*). *Nature*, **187**, 340.

Buchmann, O. L. and Grecian, E. A. (1974). Discrimination-reversal learning in the marsupial *Isoodon obesulus* (Marsupialia, Paramelidae). *Anim. Behav.*, **22**, 975 – 81.

Burns, R K. (1956). Hormones versus constitutional factors in the growth of embryonic sex primordia in the opossum. *Am. J. Anat.*, **98**, 35 - 67.

— Cone, A. L. and Cone, D. M. (1970). Operant conditioning of Virginia opossum. *Psychol. Rep.*, **26**, 83 - 6.

— Doolittle, J. H. and Weimer, J. (1968). Spatial probability learning in the Virginian opossum. *Psychon. Sci.*, **13**, 191.

Ewer, R. F. (1968). A preliminary survey of the behaviour in captivity of the dasyurid marsupial *Sminthopsis crassicaudata* (Gould). *Z. Tierpsychol.*, **26**, 151 - 88.

— Friedman, H. (1967). Colour vision in the Virginia opossum. *Nature*, **213**, 835 - 6.

— Friedman, H. and Marshall, D. A. (1965). Position reversal training in the Virginia opossum: evidence for the acquisition of a learning set. *Quart. J. Exp. Psychol.*, **17**, 250 - 4.

Glickman, S. E. and Sroges, R. W. (1966). Curiosity in zoo animals. *Behaviour,* **25**, 319 – 63.

Gossette, R. L., Kraus, G. and Speiss, J. (1968). Comparison of successive dis-crimination reversal (SDR) performances of seven mammalian species on a spatial task. *Psychon. Sci.,* **13**, 183 – 4.

Hebb, D. O. and Williams, K. (1946). A method of rating animal intelligence. *J. Gen. Psychol.,* **34**, 59 – 65.

Herring, F. H., Mason, D. J., Doolittle, J. H. and Starrett, D. E. (1966). The Virginia opossum in psychological research. *Psychol. Rep.,* **19**, 755 – 7.

James, W. T. (1937). An experimental study of the defence mechanism in the opossum, with emphasis on natural behaviour and its relation to mode of life. *J. Genet. Psychol.,* **51**, 95 – 100.

James, W. T. (1955). The behaviour of the opossum in the Guthrie Horton puzzle box. *J. Genet. Psychol.,* **87**, 203 – 6.

James, W. T. (1958). Conditioned responses in the opossum. *J. Genet. Psychol.,* **93**, 179 – 83.

James, W. T. (1959). Behaviour of the opossum in the Fink Arrow Maze. *J. Genet. Psychol.,* **94**, 199 – 203.

James, W. T. (1960). A study of visual discrimination in the opossum. *J. Genet. Psychol.,* **97**, 127 – 30.

James, W. T. and McFarland, J. (1966). A study of form discrimination in the opossum. *J. Psychol.,* **64**, 193 – 8.

James, W. T. and Turner, W. W. (1963). Experimental study of maze learning in young opossums. *Psychol. Rep.,* **13**, 921 – 2.

Kirkby, R. J. (1969a). Learning in marsupials. *Aust. Psychol.,* **4**, 155 – 66.

Kirkby, R. J. (1969b). Marsupials are a 'them' not an 'it'. *Austr Psychol.,* **4**, no page number [Abstract].

Kirkby, R. J. and Preston, A. C. (1972). The behaviour of marsupials. II. Reactivity and habituation to novelty in *Sminthopsis crassicaudata. J. Biol. Psychol.,* **14**, 21 – 4.

Livesey, P. J. and Di Lollo, V. (1969). An attempt to train the quokka (*Setonix brachyurus*) in a visual discrimination task. Paper presented at the *Fourth Annual Conference of the Australian Psychological Society*, Sydney.

Martin, P. G. (1965). The potentialities of the fat-tailed marsupial mouse, *Sminthopsis crassicaudata* (Gould), as a laboratory animal. *Aust. J. Zool.,* **13**, 559 – 62.

Mizell, M. (1968). Limb regeneration: induction in the newborn opossum. *Science,* **161**, 283 – 6.

Munn, N. L. (1964). Discrimination-reversal learning in kangaroos. *Aust. J. Psychol.,* **16**, 1-8.

Neumann, V. C-H. (1961). Die visuelle Lernfahigkeit primitiver Saugetiere. *Z. Tierpsychol.,* **18**, 71 – 83.

O'Neil, W. M. (1966). Australia. *International Opportunities for Advanced Training and Research in Psychology,* American Psychological Association, Washington, pp. 11 – 17.

Pollard, J. S. (1961). Rats and cats in the closed field test. *Aust. J. Psychol.,* **13**, 215 – 21.

Pollard, J. S. and Lysons, A. M. (1967). Possums in the closed field test. *Anim.*

Behav., 15, 129 – 33.

Pollard, J. S., Lysons, A. M. and Hughes, R. N. (1965). Size of closed field test and performance of rats. *Aust. J. Psychol.,* 17, 217 – 9.

Powell, M. R. and Doolittle, J. H. (1971). Repeated acquisition and extinction of an operant by opossums and rats. *Psychon. Sci.,* 24, 22 – 3.

Seligman, M. E. (1970). On the generality of the laws of learning. *Psychol. Rev.,* 77, 406 – 18.

Thompson, W. R. (1953). Exploratory behaviour as a function of hunger in 'bright' and 'dull' rats. *J. Comp. Physiol. Psych.,* 46, 323 – 6.

Thompson, W. R. and Kahn, A. (1955). Retroaction effects in the exploratory activity of 'bright' and 'dull' rats. *Can. J. Psychol.,* 9, 173 – 82.

Thorndike, E. L. (1898). Animal intelligence; an experimental study of the associative processes in animals. *Psychol. Monogr.,* 8, 529 – 827.

Tilley, M. W., Doolittle, J. H. and Mason, D. J. (1966). Olfactory discrimination learning in the Virginia opossum. *Percept. Mot. Skills,* 23, 845 – 6.

Twyver, H. V. and Allison, A. (1970). Sleep in the opossum *Didelphis marsupialis. Electroenceph. Clin. Neurophysiol.,* 29, 181-9;

Voronin, L. G. (1962). Some results of comparative physiological investigations of higher nervous activity. *Psychol. Bull.,* 59, 161 – 95.

Waring, H., Moir, R. J. and Tyndale-Biscoe, C. H. (1966). Comparative physiology of marsupials. In: *Advances in Comparative Physiology and Biochemistry* (ed. O. Lowenstein), Academic Press, New York, pp. 237 – 376.

Warren, J. M. (1965a). The comparative psychology of learning. *Ann. Rev. Psychol.,* 16, 95 – 118.

Warren, J. M. (1965b). Primate learning in comparative perspective. In: *Behaviour of Nonhuman Primates,* Vol. 1 (ed. A. M. Schrier, H. F. Harlow and F. Stollnitz), Academic Press, New York, pp. 249 – 83.

Wiedorn, W. S. (1954). A new experimental animal for psychiatric research: the opossum *Didelphis virginiana. Science,* 119, 360 – 1.

Williams, G. and Kirkby, R. J. (1976). Behaviour of marsupials. IV: Learning set in the brushtailed possum (*Trichosurus vulpecula*). In Press.

Wünschmann, A. (1963). Quantitative Untersuchungen zum Neugierverhalten von Wirbeltieren. *Z. Tierpsychol.,* 20, 80 – 109.

13 Corticosteroid levels and male mortality in Antechinus stuartii

Anthony K. Lee*, Adrian J. Bradley † and Richard W. Braithwaite‡

Current views on the regulation of numbers of small mammals have developed mainly from studies of murid and cricetid rodents (Christian, 1971; Krebs et al., 1973; Krebs and Myers, 1974), and particularly those species which provide spectacular examples of cyclical fluctuations. In spite of these studies, the causes of cycles and intrinsic population regulatory mechanisms remain poorly understood. Among the difficulties encountered has been the complex age structure of many populations, and problems of ageing and identifying cohorts, and of finding the causes of differing mortalities within cohorts. With these difficulties in mind we have begun to examine populations with simple, easily determined age structure, which also show abrupt changes in numbers.

Certain species of the dasyurid genus *Antechinus* - monoestrous, insectivorous marsupials — have a highly synchronous breeding season after which all males die, sometimes within two or three weeks. In one of these species, *A. stuartii*, mortality is negligible for four months before breeding but is then particularly abrupt (Wood, 1970; Braithwaite, 1973), making it possible to study age-related changes in males during a highly predictable life span. Here we review the probable causes of this mortality and discuss their wider ecological implications.

Life history and ageing

At least six *Antechinus* species — *A. stuartii, A. swainsonii, A. flavipes, A. minimus, A. bellus* and *A. bilarni* — appear to be monoestrous and show a post-mating mortality of males (Woolley, 1966a, 1973; Taylor and Calaby, personal communication). In certain of these, the post-mating mortality of males appears more abrupt than in others. For example, in *A. stuartii* all males die within three weeks of the onset of mating at an age of 11.5 months (Wood, 1970; Braithwaite, 1973), whereas Schultz (personal communication) has detected male *A. swainsonii* which,

*Anthony K. Lee, BSc, PhD (W.A.), MA, PhD (U.C.L.A.) was awarded Doctorates in Philosophy as a result of studies on the ecology and evolution of a genus of Australian burrowing frogs, and adaptations of Californian woodrats to arid environments. Currently an Associate Professor in the Department of Zoology, Monash University, He has published on aspects of the physiological ecology of small mammals, reptiles and amphibians, with a strong orientation towards unique elements in the Australian fauna.

†Adrian J. Bradley BSc (Tas.) is a PhD candidate in the Department of Physiology at Monash University studying adrenocortical function in *Antechinus* spp.

‡Richard W. Braithwaite BSc, MSc (Qld) was awarded an MSc. on the basis of an ecological and behavioural study of *A. stuartii.* He is currently a PhD student in Zoology at Monash University.

on the basis of skull dimensions, are older than one year. In *A. stuartii* there is also a mortality among females at the times of mating, gestation and lactation, but some females survive to a second, and occasionally a third, breeding season (Wood, 1970; Leonard, 1972; Robinson, personal communication). Virgin and parous females are distinguished by nipple development (Woolley, 1966a; Wood, 1970).

Mating occurs in *A. stuartii* in late winter or early spring, and although there is variation in the timing of mating between populations, individual populations seem to breed at the same time each year (Figure 13.1; Wood 1970; Braithwaite, 1973).

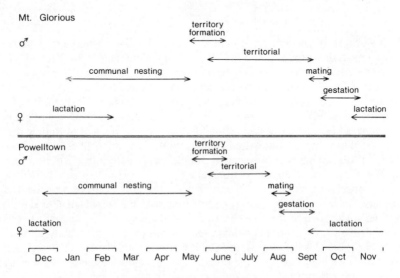

Figure 13.1 Comparison of the timing of life history events of *A. stuartii* at Mt Glorious, Queensland, and Powelltown, Victoria.

Gestation lasts 26 – 35 days (Woolley, 1966a) and all births in a population occur within a two week period (Wood, 1970). Neonates remain attached to the nipples for approximately 35 days after parturition, and are suckled in a nest for an additional two to three months (Marlow, 1961; Woolley, 1966a; Wood, 1970). Juveniles are first caught in traps in December and January but may continue to share a nest until territorial behaviour starts, approximately two months before mating (Braithwaite, 1976).

Information on the life histories of other monoestrous *Antechinus* species in which males die after mating is less complete. All breed in winter or early spring, though with some variation of timing between species and populations (Taylor and Horner, 1970; Wood, 1970; Leonard, 1972; Green, 1972; Taylor and Calaby, personal communication).

Post-mating mortality of males

Post-mating mortality of males has been studied only in *A. stuartii*, mainly in two populations – one living in subtropical rain forest at Mt Glorious, near Brisbane,

Queensland, and the other in temperate wet sclerophyll forest at Powelltown, east of Melbourne, Victoria. Queensland animals breed slightly later, and are larger (males up to 70 g, females to 35 g) than those of Victoria (males to 40 g, females to 25 g) which share their habitat with a possible competing species, *A. swainsonii.* We have no reason to suspect that these differences affect the interpretation of our present data. Three observations suggest that the mortality has an intrinsic basis: first, the precision of timing; second, the abruptness of onset; and third, the occurrence of the mortality in populations as geographically and climatically separate as those of southern Queensland and southern Victoria.

Probable causes of death in three samples of *A. stuartii* are summarised (see Arundel *et al.*, in this volume). One sample of six males was captured on 27 July 1974 in Sherbrook Forest before the onset of mating, and isolated from other animals. Others were taken from Healesville (four males) and Powelltown (seven males, five females) after the onset of mating, and during the last week of natural life of the males.

Most deaths occurring in the three samples were associated with two disease states. The first, detected as focal hepatic necrosis containing Gram-positive bacteria, was attributed to *Listeria monocytogenes* cultured from this tissue. The second, characterised by a moderately reduced haematocrit, was considered to be a consequence of pathogenic infections of a species of *Babesia,* in the absence of evidence of another specific disease. In another study Cheal, Lee and Barnett (1976) detected an anaemia in four of a sample of six males and two of a sample of nine females captured at Powelltown at the time of the natural mortality of males. Evidence of microcytosis and hypochromism in smears from anaemic animals suggested an aetiology of iron deficiency and chronic haemorrhage (Wintrobe, 1967). The frequent occurrence of gastric and duodenal haemorrhages in moribund laboratory-held *A. stuartii* and their inclusion among the probable causes of death (see table summarising causes of death in Arundel *et al.,* this volume) suggest that the anaemia may result from a dual aetiology.

These changes in red cell parameters appear moderate when compared with anaemias tolerated by larger mammals (Schalm, 1965). However, there is a correlated increase in resting metabolism (Cheal *et al.,* 1976) which may reflect an increase in the cost of oxygen transport in the presence of the anaemias. If so, it seems that even moderate decreases in the oxygen-carrying capacity of the blood are taxing for small mammals with their relatively high oxygen demands.

The development of *L. monocytogenes* and a species of *Babesia* in animals collected before mating (late July) suggests that these pathogens may invade the hosts early in life and survive as latent infections, which redevelop if the immune system is compromised. Such a compromise may occur if plasma corticosteroids rise (Eberlein, Bongiovani and Rodriguez, 1967). The association of these diseases with gastric and duodenal ulcers, as other probable causes of death, is consistent with the hypothesis that the synchronous death of males is aetiologically related to a stress response, and its physiological consequences.

Corticosteroid response

Barnett (1973) obtained evidence of increased adrenocortical activity in male *A. stuartii* in the last weeks of life, including increased plasma corticosteroid levels,

adrenal weight, cross-sectional area of the zona fasciculata and liver glycogen concentration. Other changes consistent with a hyperadrenocortical response which
also occur at this time include an increased concentration of skin hydroxyproline
(Barnett, 1974), loss of body weight (Woolley, 1966a; Wood, 1970; Braithwaite,
1973), negative nitrogen balance (Woollard, 1971) and a marked neutrophilia and
lymphopenia (Cheal *et al.*, 1976).

Bradley, McDonald and Lee (1975) also observed an increase in the plasma
corticosteroid concentration of male *A. stuartii* from a mean of 2.9 ± 0.51
μg/100 ml in late July, before mating, to a mean of 5.2 ± 0.60 μg/100 ml in mid-
and late August, during the last two weeks of natural life. However, these are
measurements of the total plasma corticosteroid concentrations. In man the total
concentration includes a free fraction, normally contributing 10 per cent of the
total; a fraction weakly bound to albumin, normally contributing another 10 per
cent; and the remainder which is firmly bound to transcortin, a globulin with a high
affinity for corticosteroids (Tait and Burstein, 1964). It is generally accepted
that only the free and albumin-bound fractions are available for biological
activity.

The maximum corticosteroid binding capacity of plasma at $4°C$, measured by
the gel filtration technique of Doe, Fernandez and Seal (1964), gives a measure of
plasma transcortin concentration. To assess the increase in plasma corticosteroid
concentrations of the male *A. stuartii* in terms of biological activity, the maximum
high affinity cortisol binding capacity was measured and compared with that of
females at the same time (Bradley, McDonald and Lee, 1976). The maximum
cortisol binding capacity decreased markedly in males between April and late
August (Figure 13.2). In fact, in August the peripheral plasma corticosteroid
concentration of males exceeded the maximum binding capacity of the plasma: at
$4°C$ at least 50 per cent of the measured corticosteroids was available for biological
activity, a proportion which would normally be higher at normal body temperature of $37°C$. The severe decrease in maximum cortisol binding capacity of
plasma from males does not result from a decrease in total protein concentration,
since this falls only slightly towards the end of life (Cheal *et al.*, 1976).

The maximum cortisol binding capacities of plasma from males and females
were most similar in April (Figures 13.2 and 13.3), but even so, the mean value for
females (9.67 ± 0.31 μg/100 ml) was significantly greater than the mean value for
males (7.95 ± 0.24 μg/100 ml, $p < 0.01$). The difference increased in late June
when the values for females increased. The binding capacity of the plasma of females then decreased until at least late August, but at no time did it fall below the
peripheral plasma corticosteroid concentration.

De Moor *et al.* (1967) provided evidence supporting the hypothesis that the
cortisol binding capacity of plasma transcortin is sex-linked. Consistently, Lohrenz,
Seal and Doe (1967) detected a familiar difference in the concentration of corticosteroid binding capacity in man. According to Seal and Doe (1966), oestrogen
therapy and pregnancy are the only conditions that have been demonstrated to
increase this capacity in man. Oestrogen may be responsible for the difference in
cortisol binding capacity between sexes observed in *A. stuartii*, and for the increase in cortisol binding capacity of the plasma of females to 13.11 ± 0.09 μg/
100 ml in late June.

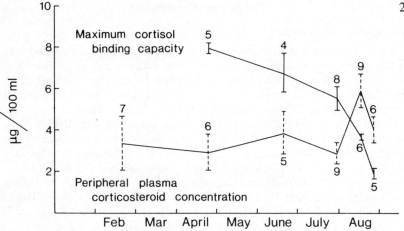

Figure 13.2 Changes in the mean maximum cortisol binding capacity and mean total peripheral plasma corticosteroid concentration of male *A. stuartii* with time of the year. Vertical lines either side of means denote ± 1 standard error. Numbers indicate sample size.

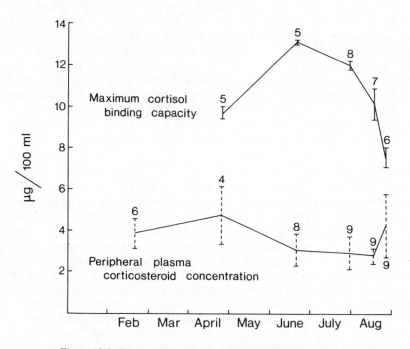

Figure 13.3 Changes in the mean maximum cortisol binding capacity and mean total peripheral plasma corticosteroid concentration of female *A. stuartii* with time of year. Symbols as in Figure 13.2.

Corticosteroids and mortality

To test the hypothesis that the increase in peripheral plasma corticosteroid concentrations observed in males captured in August could be a cause of the mortality, Bradley *et al.* (1975) injected laboratory-held males with exogenous cortisol acetate. Such males will survive at least four months beyond the time of natural mortality if captured before mating and isolated in the laboratory.

In each of two years animals kept under these conditions were divided into three groups. Two experimental groups received intramuscular injections of 1 and 10 mg/kg per day cortisol acetate for three weeks, control groups receiving the vehicle alone. Those receiving the low level of cortisol acetate (16 individuals) showed mean corticosteroid concentrations of 3.4 ± 0.78 µg per 100 ml of plasma, from samples obtained 4, 7, 12 and 24 h after injection. Samples obtained at similar intervals from 20 animals with the higher dose averaged 8.3 ± 1.86 µg per 100 ml. These values fell on either side of a mean value measured in males captured in mid- and late August (5.2 ± 0.60 µg per 100 ml; Bradley *et al.*, 1975). The maximum cortisol binding capacity of the plasma from the injected groups has yet to be measured.

The conclusion of the experiment coincided with the time of natural mortality of males. Mortality appeared to be dose-related. There were no deaths among the 11 control animals during the second and third weeks following the experiments, but three of the 13 which received the lower dose (23 per cent) and six of the 14 which received the higher dose (43 per cent) died during the same period. The relationship between cortisol injection and mortality, regardless of dose, was highly significant ($p < 0.01$).

The survivors were examined for evidence of disease, and although there was no correlation between the incidence of hepatic necrosis and cortisol dose level, there was a dose-related incidence of infection by cyto-megalo virus of the prostate gland (Barker, personal communication). Thus there is some evidence to link high levels of corticosteroids with post-mating mortality in *A. stuartii.*

Immunosuppression

Administration of large doses of anti-inflammatory steroids in other animals is known to decrease the inflammatory reaction, suppress the reticulo-endothelial system and diminish antibody formation, all of which may predispose an animal to infections and invasion by parasites (Eberlein *et al.*, 1967; Mahoney, 1972; Rose, 1972).Bradley (unpublished data) has attempted to determine the possible effects of the peripheral plasma corticosteroid concentration on the immune system of *A. stuartii,* and whether these changes occur naturally in association with the rise in endogenous plasma corticosteroid concentrations. Immunoglobulin concentrations were assessed by resolving plasma samples in two dimensions by a combination of polyacrylamide disc and gradient gel electrophoresis (Margolis and Kenrick, 1969). Using this method, three protein fractions in the plasma of *A. stuartii* were observed to behave similarly to proteins considered to be IgM, IgG and IgA in human serum, and were provisionally identified as such.

At least two of the immunoglobulins, IgG and IgA, occurred at lower concentrations in the plasma from freshly captured males in late August, when compared with females captured at the same time, and males captured in late July and held in isolation in the laboratory. A decrease in IgG and IgA concentration, as well as in IgM concentration, also occurred in laboratory-held males receiving injections of 10 mg/kg per day cortisol acetate. This is in agreement with the findings of Levy and Waldmann (1970), who have shown that serum concentrations of all subclasses of IgG, as well as of IgA and IgM, are reduced in mice by administration of cortisol acetate, although their doses were in most cases considerably higher than those used with *A. stuartii*. This evidence, and the depletion of splenic lymphoid elements in males in late August (Barker, personal communication), as well as the lymphopenia referred to earlier, are compatible with the inference that there is a suppression of the immune system in male *A. stuartii* due to increased plasma corticosteroid levels.

Behaviour and mortality

Several observations implicate the behaviour of male *A. stuartii* during the breeding season as the stimulus for the adrenocortical response. Males captured before mating and isolated in the laboratory survive beyond the period of natural mortality. In contrast, males captured during the last week of natural life and isolated from the time of capture succumb at the time of natural mortality. Finally, a mortality occurs among males captured before mating and subjected to encounter experiments or caged together, although this mortality may occur two to four weeks after the natural mortality (Wood, 1970).

Several conspicuous behavioural changes occur in males during the breeding season, of which the most evident are increased aggressiveness and activity. Braithwaite (1973, 1974, 1976) studied seasonal changes in the social behaviour of *A. stuartii* in encounters contrived in an observation cage located in the field. Animals were trapped overnight and tested in pairs on the following day. The frequency of approaches, withdrawals, attacks, grooming, cohesive behaviours, submissions, cloacal marking and three types of threat were recorded for each test. These data, analysed with a stepwise multiple linear regression program, revealed that the time of year and sex of the animals were the important predictors of aggressive behaviour. Male defensive behaviour (chee threats; see Braithwaite, 1974) increased gradually with age until about six weeks before the onset of mating, and then decreased as offensive behaviour (attacks and huffing threats) increased conspicuously (Figure 13.4). The behaviour of females contrasted with that of males, for although aggressiveness (entirely defensive) increased in females during the period in which males establish territories, it returned to a low level until birth of the young. Females were never seen to perform any element of offensive behaviour.

Changes in activity of males during the breeding season include increases both in diurnal activity and in the distance between successive captures. On the basis of a trapping study, Wood (1970) concluded that *A. stuartii* was primarily but not exclusively nocturnal. However, during the breeding season males noticeably increase diurnal activity and are observed moving about the forest floor without apparent fear (Braithwaite, 1973, 1976). Wood (1970) also found that males showed the greatest average distance between consecutive captures at this time and during territory formation, and at all times remained active longer than females. In a detailed analysis of the relationships between use of space and domin-

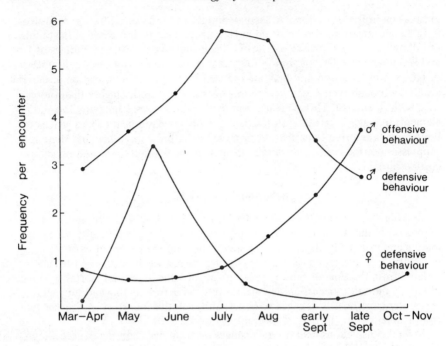

Figure 13.4 Changes in the behaviour of *A. stuartii* in contrived
encounters with time of year.

ance Braithwaite (1973) observed that dominant males undertook forays beyond
the boundaries of territories established in May or June, and that these forays
increased in frequency during the breeding season. Consistently, the number of
new animals captured on trapping grids also increases at this time (Wood, 1970;
Leonard, 1972).

Male *A. stuartii* show a prodigious appetite for copulation in the laboratory.
Copulation, consisting of short bouts of thrusting interrupted by longer periods of
rest during which the male remains mounted, usually lasts five to six hours and
may continue for 12 hours (Marlow, 1961; Woolley, 1966a). Woolley (1966b)
also observed that behavioural oestrus lasted an average of 6.2 days in females
housed with males, during which the pairs generally copulated once daily; the
longest behavioural oestrus observed was 13 days, encompassing 14 copulations.
Whether this capacity is fulfilled in the field is unknown. Females housed with
males lose hair from the neck during repeated copulations, but hair-loss has not
been observed in females in natural populations, which suggests a lower frequency
of copulation in the field.

Frequent copulations and agonistic encounters, interspersed with periods of
searching for females and feeding, seem therefore to characterise male behaviour
during the mating season. This combination of behaviour seems to be more lethal
than either element on its own, since there is little mortality among individual
males housed with females, and the mortality among males housed in pairs during

the breeding season generally occurs two to four weeks after the natural mortality (Wood, 1970).

Significance in the life history

Semelparity or single reproduction (Cole, 1954), as exemplified by the life history of male *A. stuartii*, occurs commonly among invertebrates but only rarely among lower vertebrates (for example, Pacific salmon, *Oncorhynchus* spp.; lampreys; fresh-water eels, *Anguilla* spp.). It has not previously been described in mammals or birds. Since semelparous animals are generally fecund, one reason why it seldom appears in higher vertebrates may be the reduction in litter size associated with progressive improvements in the care of litters (viviparity, placentation, lactation, etc.). In these circumstances iteroparity (repeated reproduction), facilitated by the long reproductive life of most vertebrates, increases the reproductive capacity of individuals, as well as providing insurance against reproductive failure.

The value of iteroparity in increasing reproductive capacity in a seasonally fluctuating environment depends upon the relationships between the time available for successful reproduction and the minimum period between litters. If successful reproduction can be achieved only during a short period of time each year, or if the duration of gestation and lactation is prolonged, the contribution of iteroparity to recruitment is reduced. Provided reproduction may be reliably timed to coincide with the period when most young survive, it becomes important for organisms to maximise reproductive effort during this period of the year, even to the detriment of longevity (Wilbur, Tinkle and Collins, 1974). This appears to have occurred in monoestrous species of *Antechinus*.

The minimum period between litters is particularly long in polyoestrous dasyurids when compared with other animals of similar size (Woolley, 1973; Braithwaite and Lee, 1976). It seems also that the period favourable for reproduction in small insectivorous dasyurids, in view of their relatively high energy demands, is confined to times of the year when insects are particularly abundant. This problem is exacerbated in *Antechinus* species, in which a litter at weaning may weigh three times as much as the mother (Braithwaite, 1973).

All *Antechinus* species with semelparous males mate in winter or early spring, and the litters are weaned in the period between November and January. Lactation therefore corresponds with the spring flush of insects. If a second litter followed immediately, lactation would coincide with the hot and frequently dry summers characteristic of most of Australia, and weaning with the autumn decline in air temperatures. As these circumstances appear to favour the production of only one litter in the spring of each year, and since it is highly likely that the probability of survival from year to year is low in an animal of this size, selection will favour individuals which maximise reproductive effort during their first reproduction. It is this reproductive effort by males, including searching for females, prolonged and repeated copulation, and agonistic encounters with other males, which appears to lead to their mortality.

Monoestrous species of *Antechinus* occur only in areas which receive a reliable rainfall of at least 600 mm annually. This limitation on environment is hardly surprising, as the only insurance against reproductive failure is the capacity of some females to breed in a second and very occasionally a third year. *Antechinus* species with semelparous males are replaced in arid Australia by iteroparous species of *Antechinus* and *Sminthopsis*. This model suggests where other instances of semelparity in mammals might be found. Other small marsupials, with their relatively long periods of gestation and lactation, living in highly predictable environments, appear to be suitable candidates. Other instances of semelparity in mammals may occur where the growing season is particularly short – for instance, on mountain tops in middle latitudes (Braithwaite and Lee, 1976).

Population regulation

The pathways through which behaviour may effect reproduction, growth and survival are well documented for laboratory colonies of small mammals (Christian, Lloyd and Davis, 1965; Christian, 1971), but the significance of these relationships to natural populations has not been fully ascertained. The behaviourally mediated mortality of male *A. stuartii* provides one such illustration, and suggests other circumstances where this phenomenon may apply.

Mortality of adults is frequently associated with the breeding season in small mammals, and may be particularly severe where reproduction is highly synchronised, even in iteroparous species. Examples appear in microtine rodents where, in peak populations, there is often a conspicuous mortality following a highly synchronised bout of mating in spring, particularly *Lemmus trimucronatus* (Krebs, 1964), *Microtus pennsylvanicus* (Krebs, Keller and Tamarin, 1969; Christian, 1971). Krebs (1964) estimated that this mortality, concentrated over a few days, accounted for 30-67 per cent of individuals in a peak population of the Brown lemming (*L. trimucronatus*). As in *A. stuartii*, mortality was greater among males than females, and was associated with haphazard movements by individuals, and a noticeable increase in aggressiveness when handled. A possible pathogen, *Listeria monocytogenes*, has been isolated from lemmings (Nordland, 1959), but the data relating disease and changes in physiological state with this mortality are insufficient to confirm or refute an association.

Attempts to test and understand this association must be based upon more than a few superficial indices of physiological and pathological status, as our own example illustrates. They must also recognise that diseases may develop acutely in small mammals, by virtue of their high metabolic rates, and so may be difficult to detect. For this reason it may be best studied initially in populations where reproduction occurs in bursts, or where breeding is confined to a short and precisely timed period.

We are not suggesting that behaviourally mediated disease is a major factor governing microtine cycles. Instead, we anticipate that this and other studies of mortality associated with reproduction may provide a more appropriate assessment of the contribution of this phenomenon to fluctuations in numbers of small mammals.

Acknowledgements

We wish to acknowledge the assistance of Drs I. R. McDonald, I. Barker and I. Beveridge in preparation and criticism of this manuscript.

References

Barnett, J. L. (1973). A stress response in *Antechinus stuartii* (Macleay). *Aust. J. Zool.*, **21**, 501 – 13.

Barnett, J. L. (1974). Changes in the hydroxyproline concentration of the skin of *Antechinus stuartii* with age and hormonal treatment. *Aust. J. Zool.*, **22**, 311 – 8.

Bradley, A. J., McDonald, I. R. and Lee, A. K. (1975). Effect of exogenous cortisol on mortality of a dasyurid marsupial. *J. Endocr.*, **66**, 281 – 2.

Bradley, A. J., McDonald, I. R. and Lee, A. K. (1976). Corticosteroid-binding globulin and mortality in a dasyurid marsupial. *J. Endocr.* (in press).

Braithwaite, R. W. (1973). An ecological study of *Antechinus stuartii* (Marsupialia: Dasyuridae). M.Sc. Thesis, University of Queensland, Brisbane.

Braithwaite, R. W. (1974). Behavioural changes associated with the population cycle in *Antechinus stuartii* (Marsupialia). *Aust. J. Zool.*, **22**, 45 – 62.

Braithwaite, R. W. (1976). Social organisation of *Antechinus stuartii* (Marsupialia). *Aust. J. Zool.*

Braithwaite, R. W. and Lee, A. K. (1976). A mammalian example of semelparity. *Amer. Nat.* Manuscript.

Cheal, P. D., Lee, A. K. and Barnett, J. L. (1976). Haematological adjustments with age in *Antechinus stuartii* (Marsupialia). *Aust. J. Zool.* (in press).

Christian, J. J. (1971). Population density and reproductive efficiency. *Biol. Reprod.*, **4**, 248 – 94.

Christian, J. J., Lloyd, J. A. and Davis, D. E. (1965). The role of endocrines in the self-regulation of mammalian populations. *Rec. Prog. Hormone Res.*, **21**, 501 – 78.

Cole, L. C. (1954). The population consequences of life history phenomena. *Quart. Rev. Biol.*, **29**, 103 – 37.

De Moor, P., Meulepas, E., Hendrikx, A., Heyns, W. and Vandenschrieck, H. G. (1967). Cortisol-binding capacity of plasma transcortin: a sex-linked trait? *J. Clin. Endocr. Metab.*, **27**, 959 – 65.

Doe, R. P., Fernandez, R. and Seal, U. S. (1964). Measurement of corticosteroid-binding globulin in man. *J. Clin. Endocr. Metab.*, **24**, 1029 – 39.

Eberlein, W. R., Bongiovanni, A. M. and Rodriguez, C. S. (1967). Diagnosis and treatment: the complications of steroid treatment. *Pediatrics*, **40**, 279 – 82.

Green, R. H. (1972). The murids and small dasyurids in Tasmania. Part 5. *Antechinus swainsonii* (Thomas). Part 6. *Antechinus minimus* (Geoffroy). *Rec. Queen Vict. Mus.*, **46**, 1 – 27.

Krebs, C. J. (1964). The lemming cycle at Baker Lake, Northwest Territories, during 1959 – 62. *Tech. Pap. Arct. Inst. N. Am.*, **15**, 1 – 104.

Krebs, C. J., Gains, M. S., Keller, B. L., Myers, J. H. and Tamarin, R. H. (1973). Population cycles in small rodents. *Science*, **179**, 35 – 41.

Krebs, C. J., Keller, B. L. and Tamarin, R. H. (1969). *Microtus* population biology: demographic changes in fluctuating populations of *M. ochrogaster* and *M. pennsylvanicus* in southern Indiana. *Ecology*, **50**, 587 – 607.

Krebs, C. J. and Myers, J. H. (1974). Population cycles in small mammals. In: *Advances in Ecological Research*, Vol. 8 (ed. A. Macfadyen), Academic Press, New York, pp. 267 – 399.

Leonard, B. V. (1972). The effect of fire upon selected small mammals and leaf litter fauna in sclerophyll forest in southern Australia. M.Sc. Thesis, Monash University, Melbourne.

Levy, A. L. and Waldmann, T. A. (1970). The effect of hydrocortisone on immunoglobulin metabolism. *J. Clin. Invest.*, **49**, 1679 – 84.

Lohrenz, F. N., Seal, U. S. and Doe, R. P. (1967). Adrenal function and serum protein concentrations in a kindred with decreased corticosteroid-binding globulin (CBG) concentration. *J. Clin. Endocr. Metab.*, **27**, 966 – 972.

Mahoney, D. F. (1972). Immune response to hemoprotozoa. In: *Immunity in Animal Parasites* (ed. E. J. L. Soulsby), Academic Press, New York, pp. 301 – 42.

Margolis, J. and Kenrick, K. G. (1969). Two dimensional resolution of plasma proteins by combination of polyacrylamide disc and gradient gel electrophoresis. *Nature*, **221**, 1056 – 7.

Marlow, B. J. (1961). Reproductive behaviour of the marsupial mouse, *Antechinus flavipes* (Waterhouse) (Marsupialia) and the development of the pouch young. *Aust. J. Zool.*, **9**, 203 – 18.

Nordland, O. S. (1959). Host-parasite relations in initiation of infection. 1. Occurrence of listeriosis in Arctic mammals, with a note on its possible pathogenesis. *Can. J. Comp. Med.*, **23**, 393 – 400.

Rose, M. E. (1972). Immune response to intracellular parasites. In: *Immunity to Animal Parasites* (Ed. E. J. L. Soulsby), Academic Press, New York, pp. 365 - 88.

Schalm, O. W. (1965). *Veterinary Haematology*, 2nd edn, Lea and Febiger, Philadelphia.

Seal, U. S. and Doe, R. P. (1966). Corticosteroid-binding globulin: Biochemistry, physiology and phylogeny. In: *Steroid Dynamics* (ed. G. Pincus, T. Nakao and J. F. Tait), Academic Press, New York, pp. 63 – 90.

Tait, J. F. and Burstein, S. (1964). *In vivo* studies of steroid dynamics in man. In: *The Hormones*, Vol. 5 (ed. G. Pincus, K. V. Thimann and E. B. Astwood), Academic Press, New York, pp. 441 – 557.

Taylor, J. M. and Horner, B. E. (1970). Gonadal activity in the marsupial mouse, *Antechinus bellus*, with notes on other species of the genus (Marsupialia: Dasyuridae). *J. Mammal.*, **51**, 659 - 68.

Wilbur, H. M., Tinkle, D. W. and Collins, J. P. (1974). Environmental certainty, trophic level, and resource availability in life history evolution. *Am. Nat.*, **108**, 805 - 17.

Wintrobe, M. M. (1967). *Clinical Hematology*, 6th edn, Lea and Febiger, Philadelphia.

Wood, D. H. (1970). An ecological study of *Antechinus stuartii* (Marsupialia) in a south-east Queensland rain forest. *Aust. J. Zool.*, **18**, 185 - 207.

Woollard, P. (1971). Differential mortality of *Antechinus stuartii* (Macleay): nitrogen balance and somatic changes. *Aust. J. Zool.*, **19**, 347 – 53.

Woolley, P. (1966a). Reproduction in *Antechinus* spp. and other dasyurid marsupials. *Synp. Zool. Soc. London*, **15**, 281 - 94.

Woolley, P. (1966b). Reproductive biology of *Antechinus stuartii* Macleay (Marsupialia: Dasyuridae). Ph.D. Thesis, Australian National University, Canberra.

Woolley, P. (1973). Breeding patterns, and the breeding and laboratory maintenance of dasyurid marsupials. *Exp. Animals*, **1**, 161 – 72.

14 Imbalance in the sex ratio and age structure of the Red kangaroo, Macropus rufus, in central Australia

A. E. Newsome*

Introduction

The Red kangaroo, *Macropus rufus*, inhabits the better-grassed regions of arid and semi-arid Australia, and is one of the best-studied marsupials (for summaries, see Frith and Calaby, 1969, and Newsome, 1971, 1975). The species' requirement for green herbage to eat (Griffiths and Barker, 1966; Chippendale, 1968; Storr, 1968; Bailey, Martensz and Barker, 1971) dominates its ecology in that capricious environment. Thus the Red kangaroo is nomadic (Newsome, 1962, 1965a, b; Frith, 1964; Bailey, 1971), and breeding is variable (Newsome, 1962, 1965c; Frith and Sharman, 1964; Bailey, 1967). In central Australia the proportions of females breeding and of young surviving the full eight months of pouch-life are controlled by the food supply (Newsome, 1966b). Half the females cease breeding after 3 – 5 months' drought (defined in Newsome, 1966a; and see below), and half the pouch-young die after only 1.5 – 2.5 months' drought (Newsome, 1965c). No young survive 8 months' drought. Large gaps may be expected, then, in age structures.

This study examines age structure of each sex in populations inhabiting central Australia from 1959 to 1962. It also examines the rainfall record of the past century there, to detect periods favourable for population growth, and relates their occurrence to the Red kangaroo's chance of survival as a species.

The study area of about 1050 km² lies north of the MacDonnell Ranges, 20 – 40 km north of Alice Springs (23°26'S; 133°35'E), and along the Tropic of Capricorn. The environment is arid (Meigs, 1953). Annual rainfall is erratic, averaging 26 cm; evaporation is high, at 250 cm per year. Maximum summer temperatures average 31 – 35 °C. The general environment was described in detail in Perry (1962) and specifically for the Red kangaroo by Newsome (1965a).

Methods

Samples

Samples of kangaroos were shot at night using a spotlight and a four-wheel-drive vehicle (Newsome, 1965c). In all, 16 samples were collected between October

*Dr Alan Newsome is a graduate of the Universities of Queensland and Adelaide, where he has also held teaching and research posts. Since 1966 he has been a Principal Research Scientist in the Division of Wildlife Research, C.S.I.R.O., Canberra. In 1972 he held a Senior Fulbright Scholarship at Berkeley, USA, and was a Research Associate of the Museum of Vertebrate Zoology. He is especially interested in the ecology of Australian vertebrates, including rodents, dingoes and marsupials.

221

1959 and October 1962, a period that included severe drought (1961) and good rains (1962). Samples were large, mostly about 150 each (range: 79 - 197), totalling 1991 animals, of which 1516 were females and 475 males. Victims were selected randomly from groups of kangaroos encountered along set transects about 100 km long on the open plains, except on three occasions when the hunt followed good general rains. Kangaroos were rare on the open plains then and tended to stay inaccessibly in the mulga woodlands (*Acacia aneura*) (Newsome, 1965a, b); so all animals found were hunted.

Nine of the 16 samples were collected in two parts each to study population structure and reproduction. Thus the first 100 animals were selected randomly as described above to study population structure but, to save unnecessary killing, only mature females (those with second molar fully erupted; see below) were selected for the study of reproduction to bring the complement of mature females up to 100 (Newsome, 1965c). In this way 322 of the 1516 females were collected.

Age

Age in the Red kangaroo, as with some other macropodids, can be estimated by the sequence of eruption of molars, and by their progression anteriorly along the jaw (Sharman, Frith and Calaby, 1964). Data on tooth eruption were kindly made available by Mr J. H. Calaby from captive animals held at the Division of Wildlife Research, C.S.I.R.O., Canberra, for over 16 years, and analysed by the late Mr G. McIntyre to 'age' Red kangaroos beyond 11 years, the limit in Sharman *et al.*'s (1964) data. The following equation was obtained:

$$\text{tooth eruption} = -1.813 + 2.701 \log A$$

where A was the age in months.

The eruption patterns from one molar in full eruption to the next was divided into five equal steps, each given appropriate decimal rank, 0.2, 0.4, etc. Each of the four molars was allocated a Roman numeral according to its order of eruption. The equation indicated that the first molar, MI, was fully erupted at a mean of 11 months, MII at 25.8 months, MIII at 60.6 months and MIV at 142.1 months. These values differ a little from those of Sharman *et al.* (1964) owing to the extra data, and the decimal ranking is double theirs to provide a continuous scale.

A second method of 'ageing' based on molar progression, was used once MIV was fully erupted. Progress of MIV in fifth parts past the zygomatic process (see Sharman *et al.*, 1964, for illustrations) provided another decimal ranking for the following equations:

$$\text{molar progression} = -2.849 + 2.937 \log A$$

ignoring sexual differences;

$$\text{molar progression} = -2.251 + 2.700 \log A$$

for males alone;

$$\text{molar progression} = -2.559 + 2.700 \log A$$

for females.
Old kangaroos were 'aged' by all three methods.

There are problems of extrapolation: firstly, the particular one of using data from the temperate climate of Canberra on kangaroos from the arid subtropics, and, secondly, the general problem of using data from captive animals on wild ones. The latter problem remains, but the former was partially resolved.

Eight Red kangaroos from central Australia were reared there during the study, and 32 molar patterns were obtained at known ages up to a limit of MIII at full eruption ('age': 60.6 months). There were no significant differences between this limited set of data and the larger set used here.

Periods favouring reproduction

An empirical method of analysing weather records in central Australia was devised to compare droughts of different intensities and in different seasons (Newsome, 1966a), and also to measure their effects on fertility and fecundity in Red kangaroos (Newsome, 1965c). The method is essentially a profit and loss account of daily rainfall and potential evaporation of soil moisture, as expressed in a Drought Index, and its biological basis is the response of pastures.

Pastures inhibited by drought respond to falls of rain ranging from about 2.9 cm in the month of January to about 0.7 cm in July (Slatyer, 1962). Pastures respond well should extra rain fall, keeping the soil moist for the whole month. Since water evaporates from the soil at about one-fifth the rate from a standard evaporimeter, an extra 5.9 cm rain is needed on average in January for good growth and 1.5 cm in July. Appropriate values of rainfall and evaporation per month were provided by Slatyer (1962) for Alice Springs.

When the rain falling on the study area (average of 5 – 10 recording stations) in any month exceeded the stated amount needed for pastures to respond, all female Red kangaroos were found to be breeding (Newsome, 1965c). If no more good rain fell for some months, breeding and survival of pouch-young declined predictably according to the value of the Drought Index (Newsome, 1965c). The quantity of food for kangaroos also declined predictably as the Drought Index increased (Newsome, 1966b).

Daily rainfalls from the study area since 1957 and from Alice Springs since 1875 were analysed to derive periods of pasture growth and intervening droughts. In particular, periods of pasture growth were calculated annually from December one year to November the next from 1941 onwards. A month's precession on the calendar year was allowed because pregnancy in the Red kangaroo is about five weeks long (Sharman and Pilton, 1964). Young born in January one year were therefore conceived the previous year.

Results

Sex ratio

The 1669 animals in the population sample are grouped by 'age' (based on molar eruption) and sex in Table 14.1. The 925 pouch-young and joeys-at-heel also obtained in sampling (Newsome, 1965c) have been included. Their sex ratio, 1.17 males to 1.00 females, was significantly imbalanced because there were different numbers of males and females in pouch-young specifically aged 5 – 6 months

Table 14.1 Change in the sex ratio of Red kangaroos with age (population data only)

Ascribed age (years)	<1*	1–2	2–3	3–4	4–5	5–6	6–7	7–8	8–9	9–10	10–11	11–12	>12	Totals
Males	499	47	45	44	31	30	65	45	37	14	25	9	71	962
Females	426	54	55	100	76	75	98	139	86	29	87	26	381	1632
Totals	925	101	100	144	107	105	163	184	123	43	112	35	452	2594
Sex ratios	1.17	0.87	0.82	0.44	0.41	0.40	0.66	0.24	0.43	0.48	0.29	0.35	0.19	0.59

*From Newsome (1965c)

Table 14.2 The number of Red kangaroos in each tooth class

Molar eruption	0.2	0.4	0.6	0.8	MI	L2	L4	L6	L8	MII	IL2	IL4	IL6	IL8	MIII	IIL2	IIL4	IIL6	IIL8	MIV	Totals
Ascribed age (months)	6.0	6.6	7.8	9.3	11.1	13.0	15.5	18.1	21.5	25.8	30.6	36.4	43.1	51.1	60.6	71.8	85.2	101.0	120.0	142.0	
Males — population sample	4	7	3	3	2	12	20	15	18	16	26	18	19	22	18	69	46	52	27	78	475
Females — population sample	7	6	7	8	11	12	17	22	32	18	50	36	39	72	38	118	146	100	92	363	1194
Females — extra reproductive sample										2	5	5	6	18	12	46	57	19	26	126	322
Totals	11	13	10	11	13	24	37	37	50	36	81	59	64	112	68	233	249	171	145	567	1991
No. of samples containing each class	4	4	3	5	8	12	12	10	13	10	14	15	16	15	15	16	16	16	16	16	16

(Newsome, 1965c). This difference was probably due to chance for, otherwise, the sex ratio was 1 : 1, and was so at birth.

Sex ratios could also be accepted as 1 : 1 in animals up to three years old. From then on, however, the numbers of females progressively outnumbered males, reaching five to one in old animals.

Population structure

The 'age' distribution based on molar eruption is shown in Table 14.2 for all 1991 Red kangaroos. Frequencies for the 322 supplemental mature females, shown separately, do not differ significantly from frequencies for the other females excepting the two youngest classes, which are a little low because they would normally have included some immature females.

Frequencies in each tooth class increased from young to old, quite the reverse of the expected age structure of populations. This was especially so among females. Molar eruption was truncated at MIV, so artificially enhancing the trend, but few samples contained young animals, whereas all contained old ones.

The 'ages' of all animals were also used to plot their years of birth in Figure 14.1. Every animal with MIV fully erupted was assigned two other 'ages' according to its molar progression. Because variability in age is large for such molar progressions, these two methods probably spread the data widely, so dampening real trends. Thus the mean of all 'ages' was calculated for each animal and appropriate frequencies plotted on Figure 14.1 also. An increase pro rata of 19.3 per cent in the frequencies of animals up to three years old would allow for the inclusion of the supplemental mature females in other frequencies, but would not significantly alter trends.

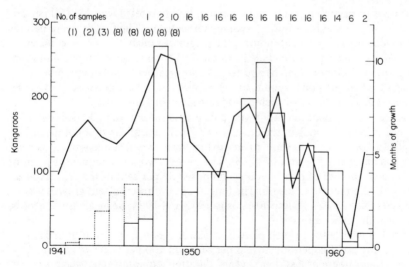

Figure 14.1. Estimated years of birth of 1991 Red kangaroos from central Australia and the estimated periods of pasture growth each year (line). The solid histogram represents kangaroos 'aged' on molar eruption, and the dotted one, old animals 'aged' by molar eruption and molar progression averaged (see text).

Results indicate three peaks in births, in 1948, 1955 and 1958. These peaks coincide with long periods of pasture growth in those years, estimated at 10.3, 8.4 and 5.6 months, respectively (see Figure 14.1). Similarly, troughs in the numbers of births coincide with droughts, the worst being the most recent, in 1961, when so little rain fell that pastures grew for about half a month only. Very few young survived that year (Newsome, 1965c). The next year, 1962, was a wet one and a large crop of young was produced (Newsome, 1965c), but not early enough to have been included in any except the two late samples that year (see Figure 14.1). In October that year 30 per cent of the population were joeys-at-heel (Newsome, 1965c).

There was an extraordinary sequence of wet years from 1942 to 1950, especially 1947 to 1949 (see Figure 14.1), and an exceptionally large number of animals in the samples was born then. Thus 599 out of 1669 (35.9 per cent) were older than 10 years, and 452 (27.8 per cent) were older than 12 years. The oldest animals were born as early as 1946, or perhaps 1942, depending on the 'ageing' technique used.

Pasture growth and drought, 1875 - 1974

There were an estimated 214 periods favouring growth of pastures during the past century (48 per cent of the time), and 213 intervening droughts. Most periods, favourable or unfavourable, were short-lived, the former being 2.7 months long on average, and the latter, 2.9 months. The longest continuous period of growth was 34 months (1920 - 3) and the longest continuous drought 14 months (1963 - 4). Very much longer periods of discontinuous dry weather prevailed, however, the longest being 6.5 years (1958 - 65) (Bureau of Meteorology, Aust., 1965).

Periods of pasture growth were separated into two sets favouring 100 per cent and 50 per cent survival of pouch-young, respectively. The first set were periods continuous for at least eight months, the length of pouch-life. The second set also spanned eight months but included short droughts no longer than one month in hot weather (October - March) and two in cool weather (April - September), enough for half the pouch-young in the population to have perished (Newsome, 1965c). The frequencies of such periods are given in Table 14.3.

There were 17 continuously favourable periods permitting 100 per cent survival of pouch-young in the century (17.7 per cent of the time) of mean length 12.5 months. Five were longer than a year, of which one was 1.6 years long and another 2.8 years (1920 - 3). The 29 less favourable sequences permitting 50 per cent survival of pouch-young comprised 31 per cent of the century and averaged 12.8 months in length. The longest span of intermittent growth was 3.1 years (also 1920 - 3). Recent rains in 1973 and 1974, being all-time records, may provide longer spans.

The intervals between such periods of pasture growth are grouped in Table 14.4. About half the periods (for 50 per cent and 100 per cent survival of pouch-young) were of less than 2 years' duration. No interval between periods favouring 50 per cent survival exceeded 13 years, and most (22 out of 28) were less than 4 years. Intervals between periods favouring 100 per cent survival of pouch-young are longer, half exceeding 4 years, four exceeding 11 years and three exceeding 13 years.

Table 14.3 Periods of pasture growth favouring survival of pouch-young in central Australia during the century 1875 – 1974

Expected survival of pouch-young (%)	Estimated length of period (months)													Totals
	8 – 9	9 – 10	10 – 11	11 – 12	12 – 13	13 – 14	15 – 16	18 – 19	19 – 20	21 – 22	23 – 24	33 – 34	37 – 38	
100	4	2	2	4	1	1	1 Mean: 12.5	0	1	0	0	1	0	17
50	9	6	3	3	1	0	0 Mean: 12.8	1	2	2	1	0	1	29

Table 14.4 Intervals between growing periods favouring survival of pouch-young in central Australia during the century 1875 – 1974

Expected survival of pouch-young during growing periods (%)	Estimated length of intervals (months)							Totals
	0 – 5	5 – 10	10 – 20	20 – 50	50 – 100	100 – 150	150 – 200	
100	1	1	4	2 Mean: 65.3	4	1	3	16
50	1	6	9	7 Mean: 28.2	4	0	1	28

Discussion

Sampling errors

Three factors may have modified the accuracy of the age distribution: avoidance behaviour of kangaroos hunted, errors in 'ageing', and intense shooting of kangaroos for skins in the 1950s.

The first problem is thought to be minimal because large males were never seen especially reacting to avoid a slowly approaching motor vehicle, by day or night, as reported in less arid regions in New South Wales and Queensland (Frith and Sharman, 1964). Kangaroos were certainly flighty when in good condition, but sex ratios in samples taken then are not significantly different from those taken in drought, when kangaroos were very easily shot. Indeed all 16 samples taken displayed the same gross imbalance in sex ratios.

The second problem is difficult to resolve. McIntyre (1964), examining the problem in captive animals, found an array of molar patterns for every age. Each array overlapped the next one, sometimes two or three of them in young animals. For example, a two year old animal could have any molar eruption pattern from MI.6 to MII.4, a three year old from MII.2 to MII.8, a five year old from MII.8 to MIII.4, and so on (ignoring sexual differences and natural mortality).

By interpolation within the probability matrix provided by McIntyre, the degree of approximation involved in nominating a given mean age was calculated for a given tooth class. The probabilities in each of the overlapping arrays at that tooth class were then summed. Thus, for the five eruption stages of MI, the over-lapping frequencies added up to a value 0.76 of the expected value 1.0 (range: 0.50 - 0.95); of MII, 0.83 (0.66 - 0.90); of MIII, 0.91 (0.78 - 1.03); and of MIV, 0.89 (0.73 - 1.23). I have therefore accepted the frequencies given in Figure 14.1 as reasonable.

The third problem is most important. Shooting for hides began commercially in central Australia in 1951 (W. T. Hare, personal communication) after an 8 - 10 year period which greatly favoured breeding and survival of Red kangaroos (see Figure 14.1). Old kangaroos of large size should have been numerous, and indeed were probably responsible for attracting six to eight professional shooters to the study area, the first time any commercial harvesting was attempted there. No official records exist but, by all accounts, harvests were very high. Some shooters at their best took 7000 and more skins in a year (G. Holt, personal communication), a harvest calculated at about 6 kangaroos/km^2. Kangaroos must have been very abundant indeed.

That there were still so many old animals 10 years later despite the heavy shooting emphasises just how great the recruitment must have been in the 1940s. The peak in birth dates for 1955 in Figure 14.1 would not have been diminished to anywhere near the same extent as earlier ones because only two to three shooters were operating then.

Sex ratio

The progressive excess of females in the population structure began after Red kangaroos reached three years old. This is the age of sexual maturity, reached in males around 36 - 43 months and in females around 30 - 36 months (Newsome,

1965c). It is also the age when sexual dimorphism becomes apparent. Growth in the male far outstrips that in the female thereafter (see Figure 14.2), such that few females were heavier than 25 kg in central Australia (though they can be elsewhere), whereas some males were double this weight. The heaviest female shot was 29 kg and the heaviest male 56.7 kg.

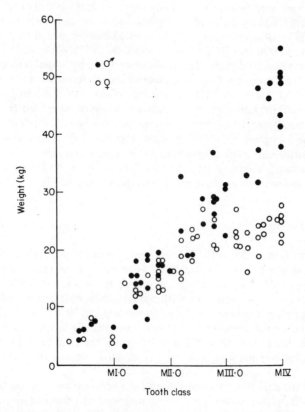

Figure 14.2   Dimorphic growth in Red kangaroos in central Australia.

The advantage of large size to the male Red kangaroo has never been studied, but presumably has some link with sexual success, as in seals (Bartholomew, 1970). If so, there must be social interactions in Red kangaroos to a degree not yet measured. Size seems to be related, however, to heavy mortality among males because of the grossly uneven sex ratios of five females to one male in old animals. Selective predation by man in the early 1950s was probably the main cause, kangaroo shooters taking the larger animals for greater reward.

Any natural mortality would have been masked by heavy human predation. It may be larger in male kangaroos just as in mountain sheep, *Ovis canadensis* and *O. dalli* (Caughley, 1966; Geist, 1971), which are sexually dimorphic also and live to about the same age. The rigours of the rut (Geist, 1971) and predation from wolves (Murie, 1944) are the main causes of mortality in the sheep; but there is no rut in kangaroos, for they breed all year round, and there is little predation from dingoes, *Canis familiaris dingo* (A. Newsome and L. Corbett, unpublished).

Differential mortality, polygamy and an even sex ratio at birth, all characteristics of the Red kangaroo, can be evolutionary associates according to Hamilton (1967) and Leigh (1970), although Emlen (1968) does not agree. Kangaroos of all species may have been vulnerable to predation in evolutionary history. Their main predators probably included aboriginals, who have been present in Australia for over 30 000 years (Mulvaney, 1969), and also dingoes, which have been present for between 3000 and 8000 years (MacIntosh, 1975). Although selection for small size in males would probably have been opposed by sexual selection, the females may have been free to grow smaller so long as longevity and reproduction were not reduced. Increased mortality among males would result in polygamy.

Although modern hunting may concentrate on males, so sparing the breeders, the full breeding potential of the population may not be realised despite opportunistic polygamy. Pregnancy rates can be quite low during drought owing to heat-induced infertility in males (Newsome, 1973). Even in good seasons when all males were potent, fecundity did not exceed about 70 per cent, possibly because there were too few adult males to go around (Newsome, 1973).

Population structure

It is clear that, from 1959 to 1962, I was studying an old population of Red kangaroos basically generated in the late 1940s and partially regenerated in 1953 - 5 and in 1958. Its inverted age structure greatly contrasts with conventional ideas of populations, and with age structures obtained from Red kangaroos in the milder temperate environment of western New South Wales about 1500 km to the south-east of central Australia (Frith and Sharman, 1964), and also with that of the other arid zone kangaroo, the Euro or Hill kangaroo (*Macropus robustus*) in arid coastal country in north-western Australia (Ealey, 1964).

Although the environments and climates of north-western and central Australia are similarly arid, Ealey (personal communication) sampled his populations in 1956 after four to five favourable rainy seasons. The contrast between age structure obtained from the two kangaroos of the arid zone highlights the importance of time and place in obtaining population samples of large animals, especially in semi-arid and arid environments. Presumably, the population structure of Red kangaroos in central Australia in the late 1940s would have appeared balanced too.

The results here do establish the great importance of unusually rainy years to the Red kangaroo. Of particular importance are the sporadic runs of flush years. They induce great pulses in the long-term population trends — the greater the run of good years, the greater the pulse. Recruitment surges similarly associated with rainfall have been found in the Elephant (*Loxodonta africana*) (Laws, 1969). Living in mild tropical environments of east Africa and being so long-lived, elephants may not need such surges to tide populations over droughts. But it appears certain that the Red kangaroo has to capitalise on the sporadic flush years for that purpose.

Nothing is known of survival rates for juveniles (9 – 12 months old) or immature and subadult kangaroos (< 3 years old) in central Australia. The climatic records indicate, however, that the probability of weather being adequate for all pouch-young in any cohort to survive pouch-life is 0.18. The probability that its first year out of the pouch will also be favourable falls greatly to 0.06, and the probability during the time taken to sexually mature is quite low, 0.03. Similar probabilities for partly favourable seasons (50 per cent mortality of young) for such survival improve to 0.31, 0.14 and 0.05. (The chance for a female kangaroo to rear an offspring is really less than the figures indicate because one such favourable season is lost to the mother itself during its own pouch-life.)

The long interval between runs of good seasons that promote survival of pouch-young is likely to be a telling ecological variable. The interval between partly favourable breeding seasons (2.3 years), does not seem too severe, but it rises to 5.4 years for totally favourable seasons.

Gaps between runs of two consecutive years containing partly or wholly favourable eight month periods may be more significant for such long-lived animals. The mean intervals were 5.5 years (range, 0 – 14) and 11 years (range, 0 – 24), respectively, and 11.9 (0 – 44) and 22.3 (0 – 44) for runs of three consecutive years. Since two of these means and an upper limit approach the longevity of wild Red kangaroos (about 15 years, 20 years at the most; McIntyre, 1964, and see Figure 14.1), and one mean and three upper limits exceed it, shorter moist periods are clearly vital to the maintenance of populations. But it is the rarer rainy excesses which regenerate populations sufficiently to tide them over long periods of intermittent drought.

It would seem, then, that runs of two or more good years may be needed every 15 – 20 years in central Australia to prevent the risk of populations declining to extinction in many places. The really wet runs of years in the past century were 1920 – 3, 1945 – 50, 1966 – 9 and 1973 – 4. Before that, between 1875 and 1919, there were two lots of consecutively favourable years and two other sets partly favourable. The environment in central Australia, therefore, does not seem far from the limits of survival for the Red kangaroo.

Acknowledgements

I gratefully thank the following: Mr D. R. Stephens and Mr M. Mahon, who did much of the field sampling, often under trying conditions; Mr W. Pryor and the late Mr J. Gorey for permission to work on their cattle stations; Professors H. G. Andrewartha and G. B. Sharman, who supervised the field programme; and Dr H. J. Frith who advised. The regular and tedious collection of molar patterns in captive Red kangaroos at Canberra was carried out by Mr J. Merchant and other members of the Division of Wildlife Research, C.S.I.R.O., Canberra under the direction of Mr J. Calaby, who generously made the extensive data available for ageing my samples. The late Mr G. McIntyre very kindly calculated equations from the data. Dr G. Caughley and Professor W. Z. Lidicker critically read the manuscript and helped considerably with many points. I am most grateful also to the Department of the Northern Territory, in particular to Mr J. H. Whittem and Mr W. T. Hare, for the opportunity to conduct this research. Mr F. Knight drew the figures and Mrs C. Todd typed the manuscript.

References

Bailey, P. T. (1967). The ecology of the red kangaroo, *Megaleia rufa* (Desmarest), in north-western New South Wales. M.Sc. Thesis, University of Sydney.

Bailey, P. T. (1971). The red kangaroo, *Megaleia rufa* (Desmarest), in north-western New South Wales. I. Movements. *C.S.I.R.O. Wildlife Res.,* **16**, 11 – 28.

Bailey, P. T., Martensz, P. N. and Barker, R. (1971). The red kangaroo, *Megaleia rufa* (Desmarest), in north-western New South Wales. II. Food. *C.S.I.R.O. Wildlife Res.,* **16**, 29 – 39.

Bartholomew, G. A. (1970). A model for the evolution of pinniped polygamy. *Evolution,* **24**, 546 – 59.

Bureau of Meteorology, Aust. (1965). Meteorological aspects of the recent drought in central Australia. Meteorological Summary (September 1965). Commonwealth Bureau of Meteorology, Melbourne.

Caughley, G. (1966). Mortality patterns in mammals. *Ecology,* **47**, 906 – 18.

Chippendale, G. M. (1968). The plants grazed by red kangaroos, *Megaleia rufa* (Desmarest), in central Australia. *Proc. Linnean Soc. N.S. Wales,* **93**, 98 – 110.

Ealey, E.H.M. (1964). L'euro, ou kangarou des collines, *Macropus robustus,* dans le nord-ouest de l'Australia. *La Terre et la Vie,* **1**, 3 – 19.

Emlen, J. M. (1968). Selection for the sex ratio. *Am. Nat.,* **102**, 589 – 91.

Frith, H. J. (1964). Mobility of the red kangaroo, *Megaleia rufa. C.S.I.R.O. Wildlife Res.,* **9**, 1 – 19.

Frith, H. J. and Calaby, J. H. (1969). *Kangaroos,* Cheshire, Melbourne.

Frith, H. J. and Sharman, G. B. (1964). Breeding in wild populations of the red kangaroo, *Megaleia rufa. C.S.I.R.O. Wildlife Res.,* **9**, 86 – 114.

Geist, V. (1971). *Mountain Sheep,* University of Chicago Press.

Griffiths, M. and Barker, R. (1966). The plants eaten by sheep and by kangaroos grazing together in a paddock in south-western Queensland. *C.S.I.R.O. Wildlife Res.,* **11**, 145 – 67.

Hamilton, W. D. (1967). Extraordinary sex ratios. *Science,* **156**, 477 – 88.

Laws, R. M. (1969). Aspects of reproduction in the African elephant, *Loxodonta africana. J. Reprod. Fert., Suppl.,* **6**, 193 – 217.

Leigh, E. G. Jr (1970). Sex ratio and differential mortality between the sexes. *Am. Nat.,* **104**, 205 – 10.

MacIntosh, N. W. G. (1975). The origin of the Dingo: an enigma. In: *The Wild Canids* (ed. M. W. Fox), Van Nostrand Reinhold, New York, pp. 87 – 106.

McIntyre, G. A. (1964). The relation of tooth development to age in the red kangaroo. Appendix in: Sharman, G. B., Frith, H. J. and Calaby, J. H. (1964).

Meigs, P. (1953). World distribution of arid and semi-arid homo-climates. In: *Reviews of Research on Arid Zone Hydrology,* UNESCO, Paris, pp. 203 – 10.

Mulvaney, D. J. (1969). *The Prehistory of Australia,* Thames and Hudson, London.

Murie, A. (1944). *The Wolves of Mt. McInley,* Fauna Nat. Parks U.S., Fauna Ser. 5.

Newsome, A. E. (1962). The biology of the red kangaroo, *Macropus rufus* (Desmarest), in central Australia. M. Sc. Thesis, University of Adelaide.

Newsome, A. E. (1965a). The abundance of red kangaroos, *Megaleia rufa* (Desmarest), in central Australia. *Aust. J. Zool.,* **13**, 269 – 87.

Newsome, A. E. (1965b). The distribution of red kangaroos, *Megaleia rufa* (Desmarest), about sources of persistent food and water in central Australia. *Aust. J. Zool.*, **13**, 268 – 99.

Newsome, A. E. (1965c). Reproduction in natural populations of the red kangaroo, *Megaleia rufa* (Desmarest), in central Australia. *Aust. J. Zool.*, **13**, 735 – 59.

Newsome, A. E. (1966a). Estimating severity of drought. *Nature*, **209**, 904.

Newsome, A. E. (1966b). The influence of food on breeding in the red kangaroo in central Australia. *C.S.I.R.O. Wildlife Res.*, **11**, 187 – 96.

Newsome, A. E. (1971). The ecology of red kangaroos. *Aust. Zool.*, **16**, 32 – 50.

Newsome, A. E. (1973). Cellular degeneration in the testis of red kangaroos during hot weather and drought in central Australia. *J. Reprod. Fert. Suppl.*, **19**, 191 – 201.

Newsome, A. E. (1975). An ecological comparison of the two arid zone kangaroos of Australia, and their anomalous prosperity since the introduction of ruminant stock to their environment. *Quart. Rev. Biol*, **50**, 389 – 424.

Perry, R. A. (ed.) (1962). *Lands of the Alice Springs Area, Northern Territory, 1956 – 57,* C.S.I.R.O. Aust. Land Res. Series No. 6.

Sharman, G. B., Frith, H. J. and Calaby, J. H. (1964). Growth of the pouch young, tooth eruption and age determination in the red kangaroo, *Megaleia rufa. C.S.I.R.O. Wildlife Res.*, **9**, 20 – 49.

Sharman, G. B. and Pilton, P. E. (1964). Life history and reproduction of the red kangaroo (*Megaleia rufa*). *Proc. Zool. Soc. London,* **142**, 29 – 48.

Slatyer, R. O. (1962). Climate of the Alice Springs area. In: *Lands of the Alice Springs Area, Northern Territory, 1956 – 57,* C.S.I.R.O. Aust. Land Res. Series No. 6, pp. 109 – 28.

Storr, G. M. (1968). Diet of kangaroos (*Megaleia rufa* and *Macropus robustus*) and merino sheep near Port Hedland, Western Australia. *J. Roy. Soc. West. Aust.*, **51**, 26 – 32.

Section 4 Anatomy of marsupials

The anatomy of marsupials was examined with characteristic thoroughness by the comparative anatomists of the 19th century, and interpreted – where interpretation was felt necessary – as representing a primitive stage in the evolutionary sequence between reptiles and higher mammals; time, it seemed, had stood still for the marsupials but moved on for the eutherians. Modern anatomists build on the work of their predecessors but interpret differently. Those who have contributed to this section are concerned with function rather than homologies; with insights into the ecology, physiology, mechanics and behaviour of their study animals, they interpret in terms of present-day adaptive advantage – or at least feel that further study of the living animal is needed if they do not fully understand the function of a structure. Dr. Barbour (Chapter 15) includes a number of key references to earlier studies of marsupials, but his contribution is mainly a review of more recent work in which form and function are considered together. Dr. Pamela Parker's anatomy-based studies of marsupial and eutherian patterns of reproduction (Chapter 16) fit this specification exactly; her conclusions are a further refutation of the concept that marsupials survive despite the inadequacies of their outmoded systems.

In Chapter 17 Professor Crompton and his colleagues show the workings of the jaw musculature of the Virginia opossum *Didelphis virginiana*. The elongate face and jaws, and tribosphenic molars of this species suggest that its jaw musculature may have changed little from those of early mammals, and so provide a better starting point for comparative study than the jaw muscles of more specialised forms. Chapter 18, contributed by Woolley and Webb, draws attention to a curious appendage of the penis found in many but not all dasyurids. Though serological and other biochemical techniques are rapidly giving taxonomy a welcome new look, diagnosis of genera and species still depends largely on combinations of minor features of this kind, and investigations of their incidence, functions, homologies and taxonomic significance are by no means outmoded.

15   Anatomy of marsupials

R. A. Barbour*

Early studies often were made with the assumption that marsupials are fundamentally more primitive than eutherian mammals. Findings, however, show more
similarities than differences and the differences do not always indicate two levels
of evolutionary advancement but sometimes just different ways of dealing with
the same biological problem. Current thought now tends to regard marsupials as
representing an alternative rather than an inferior path in evolution. Among
marsupials themselves there are marked differences related to the wide adaptive
radiation they have undergone (see Wood Jones, 1923; Sharman, 1959a; Troughton,
1959), and they show a range of degrees of specialisation in various aspects of
their morphology.

The scope of a brief review like this necessarily must be limited, and particular
attention is given to the more or less unique features of marsupial anatomy, primitive features where they exist and interesting points of individual species, genera
or families. Emphasis is placed on recent studies. The subject is treated system-
atically and the sequence has been chosen to give priority to those areas that have
achieved the greatest prominence from the point of view of taxonomy or general
research interest. Little attention, however, is given to taxonomy itself or to
theories of marsupial evolution, which are discussed elsewhere in this symposium.
Embryology also is largely ignored. For brevity's sake specific names are often
omitted: *Trichosurus* always refers to *vulpecula* and *Didelphis* unqualified can be
assumed to refer to *marsupialis virginiana*; similarly with *Setonix (brachyurus)*,
Tarsipes (spenserae), *Pseudocheirus (peregrinus)*, *Myrmecobius (fasciatus)*,
Phascolarctos (cinereus), *Notoryctes (typhlops)*, *Perameles (nasuta)*, *Petrogale*
(xanthopus) and *Thylogale (eugenii)*.

The bibliography, while extensive, is incomplete, the many omissions being
mainly older papers which the reader can find by consulting the more recent
works cited. Historically the first definitive paper on marsupial anatomy is
probably that of Tyson (1698) on *Didelphis*. Little more appeared for 100 years;
then commenced the steady and increasing stream of works of the nineteenth
century, from the early part of which stands out prominently the name of
Richard Owen, who, apart from his many papers, attempted one of the first

*Dr Robert A. Barbour graduated M.B., B.S. from the University of Adelaide in 1955
and M.D. in 1962. Since 1956 he has been on the staff of the Adelaide Department of
Anatomy and Histology, where he is now a senior lecturer and teaches histology and gross
anatomy: during two sabbatical years, spent mainly in England, he studied electron microscopy and histochemistry. He has worked in the field of marsupial anatomy at both gross
and microscopic levels and his current research interest is in the histology of endocrine and
salivary glands.

major overviews (Owen, 1866, 1868). In the last 80 – 90 years studies have tended
to concentrate more on individual organs or systems and have been more detailed,
some especially bringing to light information at the microscopic level.

Female reproductive system

This system has been studied extensively because of interest arising from some
unique features of the internal genitalia and the birth of extremely small and
immature young followed by their nurture within the pouch possessed by
females of most species. Various general accounts have summarised, and can
provide a guide to, the abundant literature (Nelsen and Maxwell, 1942; Wood
Jones, 1943; Pearson, 1945; Eckstein and Zuckerman, 1956; Sharman, 1959b;
Kean, 1961, 1967): they provide the basis for what follows.

The female marsupial possesses two ovaries, two oviducts, two separate uteri
and a unique vaginal apparatus that opens into a urogenital sinus. Morphological
changes occur in an oestrous cycle that resembles that of eutherian mammals
(well summarised by Tyndale-Biscoe, 1973).

The ovaries present most of the usual mammalian features. Maturing follicles,
however, exhibit no corona radiata (Sharman, 1959b) and the theca interna is
often not prominent and may become less obvious as the follicle matures (Hughes,
Thomson and Owen, 1965). The marsupial ovum is relatively large (Sharman,
1959b; Tyndale-Biscoe, 1973) with a diameter up to 0.25 mm (Tyndale-Biscoe,
1974). Following ovulation a corpus luteum forms from the remaining membrana
granulosa by enlargement of its cells accompanied by ingrowth of vascular connective tissue from the theca externa. The contribution from the theca interna seems
unclear; Mossman and Duke (1973) conclude that marsupials have no theca lutein
cells, although Sharman (1955) mentions some in *Setonix*, as do Kean, Marryatt
and Carroll (1964) in *Trichosurus*. Shorey and Hughes (1973), however, mention
only granulosa lutein cells in their electron microscopic study on the latter. The
corpus luteum of pregnancy is the same as that of the non-pregnant state. Interstitial cells are not prominent (Mossman and Duke, 1973); O'Donoghue (1916)
found them present only in diprotodonts. Hughes *et al.* (1965) found interstitial
cells more abundant in young animals and during oestrus and pregnancy in
Pseudocheirus. They may sometimes have an origin other than from atretic
follicles (O'Donoghue, 1916; Hughes *et al.*, 1965; Tyndale-Biscoe, 1968).

The oviducts vary considerably between species in length and degree of
convolution: for example, they are short and simple in *Tarsipes* (de Bavay, 1951)
and *Bettongia lesueur* (Tyndale-Biscoe, 1968), long and convoluted in *Setonix*
(Waring *et al.*, 1955) and *Trichosurus* (Kean *et al.*, 1964). During its passage along
the oviduct the ovum acquires an albumen coat and a thin shell membrane, the
latter being absent from eutherian eggs (Hill, 1910). Hartman (1923) noted the
shell membrane glands in the oviduct in *Didelphis*.

The uteri converge towards their caudal ends, which lie close together but
always have separate external ora. The uterine wall has the usual layers. The
myometrium is predominantly circular (Hill, 1899; Flynn, 1911; Matthews,
1947) – or rather, presumably, a close spiral – and in *Didelphis* Lierse (1965) has
shown that the arrangement is simpler than in eutherians as the main layer consists of a single spiral rather than two opposite ones. The endometrium has a

connective stroma with cuboidal or columnar epithelium covering the surface and lining numerous glands. The presence of ciliated epithelial cells here, and in the oviduct, seems to depend on species; for example, they are present in *Didelphis* (Morgan, 1946) and *Pseudocheirus* (Hughes *et al.*, 1965), but not in *Trichosurus* (Kean *et al.*, 1964). The cervical portion may lack glands in some forms – for example, *Tarsipes* (Hill, 1900; de Bavay, 1951), *Myrmecobius* (Hill, 1900). The endometrium shows typical cyclical changes, including epithelial degeneration and replacement, increased stromal vascularity and oedema, and definite (but not marked) indications of secretory activity in the luteal phase. In *Trichosurus* Shorey and Hughes (1972) evince that epithelial replacement involves complete loss followed by regeneration from stromal cells.

The vaginal apparatus consists basically of two lateral vaginae each connecting a uterus to the urogenital sinus. During development the cranial end of each comes to lie adjacent to its fellow, with which it subsequently shares a common median septum. The chambers so formed, into which the uterine cervices open, are usually termed vaginal culs-de-sac and in many species unite as the median septum is ruptured at, or even before, the first parturition. The rest of each lateral vagina originally lies with the urethra and the ends of the ureters in a mass of connective tissue, the urogenital strand, that extends from the cul-de-sac region to the anterior end of the urogenital sinus. In most species this part, called the lateral canal, loops laterally beyond the confines of the strand in the adult, sometimes looping anteriorly as well to assume the form of the handle of an urn or even becoming convoluted – as in the Caenolestidae (Osgood, 1921; Baxter, 1935a). Posteriorly the lateral canals may unite into a short common canal before joining the urogenital sinus. The lateral canals serve as channels for seminal entry but not for parturition, which, probably in all species, occurs through a median passage which opens through the connective tissue of the urogenital strand. Generally this passage is transient, but in many of the Macropodidae, in *Tarsipes* and in some specimens of *Didelphis* (according to Ratcliffe, 1941) it becomes epithelialised and remains as a permanent median vagina. The distance from the cul-de-sac region to the urogenital sinus varies according to the family concerned (see Pearson, 1945); it is long in the Didelphidae, Peramelidae and Phascogalinae (Pearson and de Bavay, 1951) but small in most Australian families. The anterior portion of each lateral canal is usually somewhat enlarged to serve as a seminal receptacle, and diverticula may extend in a cranial direction between the uteri and bladder. These occur in the Peramelidae (Pearson, 1944, 1945; Eckstein and Zuckerman, 1956) and also some Macropodidae. Tyndale-Biscoe (1968) discusses them in the Macropodidae and considers that two distinct types may coexist. The whole vaginal complex except for the median vaginal canal has a muscular layer in its wall and is largely lined with stratified squamous epithelium continuous with that of the urogenital sinus; there is a variable region of cuboidal-columnar epithelium near the uterine openings. The vaginal complex shows changes with the oestrous cycle, including enlargement up to the time of oestrus followed by subsequent shrinkage, even if pregnancy occurs; this is a marked difference from the condition in eutherians (Kean *et al.*, 1964). Oestrous changes also include cornification of the stratified epithelium in some species, while in many epithelial thickening obstructs the narrow posterior parts of the lateral canals at non-breeding times.

The persistence of two lateral vaginae instead of their fusion into a single median vagina as in eutherians is usually explained by the position of the ureters, which in marsupials develop medial to, instead of lateral to, the Müllerian and Wolffian ducts. The former give rise to the female genital pathway as far as the posterior third of the lateral canal — the last part probably having a variable origin according to species from the urogenital sinus or Müllerian/Wolffian ducts (Buchanan and Fraser, 1918; Baxter, 1935b). This theory of the ureters preventing fusion of the vaginae is attributed to van den Broek (Bolliger, 1946) and has been followed by others (Pearson, 1945; Sharman, 1954, 1959a, b, 1965, 1970, 1974; Tyndale-Biscoe, 1973). Other theories proposed by Abbie (1941) and Kean (1961) suggest derivation of the whole lateral canal from the urogenital sinus, but these do not seem to take sufficient account of the embryological evidence, the finding of vestiges of the Wolffian duct related to the posterior part of the lateral canal in some species (Flynn, 1911; de Bavay, 1951; Tyndale-Biscoe, 1965, and others cited by Sharman, 1965), or the fact that the male genital duct (derived from the Wolffian duct) also lies lateral to the ureter (see Sharman, 1965).

The reversed relationship of the ureters and genital ducts (in both sexes) is regarded by Wood Jones (1943) as *the* fundamental difference between marsupials and eutherians; Tyndale-Biscoe (1973) also takes this view.

Placentation in marsupials has been summarised by Pearson (1949), Amoroso (1952) and Sharman (1959a, b, 1961). In contrast to eutherians marsupials have a yolk-sac (choriovitelline) placenta, the yolk-sac developing vascular and non-vascular areas, and there being no erosion of endometrium so that the embryonic and maternal circulations are separated by six layers of tissue. In an ultrastructural study on *Philander opossum* Enders and Enders (1969) showed that desmosomal attachments develop between the trophoblast and the maternal epithelium. The thickness of the separating tissues and the degree of vascularisation of the endometrium vary with species; for example, in some Phalangeridae the layers are thinner and there is greater vascularisation than in some Macropodidae (Sharman, 1961). In most marsupials there is no allantochorial contact. In *Phascolarctos, Vombatus* and *Dasyurus* some contact is made but the allantois probably plays no part in placental formation. In some Peramelidae, on the other hand, a functional chorioallantoic placenta is added to the yolk-sac one; here the maternal epithelium forms a syncytium that becomes vascularised by capillaries and to which there is close attachment of, or fusion with, the trophoblast (Hill, 1897; Flynn, 1923). This brings closer vascular contact but the endothelia usually remain intact. After parturition the foetal membranes are left in the uterus to be absorbed (Hill, 1897; Flynn, 1922, 1930; Amoroso, 1952; Sharman, 1959b). The marsupial placenta is non-secretory (Sharman, 1970) and this is probably related to the short gestation period, which, with only two possible exceptions (Sharman, Calaby and Poole, 1966), is shorter than the length of the oestrous cycle.

Male reproductive system

Externally the male genitalia are notable for the prepenial position of the scrotum (Owen, 1868; Bolliger, 1946; Eckstein and Zuckerman, 1956; Sharman, 1959b; Biggers, 1966) which houses the testes permanently in most species. In *Notoryctes*, however, the testes remain undescended (Stirling, 1891) and in wombats descend only in the breeding season (Sharman, 1959b). The penis is usually bifid, except in kangaroos (Owen, 1868; Eckstein and Zuckerman, 1956; Biggers, 1966), *Tarsipes* (Eckstein and Zuckerman, 1956) and *Notoryctes* (Stirling, 1891). The urethra terminates before the bifid end but two grooves extend along the two branches presumably to direct semen into the lateral vaginae of the female (Biggers, 1966).

The testes contain the usual seminiferous tubules and interstitial cells. The latter are abundant and show relatively little lipid in *Didelphis* (Christensen and Fawcett, 1961); paucity of lipid was also noted in *Trichosurus* by Green (1963a). Spermatogenesis occurs in a cyclical fashion that appears to be similar in pattern and timing to that occurring in eutherians (Sharman, 1959c; Setchell and Carrick, 1973). Spermatozoa vary in form from family to family (Biggers, 1966). In the American marsupials three morphological types of sperm head occur (Biggers and Creed, 1962; Biggers, Creed and de Lamater, 1963; Biggers and de Lamater, 1965): the *Didelphis* type common to several genera and having a hook-shaped or inverted V-shaped head with the middle piece attached to one side in the concavity; the *Caluromys* type with a saucer-shaped head and the middle piece attached to the convex side; and the *Caenolestes* type with the middle piece attached to a depression on one side. Phillips (1970a, b) has described the fine structure and development of the sperm of *Caluromys philander*, which he likens to a carpet tack. Similarly, Rattner (1972) has described the development of the hook-shaped head in *Didelphis* and *Marmosa mitis*. In *Caluromys* (Phillips, 1970a) and *Didelphis* (Guraya, 1971) mitochondria enlarge and develop concentric cristae during spermiogenesis. Hughes (1965) reviewed much of the earlier work on marsupial sperm and studied many Australian forms, among which he found five types all with a groove on one aspect of the head for attachment of the middle piece but varying in detail in the five families he examined. In general, marsupial sperm have thick tails (Biggers, 1966) and some are very long, especially in the Dasyuridae and to a lesser degree the Peramelidae (Hughes, 1965). Sapsford, Rae and Cleland (1967, 1969, 1970) made a very detailed ultrastructural study of spermiogenesis in *Perameles* and found it to vary only in detail from the process in eutherians. In an associated study Sapsford and Rae (1969) found that Sertoli cells appear to dissect excess cytoplasm from late spermatids and help push them into the tubular lumen. In all American species (but not every animal) many of the spermatozoa undergo a peculiar phenomenon of pairing when they reach the epididymis; this does not occur in any Australian species (see Biggers and Creed, 1962; Biggers and de Lamater, 1965).

In most or all Didelphidae the parietal layer of the tunica vaginalis shows marked pigmentation (Finkel, 1945; Biggers, 1966), which may have a thermo-regulatory function. Such pigmentation has been reported in limited Australian species (Finkel, 1945; Biggers, 1966; Heddle and Guiler, 1970). The pigment is first seen in the gubernacular apparatus (of both sexes) during development and later concentrates in the tunica vaginalis (Finkel, 1945).

The efferent ductules of the testis possess ciliated and non-ciliated cells in four species of Didelphidae studied with the electron microscope (Martan, Hruban and Slesers, 1967; Ladman, 1967); intracellular granules, presumably secretory, show ultrastructural and histochemical differences between the four species (Martan *et al.*, 1967). The epididymis is large (Owen, 1868; Bolliger, 1946; Kean, 1967) and only loosely attached to the testis (Owen, 1868; Young, 1879; Parsons, 1903; Osgood, 1921). The ductus deferens has little muscle compared with eutherians (Bolliger, 1946), after entering the abdomen passes lateral instead of medial to the ureter (Eckstein and Zuckerman, 1956; Sharman, 1965; Kean, 1967) and has no terminal ampulla (Bolliger, 1946; Eckstein and Zuckerman, 1956) – although Osgood's (1921) description suggests one in *Caenolestes*. Male marsupials have no seminal vesicles (Owen, 1868; Bolliger, 1946; Eckstein and Zuckerman, 1956; Biggers, 1966; Kean, 1967; Rodger and Hughes, 1973).

The testicular vessels in the spermatic cord form a unique 'rete mirabile', both the artery and vein being divided into numerous parallel branches (Barnett and Brazenor, 1958; Guiler and Heddle, 1970). This is not seen in eutherians but is absent from only a few species of small marsupials (including *Notoryctes* with abdominal testes) and shows varying regularity of arrangement, being most regular in the kangaroos and bandicoots (Barnett and Brazenor, 1958). The thermoregulatory function originally suggested has been questioned by Heddle and Guiler (1970), who also showed the presence of lymphatics in the rete.

The marsupial prostate is markedly different from that of eutherians. It is of the disseminate or diffuse type, consisting of many small glands within the mucosa and totally internal to the urethral muscular coat (Owen, 1868; Bolliger, 1946; Eckstein and Zuckerman, 1956; Biggers, 1966; Rodger and Hughes, 1973). It is relatively large (Owen, 1868; Parsons, 1903; Osgood, 1921; Bolliger, 1946; Biggers, 1966; Kean, 1967) and usually consists of three macroscopically and microscopically distinguishable zones (Eckstein and Zuckerman, 1956; Biggers, 1966; Kean, 1967; Tyndale-Biscoe, 1973; Rodger and Hughes, 1973), although sometimes there are only two (Kean, 1967; Rodger and Hughes, 1973). Detailed histochemical and ultrastructural studies in *Didelphis* have shown a variety of cell types suggesting a complex total secretion some of which is liberated in an apocrine fashion (Ladman and Soper, 1963; Martan and Allen, 1965; Hruban *et al.*, 1965; Hardin, 1965). The work of Rodger and Hughes (1973) suggests considerable complexity in several Australian species too.

Beyond the prostate there is a short membranous urethra that receives the ducts of the bulbourethral (Cowper's) glands (Biggers, 1966). There are up to three pairs of these glands (Owen, 1868; Young, 1879; Parsons, 1903; Osgood, 1921; Eckstein and Zuckerman, 1956; Kean, 1967; Rodger and Hughes, 1973).

Teeth

Teeth, of course, are widely studied for their taxonomic value, and among the works on marsupials stand out particularly those of Bensley (1903) and Tate (1947, 1948a, b). The basic marsupial dental formula is $I_4^5\ C_1^1\ P_3^3\ M_4^4$ (Wood Jones, 1923; Gregory, 1951; Sharman, 1974), exhibited only in the Didelphidae among recent species. The nearest among Australian examples are the Peramelidae, where the incisors are reduced to I_3^{4-5}, and Dasyuridae, with I_3^4 (Bensley, 1903; Tate, 1947, 1948b; Wood Jones, 1948). In other families there is further reduction or, less commonly, an increase in part of the formula – some Macropodidae and *Myrmecobius* have more than four molars (Tate, 1948a). Ziegler (1971) regards the basic therian formula at the time of metatherian divergence from eutheria as being I_4^5 C_1^1 P_4^4 M_4^4; marsupials are nearer to this primitive condition, all having lost only P1, than the eutheria, whose basic pattern is I_3^3 C_1^1 P_4^4 M_3^3 (Gregory, 1951; Sharman, 1974).

The dental formula, particularly of the incisor teeth, has been used as a fundamental criterion in dividing marsupials into two orders, Polyprotodonta and Diprotodonta, a subdivision created by Owen (Abbie, 1937, 1941; Troughton, 1959). The polyprotodonts are near to the basic formula, with $I_3^4 - _4^5$; diprotodonts have at most I_1^3, one or both canines may be absent and premolars are reduced sometimes as far as P_1^1 (Abbie, 1937; Troughton, 1959; Tyndale-Biscoe, 1973). Some diprotodonts exhibit a gap or diastema in the region of the lost anterior premolars (Calaby, 1968; Tyndale-Biscoe, 1973); most retain four molars. Marsupials exhibit little tooth replacement, only the last premolars having deciduous precursors (Gregory, 1951); the deciduous teeth anterior to this have been reduced to developmental vestiges or lost altogether (P1)(Ziegler, 1971).

The form of marsupial teeth is similar to that of eutherians and varies similarly according to diet; thus, in general, the polyprotodonts exhibit a carnivorous pattern, the diprotodonts a herbivorous one (Abbie, 1941).

Since the mid-nineteenth century it has been known that marsupial tooth enamel is unusual in possessing fine tubules extending from the dentinal aspect towards the free surface. There was much difference of opinion as to whether these tubules are within or between the enamel rods and whether they house continuations of the dentinal fibres or contain some other sort of fibre. Moss and Applebaum (1963) reviewed the literature on this matter and from their own studies on kangaroo teeth conclude that the 'tubules' are uncalcified enamel rods in a calcified matrix and that they do not contain fibres, dentinal or otherwise. The presence of layers in the cement has been used as a means of estimating age of teeth in eutherians; Pekelharing (1970) has shown this to be applicable to *Trichosurus* also.

The enamel organ of developing teeth presents similar features in marsupials and higher mammals. Berkovitz (1967a) reports the presence of an enamel cord (a row of cells through the stellate reticulum from the stratum intermedium towards the external enamel epithelium) in *Setonix* and cites references to its presence in other marsupials; he also found an enamel knot (a thickening of the centre of the external enamel epithelium) in *Trichosurus* – previously unknown in marsupials. Vascularisation of the enamel organ occurs in many eutherians; Berkovitz (1967b) cites some previous findings in marsupials and describes his observations on the above two species, where he found marked capillary invasion of the enamel organ only in the molariform teeth and then only after considerable enamel formation – rather later than is typical of eutherians.

Skeleton and joints

The marsupial skeleton is essentially mammalian (Abbie, 1941) and has few unique features present in all species. The hard palate possesses vacuities (Abbie, 1941; Tate, 1947; Gregory, 1951) in all except *Myrmecobius* (Wood Jones, 1923) and the angle of the mandible is inflected in most (Owen, 1841b; Abbie, 1941; Tate, 1948a; Gregory, 1951; Tyndale-Biscoe, 1973) but not in *Tarsipes* (Abbie, 1939a), and the inflection is poorly developed in *Phascolarctos* (Abbie, 1939a) and *Myrmecobius* (Wood Jones, 1923). The inflected portion in diprotodonts provides extra area for insertion of their large medial pterygoid muscle (Abbie, 1941). In many diprotodonts the lateral aspect of the mandible possesses an excavation, the masseteric fossa, for insertion of the masseter; especially in Macropodidae this may become very deep as a masseteric canal (Abbie, 1939a, 1941). Other notable features of the skull are the alisphenoid contribution to the tympanic bulla (Abbie, 1941; Gregory, 1951; Tyndale-Biscoe, 1973), the extension of the lacrimals into the snout (Tyndale-Biscoe, 1973), the zygomatic bones reaching the temporomandibular articulation, the carotid canal piercing the basisphenoid, fusion of the optic canal and the foramen lacerum and the spreading nature of the proximal ends of the nasal bones (Gregory, 1951). Abbie (1939b) found a temporomandibular articular disc present only in diprotodonts and considers this to be against the theory that the disc has evolved from the lateral pterygoid tendon.

The vertebral column has 26 presacral vertebrae (Owen, 1866; Wood Jones, 1923) and always seven cervical (Owen, 1841b, 1866; Wood Jones, 1923). The number of thoracic vertebrae (and therefore the number of pairs of ribs) is usually 13 (Owen, 1866; Wood Jones, 1923) but there are 15 in wombats (Owen, 1841a, b, 1866, 1874) and *Notoryctes* (Stirling, 1891) and only 12 in *Petaurus* (Owen, 1841b, 1866). The number of bones fused to form a sacrum varies from one to seven (Owen, 1841b, 1866). In many the ventral arch of the atlas is incompletely ossified (Owen, 1841b, 1874; Parsons, 1896; Gregory, 1951), and the dens usually remains separate from the axis (Gregory, 1951).

The coracoid element of the forelimb girdle is small and does not articulate with the sternum in the adult (Wood Jones, 1923; Tyndale-Biscoe, 1973). Clavicles are present except in bandicoots (Wood Jones, 1923; Shrivastava, 1962). Most species (like many eutherians) have a humeral entepicondylar foramen (Wood Jones, 1923; Abbie, 1941; Tyndale-Biscoe, 1973). In the wrist region there is a tendency for fusion of scaphoid and lunate (Abbie, 1941), and it has been noted in *Petrogale* (Parsons, 1896) and *Trichosurus* (Lewis, 1965) that the inferior radioulnar joint is syndesmotic and that there is no disc separating the ulna from the carpus. None of these features is unique to marsupials.

The hind-limb girdle bones contribute to a pelvis with small inlet and outlet (Gregory, 1951) and associated with this are the epipubes, two bones extending forwards into the ventral abdominal wall from the pubes and giving attachment to abdominal muscles (Barbour, 1963; Kean, 1966). An epipubis is found in all marsupials, although it is rudimentary in *Thylacinus* (Wood Jones, 1923; Abbie, 1941; Gregory, 1951; Kean, 1966). It is also present in monotremes (Abbie, 1941; Gregory, 1951; Tyndale-Biscoe, 1973) and reptiles (Abbie, 1941). Many marsupials show a tendency to tibiofibular fusion (Abbie, 1941), while, on the other hand, some have rotatable tibiofibular joints (Owen, 1841a, 1866). There is often a fabella or parafibula associated with the head of the fibula (Osgood, 1921; Lewis, 1962), while only the bandicoots possess a well-ossified patella (Wood Jones, 1923). Many marsupials have a fibulotalar meniscus but few eutherians do (Lewis, 1964). A plantar sesamoid (prehallux) is common in both marsupials and eutherians (Lewis, 1964). The marsupial hallux is frequently modified as an aid to prehension; in eutheria the pollex is more commonly utilised in this way (Kean, 1966).

Skin

The pouch, perhaps often thought to be the hallmark of a marsupial, should not be so regarded as it is absent in several polyprotodont species, both Australian and American (Abbie, 1941; Tate, 1948b; Eckstein and Zuckerman, 1956; Sharman, 1959b; Kean, 1966): on the other hand, a monotreme, *Tachyglossus*, has one (Abbie, 1941). It is simply a cutaneous invagination through the panniculus carnosus with the nipples of the mammary glands on its dorsal wall (Abbie, 1941). In general, it opens forwards in diprotodonts, backwards in most polyprotodonts (Abbie, 1941; Tate, 1948b; Sharman, 1959b; Kean, 1966); in the wombat some of the pouch extends both ways from the opening (Sharman, 1959b). From his observations and experiments on *Trichosurus* Bolliger (1942, 1944) believes the pouch skin is homologous with the scrotal skin of the male. The wider significance of this is doubtful, however, as Tate (1947) considers that the marsupial and eutherian scrota are not homologous. The number of nipples varies (largely with litter size); usually there are four but some diprotodonts have only two and many polyprotodonts more than four (Abbie, 1941; Eckstein and Zuckerman, 1956; Sharman, 1959b).

Diprotodont marsupials and Peramelidae all have the second and third pedal digits reduced in size and connected by a web of skin (Abbie, 1937; Tyndale-Biscoe, 1973) – a condition known as syndactyly. This feature has been used to subdivide marsupials into two groups, Syndactyla and Didactyla, a system which of course allies the Peramelidae differently from the Di - and Polyprotodonta scheme. Wood Jones (1923) ascribes this classification to Bensley and others, and himself subscribes to the rather unpopular idea that it is more fundamental than a subdivision based on the teeth. As pointed out by Kean (1966), any method of classification based on only one criterion is far from ideal anyway.

The general structure of the skin is much the same as in other mammals. Hairs are arranged in groups consisting of a (usually larger) central hair and lateral clusters in which several hairs may share a common follicle opening (Sweet, 1907; Gibbs, 1938; Hardy, 1947; Lyne, 1957; Mykytowycz and Nay, 1964; Kean, 1966). The central hair of each group has a sebaceous and an apocrine gland; the lateral hairs have only sebaceous glands – usually one per cluster (Hardy, 1947; Green, 1963b; Mykytowycz and Nay, 1964; Kean, 1966). Bolliger and Hardy (1944) examined the sternal skin (with brown hair) in *Trichosurus* but found it only quantitatively different from other regions. Eccrine sweat glands occur only in the non-hairy skin of the paws and (in some species) the tail (Green, 1961, 1963b; Mykytowycz and Nay, 1964; Kean, 1966; Fortney, 1973). Fortney's (1973) detailed ultra-structural and histochemical study of the eccrine glands of *Didelphis* shows them to consist of light and dark cells and myoepithelial cells on a basement membrane as in eutherians, but the dark cells appear to be protein-secreting rather than mucoid in type. In Macropodidae Mykytowycz and Nay (1964) describe myoepithelial cells outside the basement membrane and also report virtual absence of the reticular layer of the dermis; one suspects their histological interpretation, however. Using four different species, Green (1963b) describes the histology of the special glands of the eyelid and cloacal region, the latter being sebaceous, sometimes with the addition of a cell-secreting type. Bolliger and Hardy (1944) and Mykytowycz and Nay (1964) were unable to identify definite arrector pili muscles in the species they studied, but Gibbs (1938) and Hardy (1947) describe the presence of one per hair group and Gibbs regards this as a difference from eutherians.

Tylotrich (tactile sensory) hair follicles, well known in eutherian mammals, have similar structure and distribution in *Didelphis* (Mann, 1968). Hollis and Lyne (1974) found the innervation of vibrissal follicles in *Trichosurus* to exhibit details as in eutherians, including the presence of Merkel-cell neurite complexes which had been described earlier in *Didelphis* (Munger, 1965). Winkelmann found Meissner's corpuscles in the hand skin of *Didelphis* (1964) and the Red kangaroo (1966); these have otherwise only been associated with primates. He found other mammalian type endings too. In these studies Winkelmann also noted some unusual histochemical reactions in nerve endings, including the unique finding of alkaline phosphatase activity in hair nerve endings and motor end plates of the Red kangaroo.

Nervous system

In many marsupials the brain is relatively small (Owen, 1868; Wood Jones, 1923), but there are, nevertheless, some where it is quite large even by general eutherian standards (Kean, 1967).

Differing little externally from the typical mammalian brain (Abbie, 1941), the brain of marsupials has one major distinction — the complete absence of a corpus callosum (Elliot Smith, 1894; Wood Jones, 1923; Abbie, 1941; Tyndale-Biscoe, 1973). The report by Masai (1960) of the presence of a corpus callosum in five diprotodont species must be viewed with considerable doubt. Various studies on *Didelphis* have shown that the large anterior commissure carries commissural fibres connecting virtually all parts of the neocortex (Ebner, 1965, 1967a; Nelson and Lende, 1965; Martin, 1967, 1968a) and so confirm Elliot Smith's (1894) theory that it has a component representing the eutherian corpus callosum. In some marsupials some fibres of the anterior commissure form a bundle termed the fasciculus aberrans that radiates to the cortex through the internal capsule instead of the external (as do the other fibres in marsupials and all in eutherians). Elliot Smith (1902) first appreciated that this fasciculus is unique to diprotodonts, and this has been further substantiated by Abbie (1937, 1941) and Masai (1960). Abbie stresses the taxonomic importance of the fasciculus, considers that it indicates that the diprotodonts are more advanced than the polyprotodonts, and suggests it as the basis for a classification of marsupials into two groups, Simplicicommissurala (= Polyprotodonta) and Duplicicommissurala (= Diprotodonta).

According to Owen (1868), the spinal cord usually extends to the sacrum, and Voris's (1928b) account of *Didelphis* shows a generally typical mammalian cord that conforms on this point.

The internal structure and pathways of the central nervous system have been studied very fully in *Didelphis* because of its position as a relatively primitive or generalised marsupial (Bodian, 1939; Lende, 1963a; Beran and Martin, 1971; Martin and Dom, 1971) and of its ready availability in North America, where much of this work has been done. Among Australian species most is known of *Trichosurus*. In general, the findings indicate only minor differences from eutherians.

Ascending pathways in *Didelphis* are similar to those of other mammals (Mehler, 1969; Hazlett, Dom and Martin, 1972; Hazlett, Martin and Dom, 1971). Hazlett *et al.* (1971) could not distinguish separate dorsal and ventral spinocerebellar tracts as did Larsell (1936) and Mehler (1969), but both are represented as fibres enter the cerebellum through both superior and inferior peduncles; cerebellar cortical terminations are also similar to those of eutherians. The deep nuclei of the cerebellum are dentate, fastigial and interpositus, the last showing signs of division into the emboliform and globose of higher mammals (Larsell, 1935). The spinothalamic tract is small as in many other non-primates (Hazlett *et al.*, 1972). Clezy, Dennis and Kerr (1961) demonstrated the medial lemniscus and spinocerebellar tracts but found no spinothalamic or spinotectal tracts in *Trichosurus*, and Blumer (1963) could find no spinothalamic tract in *Setonix*; it should be emphasised, however, that such discrepancies between recent and somewhat older studies using anterograde degeneration methods may depend on the greater sensitivity of the newer silver techniques compared with the older Marchi staining, which detects only myelinated fibres. Hazlett *et al.* (1972) point out the preponderance of spinal fibres ending in the hind-brain in *Didelphis*, many fibres going to the reticular formation, and they suggest the existence of major indirect relays to the forebrain. Physiological studies indicate the dominance of such primitive indirect paths in *Trichosurus* too (Dennis and Kerr, 1961a, b).

In the brainstem of *Didelphis* the basilar pons is totally pretrigeminal and is small; the pontine nuclei are not distinguishable on the basis of neurone size as they are in some eutherians (King, Martin and Biggert, 1968). King, Bowman and Martin (1971) classified the neurones of the red nucleus into three morphological types but could not distinguish magnocellular and parvocellular parts of the nucleus; Woodburne (1943), however, mentions a magnocellular portion with resemblances to some submammalian forms. Efferent paths from the tectum resemble those of other mammals (Morest, 1965; Martin, 1969; Rafols and Matzke, 1970) and include some fibres to oculomotor nuclei whose presence has been disputed in eutherians (Rafols and Matzke, 1970). Martin (1969) points out that the tectum has apparently already become an important extrapyramidal centre as in eutherians rather than just a sensory relay as in lower forms.

Bodian (1939, 1940) found the diencephalon of *Didelphis* typically mammalian but showing primitive features particularly in the dorsal thalamic region, while the hypothalamus appears more advanced and shows clear nuclear differentiation. Oswaldo-Cruz and Rocha-Miranda (1967a, b, 1968) have studied this region fully in *Didelphis virginiana* and *D. aurita* and presented the details in sterotaxic co-ordinates. Detailed study of the hypothalamic neuroglia of *Didelphis* by Royce (1971) showed no special marsupial characteristics. In *Trichosurus* the thalamus is somewhat more advanced (Goldby, 1941); the lateral geniculate body is larger (as is the superior colliculus), suggesting greater development of the optic system. Unlike *Didelphis* and some eutherians (for example, rat), *Trichosurus* has a laminated lateral geniculate body; cytologically there are four layers but degeneration experiments show six on the basis of nerve fibre terminals and the first layer is unusual in receiving mostly uncrossed fibres, a condition previously known only in the primitive primate *Tupaia* (Hayhow, 1967). Somatotopical representation in the thalamus similar to that of eutherians has been reported in *Didelphis virginiana* (Pubols and Pubols, 1966) and *Didelphis aurita* (Sousa, Oswaldo-Cruz and Gattass, 1971); on the other hand, Hazlett *et al.* (1972) found no such arrangement in the former species and Erickson *et al.* (1964) report some neurones in the ventroposterior thalamus with wide-area somatic fields which they suggest are a primitive feature. The thalamus projects to all areas of neocortex (Bodian, 1942; Ebner, 1967a) but there is no sign of 'essential' projections that occur in eutherians (Diamond and Utley, 1963), nor is there much evidence of the specialised endings seen in cortical layer IV in eutherians (Ebner, 1967a). Pubols (1968) found topographic representation of the ventro-basal thalamus in the sensory cortex. A study of the development of the dien-cephalon of *Trichosurus* showed no very unusual features (Warner, 1969).

Didelphis has areas of sensory cerebral cortex (somatic I and II, auditory and visual) disposed very like those of rodents, and the somatic I area shows typical eutherian somatotopic localisation (Lende, 1963a). Similar somatic localisation exists in *Trichosurus* (Adey and Kerr, 1954) and *Thylogale* (Lende, 1963c). The main visual cortex in *Didelphis* is typical striate cortex (Gray, 1924) and receives geniculocortical projection fibres much as in most eutherians (Bodian, 1935; Benevento and Ebner, 1969, 1971a, b); the same applies to *Trichosurus* (Packer, 1941). *Didelphis* also has projections from other thalamic regions to striate cortex not known in other mammals (Benevento and Ebner, 1971b) and striate to striate commissural fibres known only in cat among eutherians (Kunze, Putnam and Manning, 1968). Other aspects of the optic system in these two species show the basic mammalian arrangement; in *Trichosurus* (Packer, 1941) there is about 75 per cent chiasmatic decussation, in *Didelphis* 80 per cent (Bodian, 1937). In the accessory optic system *Didelphis* shows a simple arrangement similar to that in reptiles and birds with only one tract terminating in the nucleus of the interpeduncular tract (Giolli, 1965); *Trichosurus* has the typical mammalian pattern of three tracts ending in the midbrain, although the transpeduncular tract is poorly developed (Hayhow, 1966). The medial geniculate body in *Didelphis* projects to the temporal cortex and the lentiform nucleus and amygdaloid (Ebner, 1967b).

Electrical stimulation studies have shown an area of motor cortex mostly behind the orbital sulcus, and it is of the granular type (Gray, 1924; Abbie, 1940; Martin and Megirian, 1972). There has, however, been disagreement on whether there is any representation of the hinder part of the body. In *Didelphis* most early studies found no hind-limb or tail areas; in more recent studies Lende (1963b) found such areas but Bautista and Matzke (1965) did not. In studies on five Australian species Goldby (1939) and Abbie (1940) found hind-limb and tail areas in diprotodonts but not polyprotodonts; this was not in full accord with older works, but more recent studies on two diprotodonts, *Thylogale* (Lende, 1963c) and *Trichosurus* (Rees and Hore, 1970), agree. Lende has shown that in *Didelphis* (1963b) and *Thylogale* (1963c) there is complete overlap and topological correspondence of sensory and motor areas, and he refers to the cortical region as a sensorimotor amalgam. In *Trichosurus* the overlap is probably not quite complete (Martin, Megirian and Roebuck, 1970). The marsupial pyramidal tract follows a typical course to the medulla, where its fibres mostly decussate to form a major tract in the dorsal white column of the cord and a lesser one in the lateral column, an arrangement found also in some eutherians (Turner, 1924; Goldby, 1939). The caudal limit of the tracts varies with species, is further caudal for the dorsal tract than the lateral one, and estimates have extended further in more recent studies compared with older ones; the dorsal tract has been traced to the fifth thoracic segment in *Didelphis* (Martin and Fisher, 1968), but most fibres end by the last cervical segment, to the seventh thoracic segment in *Setonix* (Watson, 1971) and to the tenth thoracic segment in *Trichosurus* (Martin *et al.*, 1970; Rees and Hore, 1970). The cellular lamination pattern of the spinal grey matter in *Didelphis* is similar to that in cat, but the pyramidal fibre terminations are rather more dorsally placed than in eutherians, where the tract arises in agranular cortex (Martin and Fisher, 1968); the pattern in *Trichosurus* (Martin *et al.*, 1970; Rees and Hore, 1970) and *Setonix* (Watson, 1971) resembles that in *Didelphis*. The pyramidal tract in *Didelphis* is a fine fibre system, having no large axons as in some eutheria (Biedenbach and Towe, 1970). In the newborn *Didelphis* the brain is extremely immature (Ulinski, 1971); there are no cerebral cortex or pyramidal fibres (Riese, 1945; Ward, 1954), so early movements of the young animal must depend on some primitive neurological mechanism. Descending paths from the cortex to the corpus striatum and brainstem have been studied in detail in *Didelphis*, mainly by Martin and his co-workers (Martin and West, 1967; Martin, 1968a, b; Martin and King, 1968; Martin and Fisher, 1968), and to some extent in *Trichosurus* (Rees and Hore, 1970; Martin, Megirian and Roebuck, 1971; Martin and Megirian, 1972). In the main both present a typical mammalian pattern. As in other non-primates, corticobulbar fibres do not make direct contact with cranial motor nuclei. Unlike the condition in some eutherians, all cortical areas project to the basilar pons. There are abundant projections to the red nucleus and reticular formation. Topological projection has been noted from the sensory cortex to the spinal trigeminal, cuneate and gracile sensory nuclei (Zimmerman and Chambers, 1963; Martin *et al.*, 1971). In the corpus striatum of *Didelphis* the tail of the caudate nucleus is small; there are no definite cortical projections to the globus pallidus nor are there any neocortical projections to the cytologically different ventro-medial part of the caudate head (Martin and Hamel, 1967). The amygdala of *Didelphis* has been described by Johnstone (1923) and van der Sprenkel (1926); and Ebner (1967a) has found an interamygdaloid bundle in the anterior commissure not known in eutherians.

Descending pathways from the brainstem in *Didelphis* are similar to those of other mammals. The rubrospinal tract is mainly crossed and extends to the end of the cord (Martin and Dom, 1970a); rubrobulbar fibres end in various brainstem nuclei but the facial is the only motor one (Martin and Dom, 1970b). Reticulospinal fibres also extend the full length of the cord (Martin and Dom, 1971); this and the rubrospinal tract probably play an important part in motor control (Martin and Dom, 1970a, 1971), especially to the caudal regions beyond direct pyramidal influence. Tectospinal fibres are sparse (Martin, 1969; Rafols and Matzke, 1970).

The rhinencephalon has been studied in *Didelphis* (Herrick, 1924) and *Trichosurus* (Adey, 1953), the latter work showing that, as in eutheria, there are no connections to certain traditional 'rhinencephalic' areas such as the hippocampus and the entorhinal area of the pyriform cortex. Also in *Trichosurus* Adey, Dunlop and Sunderland (1958) describe connections between rhinencephalon and brain-stem centres.

The arteries of the brain of *Didelphis* have been studied by Voris (1928a) and Gillilan (1972); they come, according to typical mammalian pattern, from two internal carotid and two vertebral arteries, but each carotid gives off a posterior communicating soon after entering the skull and there is no anterior communicating; this communication is present, however, in *Trichosurus* (Teo, 1974). As in submammalian forms, there is only one pair of inferior cerebellar arteries and the superior cerebellars supply most of the cerebellum (Gillilan, 1972). Gillilan does not agree with Voris's contention that the internal carotids supply all the forebrain. The spinal cord shows a primitive arrangement of six instead of three longitudinal arterial anastomoses (Voris, 1928a). Inside the brain and cord a more surprising arrangement is found, first described in *Didelphis* (Wislocki and Campbell, 1937), further confirmed in that species (Scharrer, 1940; Gillilan, 1972) and noted in several Australian marsupials (Craigie, 1938; Sunderland, 1941). While the arteries and veins at the surface run apart from one another and anastomose freely in typical mammalian fashion (Voris, 1928a; Wislocki and Campbell, 1937; Scharrer, 1940), within the nervous tissue the arteries and veins run together, the arteries being end arteries without any anastomoses and giving off discrete capillary loops instead of the usual mammalian capillary network. Such arrangement is otherwise seen only in submammalian classes (Sunderland, 1941). Some regions of the brain show the more usual capillary net and appear to be those outside the blood - brain barrier (Wislocki, 1940).

In the peripheral nervous system the main limb plexuses have received some attention: Voris (1928b) described those of *Didelphis*; Barbour (1963) gave an account of those of *Trichosurus* and reviewed most of the relevant marsupial literature to that date. No outstanding special features were noted. Sonntag (1921b) noted that the vagus and cervical sympathetic were fused in the neck in most marsupials, being separate but connected by branches in some diprotodonts.

The eye

There is little literature on marsupial eyes. O'Day (1938) reports on four species, noting the presence of oil droplets in the retinal cones (not present in eutherians) and the occurrence of twin cones in three of the four. He also comments that most marsupials except *Didelphis* and the Dasyuridae have an 'avascular retina'; this is in keeping with the note that no retinal vessels are visible ophthalmoscopically in *Trichosurus* (Bolliger and Macindoe, 1950). In the retina of *Didelphis* there are eight to ten pairs of arteries and veins with capillary loops as in the brain (Wislocki, 1940). Green (1963b) was able to find a lacrimal gland in only one species (*Trichosurus*) out of four studied. *Notorcytes,* the blind marsupial mole, has a vestigial eye not visible externally and containing no nervous tissue (Stirling, 1891); correspondingly it lacks optic nerves, chiasma and tracts and also the third, fourth and sixth cranial nerves (Schneider, 1968).

Skeletal muscular system

According to Abbie (1941), the muscles conform to the general mammalian pattern; the fusion of the temporalis with the small lateral pterygoid he considers a primitive feature and he also comments on the large size of the medial pterygoid in diprotodonts, the extension of the masseter around the base of the mandible to reach the inflected angle, and extensive nature of the cutaneous musculature, which has a tendency to fuse with the deep muscles especially in the pectoral region. Lewis (1962) points out some primitive features of the hindlimb muscles based on his study of 12 species — fusion of the soleus with the lateral head of gastrocnemius, insertion of plantaris into the plantar aponeurosis and insertion of flexor fibularis to all toes. He also regards as primitive the crural origin of flexor digitorum brevis that occurs in some species (in spite of its having a recurrent nerve supply from the foot) and comments on the difference from eutheria seen in the common insertion of flexor tibialis into the plantar aponeurosis and medial cuneiform. Shrivastava (1962) records the continuation of some of the trapezius into the deltoid seen in some species as a unique marsupial feature. Barbour (1963) gives a very detailed account of the muscles of *Trichosurus* and reviews the descriptive literature up to 1960 but makes no specific eutherian comparisons. The presence of the epipubis, of course, causes modifications to the abdominal muscles, and the cutaneous musculature has an opening for the pouch — an opening which is present in pouchless forms (Abbie, 1941) and in males (Barbour 1963), where it transmits the spermatic cords and the cremaster muscles. In females the cremasters are inserted into the mammary glands deep to this muscle layer.

Muscle spindles in *Trichosurus* (Jones, 1966a, b) are like those of other mammals in such details as the presence of two types of intrafusal fibres and their innervation. *Didelphis* was one of the first mammals, with some eutherians, in which spindles were first found in extrinsic lingual muscles (Langworthy, 1924).

Cardiovascular system

The erythrocytes of *Didelphis* are nucleated at birth (McCrady, 1938); but in adults only a few species have been noted as having nucleated cells and then only in very small numbers (Parsons *et al.*, 1971). Leukocytes and platelets are typically mammalian; in *Trichosurus* the only really unique feature is the presence of alkaline phosphatase in the granules of basophils (Barbour, 1972).

The marsupial heart, reviewed by Dowd (1969), presents a number of unusual features and, in *Trichosurus* at least, resembles the avian condition more than the mammalian (Dowd, 1974). The pericardium is generally adherent to the sternum and diaphragm (Sonntag, 1921b) but there are exceptions (Sonntag, 1921a, b; Boardman, 1941; Dowd, 1969). Often the right atrium has a bilobed auricle, the two parts of which surround the base of the aorta (Dowd, 1969), but again some exceptions have been noted (Sonntag, 1921b; Boardman, 1941; Dowd, 1969); this feature is practically unknown in other classes (Dowd, 1969). The interatrial septum has no fossa or annulus ovalis (Wade, 1968; Dowd, 1969), possibly owing to closure of the foramen ovale in a manner such as occurs in birds (Dowd, 1969). The right atrioventricular valve is unlike that of eutherians in having no papillary muscles attached to the nonseptal wall of the ventricle and usually it has only one large cusp, that being on the nonseptal aspect of the orifice (Wade, 1968; Dowd, 1969). Dowd (1974) found in *Trichosurus* that the coronary arteries arise in the usual way but have their distributing branches running in the myocardium as in birds, and the septal (interventricular) branches run in a unique position almost in the endocardium of the right ventricle covered by only a few muscle fibres. The coronary venous system is notable for the fact that the great cardiac vein turns to the right behind the pulmonary trunk and aorta to enter the right atrium (Dowd, 1974), a course similar to that of a vein in the avian heart. The conducting system of the heart in three species of kangaroos studied by Blair, Davies and Francis (1942) is essentially like that of eutherians.

The aorta follows the usual course. Branching of the aortic arch is very variable; Sonntag (1921b) recognises six different patterns, in all of which the left sub-clavian artery arises independently; Pearson (1940) says the pattern (which has no taxonomic value) may range from the primitive condition of four separate branches to that where all arise from a common trunk, the commonest form being two branches with the left subclavian independent. Two more recent reports conform to this commonest pattern (Boardman, 1941; Wade and Neely, 1949). In the abdomen the anterior and posterior mesenteric arteries arise by a common trunk (usually arising separately in eutheria) (Pearson, 1940); Sonntag (1921a) found a common coeliacomesenteric trunk in *Phascolarctos*. According to Pearson (1940), there are two pairs of gonadal arteries that join together before reaching the gonads. External and internal iliac arteries arise separately from the aorta, there being no common iliac as is usual in eutheria (Pearson, 1940).

All marsupials except *Petaurus breviceps* possess right and left anterior venae cavae (Pearson, 1940). The posterior vena cava lies ventral to the abdominal aorta in most species (Sonntag, 1921b; Pearson, 1940), a position rare in eutheria (McClure, 1903), and the mode of formation of this vessel is extremely variable in *Didelphis* (McClure, 1900, 1903, 1906). Most marsupials have two azygos veins but in some there is only one, usually the left (Pearson, 1940). In *Didelphis* the external jugular vein is very large and drains superficial structures of the head and neck; the internal jugular is small, draining only deep parts of the neck; the blood from the brain goes mainly into the internal vertebral plexus (Dom, Fisher and Martin, 1970). A large external jugular has been found in *Petrogale* too (Parson, 1896), so this pattern may be general. The pulmonary veins of the few species reported enter the left atrium by two trunks (Owen, 1836; Parsons, 1896) or a single common trunk (McClure, 1903; Wade and Neely, 1949; Wade, 1968).

Lymphatic system

The occurrence of the thymus in marsupials has been studied quite extensively, particularly by Yadav (1973), who looked at 93 species and reviewed most of the older literature except Boardman (1941), Wood Jones (1948) and Kathiresan (1969), whose findings agree with his general conclusions. Polyprotodonts have only a thoracic thymus, divisible into two or four lobes; diprotodonts have in addition a pair of superficial cervical thymi (not common in eutheria) closely associated with the salivary glands. Exceptions to this are found among the Vombatidae, some of which lack the thoracic members. Some accounts report similar lack in *Phascolarctos* (Symington, 1900; Fraser, 1915). The thoracic thymus develops from the endoderm of the third and fourth pharyngeal pouches, the cervical mainly from the ectoderm of the cervical sinus (Yadav, 1973); Kingsbury (1940) reports finding a microscopic amount of thymic tissue in relation to the cervical sinus in *Didelphis*. In *Trichosurus*, according to Adams (1955), the thoracic thymus lies in company with parathyroid IV, while parathyroid III may have some accessory thymic nodules associated with it in the neck.

The spleen is triangular or triradiate in form (Martin and Jones, 1834; Forbes, 1881; Parsons, 1896, 1903; Windle and Parsons, 1897; Mackenzie, 1916, 1918a; Osgood, 1921; Sonntag, 1921a, b; Wood Jones, 1948, Perrott, 1966). The internal structure of the spleen in *Didelphis* is typically mammalian, although (as in a few eutherians) it lacks ellipsoids and sinusoids (Hayes, 1968).

In two species of *Didelphis* studied by Azzali and DiDio (1965) lymph nodes are less developed than in rodents of the same size and some stations are missing. The cisterna chyli and thoracic duct are partly duplicated and some abdominal and diaphragmatic lymph passes to a mediastinal node not observed in other mammals. All lymph, however, passes through a node before reaching the blood.

Respiratory system

Lobation of the lungs is somewhat variable; the general comment of Owen (1868) seems to fit, as well as any, the specific observations made since (Forbes, 1881; Parsons, 1896, 1903; Osgood, 1921; Sonntag, 1921a, b; McCrady, 1938; Boardman, 1941; Wood Jones, 1948) — the left lung is often undivided but may have two lobes; the right lung has up to four lobes; an azygos lobe is sometimes present.

The structure of the lung (interpreting from studies on *Didelphis*) is reptilian at birth but becomes mammalian in the adult (Sorokin, 1962). The reptilian features are the presence of a separate capillary bed for each air space (Sorokin, 1962) and a lining for the air spaces consisting of cuboidal epithelium with squamous epithelium only over the capillaries (Bremer, 1904). The terminal air spaces in the newborn are not true alveoli but represent bronchi and bronchioles (Bremer, 1904, 1935).

In the adult *Didelphis* the bronchial walls contain many glands but there are few goblet cells; there is a relatively large amount of fat in alveolar walls (Sorokin, 1962). More detailed study of the glands shows them to be individually small and numerous and to extend into the muscular layer of the wall; they contain both mucous and serous cells, the latter being unusual in that histochemically and ultrastructurally they appear to be of ionic secretory type rather than zymogenic (Sorokin, 1965). Great alveolar cells are typically mammalian (Sorokin, 1967).

Alimentary system

Alimentary morphology in marsupials shows a wide range of variation, as in eutherians, related to diet (Owen, 1868; Abbie, 1941; Waring, Moir and Tyndale-Biscoe, 1966), the carnivorous and insectivorous polyprotodonts having, in general, a shorter and simpler gut than the largely herbivorous diprotodonts.

Doran and Baggett (1971) have noted structural differences in mammalian tongues correlated with predominant extraoral or intraoral function: marsupials conform to their two classes just as eutherians do. The papillae on the tongue include a compound filiform type termed 'coronate' (Poulton, 1883); there are usually three vallate papillae (Poulton, 1883; Parsons, 1896; Windle and Parsons, 1897; Osgood, 1921; Sonntag, 1921a; Wood Jones, 1948). The ultrastructure of the oral epithelium in *Trichosurus* reveals an abundance and arrangement of cytoplasmic fibrils somewhat similar to the condition in some birds (Tucker, 1969).

The three pairs of major salivary glands are usually present (Owen, 1868; Parsons, 1896; Osgood, 1921; Sonntag, 1921b; Wood Jones, 1948; Forbes and Tribe, 1969), although some older accounts apparently overlooked the sublinguals and Forbes (1881) could find no parotids in *Phascolarctos*. In general, the parotids are small in polyprotodonts, larger in many diprotodonts, especially Macropodidae (Parsons, 1896; Forbes and Tribe, 1969); the submandibular glands show the opposite relationship and the sublinguals are very variable (Sonntag, 1921b). Histological study of the glands in three macropod species by Forbes and Tribe (1969) showed that the parotid is serous, the submandibular has two serous cell types and occasional mucous acini, while the sublingual is almost exclusively mucous. Detailed studies on *Didelphis* show the parotid to be of serous type (Quintarelli and Dellovo, 1969), while the submandibular has typical mucous cells and a somewhat unusual type of serous cells (Quintarelli and Dellovo, 1969; Wilborn and Shackleford, 1969).

The stomach is unremarkable in most species but in two groups presents some interesting special features. Firstly, in the Macropodidae the stomach has a decidedly ruminant character (Kean, 1967) and the features have been described in several species (Schäfer and Williams, 1876; Mackenzie, 1918b; Moir *et al.*, 1956; Griffiths and Barton, 1966). The macropod stomach is very large, has sacculations along the proximal part of the greater curvature and sometimes may have longitudinal muscle bands resembling taeniae and may assume a spiral form. Internally near the cardia there is a considerable area lined by stratified squamous epithelium and representing a rumen; beyond this much of the mucosa, including an area towards the pylorus, has mucus secreting glands and a columnar surface, while an area proximal to the pylorus is of typical acid and enzyme secreting form (representing abomasum of ruminants). In *Macropus rufus* before birth a single cell type produces both acid and enzyme (Griffiths and Barton, 1966), a condition known in some frogs. The other gastric modification of note is the cardiogastric gland patch of the wombat and koala, known only in the beaver among the eutheria. Noted often over many years, this has been most fully described by Hingson and Milton (1968) (wombat) and Krause and Leeson (1973) (koala). The patch consists of some 25 - 30 branching evaginations of the mucosal lining near the cardia and adjacent to the lesser curvature; the mucosa contains glands similar to those of the corpus of the stomach. Johnstone (1899) (on both species) and Hingson and Milton describe the muscularis mucosae as folding around the evaginations and limiting the patch from submucosa, but Krause and Leeson say they extend through it into the submucosa. Johnstone (1899) noted some skeletal muscle fibres in the adjacent muscularis externa in wombat. Ultrastructurally the glandular cells of the patch in koala are like those of eutherian gastric glands but there are no argentaffin cells present (Krause and Leeson, 1973). In *Didelphis* and macropods there is a considerable length of abdominal oesophagus before the stomach is reached (Owen, 1868; Mackenzie, 1918b). Most marsupials have typical omenta attached to the stomach (Mackenzie, 1918b; Perrott, 1966).

The small intestine is usually a distinct part of the gut with a mesentery; in *Sarcophilus ursinus* there is no distinction between small and large intestines and the whole has a single dorsal mesentery, while in *Didelphis* there is a primitive mesentery continuous with the greater omentum (Mackenzie, 1918b). Perrott (1966) describes a free dorsal border to the mesentery in *Trichosurus*. Villi have been noted in the small intestine in a number of species (Forbes, 1881; Lönnberg, 1902; van Lennep, 1962); they are tongue-shaped in *Petaurus breviceps* and *Acrobates* (Lönnberg, 1902), while in *Perameles* van Lennep (1962) has described zig-zag folds replacing villi – an arrangement otherwise known only in bats. Branching crypts have been found in some and have no lumen at their ends in *Perameles* (van Lennep, 1962); crypts are very short in *Trichosurus* (van Lennep, 1962) and *Didelphis* (Klein, 1906). Paneth cells are absent in *Perameles* (van Lennep, 1962) and are scattered right to the tips of the long villi in *Didelphis* (Klein, 1906). Peyers patches have been noted in a few marsupials (Lönnberg, 1902; Azzali and DiDio, 1965). Brunner's glands have been studied in 55 Australian species (Krause, 1972) and typically form small clumps arranged into a narrow collar just beyond the pylorus; in the larger Phalangeridae and the Phascolomidae there is also some distal scatter most marked in just those species with cardiogastric glands; the ducts open onto the lumenal surface, not into crypts as in eutheria. In *Didelphis* Brunner's glands have a similar arrangement but open into mucosal depressions termed stomata (Krause and Leeson, 1969a). In *Didelphis* Krause and Leeson (1969b) have also described a basement membrane-like layer limiting both the deep and superficial surfaces of the mucosal lamina propria; the deep layer seems reminiscent of the stratum compactum of some eutherians and lower vertebrates (Andrew and Hickman, 1974).

The large intestine has a caecum in most marsupials, but not the Dasyuridae, Notoryctidae, *Tarsipes* and *Dromiciops australis*. The size varies much; some are quite large, especially in some diprotodonts and most notably in *Phascolarctos* – see Hill and Rewell (1954), who did an extensive study and reviewed most of the literature. Although it has sometimes been reported, Hill and Rewell (1954) say a true vermiform appendix is never present. The colon is usually divided into a proximal or right portion and a somewhat longer distal or left portion joined by a relatively fixed part bound to the pylorus or by a short transverse colon; in polyprotodonts the colon is shorter and in some there is almost no right colon (Mackenzie, 1918b; Hill and Rewell, 1954). In some polyprotodonts there is no clear distinction of colon from rectum (Mitchell, 1905; Osgood, 1921). The right colon may have its own mesocolon or share the mesentery with the small intestine the left colon typically has its own mesentery (Mackenzie, 1918b). Perrott (1966) describes a free dorsal border for the right mesocolon in *Trichosurus*. A pelvic colon has only been reported in macropods (Hill and Rewell, 1954).

Lobation of the liver has been noted in various species, the number of lobes varying with species (and probably with interpretation too) from three in *Didelphis azarae* (Martin and Jones, 1834) to seven in *Trichosurus* (Perrott, 1966). All have a gall bladder (Owen, 1868). The pancreas is usually diffuse (Mackenzie, 1918a; Osgood, 1921; Sonntag, 1921b; Wood Jones, 1948) but is more compact in koala (Sonntag, 1921b) and wombat (Mackenzie, 1918a). The usual position for the pancreas is in the mesoduodenum sometimes extending to the dorsal abdominal wall; it may also extend into the lesser omentum with the common bile duct which its duct joins in some species a considerable distance from the duodenal wall (Mackenzie, 1918a).

Renal system

Marsupial kidneys are unipapillary (Martin and Jones, 1834; Owen, 1836, 1868; Sonntag, 1921a, b; Wood Jones, 1948; Pfeiffer, 1968) and usually the right lies somewhat more anteriorly than the left (Owen, 1868; Parsons, 1903; Sonntag, 1921b; Fordham, 1928; Wood Jones, 1948). Like some eutherians, *Didelphis* has complex renal pelvic fornices and secondary pyramids of the outer zone of the medulla (Pfeiffer, 1968) and also medullary zonation with looped vasa recta (Plakke and Pfeiffer, 1964). Ultrastructurally the renal glomerulus in *Trichosurus* is like that of eutheria (Pak Poy, 1957). Limited studies suggest that at birth the mesonephros is still the functional kidney and continues as such for a time while the metanephros develops (McCrady, 1938; Bancroft, 1973).

Endocrines

The hypophysis has the usual parts. Commonly the pars intermedia is only a single layer of cells separated from the pars distalis by an extensive cleft on the one hand, applied to and almost completely surrounding the neural lobe on the other (Dawson, 1938; Green, 1951; Roth and Luse, 1964; Hanström, 1966; Teo, 1974). In *Didelphis* (Dawson, 1938) and *Trichosurus* (Dorsch, 1974) the cells of the intermedia are largely chromophobes and the pars tuberalis is poorly developed. In *Setonix* (Hanström, 1954) the pars intermedia is thicker and the cleft reduced to some cysts near the midline; in *Setonix* also the pars tuberalis consists of internal and external parts and has similarities with monotremes (Hanström, 1954, 1966). In the pars distalis occur chromophobes and chromophils; of the latter four types have been identified in *Didelphis* (Wheeler, 1943; Green, 1951), five in *Trichosurus*, where they have been functionally correlated with their hormonal secretions (Dorsch, 1974), and seven, including a unique 'acid fuchsin cell', in *Wallabia rufogrisea* (Ortman and Griesbach, 1958). In *Didelphis* (Dawson, 1938) and *Trichosurus* (Dorsch, 1974) acidophils show a marked concentration postero-laterally but Hanström (1954) notes no such localisation in *Setonix*. Blood supply through a portal system exists as in all vertebrates above the Anura (Green, 1951).

The ventricular recess does not extend into the neural lobe (Hanström, 1966), which commonly shows primitive lobulation. This is seen at its most extreme in *Didelphis virginiana* (Bodian, 1951), where there are lobules throughout, each with a central core of nerve fibres and a peripheral palisade zone of parallel nerve endings and pituicyte fibres perpendicular to the vascular interlobular connective tissue septa. This lobulation is seen also in *Didelphis aurita* (Hanström, 1966), is rather less well developed centrally in *Trichosurus* (Teo, 1974), and less still in *Setonix*, where it is peripheral only (Hanström, 1954). Ultrastructural study of *Didelphis* has shown unusual features in the high fibrillar content of pituicytes and the presence of glycogen in neurohypophyseal nerve fibres (Roth and Luse, 1964).

The pineal in *Didelphis* presents similarities to submammalian forms (Jordan, 1911), and its histology in macropods has been described briefly (Quay, 1966).

The thyroid lies beside the larynx and trachea, usually as two separate lobes, although a thin isthmus has been noted sometimes in *Trichosurus* (Fraser and Hill, 1915; Sonntag, 1921a, b; Adams, 1955), *Didelphis* (Kingsbury, 1940) and two species of *Perameles* (Fraser, 1915). Azzali (1964) found the parafollicular cells in two species of *Didelphis* to be similar to those of eutheria.

One parathyroid (III) typically lies at the commencement of the internal carotid artery (Kingsbury, 1940; McCrady, 1941; Adams, 1955); the other (IV) is inconstant in *Didelphis* (Kingsbury, 1940; McCrady, 1941) and lies with the thoracic thymus in *Trichosurus* (Adams, 1955).

Much information on marsupial adrenals comes from Bourne (1949), who performed an extensive study and review; they are typically mammalian and show considerable variation in such details as corticomedullary ratio, prominence of cortical zones, the presence of a corticomedullary septum, compactness of medulla and size of medullary sinuses. In some the medulla comes to the surface in places, and in many the right gland is intimately connected with the posterior vena cava and/or liver. Chester Jones (1957) found cortical zones clearly defined in five macropod species he studied, as did Bolliger (1953) in koala. Bourne (1934) describes an unusual portion of the cortex he terms the delta zone in *Trichosurus*. This is found only in mature females, bulges into the medulla from the inner aspect of the cortex on one side, and shows changes during pregnancy. Vinson *et al.* (1971) produce evidence that it may be associated with the relatively high level of circulating testosterone these females possess. As in many eutherian species, the medulla of *Trichosurus* contains both adrenalin and noradrenalin storing cells (Barbour, unpublished observations). Ultrastructural studies on cortical cells in *Didelphis* have shown a typical mammalian picture (Long and Jones, 1967).

In the pancreatic islets of *Didelphis* B cells lie centrally, the rare D cells mainly around them and the A cells peripherally; the cells resemble those of some eutherians except that the A cells contain considerable glycogen (Munger, Caramia and Lacy, 1965). There is also a unique E cell type first recognised by Thomas (1937).

References

Abbie, A. A. (1937). Some observations on the major subdivisions of the Marsupialia with especial reference to the position of the Peramelidae and Caenolestidae. *J. Anat.* **71**, 429 – 36.

Abbie, A. A. (1939a). A masticatory adaptation peculiar to some diprotodont marsupials. *Proc. Zool. Soc. London*, **B109**, 261 – 79.

Abbie, A. A. (1939b). The mandibular meniscus in monotremes and marsupials. *Aust. J. Sci.*, **2**, 86 – 8.

Abbie, A. A. (1940). The excitable cortex in *Perameles, Sarcophilus, Dasyurus, Trichosurus* and *Wallabia (Macropus)*. *J. Comp. Neurol.*, **100**, 469 – 87.

Abbie, A. A. (1941). Marsupials and the evolution of mammals. *Aust. J. Sci.*, **4**, 77 – 92.

Adams, W. E. (1955). The carotid sinus complex, 'parathyroid' III and thymo-parathyroid bodies, with special reference to the Australian opossum, *Trichosurus vulpecula. Am. J. Anat.*, **97**, 1 – 57.

Adey, W. R. (1953). An experimental study of the central olfactory connexions in a marsupial (*Trichosurus vulpecula*). *Brain*, **76**, 311 – 30.

Adey, W. R., Dunlop, C. W. and Sunderland, S. (1958). A survey of rhinencephalic interconnections with the brain stem. *J. Comp. Neurol.*, **110**, 173 – 203.

Adey, W. R. and Kerr, D. I. B. (1954). The cerebral representation of deep somatic sensibility in the marsupial phalanger and the rabbit; an evoked potential and histological study. *J. Comp. Neurol.*, **100**, 597 – 625.

Amoroso, E. C. (1952). Placental phenomena in marsupials. In: *Marshall's Physiology of Reproduction*, Vol. 2 (ed. A. S. Parkes), Longmans, Green, London, pp. 159 – 64.

Andrew, W. and Hickman, C. P. (1974). *Histology of the Vertebrates. A Comparative Text*, Mosby, St. Louis.

Azzali, G. (1964). Ultrastructure des cellules parafolliculaires de la thyroide chez quelques mammifères. *Ann. Endocrinol. (Paris)*, **25**, Suppl., 8 – 13.

Azzali, G. and DiDio, J. A. (1965). The lymphatic system of *Didelphys azarae* and *Didelphys marsupialis. Am. J. Anat.*, **116**, 449 – 70.

Bancroft, B. J. (1973). Embryology of *Schoinobates volans* (Kerr) (Marsupialia; Petauridae). *Aust. J. Zool.*, **21**, 33 – 52.

Barbour, R. A. (1963). The musculature and limb plexuses of *Trichosurus vulpecula. Aust. J. Zool.*, **11**, 488 – 610.

Barbour, R. A. (1972). The leukocytes and platelets of a marsupial, *Trichosurus vulpecula*. A comparative morphological, metrical and cytochemical study. *Arch. Histol. Jap.*, **34**, 311 – 60.

Barnett, C. H. and Brazenor, C. W. (1958). The testicular rete mirabile of marsupials. *Aust. J. Zool.*, **6**, 27 – 32.

Bautista, N. S. and Matzke, H. A. (1965). A degeneration study of the course and extent of the pyramidal tract of the opossum. *J. Comp. Neurol.*, **124**, 367 – 75.

Baxter, J. S. (1935a). On the female genital tract of the Caenolestids (Marsupialia). *Proc. Zool. Soc. London*, 157 – 62.

Baxter, J. S. (1935b). Development of the female genital tract in the American opossum. *Carneg. Inst. Contr. Embryol.*, **25**, 15 – 36.

Benevento, L. A. and Ebner, F. F. (1969). Lateral geniculate nucleus projections to neocortex in the opossum. *Anat. Rec.*, **163**, 294.

Benevento, L. A. and Ebner, F. F. (1971a). The areas and layers of corticocortical terminations in the visual cortex of the Virginia opossum. *J. Comp. Neurol.*, **141**, 157 – 89.

Benevento, L. A. and Ebner, F. F. (1971b). The contribution of the dorsal lateral geniculate nucleus to the total pattern of thalamic terminations in striate cortex of the Virginia opossum. *J. Comp. Neurol.*, **143**, 243 – 60.

Bensley, B. A. (1903). On the evolution of the Australian Marsupialia; with remarks on the relationships of the marsupials in general. *Trans. Linnean Soc., Ser. 2, Zool.*, **9**, 83 – 217.

Beran, R. L. and Martin, G. F. (1971). Reticulospinal fibres of the opossum, *Didelphis virginiana.* I. Origin. *J. Comp. Neurol.*, **141**, 453 – 65.

Berkovitz, B. K. B. (1967a). An account of the enamel cord in *Setonix brachyurus*

(Marsupialia) and on the presence of an enamel knot in *Trichosurus vulpecula. Arch. Oral Biol.,* **12**, 49 – 59.

Berkovitz, B. K. B. (1967b). The vascularity of the enamel organ in the developing teeth of *Setonix brachyurus* and *Trichosurus vulpecula* (Marsupialia). *Arch. Oral Biol.,* **12**, 1299 – 305.

Biedenbach, M. A. and Towe, A. L. (1970). Fiber spectrum and functional properties of pyramidal tract neurones in the American opossum. *J. Comp. Neurol.,* **140**, 421 – 9.

Biggers, J. D. (1966). Reproduction of male marsupials. *Symp. Zool. Soc. London,* **15**, 251 – 80.

Biggers, J. D. and Creed, R. F. S. (1962). Conjugate spermatozoa of the North American opossum. *Nature,* **196**, 1112 – 3.

Biggers, J. D., Creed, R. F. S. and de Lamater, E. D. (1963). Conjugated spermatozoa in American marsupials. *J. Reprod. Fert.,* **6**, 324.

Biggers, J. D. and de Lamater, E. D. (1965). Marsupial spermatozoa pairing in the epididymis of American forms. *Nature,* **208**, 402 – 4.

Blair, D. M., Davies, F. and Francis E. T. B. (1942). The conducting system of the marsupial heart. *Trans. Roy. Soc. Edinburgh,* **60**, 629 – 38.

Blumer, W. F. C. (1963). Ascending and descending spinal tracts of the quokka (*Setonyx brachyurus*). *J. Anat.,* **97**, 490.

Boardman, W. (1941). On the anatomy and functional adaptation of the thorax and pectoral girdle in the wallaroo (*Macropus robustus*). *Proc. Linnean Soc. N. S. Wales,* **66**, 349 – 87.

Bodian, D. (1935). The projection of the lateral geniculate body on the cerebral cortex of the opossum, *Didelphis virginiana. J. Comp. Neurol.,* **62**, 469 – 94.

Bodian, D. (1937). An experimental study of the optic tracts and retinal projection of the Virginia opossum, *J. Comp. Neurol.,* **66**, 113 – 44.

Bodian, D. (1939). Studies on the diencephalon of the Virginia opossum. Part I. The nuclear pattern in the adult. *J. Comp. Neurol.,* **71**, 259 – 323.

Bodian, D. (1940). Studies on the diencephalon of the Virginia opossum. Part II. The fiber connections in normal and experimental animals. *J. Comp. Neurol.,* **72**, 207 – 97.

Bodian, D. (1942). Studies on the diencephalon of the Virginia opossum. Part III. the thalamo-cortical projection. *J. Comp. Neurol.,* **77**, 525 – 75.

Bodian, D. (1951). Nerve endings, neurosecretory substance and lobular organization of the neurohypophysis. *Bull. Johns Hopkins Hosp.,* **89**, 354 – 79.

Bolliger, A. (1942). Functional relations between scrotum and pouch and the experimental production of a pouch-like structure in the male of *Trichosurus vulpecula. J. Proc. Roy. Soc. N. S. Wales,* **76**, 283 – 93.

Bolliger, A. (1944). An experiment on the complete transformation of the scrotum into a marsupial pouch in *Trichosurus vulpecula. Med. J. Aust.,* **2**, 56 – 8.

Bolliger, A. (1946). Some aspects of marsupial reproduction. *J. Proc. Roy. Soc. N. S. Wales,* **80**, 2 – 13.

Bolliger, A. (1953). The adrenals of the koala (*Phascolarctos cinereus*) and their alleged relationship to eucalyptus leaf diet. *Med. J. Aust.,* **1**, 917 – 9.

Bolliger, A. and Hardy, M. H. (1944). The sternal integument of *Trichosurus vulpecula. J. Proc. Roy. Soc. N. S. Wales,* **78**, 122 – 33.

Bolliger, A. and Macindoe, N. M. (1950). Eye changes in a marsupial experimentally infected with kala-azar and trypanosomiasis. *Am. J. Ophthalmol,* 33, 1871 - 7.

Bourne, G. H. (1934). Unique structure in the adrenal of the female opossum. *Nature,* 134, 664 - 5.

Bourne, G. H. (1949). *The Mammalian Adrenal Gland,* Oxford University Press.

Bremer, J. L. (1904). On the lung of the opossum. *Am. J. Anat.,* 3, 67 - 73.

Bremer, J. L. (1935). Postnatal development of alveoli in the mammalian lung in relation to the problem of the alveolar phagocyte. *Carneg. Inst. Contr. Embryol.,* 25, 83 - 111.

Buchanan, G. and Fraser, E. A. (1918). The development of the uro-genital system in the Marsupialia, with special reference to *Trichosurus vulpecula.* Part I. *J. Anat.,* 53, 35 - 95.

Calaby, J. H. (1968). Observations of the teeth of marsupials, especially kangaroos. *Dent. Mag.,* 85, 23 - 7.

Chester Jones, I. (1957). *The Adrenal Cortex,* Cambridge University Press.

Christensen, A. K. and Fawcett, D. W. (1961). The normal fine structure of opossum testicular interstitial cells. *J. Cell Biol.,* 9, 653 - 70.

Clezy, J. K. A., Dennis, B. J. and Kerr, D. I. B. (1961). A degeneration study of the somaesthetic afferent systems in the marsupial phalanger, *Trichosurus vulpecula. Aust. J. Exp. Biol. Med. Sci.,* 39, 19 - 28.

Craigie, E. H. (1938). The blood vessels of the central nervous system of the kangaroo. *Science,* 88, 359 - 60.

Dawson, A. B. (1938). The epithelial components of the pituitary gland of the opossum. *Anat. Rec.,* 72, 181 - 93.

de Bavay, J. M. (1951). Notes on the female urogenital system of *Tarsipes spenserae* (Marsupialia). *Papers Roy. Soc. Tasmania,* 1950, 143 - 9.

Dennis, B. J. and Kerr, D. I. B. (1961a). Somaesthetic pathways in the marsupial phalanger, *Trichosurus vulpecula. Aust. J. Exp. Biol. Med. Sci.,* 39, 29 - 42.

Dennis, B. J. and Kerr, D. I. B. (1961b). An afferent system in the central tegmental facsiculus. *Aust. J. Exp. Biol. Med. Sci.,* 39, 43 - 56.

Diamond, I. T. and Utley, J. D. (1963). Thalamic retrograde degeneration study of sensory cortex in opossum. *J. Comp. Neurol.,* 120, 129 - 60.

Dom, R., Fisher, B. L. and Martin G. F. (1970). The venous system of the head and neck of the opossum (*Didelphis virginiana*). *J. Morphol.* 132, 487 - 96.

Doran, G. A. and Baggett, H. (1971). A structural and functional classification of mammalian tongues. *J. Mammal.,* 52, 427 - 9.

Dorsch, M. (1974). A preliminary study of the cytology of the adenohypophysis of the brush-tail possum (*Trichosurus vulpecula*) with particular emphasis on the special cytology of the pars anterior. Thesis, University of Adelaide.

Dowd, D. A. (1969). Gross features of the heart of a marsupial, *Trichosurus vulpecula. Acta Anat.,* 74, 454 - 71.

Dowd, D. A. (1974). The coronary vessels in the heart of a marsupial, *Trichosurus vulpecula. Am. J. Anat.,* 140, 47 - 56.

Ebner, F. F. (1965). Projections of the forebrain commissures in the opossum. *Anat. Rec.,* 151, 345.

Ebner, F. F. (1967a). Afferent connections to neocortex in the opossum (*Didelphis virginiana*). *J. Comp. Neurol.,* 129, 241 - 67.

Ebner, F. F. (1967b). Medial geniculate nucleus projections to telencephalon in the opossum. *Anat. Rec.,* 157, 238 - 9.

Eckstein, P. and Zuckerman, S. (1956). Morphology of the reproductive tract. In: *Marshall's Physiology of Reproduction,* Vol. 1, Part 1 (ed. A. S. Parkes), Longmans, Green, London, pp. 43 - 155.

Elliott Smith, G. (1894). A preliminary communication upon the cerebral commissures of the Mammalia, with special reference to the Monotremata and Marsupialia. *Proc. Linnean Soc. N. S. Wales,* 9, 653 - 57.

Elliott Smith, G. (1902). On a peculiarity of the cerebral commissures in certain Marsupialia, not hitherto recognised as a distinctive feature of the Diprotodontia. *Proc. Roy. Soc. London,* 70, 226 - 31.

Enders, A. C. and Enders, R K. (1969). The placenta of the four-eyed opossum *(Philander opossum). Anat. Rec.* 165, 431 - 49.

Erickson, R. P., Jane, J. A., Waite, R. and Diamond, I. T. (1964). Single neurone investigation of sensory thalamus of the opossum. *J. Neurophysiol.,* 27, 1026 - 47.

Finkel, M. P. (1945). The relation of sex hormones to pigmentation and to testis descent in the opossum and ground squirrel. *Am. J. Anat.,* 76, 93 - 151.

Flynn, T. T. (1911). Notes on marsupial anatomy. II. On the female genital organs of a virgin *Sarcophilus satanicus. Papers Roy. Soc. Tasmania,* 1911, 144 - 56.

Flynn, T. T. (1922). Notes on certain reproductive phenomena in some Tasmanian marsupials. *Ann. Mag. Nat. Hist., Ser. 9,* 10, 225 - 31.

Flynn, T. T. (1923). The yolk-sac and allantoic placenta in *Perameles. Quart. J. Microscop. Sci.,* 67, 123 - 82.

Flynn, T. T. (1930). The uterine cycle of pregnancy and pseudo-pregnancy as it is in the diprotodont marsupial *Bettongia cuniculus.* With notes on other reproductive phenomena in this marsupial. *Proc. Linnean Soc. N. S. Wales,* 55, 506 - 31.

Forbes, D. K. and Tribe, D. E. (1969). Salivary glands of kangaroos. *Aust. J. Zool.,* 17, 765 - 75.

Forbes, W. A. (1881). On some points in the anatomy of the koala *(Phascolarctos cinereus). Proc. Zool. Soc. London,* 1881, 180 - 95.

Fordham, M. G. C. (1928). The anatomy of the urogenital organs of the male *Myrmecobius fasciatus. J. Morphol.,* 46, 563 - 83.

Fortney, J. A. (1973). Cytology of eccrine sweat glands in the opossum. *Am. J. Anat.,* 136, 205 - 19.

Fraser, E. A. (1915). The development of the thymus, epithelial bodies, and thyroid in the Marsupialia. Part II. *Phascolarctos, Phascolomys and Perameles. Phil. Trans. Roy. Soc.,* B207, 87 - 112.

Fraser, E. A. and Hill, J. P. (1915). The development of the thymus, epithelial bodies, and thyroid in the Marsupialia. Part I. *Trichosurus vulpecula. Phil. Trans. Roy. Soc.,* 207, 1 - 85.

Gibbs, H. F. (1938). A study of the development of the skin and hair of the Australian opossum, *Trichosurus vulpecula. Proc. Zool. Soc. London,* B108, 611 - 48.

Gillilan, L. A. (1972). Blood supply to primitive mammalian brains. *J. Comp. Neurol.,* 145, 209 - 21.

Giolli, R. A. (1965). An experimental study of the accessory optic system and of other optic fibers in the opossum, *Didelphis virginiana. J. Comp. Neurol.,* **124**, 229 - 42.

Goldby, F. (1939). An experimental investigation of the motor cortex and its connexions in the phalanger, *Trichosurus vulpecula. J. Anat.,* **74**, 12 - 33.

Goldby, F. (1941). The normal histology of the thalamus in the phalanger, *Trichosurus vulpecula. J. Anat.,* **75**, 197 - 224.

Gray, P. A. (1924). The cortical lamination pattern of the opossum, *Didelphys virginiana. J. Comp. Neurol.,* **37**, 221 - 63.

Green, J. D. (1951). The comparative anatomy of the hypophysis, with special reference to its blood supply and innervation. *Am. J. Anat.,* **88**, 225 - 311.

Green, L. M. A. (1961). Sweat glands in the skin of the quokka of Western Australia. *Aust. J. Exp. Biol. Med. Sci.,* **39**, 481 - 6.

Green, L. M. A. (1963a). Interstitial cells in the testis of the Australian phalanger (*Trichosurus vulpecula*). *Aust. J. Exp. Biol. Med. Sci.,* **41**, 99 - 104.

Green, L. M. A. (1963b). Distribution and comparative histology of cutaneous glands in certain marsupials. *Aust. J. Zool.,* **11**, 250 - 72.

Gregory, W. K. (1951). *Evolution Emerging. A Survey of Changing Patterns from Primeval Life to Man,* Macmillan, New York.

Griffiths, M. and Barton, A. A. (1966). The ontogeny of the stomach in the pouch young of the red kangaroo. *C.S.I.R.O. Wildlife Res.,* **11**, 169 - 85.

Guiler, E. R. and Heddle, R W. L. (1970). Testicular and body temperatures in the Tasmanian devil and three other species of marsupial. *Comp. Biochem. Physiol.,* **33**, 881 - 91.

Guraya, S.S. (1971). Morphological and histochemical changes in the mitochondria during spermiogenesis in the opossum. *Acta Anat.,* **79**, 120 - 5.

Hanström, B. (1954). The hypophysis in a wallaby, two tree-shrews, a marmoset, and an orang-utan. *Ark. Zool.,* **6**, 97 - 154.

Hanström, B. (1966). Gross anatomy of the hypophysis in mammals. In: *The Pituitary Gland,* Vol. 1 (ed. G. W. Harris and B. T. Donovan), Butterworths, London, pp. 1 - 57.

Hardin, J. H. (1965). Cytological features of glycogen-containing cells in the third part of the prostate gland of the opossum, *Didelphis marsupialis. Anat. Rec.,* **151**, 358.

Hardy, M. H. (1947). The group arrangement of hair follicles in the mammalian skin. Part I. Notes on follicle group arrangement in thirteen Australian marsupials. *Proc. Roy. Soc. Queensland,* **58**, 125 - 48.

Hartman, C. (1923). The oestrus cycle in the opossum. *Am. J. Anat.,* **32**, 353 - 421.

Hayes, T. G. (1968). Studies of a primitive mammalian spleen, the opossum (*Didelphis virginiana*). *J. Morphol,* **124**, 445 - 50.

Hayhow, W. R. (1966). The accessory optic system in the marsupial phalanger, *Trichosurus vulpecula.* An experimental degeneration study. *J. Comp. Neurol.,* **126**, 653 - 71.

Hayhow, W. R. (1967). The lateral geniculate nucleus of the marsupial phalanger, *Trichosurus vulpecula.* An experimental study of cytoarchitecture in relation to the intranuclear optic nerve fields. *J. Comp. Neurol.,* **131**, 571 - 603.

Hazlett, J. C., Dom, R. and Martin, G. F. (1972). Spino-bulbar, spino-thalamic and medial lemniscal connections in the American opossum, *Didelphis marsupialis virginiana. J. Comp. Neurol.,* 146, 95 - 117.

Hazlett, J. C., Martin, G. F. and Dom, R. (1971). Spino-cerebellar fibers of the opossum, *Didelphis marsupialis virginiana. Brain Res.,* 33, 257 - 71.

Heddle, R. W. L. and Guiler, E. R. (1970). The form and function of the testicular rete mirabile of marsupials. *Comp. Biochem. Physiol.,* 35, 415 - 25.

Herrick, C. J. (1924). The nucleus olfactorius anterior of the opossum. *J. Comp. Neurol.,* 37, 317 - 59.

Hill, J. P. (1897). The placentation of *Perameles* (Contributions to the embryology of the Marsupialia — I). *Quart. J. Microscop. Sci.,* 40, 385 - 446.

Hill, J. P. (1899). Contributions to the morphology and development of the female urogenital organs in the Marsupialia. I. On the female urogenital organs of *Perameles,* with an account of the phenomena of parturition. *Proc. Linnean. Soc. N. S. Wales,* 24, 42 - 82.

Hill, J. P. (1900). Contributions to the morphology and development of the female urogenital organs of the Marsupialia. II. On the female urogenital organs of *Myrmecobius fasciatus.* III. On the female genital organs of *Tarsipes rostratus.* IV. Notes on the female genital organs of *Acrobates pygmaeus* and *Petaurus breviceps.* V. On the existence at parturition of a pseudovaginal passage in *Trichosurus vulpecula. Proc. Linnean Soc. N. S. Wales,* 25, 519 - 32.

Hill, J. P. (1910). The early development of the Marsupialia with special reference to the native cat (*Dasyurus viverrinus*). *Quart. J. Microscop. Sci.,*56 1 - 134.

Hill, W. C. O. and Rewell, R. E. (1954). The caecum of monotremes and marsupials. *Trans. Zool. Soc. London,* 28, 185 - 240.

Hingson D. J. and Milton, G. W. (1968). The mucosa of the stomach of the wombat (*Vombatus hirsutus*) with special reference to the cardiogastric gland. *Proc. Linnean Soc. N. S. Wales,* 93, 69 - 75.

Hollis, D. E. and Lyne, A. G. (1974). Innervation of vibrissa follicles in the marsupial *Trichosurus vulpecula. Aust. J. Zool.,* 22, 263 - 76.

Hruban, Z., Martan, J., Slesers, A., Steiner, D. F., Lubran, M. and Rechcigl, M. (1965). Fine structure of the prostatic epithelium of the opossum (*Didelphis virginiana* Kerr). *J. Exp. Zool.,* 160, 81 - 105.

Hughes, R. L. (1965). Comparative morphology of spermatozoa from five marsupial families. *Aust. J. Zool.,* 13, 533 - 43.

Hughes, R. L., Thomson, J. A. and Owen, W. H. (1965). Reproduction in natural populations of the Australian ringtail possum, *Pseudocheirus peregrinus* (Marsupialia: Phalangeridae), in Victoria. *Aust. J. Zool.,* 13, 383 - 406.

Johnstone, J. (1899). On the gastric glands of the Marsupialia. *J. Linnean Soc.,* 27, 1 - 14.

Johnstone, J. B. (1923). Further contributions to the study of the evolution of the forebrain. *J. Comp. Neurol.,* 35, 337 - 481.

Jones, E. G. (1966a). The innervation of muscle spindles in the Australian opossum, *Trichosurus vulpecula,* with special reference to the motor nerve endings. *J. Anat.,* 100, 733 - 59.

Jones, E. G. (1966b). Structure and distribution of muscle spindles in the fore-paw lumbricals of the phalanger, *Trichosurus vulpecula. Anat. Rec.,* **155**, 287 - 303.

Jordan, H. E. (1911). The microscopic anatomy of the epiphysis of the opossum. *Anat. Rec.,* **5**, 325 - 38.

Kathiresan, S. (1969). Morphological studies on the thymus. Part I. Thymus of the marsupial — *Trichosurus vulpecula. Ind. J. Med. Res.,* **57**, 939 - 43.

Kean, R. I. (1961). The evolution of marsupial reproduction. *New Zealand Forest Res. Inst. Tech. Papers,* **35**, 1 - 40.

Kean, R. I. (1966). Marsupials (Part I). *Tuatara,* **14**, 105 - 20.

Kean, R. I. (1967). Marsupials (Part 2). *Tuatara,* **15**, 25 - 45.

Kean, R. I., Marryatt, R. G. and Carroll, A. L. K. (1964). The female urogenital system of *Trichosurus vulpecula* (Marsupialia). *Aust. J. Zool.,* **12**, 18 - 41.

King, J. S., Bowman M. H. and Martin G. F. (1971). The red nucleus of the opossum (*Didelphis marsupialis virginiana*): a light and electron micro-scopic study. *J. Comp. Neurol.,* **143**, 157 - 83.

King, J. S., Martin, G. F. and Biggert, T. P. (1968). The basilar pontine gray of the opossum (*Didelphis virginiana*). I. Morphology. *J. Comp. Neurol.,* **133**, 439 - 45.

Kingsbury, B. F. (1940). The development of the pharyngeal derivatives of the opossum (*Didelphis virginiana*), with special reference to the thymus. *Am. J. Anat.,* **67**, 393 - 435.

Klein, S. (1906). On the nature of the granule cells of Paneth in the intestinal glands of mammals. *Am. J. Anat.,* **5**, 315 - 30.

Krause, W. J. (1972). The distribution of Brunner's glands in 55 marsupial species native to the Australian region. *Acta Anat.,* **82**, 17 - 33.

Krause, W. J. and Leeson, C. R. (1969a). Studies of Brunner's glands in the opossum. I. Adult morphology. *Am. J. Anat.,* **126**, 255 - 73.

Krause, W. J. and Leeson, C. R. (1969b). Limiting membranes of intestinal lamina propria in the opossum. *J. Anat.,* **104**, 467 - 80.

Krause, W. J. and Leeson, C. R. (1973). The stomach gland patch of the koala (*Phascolarctos cinereus*). *Anat. Rec.,* **176**, 475 - 87.

Kunze, D. L., Putnam, S. J. and Manning, J. W. (1968). Transcortical striate connections in the opossum. *J. Comp. Neurol.,* **132**, 463 - 8.

Ladman, A. J. (1967). The fine structure of the ductuli efferentes of the opossum. *Anat. Rec.,* **157**, 559 - 75.

Ladman, A. J. and Soper, E. H. (1963). Cytological observations on intra-granular inclusion bodies in the prostate of the opossum. *Anat. Rec.,* **145**, 364.

Langworthy, O. R. (1924). A study of the innervation of the tongue musculature with particular reference to the proprioceptive mechanism. *J. Comp. Neurol.,* **36**, 273 - 97.

Larsell, O. (1935). The development and morphology of the cerebellum in the opossum. Part I. Early development. *J. Comp. Neurol.,* **63**, 65 - 94.

Larsell, O. (1936). The development and morphology of the cerebellum in the opossum. Part II. Later development and adult. *J. Comp. Neurol.,* **63**, 251 - 91.

Lende, R. A. (1963a). Sensory representation in the cerebral cortex of the opossum (*Didelphis virginiana*). *J. Comp. Neurol.,* 121, 395 – 403.

Lende, R. A. (1963b). Motor representation in the cerebral cortex of the opossum (*Didelphis virginiana*). *J. Comp. Neurol.,* 121, 405 – 15.

Lende, R. A. (1963c). Cerebral cortex: a sensorimotor amalgam in the Marsupialia. *Science,* 141, 730 – 2.

Lewis, O. J. (1962). The phylogeny of the crural and pedal flexor musculature. *Proc. Zool. Soc. London,* 138, 77 – 109.

Lewis, O. J. (1964). The homologies of the mammalian tarsal bones. *J. Anat.,* 98, 195 – 208.

Lewis, O. J. (1965). Evolutionary change in the primate wrist and inferior radio-ulnar joints. *Anat. Rec.,* 151, 275 – 85.

Lierse, W. (1965). Die Wandkonstruktion und das Gefäss-system des Uterus von Opossum (*Didelphis virginiana*). *Acta Anat.,* 60, 152 – 63.

Long, J. A. and Jones, A. L. (1967). The fine structure of the zona glomerulosa and the zona fasciculata of the adrenal cortex of the opossum. *Am. J. Anat.,* 120, 463 – 87.

Lönnberg, E. (1902). On some remarkable digestive adaptations in diprotodont marsupials. *Proc. Zool. Soc. London,* 1902, 12 – 31.

Lyne, A. G. (1957). The development and replacement of pelage hairs in the bandicoot *Perameles nasuta* Geoffroy (Marsupialia: Peramelidae). *Aust. J. Biol. Sci.,* 10, 197 – 216.

McClure, C. F. W. (1900). The variations of the venous system of *Didelphis virginiana. Anat. Anz.,* 18, 441 – 60.

McClure, C. F. W. (1903). A contribution to the anatomy and development of the venous system of *Didelphis marsupialis* (L.). Part I. Anatomy. *Am. J. Anat.,* 2, 371 – 404.

McClure, C. F. W. (1906). A contribution to the anatomy and development of the venous system of *Didelphis marsupialis* (L.). Part II. Development. *Am. J. Anat.,* 5, 163 – 226.

McCrady, E. (1938). The embryology of the opossum. *Am. Anat. Mem.,* 16, 1 – 233.

McCrady, E. (1941). Parathyroids in adult opossum. *Anat. Rec.,* 79, Suppl., 45.

Mackenzie, W. C. (1916). The shape and peritoneal relationships of the spleen in monotremes and marsupials. *J. Anat.,* 51, 1 – 18.

Mackenzie, W. C. (1918a). *The Liver, Spleen, Pancreas, Peritoneal Relations, and Biliary System in Monotremes and Marsupials,* Critchley Parker, Australia.

Mackenzie, W. C. (1918b). *The Gastro-intestinal Tract in Monotremes and Marsupials,* Critchley Parker, Australia.

Mann, S. J. (1968). The tylotrich (hair) follicle of the American opossum. *Anat. Rec.,* 160, 171 – 9.

Martan, J. and Allen, J. M. (1965). The cytological and chemical organization of the prostatic epithelium of *Didelphis virginiana* Kerr. *J. Exp. Zool.,* 159, 209 – 29.

Martan, J., Hruban, Z. and Slesers, A. (1967). Cytological studies of the ductuli efferentes of the opossum. *J. Morphol.,* 121, 81 – 101.

Martin, G. F. (1967). Interneocortical connections in the opossum, *Didelphis virginiana. Anat. Rec.,* 157, 607 – 15.

Martin, G. F. (1968a). Some efferent cortical pathways of the opossum. *J. Hirnforsch.*, **10**, 55 - 78.

Martin, G. F. (1968b). The pattern of neocortical projections to the mesencephalon of the opossum, *Didelphis virginiana. Brain Res.*, **11**, 593 - 610.

Martin, G. F. (1969). Efferent tectal pathways of the opossum (*Didelphis virginiana*). *J. Comp. Neurol.*, **135**, 209 - 24.

Martin, G. F. and Dom, R (1970a). The rubro-spinal tract of the opossum (*Didelphis virginiana*). *J. Comp. Neurol.*, **138**, 19 - 29.

Martin, G. F. and Dom, R. (1970b). Rubrobulbar projections of the opossum (*Didelphis virginiana*). *J. Comp. Neurol.*, **139**, 199 - 213.

Martin, G. F. and Dom, R. (1971). Reticulospinal fibers of the opossum, *Didelphis virginiana.* II. Course, caudal extent and distribution. *J. Comp. Neurol.*, **141**, 467 - 83.

Martin, G. F. and Fisher, A. M. (1968). A further evaluation of the origin, the course and the termination of the opossum corticospinal tract. *J. Neurol. Sci.*, **7**, 177 - 87.

Martin, G. F. and Hamel, E. G. (1967). The striatum of the opossum, *Didelphis virginiana.* Description and experimental studies. *J. Comp. Neurol.*, **131**, 491 - 515.

Martin, G. F. and King, J. S. (1968). The basilar pontine gray of the opossum (*Didelphis virginiana*). II. Experimental determination of neocortical input. *J. Comp. Neurol.*, **133**, 447 - 61.

Martin, G. F. and Megirian, D. (1972). Corticobulbar projections of the marsupial phalanger (*Trichosurus vulpecula*). II. Projections to the mesencephalon. *J. Comp. Neurol.*, **144**, 165 - 91.

Martin, G. F., Megirian, D. and Roebuck, A. (1970). The corticospinal tract of the marsupial phalanger (*Trichosurus vulpecula*). *J. Comp. Neurol.*, **139**, 245 - 57.

Martin, G. F., Megirian, D. and Roebuck, A. (1971). Corticobulbar projections of the marsupial phalanger (*Trichosurus vulpecula*). I. Projections to the pons and medulla oblongata. *J. Comp. Neurol.*, **142**, 275 - 95.

Martin, G. F. and West, H. J. (1967). Efferent neocortical projections to sensory nuclei in the brainstem of the opossum (*Didelphis virginiana*). *J. Neurol. Sci.*, **5**, 287 - 302.

Martin, W. and Jones, R. (1834). Notes on the dissection of Azara's opossum (*Didelphis azarae* Temm). *Proc. Zool. Soc. London*, **1834**, 101 - 4.

Masai, H. (1960). A further study on the fasciculus aberrans in the anterior commissure of the Diprotodontia. *Med. J. Osaka Univ.*, **10**, 309 - 11.

Matthews, L. H. (1947). A note on the female reproductive tract in the tree kangaroos (*Dendrolagus*). *Proc. Zool. Soc. London*, **117**, 313 - 33.

Mehler, W. R. (1969). Some neurological species differences — *a posteriori. Ann. N. Y. Acad. Sci.*, **167**, 424 - 68.

Mitchell, P. C. (1905). On the intestinal tract of mammals. *Trans. Zool. Soc. London*, **17**, 437 - 536.

Moir, R. J., Somers, M. and Waring, H. (1956). Studies on marsupial nutrition. I. Ruminant-like digestion in a herbivorous marsupial (*Setonix brachyurus* Quay and Gaimard). *Aust. J. Biol. Sci.*, **9**, 293 - 304.

Morest, D. K. (1965). Identification of homologous neurons in the postero-lateral thalamus of cat and Virginia opossum. *Anat. Rec.,* **151,** 390.

Morgan, C. F. (1946). Sexual rhythms in the reproductive tract of the adult female opossum and effects of hormonal treatments. *Am. J. Anat.,* **78,** 411 - 63.

Moss, M. L. and Applebaum, E. (1963). The fibrillar matrix of marsupial enamel. *Acta Anat.,* **53,** 289 - 97.

Mossman, H. W. and Duke, K. L. (1973). *Comparative Morphology of the Mammalian Ovary,* University of Wisconsin Press.

Munger, B. L. (1965). The intraepidermal innervation of the snout skin of the opossum. A light and electron microscope study, with observations on the nature of Merkel's *Tastzellen. J. Cell Biol.,* **26,** 79 - 97.

Munger, B. L., Caramia, F. and Lacy, P. E. (1965). The ultrastructural basis for the identification of cell types in the pancreatic islets. II. Rabbit, dog and opossum. *Z. Zellforsch.,* **67,** 776 - 98.

Mykytowycz, R. and Nay, T. (1964). Studies of the cutaneous glands and hair follicles of some species of Macropodidae. *C.S.I.R.O. Wildlife Res.,* **9,** 200 - 17.

Nelsen, O. E. and Maxwell, N. (1942). The structure and function of the urogenital region in the female opossum compared with the same region in other marsupials. *J. Morphol.,* **71,** 463 - 91.

Nelson, L. R. and Lende, R A. (1965). Interhemispheric responses in the opossum. *J. Neurophysiol.,* **28,** 189 - 99.

O'Day, K. (1938). The retina of the Australian mammal. *Med. J. Aust.,* **1,** 326 - 8.

O'Donoghue, C. H. (1916). On the corpora lutea and interstitial tissue of the ovary in the Marsupialia. *Quart. J. Microscop. Sci.,* **61,** 433 - 73.

Ortman, R. and Griesback, W. E. (1958). The cytology of the pars distalis of the wallaby pituitary. *Aust. J. Exp. Biol. Med. Sci.,* **36,** 609 - 18.

Osgood, W. H. (1921). A monographic study of the American marsupial, *Caenolestes. Field Mus. Nat. Hist., Chicago, Publ. Zool.,* **14,** 3 - 156.

Oswaldo-Cruz, E. and Rocha-Miranda, C. E. (1967a). The diencephalon of the opossum in stereotaxic coordinates. I. The epithalamus and dorsal thalamus. *J. Comp. Neurol.,* **129,** 1 - 37.

Oswaldo-Cruz, E. and Rocha-Miranda, C. E. (1967b). The diencephalon of the opossum in stereotaxic coordinates. II. The ventral thalamus and hypothalamus. *J. Comp. Neurol.,* **129,** 39 - 48.

Oswaldo-Cruz, E. and Rocha-Miranda, C.E. (1968). *The Brain of the Opossum* Didelphis marsupialis. *A Cytoarchitectonic Atlas in Stereotaxic Coordinates.* Instituto de Biofísica, Universidade Federal do Rio de Janeiro.

Owen, R. (1836). Notes on the anatomy of the wombat, *Phascolomys wombat,* Per. *Proc. Zool. Soc. London,* **1836,** 49 - 53.

Owen, R. (1841a). Outlines of a classification of the Marsupialia. *Trans. Zool. Soc. London,* **2,** 315 - 33.

Owen, R. (1841b). On the osteology of the Marsupialia. *Trans. Zool. Soc. London,* **2,** 379 - 408.

Owen, R. (1866) *On the Anatomy of Vertebrates,* Vol. II, Longmans, Green, London.

Owen, R. (1868). *On the Anatomy of Vertebrates,* Vol. III, Longmans, Green, London.

Owen, R. (1874). On the osteology of the Marsupialia. (Part IV) Bones of the trunk and limbs, *Phascolomys. Trans. Zool. Soc. London,* 81, 483 – 500.

Packer, A. D. (1941). An experimental investigation of the visual system in the phalanger, *Trichosurus vulpecula. J. Anat.,* 75, 309 – 29.

Pak Poy, R. K. F. (1957). Electron microscopy of the marsupial renal glomerulus. *Aust. J. Exp. Biol. Med. Sci.,* 35, 437 - 47.

Parsons, F. G. (1896). On the anatomy of *Petrogale xanthopus,* compared with that of other kangaroos. *Proc. Zool. Soc. London,* 1896, 683 – 714.

Parsons, F. G. (1903). On the anatomy of the pig-footed bandicoot (*Choeropus castanotis*). *J. Linnean Soc., Zool.,* 29, 64 – 80.

Parsons, R. S., Atwood, T., Guiler, E. R. and Heddle, R. W. L. (1971). Comparative studies on the blood of monotremes and marsupials – I. Haematology. *Comp. Biochem. Physiol.,* 39B, 203 – 8.

Pearson, J. (1940). Notes on the blood system of the Marsupialia. *Papers Roy. Soc. Tasmania,* 1939, 77 – 94.

Pearson, J. (1944). The vaginal complex of the rat-kangaroos. *Aust. J. Sci.,* 7, 80 – 3.

Pearson, J. (1945). The female urogenital system of the Marsupialia with special reference to the vaginal complex. *Papers Roy. Soc. Tasmania,* 1944, 71 – 98.

Pearson, J. (1949). Placentation of the Marsupialia. *Proc. Linnean Soc.,* 161, 1 – 9.

Pearson, J. and de Bavay, J. (1951). The female urogenital system of *Antechinus* (Marsupialia). *Papers Roy. Soc. Tasmania,* 1950, 137 – 42.

Pekelharing, C. J. (1970). Cementum deposition as an age indicator in the brush-tailed possum, *Trichosurus vulpecula* Kerr (Marsupialia). *Aust. J. Zool.,* 18, 71 – 6.

Perrott, J. W. (1966). The peritoneum of *Trichosurus vulpecula. Anat. Rec.,* 154, 295 – 304.

Pfeiffer, E. W. (1968). Comparative anatomical observations of the mammalian renal pelvis and medulla. *J. Anat.,* 102, 321 – 31.

Phillips, D. M. (1970a). Development of spermatozoa in the wooly opossum with special reference to the shaping of the sperm head. *J. Ultrastruct. Res.,* 33, 369 – 80.

Phillips, D. M. (1970b). Ultrastructure of spermatozoa of the wooly opossum *Caluromys philander. J. Ultrastruct. Res.,* 33, 381 – 97.

Plakke, R. K. and Pfeiffer, E. W. (1964). Blood vessels of the mammalian renal medulla. *Science,* 146, 1683 – 6.

Poulton, E. B. (1883). On the tongues of the Marsupialia. *Proc. Zool. Soc. London,* 1883, 599 – 628.

Pubols, B. H. (1968). Retrograde degeneration study of somatic sensory thalamocortical connections in brain of Virginia opossum. *Brain Res.,* 7, 232 – 51.

Pubols, B. H. and Pubols, L. M. (1966). Somatic sensory representation in the thalamic ventrobasal complex of the Virginia opossum. *J. Comp. Neurol.,* 127, 19 – 33.

Quay, W. B. (1966). Pineal structure and composition in red and grey kangaroos. *Anat. Rec.,* 154, 405.

Quintarelli, G. and Dellovo, M. C. (1969). Studies on the exocrine secretions. Histochemical investigations on major salivary glands of exotic animals. *Histochemie,* 19, 199 – 223.

Rafols, J. A. and Matzke, H. A. (1970). Efferent projections of the superior colliculus in the opossum. *J. Comp. Neurol.*, **138**, 147 - 60.

Ratcliffe, H. L. (1941). A median vaginal canal and other anomalies of the genital tract of the opossum, *Didelphis virginiana. Anat. Rec.*, **80**, 203 - 9.

Rattner, J. B. (1972). Nuclear shaping in marsupial spermatids. *J. Ultrastruct. Res.*, **40**, 498 - 512.

Rees, S. and Hore, J. (1970). The motor cortex of the brush-tailed possum (*Trichosurus vulpecula*): motor representation, motor function and the pyramidal tract. *Brain Res.*, **20**, 439 - 51.

Riese, W. (1945). Structure and function of the brain of the opossum (*Didelphis virginiana*) at the time of birth. *J. Mammal.*, **26**, 148 - 53.

Rodger, J. C. and Hughes, R. L. (1973). Studies of the accessory glands of male marsupials. *Aust. J. Zool.*, **21**, 303 - 20.

Roth, L. M. and Luse, S. A. (1964). Fine structure of the neurohypophysis of the opossum (*Didelphis virginiana*). *J. Cell. Biol.*, **20**, 459 - 72.

Royce, G. J. (1971). Morphology of neuroglia in the hypothalamus of the opossum (*Didelphis virginiana*), armadillo (*Dasypus novemcinctus mexicanus*) and cat (*Felis domestica*). *J. Morphol.*, **134**, 141 - 80.

Sapsford, C. S. and Rae, C. A. (1969). Ultrastructural studies on Sertoli cells and spermatids in the bandicoot and ram during the movement of mature spermatids into the lumen of the seminiferous tubule. *Aust. J. Zool.*, **17**, 415 - 45.

Sapsford, C. S., Rae, C. A. and Cleland, K. W. (1967). Ultrastructural studies on spermatids and Sertoli cells during early spermiogenesis in the bandicoot *Perameles nasuta* Geoffroy (Marsupialia). *Aust. J. Zool.*, **15**, 881 - 909.

Sapsford, C. S., Rae., C. A. and Cleland, K. W. (1969). Ultrastructural studies on maturing spermatids and on Sertoli cells in the bandicoot *Perameles nasuta* Geoffroy (Marsupialia). *Aust. J. Zool.*, **17**, 195 - 292.

Sapsford, C. S., Rae, C. A. and Cleland, K. W. (1970). Ultrastructural studies on the development and form of the principal piece sheath of the bandicoot spermatozoon. *Aust. J. Zool.*, **18**, 21 - 48.

Schäfer, E. A. and Williams, D. J. (1876). On the structure of the mucous membrane of the stomach in the kangaroos. *Proc. Zool. Soc. London*, **1876**, 165 - 77.

Scharrer, E. (1940). Arteries and veins in the mammalian brain. *Anat. Rec.*, **78**, 173 - 96.

Schneider, C. (1968). Beitrag zur Kenntnis des Gehirnes von *Notoryctes typhlops. Anat. Anz.*, **123**, 1 - 24.

Setchell, B. P. and Carrick, F. N. (1973). Spermatogenesis in some Australian marsupials. *Aust. J. Zool.*, **21**, 491 - 9.

Sharman, G. B. (1954). The relationships of the quokka (*Setonix brachyurus*). *W. Aust. Nat.*, **4**, 159 - 68.

Sharman, G. B. (1955). Studies on marsupial reproduction. II. The oestrus cycle of *Setonix brachyurus. Aust. J. Zool.*, **3**, 44 - 55.

Sharman, G. B. (1959a). Evolution of marsupials. *Aust. J. Sci.*, **22**, 40 - 5.

Sharman, G. B. (1959b). Marsupial reproduction. *Monogr. Biol.*, **8**, 332 - 68.

Sharman, G. B. (1959c). Some effects of x-rays on dividing cells in the testis and bone marrow of the marsupial *Potorous tridactylus. Int. J. Radiat. Biol.,* 1, 115 - 30.

Sharman, G. B. (1961). The embryonic membranes and placentation in five genera of diprotodont marsupials. *Proc. Zool. Soc. London,* 137, 197 - 220.

Sharman, G. B. (1965). Marsupials and the evolution of viviparity. In: *Viewpoints in Biology,* Vol. 4 (ed. J. D. Carthy and C. L. Duddington), Butterworths, London, pp 1 - 28.

Sharman, G. B. (1970). Reproductive physiology of marsupials. *Science,* 167, 1221 - 8.

Sharman, G. B. (1974). Marsupial taxonomy and phylogeny. *Aust. Mammal.,* 1, 137 - 54.

Sharman, G. B., Calaby, J. H. and Poole, W. E. (1966). Patterns of reproduction in female diprotodont marsupials. *Symp. Zool. Soc. London,* 15, 205 - 32.

Shorey, C. D. and Hughes, R. L. (1972). Uterine glandular regeneration during the follicular phase in the marsupial *Trichosurus vulpecula. Aust. J. Zool.,* 20, 235 - 47.

Shorey, C. D. and Hughes, R. L. (1973). Development, function, and regression of the corpus luteum in the marsupial *Trichosurus vulpecula. Aust. J. Zool.,* 21, 477 - 89.

Shrivastava, R. K. (1962). The deltoid musculature of the Marsupialia. *Amer. Midl. Nat.,* 67, 305 - 20.

Sonntag, C. F. (1921a). The comparative anatomy of the koala (*Phascolarctos cinereus*) and vulpine phalanger (*Trichosurus vulpecula*). *Proc. Zool. Soc. London,* 1921, 547 - 77.

Sonntag, C. F. (1921b). Contributions to the visceral anatomy and myology of the Marsupialia. *Proc. Zool. Soc. London,* 1921, 851 - 82.

Sorokin, S. (1962). A note on the histochemistry of the opossum's lung. *Acta Anat.,* 50, 13 - 21.

Sorokin, S. P. (1965). On the cytology and cytochemistry of the opossum's bronchial glands. *Am. J. Anat.,* 117, 311 - 37.

Sorokin, S. P. (1967). A morphological and cytochemical study on the great alveolar cell. *J. Histochem. Cytochem.,* 14, 884 - 97.

Sousa, A. P. B., Oswaldo-Cruz, E. and Gattass, R. (1971). Somatotopic organization and response properties of neurons of the ventrobasal complex in the opossum. *J. Comp. Neurol.,* 142, 231 - 47.

Stirling, E. C. (1891). Description of a new genus and species of Marsupialia, '*Notoryctes typhlops'. Trans. Roy. Soc. S. Australia,* 14, 154 - 87.

Sunderland, S. (1941). The vascular pattern in the central nervous system of the monotremes and Australian marsupials. *J. Comp. Neurol.,* 75, 123 - 9.

Sweet, G. (1907). The skin, hair and reproductive organs of *Notoryctes.* Contributions to our knowledge of the anatomy of *Notorcytes typhlops,* Stirling — Parts IV and V. *Quart. J. Microscop. Sci.,* 51, 325 - 44.

Symington, J. (1900). A note on the thymus gland in the koala (*Phascolarctos cinereus*). *J. Anat. Physiol.,* 34, 226 - 7.

Tate, G. H. H. (1947). On the anatomy and classification of the Dasyuridae (Marsupialia). *Bull. Am. Mus. Nat. Hist.,* 88, 97 - 155.

Tate, G. H. H. (1948a). Studies on the anatomy and phylogeny of the Macropodidae (Marsupialia). *Bull. Am. Mus. Nat. Hist.,* 91, 233 - 351.

Tate, G. H. H. (1948b). Studies in the Peramelidae (Marsupialia). *Bull. Am. Mus. Nat. Hist.,* **92**, 313 - 46.

Teo, E. H. (1974). The neurohypophysis of *Trichosurus vulpecula*: a morphological study relating to function. Thesis, University of Adelaide.

Thomas, T. B. (1937). Cellular components of the mammalian islets of Langerhans. *Am. J. Anat.,* **62**, 31 - 57.

Troughton, E. LeG. (1959). The marsupial fauna: its origin and radiation. *Monogr. Biol.,* **8**, 69 - 88.

Tucker, R. (1969). The submicroscopic structure of the oral mucosa of the phalanger (*Trichosurus vulpecula*). *Proc. Linnean Soc. N. S. Wales,* **94**, 55 - 60.

Turner, E. L. (1924). The pyramidal tract of the Virginia opossum (*Didelphis virginiana*). *J. Comp. Neurol.,* **36**, 387 - 97.

Tyndale-Biscoe, C. H. (1965). The female urogenital system and reproduction of the marsupial *Lagostrophus fasciatus. Aust. J. Zool.,* **13**, 255 - 67.

Tyndale-Biscoe, C. H. (1968). Reproduction and postnatal development in the marsupial *Bettongia lesueur* (Quoy and Gaimard). *Aust. J. Zool.,* **16**, 577 - 602.

Tyndale-Biscoe, C. H. (1973). *Life of Marsupials,* Edward Arnold, London.

Tyndale-Biscoe, C. H. (1974). Reproduction in marsupials. *Aust. Mammal.,* **1**, 175 - 80.

Tyson, E. (1698). Carigueya, *seu* Marsupiale americanum OR The anatomy of an opossum, dissected at Gresham College. *Phil. Trans. Roy. Soc.,* **20**, 105 - 64.

Ulinski, P. S. (1971). External morphology of pouch young opossum brains: a profile of opossum neurogenesis. *J. Comp. Neurol.,* **142**, 33 - 58.

van Lennep, E. W. (1962). The histology of the mucosa of the small intestine of the long-nosed bandicoot (Marsupialia: *Perameles nasuta* Geoffroy) with special reference to intestinal secretion. *Acta Anat.,* **50**, 73 - 89.

van der Sprenkel, H. B. (1926). Stria terminalis and amygdala in the brain of the opossum (*Didelphis virginiana*). *J. Comp. Neurol.,* **42**, 211 - 54.

Vinson, G. P., Phillips, J. G., Chester Jones, I. and Tsang, W. N. (1971). Functional zonation of adrenocortical tissue in the brush possum (*Trichosurus vulpecula*). *J. Endocrinol.,* **49**, 131 - 40.

Voris, H. C. (1928a). The arterial supply of the brain and spinal cord of the Virginia opossum (*Didelphis virginiana*). *J. Comp. Neurol.,* **44**, 403 - 23.

Voris, H. C. (1928b). The morphology of the spinal cord of the Virginia opossum (*Didelphis virginiana*). *J. Comp. Neurol.,* **46**, 407 - 59.

Wade, O. (1968). Structural characteristics of the heart of the wombat, *Lasiorhinus latifrons* Owen. *Amer. Midl. Nat.,* **80**, 266 - 8.

Wade, O. and Neely, P. (1949). The heart and attached vessels of the opossum, a marsupial. *J. Mammal.,* **30**, 111 - 6.

Ward, J. W. (1954). The development of the corticospinal tract in the pouch-young of the Virginia opossum, *Didelphis virginiana. J. Comp. Neurol.,* **101**, 483 - 94.

Waring, H., Moir, R. J. and Tyndale-Biscoe, C. H. (1966). Comparative physiology of marsupials. *Advan. Comp. Physiol. Biochem.,* **2**, 237 - 376.

Waring, H., Sharman, G. B., Lovat, D. and Kahan, M. (1955). Studies on marsupial reproduction. I. General features and techniques. *Aust. J. Zool.,* **3**, 34 - 43.

Warner, F. J. (1969). The development of the diencephalon in *Trichosurus vulpecula. Okajimas Fol. Anat. Jap.,* **46**, 265 -95.

Watson, C. R. R. (1971). The corticospinal tract of the quokka wallaby (*Setonix brachyurus*). *J. Anat.,* **109**, 127 - 33.

Wheeler, R S. (1943). Normal development of the pituitary in the opossum and its responses to hormonal treatments. *J. Morphol.,* **73**, 43 - 87.

Wilborn, W. H. and Shackleford, J. M. (1969). The cytology of submandibular glands of the opossum. *J. Morphol.,* **128**, 1 - 33.

Windle, B. C. A. and Parsons, F. G. (1897). On the anatomy of *Macropus rufus. J. Anat. Physiol.,* **32**, 119 - 34.

Winkelmann, R. K. (1964). Nerve endings of the North American opossum (*Didelphis virginiana*): a comparison with nerve endings of primates. *Am. J. Phys. Anthropol.,* **22**, 253 - 8.

Winkelmann, R. K. (1966). Some unusual histochemical properties of kangaroo skin. *J. Invest. Dermatol.,* **46**, 446 - 52.

Wislocki, G. B. (1940). Peculiarities of the cerebral blood vessels of the opossum: diencephalon, area postrema and retina. *Anat. Rec.,* **78**, 119 - 37.

Wislocki, G. B. and Campbell, A. C. P. (1937). The unusual manner of vascularization of the brain of the opossum (*Didelphis virginiana*). *Anat. Rec.,* **67**, 177 - 91.

Woodburne, R. T. (1943). The nuclear pattern of the non-tectal portions of the midbrain and isthmus in the opossum. *J. Comp. Neurol.,* **78**, 169 - 90.

Wood Jones, F. (1923). *The Mammals of South Australia,* Govt. Printer, Adelaide.

Wood Jones, F. (1943). *Habit and Heritage,* Kegan Paul, Trench and Trubner, London.

Wood Jones, F. (1948). The study of a generalized marsupial (*Dasycercus cristicauda* Krefft). *Trans. Zool. Soc. London,* **26**, 409 - 501.

Yadav, M. (1973). The presence of the cervical and thoracic thymus lobes in marsupials. *Aust. J. Zool.,* **21**, 285 - 301.

Young, A. H. (1879). On the male generative organs of the koala (*Phascolarctos cinereus*). *J. Anat. Physiol.,* **13**, 305 - 17.

Ziegler, A. C. (1971). A theory of the evolution of therian dental formulas and replacement patterns. *Quart. Rev. Biol.,* **46**, 226 - 49.

Zimmerman, E. A. and Chambers, W. W. (1963). Cortical projections to sensory relay nuclei in the brainstem of the opossum and rat. *Anat. Rec.,* **145**, 304.

16 An ecological comparison of marsupial and placental patterns of reproduction

Pamela Parker*

The traditional view of the relationship of marsupials is that of Huxley (1880), who held that Metatheria represent a stage of evolutionary development between the Prototheria (for example, platypus and spiny anteater) and the Eutheria (placental mammals). Huxley's view implies that the living mammals thus represent three groups which evolved one from another, in the sequence Prototheria – Metatheria – Eutheria. It is this background against which features of marsupial anatomy and physiology may be interpreted as primitive in comparison with placental equivalents; thus several authors (for example, Lillegraven, 1969; Sharman, 1970) see certain aspects of marsupial reproductive biology as primitive, restrictive or inefficient when compared with similar features of placental reproduction.

Fossil evidence, however, indicates that the ancestors of the viviparous mammals separated early from the prototherian lineage; marsupials and placentals thus together represent a single lineage co-ordinate with that of Prototheria, and not two successive derivations from early mammals. By the early Cretaceous marsupials and placentals were distinguishable from each other (Slaughter, 1968, 1971), yet very similar in dentition and probably in many other anatomical and physiological features. Lillegraven (1974) concluded that the two diverging groups were isolated from each other for long periods. If so, this separation would have afforded many opportunities for distinctive specialisation in adaptation to different ecological conditions. It is therefore possible to interpret divergent characters of the two groups as responses to differing selective pressure rather than as primitive or advanced characteristics. In this paper I argue that the features which distinguish marsupial from placental reproduction in particular represent not persistent evolutionary artifacts but separate protocols evolved in response to different ecological situations. The main body of this paper discusses some of Williams's (1966a, 1966b, 1975) theoretical work in relation to marsupial – placental reproductive differences and compares specific features of reproduction in the two groups.

In a short final section aspects of brain structure and behaviour, similarly cited as primitive in many older works, are now reinterpreted. This is necessary because biological assessments of marsupials have too often been based on data from *Didelphis virginiana* (for example, Jerison, 1973; Enlow and Brown, 1958), which

*Pamela Parker has studied at Smith College, Yale School of Forestry, and Yale University Graduate School. She is currently a post-doctoral fellow at the Museum of Comparative Zoology, Harvard University, working with A. W. Crompton.

273

is equivalent to basing an interpretation of placental biology on a primitive insectivore such as *Tenrec ecaudatus*. Moreover, this practice of using a single, relatively primitive species as representative of all marsupials perhaps stems from the persistent tradition in taxonomy of placing all marsupials within a single order (Vaughan, 1972; Walker, 1964). Since marsupials show adaptations which are arguably as diverse as those of placentals, the single order classification should perhaps be replaced with a multi-ordinal one (Ride, 1964; Kirsch, 1968, 1976; Kirsch and Calaby, this volume), which might well have the effect of forestalling easy generalisations from one species.

Reproductive differences of marsupials and placentals

Reproductive patterns of marsupial and placental males show no apparent functional differences (Tyndale-Biscoe, 1973): in either group males commit few resources to the production of each offspring, and males with little pre-mating behaviour enhance their chances of producing offspring by being as promiscuous as opportunity permits (Williams, 1966a). Females of both groups by contrast invest heavily in reproduction. Their reproductive biology is therefore more readily subject to forces of natural selection imposed by the environment. Selection has produced contrasting patterns of reproduction which may be summarised in the following terms. Placental reproduction confers advantages upon the young at the expense of the mother. In placentals events in the reproductive cycle which follow implantation are relatively insensitive to changing external conditions. Reproductive investment is high, and repro-duction is difficult to terminate through much of the offspring's dependency period, even if chances of later survival are low. In contrast, marsupial females make a smaller reproductive investment in each young, at least until later phases of the offspring's dependency period when chances of survival to reproductive age are relatively higher. Marsupials can, in addition, terminate their reproductive investment readily throughout most of the dependency period, often in response to unfavourable environmental changes.

Fisher (1930) and Williams (1966a, b, 1975) provide the theoretical framework for the following discussion. In general, there is at any time in a female's life history an optimum fraction of available resources which natural selection will direct toward reproduction, in order to maximise her ultimate genetic contribu-tion to future generations. Each female of a specified population and age has a *reproductive value*, defined as the mean amount of future reproductive contri-butions for all females of this category. Selection will always maximise reproduc-tive value (Williams, 1966b; Cody, 1971).

Any female in reproductive condition faces a number of decisions which affect her reproductive value. The first of these are whether to breed at a particular time and with whom, interrelated choices. If she chooses not to breed, this lowers the fraction of her reproductive value which is immediately at risk, and increases her future reproductive value. If she chooses to breed, the risk on her current reproductive value rises and her potential reproductive value falls. Only when the rise in immediate value exceeds the rise in future value will selection favour a decision to breed. Attendant on a positive decision is a further series of choices which are the subject of this paper, involving the distribution of the mother's resources among the component activities of breeding in a way which maximises return on the investment – that is, produces the greatest possible number of viable young in her lifetime. Thus resources are partitioned within each litter, and between successive litters, throughout her reproductive life. Substantial over-investment in any one litter must detract from the parent's potential for producing further young. Successively larger fractions of parental resources may be appropriate for each new attempt at reproduction, as the remaining life expectancy of the parent dwindles (Pianka and Feener, quoted in Smith and Fretwell, 1974). Withdrawal of investment, involving the death of a litter at some time during the dependency period, may safeguard the balance of resources; it is clearly better to abandon a reproductive attempt which has a high probability of failure if the investing parent thereby increases her likelihood of surviving to reproduce again.

At most stages of the reproductive cycle female marsupials and placentals differ in significant ways with respect to the concept of reproductive value. Most of these differences can be examined in terms of patterns of investment in the young and selective pressures on the mothers; they may be treated under the headings of development and anatomy, immunology and energy allocations.

Developmental and anatomical differences

The marsupial ovum differs from that of the placental in being larger, containing more yolk and having a shell membrane. These characteristics, which make the marsupial ovum more like the ova of other amniotes (for example, reptiles) than of placental mammals, are perhaps made necessary by the subsequent events of early development: most of marsupial embryogenesis is fueled by the yolk and by secretions of the uterine lining which occur independently of mating (Tyndale-Biscoe, 1973). Early divisions of the marsupial ovum give rise to a laminar blastocyst, and most of the dividing cells of the embryo are components of developing germ layers. The placental ovum by contrast lacks many features of the typical amniote ovum. Virtually without yolk, it divides to form a solid clump of cells (the morula) which later becomes a blastocyst. From this develops the unique, invasive trophoblast which implants in the uterine wall, allowing capillary intimacy with the maternal blood stream and providing for the needs of the embryo throughout later development (Lillegraven, 1969). Since by the time of their formation the outer cells of the morula are probably determined as trophoblastic cells (Tarkowski and Wroblewska, 1967; Alexandre, 1974), this distinction in patterns of cleavage may be the significant precursor of later immunological differences in maternal – foetal relationships. To this point the marsupial and placental resource investments in reproduction are low and therefore the quantitative difference between them may be disregarded.

Energy commitment remains low in most marsupial mothers during gestation; differences between pregnant and non-pregnant females are minimal. Gestations take place within oestrous cycles, and the luteal phase of the uterus is timed to occur when the embryo is developing (Tyndale-Biscoe, 1973). In non-macropodids the uterine cycle is not interrupted by pregnancy; in macropodids it is only slightly extended (Tyndale-Biscoe, Hearn and Renfree, 1974); and even in kangaroos the secretory phase of the uterus also terminates whether or not it facilitates the development of young. During the pre-implantation phase

Table 16.1 Comparisons of gestation lengths and relative birth weights in marsupials and placentals

Species	Gestation (days)	Possible litter size	Individual birth weight (g)	Maternal weight (kg)	Birth weight/maternal weight × per cent	
Marsupials:						
Didelphis virginiana	13	14	0.161	1.4	0.01	0.14 for litter
Dasyurus viverrinus	9	7	0.013	0.7	0.002	0.014 for litter
Isoodon macrourus	14	6	0.180	1.1	0.016	0.096 for litter
Antechinus flavipes	32	4	0.016	0.024	0.07	0.28 for litter
Trichosurus vulpecula	17	4	0.220	4.3	0.005	0.020 for litter
Bettongia penicillata	21	1	0.290	1	0.03	
Setonix brachyurus	27	1	0.343	4	0.009	
Wallabia bicolor	35	1	0.430	10	0.004	
Macropus eugenii	29	1	0.398	5.2	0.008	
M. rufus	33	1	0.817	30	0.003	
Placentals:						
Mesocricetus auratus	16	5	2.2	0.096	2.3	11.5 for litter
Rattus norvegicus	22	7	4.5	0.2	2.3	16.1 for litter
Oryctolagus cuniculus	32	6	57	1.9	3.0	18.0 for litter
Sus scrofa	112	10	2500	100	2.5	25.0 for litter
Ovis aries	150	1	5000	62.5	8.0	
Macaca mulatta	164	1	450	6	7.5	
Papio cynocephalus	182	1	900	20	4.5	

Data from Tyndale-Biscoe, 1973; Sharman, 1973; Walker, 1964; Vaughan, 1972; Hendrickx, 1971.

(more than half of gestation) nutrients are actively taken up through the shell membrane (Renfree, 1970), which remains intact. Since this membrane is capable of only slight attenuation, the foetus has a fixed maximum size within it. After dissolution of the shell membrane, the degree of maternal – foetal capillary intimacy established is variable among marsupials, but generally slight, and never so long-lasting as in placentals. The period of placental contact with the mother is short, with a narrow range of variation; even in bandicoots, with the most extensive implantation of all marsupials, total gestation is only about twelve days (see Stodart, this volume). Having only a short period for growth without the shell membrane (Tyndale-Biscoe, 1973), young therefore are small at birth (Table 16.1). Hence, in most marsupial species there are no substantial differences between pregnant and non-pregnant females throughout the gestation period.

In placental mammals pregnancy interrupts the uterine cycle and the endocrine environment changes. While extra-ovarian hormones are not known to regulate pregnancy of non-macropodid marsupials and are only slightly implicated in that of macropodids (Tyndale-Biscoe *et al.*, 1974), the eutherian placenta becomes an important endocrine organ which redirects maternal physiology presumably to better meet the requirements of the foetus. It thus temporarily lowers the fitness of the female herself, although not necessarily that of her genes. Placental gestation is relatively long and the young are larger at birth. Associated with the parturition of large young are both transitory changes (developing during gestation) and permanent osteological features of female pelvic girdles contributing to the enlargement of the birth canal. The ontogeny of the permanent features of this pelvic dimorphism probably results from the effects of testicular hormones altering the basic female pattern, as shown in mice (Crelin, 1960). Hence, it appears that selection favours the male pelvic type apart from the requirements of delivering a large foetus, indicating another cost of reproduction to the placental female. This dimorphism is not present in marsupials, at least as judged from museum specimens (Figure 16.1).

The events preceding birth, and birth itself, apparently have less impact on the marsupial than the placental mother. Marsupial mothers appear not to be aware of the moment of birth or to recognise their new-born young; birth is rapid, and does not require a long lapse of maternal vigilance. New-born young break free of the amnion at birth, and usually crawl unaided to the pouch, where they locate and attach themselves to a teat. Not all female marsupials are pouched, but all marsupial young are attached for several weeks after birth and protected from environmental influence by close contact with the mother's body. In the sense of Williams (1966a), this period of attachment corresponds directly to the uterine development of the placental; it is not, however, physiologically equivalent. Despite the complexity of the journey from birth canal to teat, the percentage of new-born kangaroos (*Macropus rufus*) reaching the pouch compares favourably with survival rates of uniparous placental young (Sharman, 1973). Just as placental young have recognisable developmental stages (Weir and Rowlands, 1973), including specialisations for intra-uterine life, marsupial young develop adaptations for their existence outside the parental body. Among these are the precocious development of the olfactory region of the brain, of the musculature and innervation of the mouth, neck, and arms and of the digestive tract (Sharman, 1973).

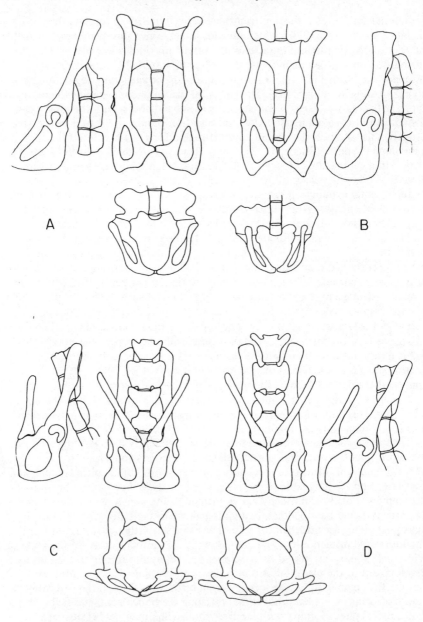

Figure 16.1 Comparisons of the pelvic girdles of the mouse and
the opossum: A. Female mouse: lateral, ventral, and posterior;
B. male mouse: lateral, ventral, and posterior (drawings based on
Crelin, 1960). C. Female opossum: lateral, ventral, and posterior;
D. male opossum: lateral, ventral, and posterior (drawings based on
specimens from Bedford, Mass).

Immunology

Selection for large size at birth and therefore prolonged gestation would seem to
have increased the danger of immunological rejection of the foetus posed by the
presence, within the mother, of tissues derived genetically from both parents;
paternal transplantation antigens are present as early as the two-cell stage in *Mus
musculus* (Palm, Heyner and Brinster, 1971). The syncytial trophoblast with its
external deposits (a structure unique to placentals) is widely believed to protect
the developing foetus from attack by maternal antibodies (Behrman, 1971; Beer
and Billingham, 1971), through some incompletely understood mechanisms which
diminish both the mother's responsiveness and her progeny's antigenicity (Jacobs
and Uphoff, 1974). Tyndale-Biscoe (1973) and Moors (1974) believe that the
shell membrane of marsupials, which is entirely of maternal origin (Boyd and
Hamilton, 1952), shields the embryo from potential rejection while it is intact
during gestation (Hughes, 1974). Very little growth occurs while the shell
membrane is in place (Tyndale-Biscoe, 1973); when it breaks down on implantation, maternal exposure to embryonic antigens begins, and this is the period of
most rapid embryogenesis. Since maternal leucocytes are common in the uterine
mucosa (Flynn, 1923; Sharman, 1961), especially about the placenta, it is reasonable to suppose that marsupials do not have an immune-suppressant system
associated with the foetal membranes comparable with that of eutherians.
Tyndale-Biscoe (1973) suggests that the efficient placenta of *Perameles* can
function only a short time before rejection mechanisms are strong enough to
damage the foetus. The short period of contact allowed between mother and
foetus in the absence of suppressant mechanisms may explain the relative developmental uniformity of all marsupials at birth (Sharman, 1973).

Gestation in some species of Macropodidae is probably extended by an
increase in length of the luteal phase of the uterus, and in other species by a
combined increase in the length of the luteal phase and protracted maintenance
of gestation into the next follicular phase; in both cases extension is probably
mediated by endocrine activity of the placenta (Tyndale-Biscoe *et al.*, 1974).
Even in this group, where gestation lasts longer than the time required for allo-
graft rejection, the period between disruption of the shell membrane and birth
is relatively short.

The nature of marsupial gestation is traditionally regarded as evidence of evolutionary stagnation, which could be circumvented by the acquisition of an advanced immunological mechanism similar to that of placentals. It can, however, be argued that
the marsupial mother may remain at an advantage in comparison with the placental by
retaining intrinsic control of her own reproductive mechanisms, rather than surrender-
ing a degree of control to an invasive organism which constitutionally includes only
half her genes. Lack of an immune-suppressant system like that of placentals is not
necessarily an evolutionary handicap which marsupials have been unable to transcend.
It may simply imply that selection has favoured an alternative pattern in which females retain control of their own physiological processes, and keep open the reproductive options which are closed when a foetus remains long in the uterus.

Allocations of energy

The decision to reproduce has very different effects on the relative fitness of placental and marsupial females. Following implantation, the placental is committed to reproduction. Indirect evidence suggests that maternal resources must fall very low before pregnancy is abandoned. Such resources as calcium are mobilised from maternal stores if the mother's diet is inadequate. Thus placental pregnancy is relatively insensitive to post-conception environmental changes, and continues to parturition if necessary at the expense of the mother's fitness. Some species are able to recall a decision to reproduce (Bruce, 1959, 1970). In this respect they resemble the marsupial pattern of reproduction. Young at parturition represent a sizable investment of maternal resources which remain at risk through lactation. During lactation the mother is more sensitive to environmental changes and reproductive effort is readily terminated; many mammal populations suffer losses of young at this stage, after further maternal resources of time and energy have been committed (Brown, 1970). By contrast, gestation in marsupials re- presents little drain of resources over those already committed in the oestrous cycle. Maternal supply of resources to the young — mainly through lactation — is continually sensitive to environmental conditions, and reproductive effort can readily be terminated by abandoning the young at any time from birth onward. Extent of maternal losses of energy then depends on the size of the young at the time when lactation fails.

Growth rate of young mammals varies with availability of resources in ways which are not clearly understood. In experiments to investigate growth rates, Merchant and Sharman (1966) fostered a young Swamp wallaby, *Wallabia bicolor*, to a female of the larger Red kangaroo, *Macropus rufus*. The growth rate of the young wallaby matched that of a young Red kangaroo, although duration of pouch life was typical of Swamp wallabies. Similarly, Smith (in press) fostered young placental Meadow voles, *Microtus pennsylvanicus*, to a female of the smaller White-footed mouse, *Peromyscus leucopus*. The young voles developed at rates normal for their species, but gained weight at rates matching those of White-footed mice. These findings suggest that somatic growth may vary more than and be separate from developmental rate; in the juveniles of both species the growth

Table 16.2 Growth rates of *Bettongia penicillata* raised in Australia and the US

	Australia	USA
Duration of pouch life	100 + days	65 – 70 days
Remain closely with mother	130 days	120 days
Active rejection by mother	180 – 200 days	180 days or less
Subadult weight	750 – 800 g	820 – 1015 g*
Age of sexual maturity	at the end of the first year in both sexes	females breed as early as 290 days
Maximum female reproduction per year	3	4
Adult weight	approximately 1 kg	1.2 – 1.6 kg†

Australian data from Sampson, 1971.
*Based on 7 individuals.
†Based on 8 individuals.

rate differed from normal by 30 per cent. Larger kangaroos also show variability
in growth rate under different environmental conditions (Frith and Calaby, 1969).
Generation length is also variable; Rat kangaroos, *Bettongia penicillata*, studied in
captivity and in the wild by Sampson (1971) raised one young every four months,
while animals of the same species which I kept captive in the United States took
only three months (Table 16.2). Weights of adults in the US averaged higher than
those recorded by Sampson, and the age of females at first reproduction was
lower. In neither case was any attempt made to alter the duration of pouch life.

It seems probable, therefore, that rates of growth and generation length are
flexible in both placentals and marsupials (Racey, 1973; Brown, 1970). However,
the ability to adjust reproduction and growth rates to environmental conditions
can more fully be exploited by marsupials, with their very short gestation periods
and lower rates of investment of energy. Macropodids show an additional specialisation – embryonic diapause, which helps them even further to match their reproductive capacity to environmental conditions. Conception may occur during a post-
partum oestrus; the zygote develops to a blastocyst and remains at that stage while
the young in the pouch continues to suckle. As the suckling stimulus diminishes,
embryogenesis resumes and birth follows. Thus a female in a good environment
may have a joey at heel, a young in the pouch and a stored blastocyst. If food and
water become scarce, competition for maternal resources results in the larger
young being left to fend for itself, while the pouch-young continues to draw its
relatively smaller rations of milk (Tyndale-Biscoe, 1973). If poor conditions
persist and all young are lost, Red kangaroos enter anoestrus, resuming oestrus
at the first rains. Thus the food supply will be growing when lactational demands
of the young becomes significant (Newsome, 1964, 1965).

Marsupials generally grow more slowly than placentals during early life, their
growth during lactation being compared unfavourably with the growth of
placental young in the uterus (Tyndale-Biscoe, 1973). Factors other than phylogenetic incompetence may, however, be involved. Ricklefs (1973) suggested that
some variation in the growth rates of nestling birds may be attributable to
differences in proportions of embryonic and differentiated tissues, embryonic
tissue having a higher rate of cell division. Clearly pouch life requires a high
percentage of differentiated tissues for competence of the digestive, respiratory
and excretory systems, and implementing this need may result in a lower growth
rate. Because of the slow and prolonged growth, more of the marsupial mother's
total resource allocation may be lost in maintenance of the young; placentals, on
the other hand, bear the energy costs of developing and sustaining the placenta. In
some marsupials the growth period extends far into adult life, epiphyses remaining
unossified longer than in most placentals (Frith and Calaby, 1969). This permits
resource utilisation at a slower rate, just as may be possible in the pouch.

In summary, marsupials invest fewer resources in their young during gestation
than do placentals, and can simply terminate their reproductive attempts over the
remainder of the long suckling period should the chances of their offspring's
survival become low. The marsupial uterine cycle is not interrupted by gestation,
and even in kangaroos, where the placenta may have some endocrine function,
there is little control of maternal physiology by non-maternal alleles during
gestation. Apart from lactation (which varies facultatively with environmental

changes), the marsupial cycle of reproduction is also less costly in energy than
that of the placental. Placentals, by contrast, make extensive investments in their
young over a long gestation period, during which the trophoblast and placenta
alter the female's endocrine environment. Whereas the selective pressures which
favoured the maternal genotype continue to maximise her reproductive value
(which includes both present and future reproductive success), those shaping the
paternal genes work towards the maximum success of the current attempt at
reproduction, future reproductive success of the mother being irrelevant. (This
analysis does not, of course, entirely apply to species forming long-term pair
bonds.) The marsupial mother invests in her offspring at a slower rate than the
placental. At the time of marsupial - placental divergence, the environments in
which each group arose (Lillegraven, 1974) must surely have selected for the
relative benefits of their distinctinve modes of reproduction. Viewed in this context, the reproductive biology of marsupials cannot be deemed less efficient than
that of placentals and may indeed have considerable advantages in the range of
environments in which marsupials are found.

Brain structure and behaviour in marsupials and placentals

Another way in which marsupials are said to be primitive relative to placentals
concerns the comparative structure of the brains in the two·groups. Marsupials
lack the corpus callosum of placentals, which functions in the inter-hemispheric
transfer of learning derived from sensory information (Myers 1962a, b). The
cerebral hemispheres of marsupials are linked, but via different structures.
Didelphis virginiana and the other polyprotodonts use the anterior commissure
as do monotremes and reptiles (Ebner, 1969). The diprotodontan marsupials have
a unique neocortical commissure, the fasciculus aberrans, paralleling the placental
corpus callosum, in addition to the anterior commissure. Despite the diverse
pathways characteristic of the different subclasses, the fibres connecting the
neocortices appear to connect the same elements in the cortices in the same way
in terms of function (Johnson, 1976; Heath and Jones, 1971). In addition to the
range of marsupial brain structure paralleling that found among placentals, there
is a similar range of brain size to body weight ratios and of neocortical expansion,
although no marsupial expansions extend as far as do those of some primates and
cetaceans (Johnson, 1976; Moeller, 1973).

Marsupials are commonly regarded as being less intelligent than placentals. The
precise relationship between brain size and intelligence in mammals is not clear,
nor is the nature of selection for intelligence (Jerison, 1973). Williams (1966a)
argues that selection should favour instinctive reactions over learned behaviour in
any situation in which a particular behaviour is universally adaptive. Instinct is
biologically efficient because less information must be maintained, integrated
and selectively assimilated over a period of time. Thus instinct will not always be
a more primitive character preceding the evolution of learned behaviour, and
intelligence is not invariably the concomitant of an advanced evolutionary state.
Williams suggests that early accidental death of the young is the selective pressure
for intelligence. If the young are lost after the mother has made extensive energetic
and time investments in the young and before they have reached reproductive
age, the mother's entire reproductive investment is lost. Hence, in those cases

where the loss is avoidable by learned behaviour, selection should favour intelligence in the young and protective behaviour in the mother.

Selective pressures for the evolution of intelligence may have been quite different in marsupials and placentals because the marsupial mode of reproduction is uniquely suited to reducing the risk of early avoidable death. While the young are in the pouch, they are not subject to external sources of mortality independent of those faced by the mother. Many species retain the well-developed young in close physical association with the mother, exposing them to the food the mother is consuming. The young can also observe the environment and appropriate maternal reactions without undertaking independent activity. In contrast, many placentals with nidicolous young must leave them unattended in a nest or den for periods of time while the parent forages, conditions that do not favour rapid learning from the parent.

There are no marsupial species for which strong maternal protective behaviour towards the young has been reported, perhaps owing to the pouch habit. On the contrary, Serventy (1970) gives examples of stressed female kangaroos spilling pouch-young while continuing to run, leaving the potential predator with an easier target than that offered by the adult. Whether this behaviour is involuntary or not is unknown. It has the effect of instantaneously removing the mechanical burden of reproduction, a behaviour impossible for placentals during pregnancy. It conserves the environmentally tested genotype of the mother while sacrificing the untried genotype of the young, and may succeed in cutting losses from two to only one individual. Because marsupials are seldom required to show maternal protective behaviour and can rapidly replace the lost young, this sacrifice may have been favoured by selection. If marsupials are ever shown to be less intelligent as a group than placentals, then there is a plausible evolutionary explanation for it.

Acknowledgements

I thank R. Cook, K. S. Thomson, B. E. Horner, B. Patterson, B. H. Slaughter, E. S. Crelin, A. W. Crompton, and especially J. A. W. Kirsch for their comments, assistance and corrections. I thank L. Radinsky and G. W. Osbaldiston for suggesting references, S. Anderson for use of collections in the American Museum of Natural History, and B. Stonehouse for his Jobean patience. This chapter is derived from part of my Ph. D. dissertation at Yale University.

References

Alexandre, H. L. (1974). Effects of x-irradiation on preimplantation mouse embryos cultured *in vitro. J. Reprod. Fert.*, **36**, 417-20.

Beer, A. E. and Billingham, R. E. (1971). Immunobiology of mammalian reproduction. *Advan. Immunol.*, **14**, 1-84.

Behrman, S. J. (1971). Implantation as an immunological phenomenon. In: *The Biology of the Blastocyst* (ed. R. J. Blandau), University of Chicago Press, pp. 480-94.

Boyd, J. D. and Hamilton, W. J. (1952). Cleavage, early development, and implantation of the egg. In: *Marshall's Physiology of Reproduction*, Vol. 2

(ed. A. S. Parkes), pp. 1 - 126.

Brown, L. (1970). Population control among large mammals. In: *Population Control* (ed. A. Allison), Penguin, Baltimore, pp. 93 - 109.

Bruce, H. M. (1959). An exteroceptive block to pregnancy in the mouse. *Nature*, **184**, 105.

Bruce, H. M. (1970). Pheremones. *Brit. Med. Bull.*, **26**, 10 - 13.

Cody, M. L. (1971). Ecological aspects of reproduction. In: *Avian Biology*, Vol. 1 (ed. D. S. Farner and J. R. King), Academic Press, New York, pp. 461 - 512.

Crelin, E. S. (1960). The development of bony pelvic sexual dimorphism in mice. *Ann. N. Y. Acad. Sci.*, **84**, 479 - 512.

Ebner, F. F. (1969). A comparison of primitive forebrain organisation in metatherian and eutherian mammals. *Ann. N. Y. Acad. Sci.*, **167**, 241 - 57.

Enlow, D. H. and Brown, S. O. (1958). A comparative histological study of fossil and recent bone tissue. Pt. 3. *Texas J. Sci.*, **10**, 187 - 230.

Fisher, R. A. (1930). *The Genetical Theory of Natural Selection*, Clarendon Press, Oxford.

Flynn, T. T. (1923). The yolk-sac and allantoic placenta in *Perameles. Quart. J. Microscop. Sci.*, **67**, 123 - 82.

Frith, H. J. and Calaby, J. H. (1969). *Kangaroos*, Humanities Press, New York.

Heath, C. J. and Jones, E. G. (1971): Interhemispheric pathways in the absence of a corpus callosum. An experimental study of commissural connexions in the marsupial phalanger. *J. Anat.*, **109**, 252 - 70.

Hendrickx, A. G. (1971). *Embryology of the Baboon*, University of Chicago Press.

Hughes, R. L. (1974). Morphological studies in implantation in marsupials. *J. Reprod. Fert.*, **39**, 173 - 86.

Huxley, T. H. (1880). On the application of the laws of evolution to the arrangement of the Vertebrata, and more particularly of the Mammalia. *Proc. Zool. Soc. London*, **1880**, 649 - 62.

Jacobs, B. B. and Uphoff, D. E. (1974). Immunologic modification: A basic survival mechanism. *Science*, **185**, 582 - 7.

Jerison, H. (1973). *Evolution of the Brain and Intelligence*, Academic Press, New York.

Johnson, J. I. (1976). Central nervous system of marsupials. In: *The Biology of Marsupials* (ed. D. Hunsaker), Academic Press, New York (in press).

Kirsch, J. A. W. (1968). Prodromus of the comparative serology of Marsupialia. *Nature*, **217**, 418 - 20.

Kirsch, J. A. W. (1976). The classification of marsupials with special reference to karyotypes and serum proteins. In: *The Biology of Marsupials* (ed. D. Hunsaker), Academic Press, New York (in press).

Lillegraven, J. A. (1969). Latest Cretaceous mammals of upper part of Edmonton Formation of Alberta, Canada, and review of marsupial-placental dichotomy in mammalian evolution. *Univ. Kansas Paleon. Contrib.* Art. 50 (Vertebrata 12), 1 - 122.

Lillegraven, J. A. (1974). Biogeographical considerations of the marsupial-placental dichotomy. *Ann. Rev. Ecol. Syst.*, **5**, 263 - 83.

Merchant, J. C. and Sharman, G. B. (1966). Observations on the attachment of marsupial pouch young to the teats and on the rearing of pouch young by foster-mothers of the same or different species. *Aust. J. Zool.*, **14**, 593 - 609.

Moeller, H. (1973). Zur Evolutionshöhe des Marsupialiagehirns. *Zool. Jhb. Anat.*, **91**, 434 – 48.

Moors, P. J. (1974). The foeto-maternal relationship and its significance in marsupial reproduction: A unifying hypothesis. *J. Aust. Mamm. Soc.*, **1**, 263 – 6.

Myers, R. E. (1962a). Transmission of visual information within and between the hemispheres: A behavioral study. In: *Interhemispheric Relations and Cerebral Dominance* (ed. V. B. Mountcastle), Johns Hopkins University Press, Baltimore, pp. 51 – 73.

Myers, R. E. (1962b). Commissural connections between occipital lobes of the monkey. *J. Comp. Neurol.*, **118**, 1 – 16.

Newsome, A. E. (1964). Anoestrus in the red kangaroo *Megaleia rufa* (Desmarest). *Aust. J. Zool.*, **12**, 9 – 17.

Newsome, A. E. (1965). Reproduction in natural populations of the red kangaroo, *Megaleia rufa* (Desmarest), in Central Australia. *Aust. J. Zool.*, **13**, 735 – 59.

Palm, J., Heyner, S. and Brinster, R. L. (1971). Differential immunofluorescence of fertilised mouse eggs with H-2 and non-H-2 antibody. *J. Exp. Med.*, **133**, 1282 – 93.

Racey, P. A. (1973). Factors affecting the length of gestation in heterothermic bats. *J. Reprod. Fert. Suppl.*, **19**, 175 – 89.

Renfree, M. B. (1970). Protein, amino acids, and glucose in the yolk-sac fluids and maternal blood sera of the tammar wallaby, *Macropus eugenii* (Desmarest). *J. Reprod. Fert.*, **22**, 483 – 92.

Ricklefs, R. E. (1973). Patterns of growth in birds. II. Growth rate and mode of development. *Ibis*, **115**, 177 – 201.

Ride, W. D. L. (1964). A review of Australian fossil marsupials. *J. Proc. Roy. Soc. W. Australia*, **47**, 97 – 131.

Sampson, J. C. (1971). The biology of *Bettongia penicillata* (Gray, 1837). Ph. D. Thesis, University of Western Australia.

Serventy, V. (1970). *Dryandra. The Story of an Australian Forest*, Reed, Sydney.

Sharman, G. B. (1961). The embryonic membranes and placentation of five genera of diprotodont marsupials. *Proc. Zool. Soc. London*, **137**, 197 – 220.

Sharman, G. B. (1970). Reproductive physiology of marsupials. *Science*, **167**, 1221 – 8.

Sharman, G. B. (1973). Adaptations of marsupial pouch young for extrauterine existence. In: *The Mammalian Fetus in Vitro* (ed. C. R. Austin), Chapman and Hall, London, pp. 67 – 89.

Slaughter, B. H. (1968). Earliest known marsupials. *Science*, **162**, 254 – 5.

Slaughter, B. H. (1971). Mid-Cretaceous (Albian) therians of the Butler Farm local fauna, Texas. *J. Linnean Soc. London Zool.*, **50**, Suppl. 1 131 – 143.

Smith, C. C. and Fretwell, S. D. (1974). The optimal balance between size and number of offspring. *Am. Nat.*, **108**, 499 – 506.

Tarkowski, A. K. and Wroblewska, J. (1967). Development of blastomeres of mouse eggs isolated at the 4- and 8-cell stage. *J. Embryol. Exp. Morphol.*, **18**, 155 – 63.

Tyndale-Biscoe, C. H. (1973). *Life of Marsupials*, American Elsevier, New York. pp. 1 – 254.

Tyndale-Biscoe, C. H., Hearn, J. P. and Renfree, M. B. (1974). Control of reproduction in macropodid marsupials. *J. Endocrinol.*, **63**, 589 - 614.

Vaughan, T. A. (1972). *Mammalogy*, Saunders, Philadelphia.

Walker, E. P. (1964). *Mammals of the World*, Johns Hopkins University Press, Baltimore.

Weir, B. J. and Rowlands, I. W. (1973). Reproductive strategies of mammals. *Ann. Rev. Ecol. Syst.*, **4**, 139 - 63.

Williams, G. C. (1966a). *Adaptation and Natural Selection*, Princeton University Press.

Williams, G. C. (1966b). Natural selection, the costs of reproduction, and a refinement of Lack's principle. *Am. Nat.*, **100**, 687 - 90.

Williams, G. C. (1975). *Sex and Evolution*, Princeton University Press.

17 The activity of the jaw and hyoid musculature in the Virginian opossum, Didelphis virginiana

A. W. Crompton*, A. J. Thexton†, Pamela Parker‡ and Karen Hiiemae∮

Introduction

A broad study of the patterns of mastication in mammals is currently being undertaken by the authors of this paper, using synchronised electromyography and cinefluorography. Several recent studies of mastication in a variety of mammals (Kallen and Gans, 1972; de Vree and Gans, 1973; Luschei and Goodwin, 1974; Weijs and Dantuma, 1975) have used roughly similar techniques. However, these studies, like others (Herring and Scapino, 1973; and see Hiiemae, 1976b) have concentrated on the relationship between activity in the adductors and the movements of the lower jaw; the Digastric has been considered to be the main depressor of the mandible. No attention has been given to the hyoid musculature as a whole or to the movements of the hyoid bone. This paper will show that the pattern of jaw and hyoid movements during chewing, and the characteristic features of the adductor and hyoid EMG activity, are so interrelated that neither can be explained without consideration of the other

The Virginian (American) opossum, *Didelphis virginiana* Kerr 1792¶ retains many of the cranio-facial features of primitive mammals, including a tribosphenic molar and is, therefore, a good structural analogue for the jaw apparatus of early mammals. The adductor and some of the hyoid musculature of this animal have been described by Coues (1872) and Hiiemae and Jenkins (1969), but in view of their importance in the production of both jaw and hyoid movement, a brief description of the hyoid musculature in the opossum, based on new dissections, has been included to clarify the experimental studies. As the physical nature of the food affects the patterns of EMG activity during mastication (Thexton, 1974; Thexton and Hiiemae, 1975; Hiiemae and Thexton, 1975), this paper is confined

---

*Dr A. W. Crompton is Director of the Museum of Comparative Zoology, Harvard University.

†Dr A. J. Thexton is at the Royal Dental Hospital School of Dental Surgery, London.

‡Dr Pamela Parker is a Post-doctoral Fellow at the Museum of Comparative Zoology, Harvard University.

∮Dr Karen Hiiemae is in the Unit of Anatomy with special relation to Dentistry, at Guy's Hospital Medical School, London.

¶In a recent review of the taxonomy of the genus *Didelphis*, Gardner (1973) states that the Northern American opossum referred to as *Didelphis marsupialis virginiana* in previous papers (Crompton and Hiiemae, 1970; et seq.) should, in fact, be raised to specific status as *Didelphis virginiana*.

to a consideration of those patterns of activity which arise when feeding on soft food. The nature of the patterns of activity associated with mastication of other foods will be reported in papers in preparation.

The hyoid musculature

The attachments of the hyoid muscles are shown in Figure 17.1. The outline of the head, neck, scapula, clavicle and sternum are based on a cinefluorographic record of a feeding opossum. An enlarged drawing of the hyoid bone (stippled) and the thyroid cartilage are included to show the arrangement of the many muscle attachments on the former. In the opossum the hyoid bone may be unossified but is, as is the case in all mammals, connected to the thyroid cartilage by means of a membrane and joints. Some of the hyoid muscles, in particular the Digastric and the Mylohyoid, are indirectly attached to the bone by means of a common tendon. This is attached to the antero-ventral margin of the bone and extends in front of it as well as laterally towards the angles of the lower jaw. The position of the common tendon and the arrangement of the muscle attachments on it are shown in a diagrammatic cross-section through the tendon.

The suprahyoid muscles

The anterior and posterior bellies of the *Digastric* are the most superficial of the suprahyoid muscles. The anterior belly is a flat sheet of muscle attaching pos-teriorly along the length of the transversely orientated common tendon, its fibres passing anteriorly and in parallel to attach to the infero-medial border of the lower jaw along a line extending from below the last molar to below the first molar. In contrast, the posterior belly of this muscle is almost cylindrical, its fibres passing antero-ventrally from their attachment on the styloid process and just lateral to the origin of the Stylohyoid to insert into the lateral part of the common tendon (Figure 17.1; and see Hiiemae and Jenkins, 1969, Figures 13 and 14).

The *Mylohyoid* has a linear attachment along the internal surface of the body of the lower jaw just above that of the anterior belly of the Digastric and extend-ing posteriorly from below the level of the first molar to approximately the mid-point of the attachment of the Superficial Masseter on the angular process (Figure 17.1). The bulk of its fibres run transversely and form a thin sheet connecting the two halves of the lower jaw through a mid-line tendon but the most posterior fibres run antero-medially to attach to the central part of the common tendon. Part of the Mylohyoid is visible between the anterior bellies of the Digastric but the lateral portions of the muscle are deep to those bellies (Figures 13 and 14, Hiiemae and Jenkins, 1969).

The *Geniohyoid* muscles have a tendinous attachment onto the internal surface of the lower jaw just postero-lateral to the internal border of the mandibu-lar symphysis (Figure 17.1). The muscle expands from this attachment to form a large belly at the level of the second premolars; these paired bellies pass posteriorly to attach to the ventro-lateral surface of the hyoid. The Geniohyoids are the largest of the suprahyoid muscles. Although their fibres run parallel to those of

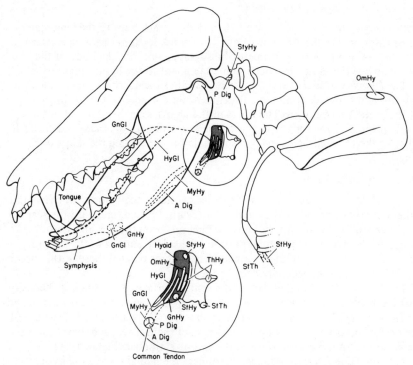

Figure 17.1 Lateral view of the head, cervical spine, scapula, sternum
and hyoid as seen in cinefluorographs, to show the attachments of
the hyoid musculature. The hyoid (see encircled enlargement at
bottom of figure) is a 'U'-shaped flat band of cartilage (stippled)
with its open end facing upwards and backwards and its somewhat
thicker base facing downwards and slightly forwards. The anterior
margin of the band at the base of the 'U' extends forwards to form
a lip (shown white) which provides attachment for the common
tendon of the Digastric and Mylohyoid. The tendon extends transversely across the base of the hyoid in front and continues, in the
same plane, laterally towards the angles of the lower jaw. The
tendon (dotted line) is shown in sagittal section. A diagrammatic
cross-section is also shown to illustrate the pattern of insertion of
the fibres of the Digastric and Mylohyoid into the tendon. The
thyroid cartilage (outlined) is linked to the hyoid through a membrane and joints and can be pivoted on the hyoid by contraction of
the Thyrohyoid muscle. Abbreviations: A. Dig., anterior belly of
Digastric; P. Dig., posterior belly of Digastric; GnGl, Genioglossus;
GnHy, Geniohyoid; HyGl, Hyoglossus; MyHy, Mylohyoid; OmHy,
Omohyoid; StHy, Sternohyoid; StyHy, Stylohyoid; StTh, Sternothyroid; ThHy, Thyrohyoid

the anterior belly of the Digastric, they are separated by the Mylohyoid, which is
superficial to the Geniohyoid but deep to the Digastric.

The *Stylohyoid* is a small cylindrical muscle which attaches on the tip of the
styloid process, just medial to the posterior belly of the Digastric and passes
antero-ventrally to insert on the dorsal tip of the hyoid (Figure 17.1).

The extrinsic muscles of the tongue

The *Genioglossus* is a large muscle with a fleshy attachment to the lower jaw immediately above the attachment of Geniohyoid. Each muscle runs postero-dorsally in the mid-line, its deepest fibres fanning out into the body of the tongue, while the most superficial gain attachment to the hyoid (Figure 17.1).

The *Hyoglossus* is also a large muscle. Its fibres form a thick sheet which runs upwards and forwards from their attachment on the sides of the hyoid passing lateral to the Genioglossus. The fibres of the Hyoglossus then fan out in the lateral portion of the tongue and, with the longitudinal, transverse and vertical fibres of the Lingualis, are largely responsible for controlling the shape of the tongue.

The infrahyoid muscles

The infrahyoid or 'strap' muscles of the neck connect the hyoid with the sternum either directly (Sternohyoid) or indirectly (Sternothyroid) and with the scapula (Omohyoid). The paired *Sternohyoids* which run from the postero-ventral margin of the hyoid to the lateral portion of the deep surface of the second sterneber are almost cylindrical muscles. Their fibres parallel those in the Sterno-thyroids which lie immediately deep to them.

The *Sternothyroid* muscles lie on the ventral surface of the trachea. The muscles of right and left sides, although separate throughout most of their length, are fused at their attachments: these are, above, to the ventro-medial surface of the thyroid cartilage and the uppermost part of the trachea, and, below, to the medial aspect of the deep surface of the second sterneber (Figure 17.1).

The small flat *Thyrohyoid* muscles connect the lateral part of the hyoid to the dorso-lateral surface of the thyroid cartilage.

The *Omohyoid* is a broad, thin band of muscle running from the lateral surface of the hyoid, where it has a linear attachment, to a tendinous insertion on the antero-lateral aspect of the cranial angle of the scapula just lateral to the attachment of the Levator anguli scapulae. During its course, the muscle runs superficial to the Sternothyroid and Thyrohyoid near the hyoid and then passes deep to the Atlantoacromialis and Sterno-cleido-mastoid before reaching the scapula.

Experimental studies

Studies of mastication in the opossum currently in progress at the Museum of Comparative Zoology use a technique in which the movements of the cranium, lower jaw, hyoid and tongue, as seen cinefluorographically, are recorded on film; the EMG activity of the jaw and hyoid muscles, together with a frame-by-frame synchronisation pulse from the cine camera are simultaneously recorded on magnetic tape. The movements of each structure can be analysed by measuring their position in sequential single frames of the cinefluorographic recordings (taken at 60 frames/sec. on 16 mm film). These positions can then be plotted against the EMG records using the synchronising signals (Thexton and Hiiemae, 1974; 1975a, b). This technique will be fully described in a forthcoming paper (Thexton and Hiiemae, in preparation). In the present study all animals were fed the same brand of canned minced dog food for the reasons cited in the introduction.

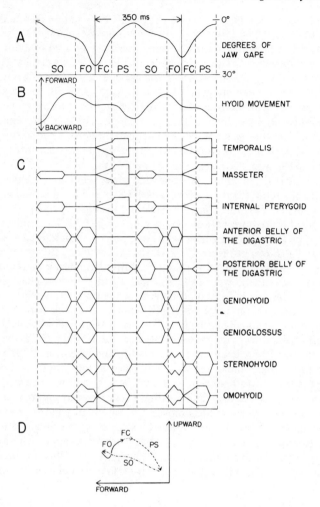

Figure 17.2 Jaw and hyoid movement (A, B) in the opossum during part of a typical sequence of chewing on soft food. In A the relative movement of upper and lower jaws has been plotted as degrees of gape (vertical axis) against time (horizontal axis). Each chewing cycle has four stages: FC – Fast Closing: PS – Power Stroke, in this case consisting almost entirely of a slow closing movement; SO – Slow Opening; and FO – Fast Opening. In B the movements of the hyoid bone have been plotted on a plane parallel to the secondary palate: the range of antero-posterior movements is of the order of 2 mm. The associated EMG activity in adductor and representative hyoid muscles is shown in C as the envelope of the electrical signal. C is a composite diagram based on data from a large number of experiments. The path of movement of a point on the hyoid during chewing is shown in D. The diagram was obtained by plotting the position of the point in the 'vertical' and 'horizontal' planes (with reference to the hard palate, see text) at equal intervals of time during the cycle shown in A

Work on mastication in general (see Hiiemae, 1977) and on this behaviour in the opossum in particular (Hiiemae and Crompton, 1971; Hiiemae, 1976 has shown that the movements of the chewing cycle are produced by both cranial and lower jaw movement. The cycle can be divided into four distinct stages based on both the direction and rate at which the upper and lower tooth rows converge or diverge. This is the best shown if 'degrees of gape' — that is, the angular separation of the cheek teeth as seen in lateral view — are plotted (for each frame of film) on the x-axis and time on the y-axis (Figure 17.2A). The actual duration of a single cycle in the opossum seems to depend on both the food supplied and the extent to which it has been reduced, but the usual range for soft food is 300 – 400 ms (Hiiemae, 1976). The four stages of the cycle in the opossum are: (1) *Fast Closing* (FC), in which the upper and lower teeth are brought into contact with the food; (2) the *Power Stroke* (PS), in which the food is punctured, crushed or sheared; (3) *Slow Opening* (SO); and (4) *Fast Opening*. Slow Opening and Fast Opening are both part of the opening stroke*, which terminates at maximum gape (the point at which cycles are, by convention, taken as beginning and ending). As shown in Figure 17.2(A), the rate at which the upper and lower teeth converge in the Power Stroke is much slower than in the preceding closing stroke. In all cycles the Power Stroke includes a period of slow closing which is normally, although not invariably, followed by a period during which no vertical movement of the jaws can be observed (Hiiemae, 1976a); this period usually lasts 40 – 50 ms, but occasionally extends to 100 ms. If the food has been completely sheared during the slow closing movement, then the teeth on the active side achieve a unilateral centric occlusion (Crompton and Hiiemae, 1970) with the protocone of the upper molar in the talonid basin of the corresponding lower.

During the first 30 – 50 per cent of Slow Opening the jaws may separate more rapidly, so separating the teeth and creating a clear freeway space, than during the rest of this stage, when vertical movement may be suspended or even transitorily reversed into slight closure. During Slow Opening the tongue is protruded so that its tip often extends beyond the lower incisors; a rapid withdrawal normally occurred during Fast Opening (Hiiemae, 1977). This tongue movement serves to move the food within the oral cavity (Hiiemae and Crompton, 1971; Hiiemae, 1977).

Movements of the hyoid

The movements of the hyoid relative to the plane of the hard palate during a typical chewing cycle on soft food are shown in Figures 17.2(B) and 17.2 (D). In lateral projection cinefluorographs the hard palate shows as a clearly defined line which can be used as a reference plane so that movements of the hyoid towards the palate or in parallel with it are readily measured. This avoids the difficulties inherent in determining hyoid position in relation to other mobile structures. In the following description movement of the hyoid in a 'vertical' plane — that is,

*In previous papers the opening stroke has been referred to as the 'recovery stroke' and Fast Closing (the closing stroke) as the 'preparatory stroke' (Hiiemae and Ardran, 1968; Crompton and Hiiemae, 1970). This terminological revision is fully discussed in Hiiemae (1977).

towards or away from the hard palate – is termed 'upward' or 'downward', respectively. Anterior movement of the hyoid parallel to the plane of the hard palate – that is, in a 'horizontal' plane – is termed 'forward' and posterior movement 'backward'.

The antero-posterior movements of the hyoid and the simultaneous movements of the jaws occurring during a typical sequence of chewing on soft food are shown in Figure 17.2(B) and (A). Figure 17.2D shows the path of movement of a point on the hyoid during a typical chewing cycle; its position in both the 'vertical' and 'horizontal' planes has been plotted at equal intervals of time. Comparison of A and B of Figure 17.2 demonstrates that the movements of the hyoid are intimately correlated with those of the jaws. The hyoid is at its furthest backward and downward position at the beginning of the Slow Opening stage of the cycle. As the jaws slowly open, the hyoid is always drawn rapidly forwards and upwards to reach its most anterior position (see also Figure 17.6C). This may be reached before the end of Slow Opening, in which case the hyoid remains virtually stationary until the commencement of Fast Opening. The rate of upward movement of the hyoid is usually greater earlier in Slow Opening and this appears to be correlated with the more rapid rate of jaw opening occurring at the same time (see second SO stage in Figure 17.2A and B). During Fast Opening, the hyoid may be drawn upwards, downwards or remain stationary in the 'vertical' plane but is almost always drawn backwards (Figure 17.2D; also Figure 17.6D). As the jaws converge in Fast Closing, the hyoid moves very slowly slightly backwards or remains stationary in the 'horizontal' plane (Figure 17.2D; also Figure 17.6D). Although movements in the 'vertical' plane are variable during Fast Closing there is an overall tendency to lift the hyoid slightly during this period. During the Power Stroke, the hyoid always moves downwards and backwards to reach its most posterior and depressed position (Figures 17.2A, and B, D; also Figure 17.6B).

In a pilot experiment that part of the common tendon which connects the anterior and posterior bellies of the Digastric was severed bilaterally to free the hyoid attachment of both bellies. There was no observable effect on the movement profile of either the jaws or the hyoid during the opening stroke. This suggests that the Digastric may not be the main depressor of the jaw, at least in the opossum, but that the suprahyoids coupled with the infrahyoids perform this function.

In summary, during Fast Closing the hyoid usually moves slightly backwards; during the Power Stroke it moves downwards and backwards; in Slow Opening it moves upwards and forwards; and in Fast Opening the hyoid tends to move only slightly backwards.

Electromyographic (EMG) activity in the adductor and hyoid musculature

EMG activity during feeding on soft food was recorded using indwelling bipolar wire electrodes (Basmaijan and Stecko, 1962) from the following muscles: various parts of the 'Temporalis' muscle mass, the Masseter, the Internal Pterygoid, the anterior and posterior bellies of Digastric and the Geniohyoid, Genioglossus, Omohyoid and Sternohyoid. These muscles were chosen for two reasons: firstly, they represent the principal muscles controlling the movements of both the lower jaw and hyoid; secondly, electrodes can be inserted into these muscles under direct vision and in such a way that the signals recorded could be guaranteed to be from that muscle. No recordings were made from (a) the External Pterygoid, which is both very small and inaccessible in the opossum, making reliable recording extremely difficult; (b) the Sternothyroid, since it can be assumed to have the same action as the Sternohyoid which overlies it; (c) the Thyrohyoid, whose functions are associated with the regulation of the angulation of the thyroid cartilage on the hyoid; (d) the Stylohyoid, since this originates close to and runs in parallel with the posterior belly of the digastric, and probably has a closely similar action; (e) the Mylohyoid, since its function is in elevating the floor of the mouth and in helping to control the horizontal distance between the bodies of the independently mobile halves of the lower jaw; and (f) the Hyoglossus, since this appears to retract the lateral part of the base of the tongue and, with the intrinsic tongue musculature, control and change the shape of the tongue rather than effect hyoid position.

It has been clearly established that the opossum only chews on one side at any one time (Crompton and Hiiemae, 1970). During the chewing of soft food, the periods in which the muscles of right and left sides showed EMG activity were similar (Figures 17.4C, 17.5A and B), although the levels of that activity could vary as between the muscles of the working and balancing sides (Figure 17.4C). This observation is in sharp contrast to that of Kallen and Gans (1972), who reported alternation of activity between right and left sides in some jaw muscles, notably the Pterygoids, in the Little brown bat (*Myotis lucifugus*).

Figures 17.3, 17.4 and 17.5 are EMG records of individual muscles during chewing; the stages of the chewing cycles are indicated. Figure 17.2C is a composite diagram to show the overall pattern of EMG activity in all the muscles examined. It must be emphasised that the pattern of EMG activity described here applies only to middle sequence chewing cycles when the animals were feeding on a single type of soft food. Both the EMG and movement patterns are known to change as a single lump of food is progressively chewed (Thexton and Hiiemae, 1975a, b) and with changes in the type of food (Thexton and Hiiemae, in preparation).

As the jaws reach maximum gape at the end of Fast Opening and then begin to close, EMG activity is seen in the Temporalis, Masseter and Internal Pterygoid. As the jaws reach tooth – food – tooth contact, the amplitude of this activity increases reaching its maximum levels during the slow close part of the Power Stroke (Figures 17.3A and C, 17.4B and C). The transition from Fast Closing to the Power Stroke is shown by a deceleration of the jaws as they close forcibly on the food between the teeth. All adductor EMG activity ceases abruptly before completion of the Power Stroke (Figure 17.4C), but the teeth remain in occlusion or in contact with the food for a further 40 – 50 ms. This can be explained by the time taken for the mechanical activity of the adductor musculature to be expressed.

Figure 17.3 Three sets of EMG records obtained from the jaw and hyoid muscles of the same side in an opossum feeding on soft food. The duration of the stages of each chewing cycle (see Figure 17.2) have been measured from the synchronous cinefluorographic recordings and are shown below each record. FC, Fast Closing; PS, Power Stroke; SO, Slow Opening; FO, Fast Opening. The total time elapsed in each record is 1 s

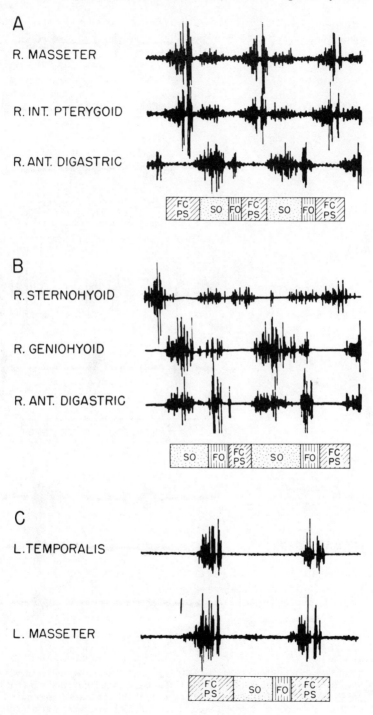

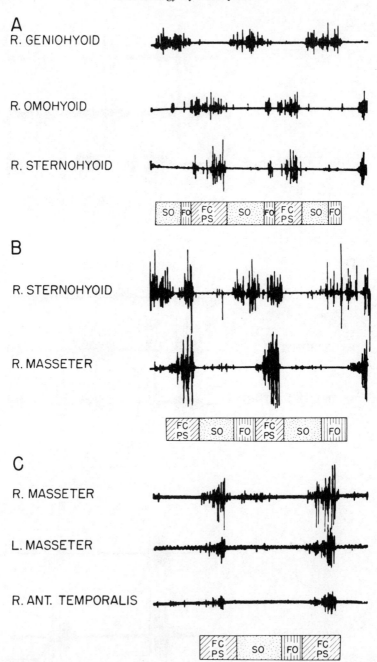

Figure 17.4 Three sets of EMG records from adductor and hyoid muscles during chewing. Record C shows the relative levels of activity found in the Masseters and the Anterior part of Temporalis when the animal is chewing on the left (first cycle) and then on the right (second cycle). The total time elapsed in each record is 1 s. The conventions and abbreviations are as in Figure 17.3

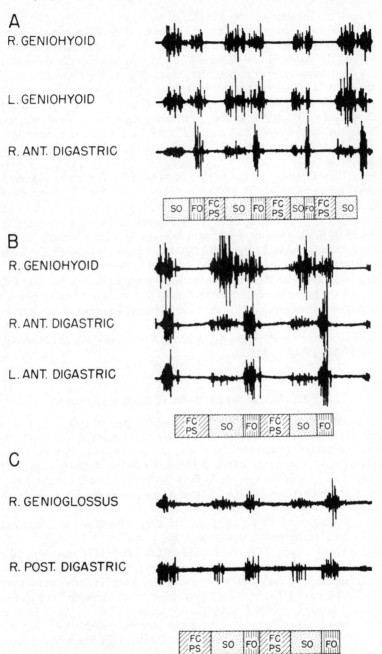

Figure 17.5 Three sets of EMG records obtained from the suprahyoid muscles of the opossum when chewing soft food. The electrical signals recorded in paired muscles are seen to be bilaterally symmetrical. The total time elapsed for each record is 1 s. The conventions and abbreviations are as in Figure 17.3

In the opossum the time to peak and half decay times of isometric twitch contractions in these muscles are both of the order of 20 ms (Thexton and Hiiemae, 1975a, b). There is very little activity in the suprahyoid musculature during Fast Closing and the Power Stroke (Figures 17.3A, 17.5A and B); only the posterior belly of the Digastric shows any activity during closing of the jaws (Figure 17.5C). The Omohyoid is active throughout Fast Closing but shows increasing activity during the slow close part of the Power Stroke (Figure 17.4A). The Sternohyoid, usually silent for part of Fast Closing, reaches its highest levels of activity during the Power Stroke (Figures 17.4A and B, 17.3B). Activity in the infrahyoid muscles ceases at approximately the same time as that of the elevators (Figure 17.4B).

During Slow Opening there is a major burst of activity in all the suprahyoid muscles (Figures 17.3A and B, 17.5). Characteristically, the Digastric, Geniohyoid and Genioglossus have two distinct bursts of activity during the chewing of soft food: the first during Slow Opening and the second during Fast Opening (Figures 17.3A and B, 17.5). The first burst in the Digastric is often of slightly lower amplitude than the second (Figure 17.3B). The infrahyoid musculature is silent during slow opening (Figures 17.3B, 17.4A), but there is a regular burst of low-level activity in the Masseter and Internal Pterygoid during this stage (Figure 17.3A). All the hyoid muscles examined show activity during Fast Opening but this is greater in the suprahyoid than in the infrahyoid muscles. The suprahyoid muscles cease firing as the jaws reach maximum gape but the Sternohyoid and Omohyoid show activity during Fast Opening, Fast Closing and the Power Stroke (Figures 17.3B, 17.4A and B).

A model for the mechanics of the hyoid apparatus

As the hyoid, and the attached larynx, are mobile, their position and orientation depend on the level of contraction of the muscles acting on the bone, namely the anterior and posterior suprahyoid muscles and the infrahyoid muscles. The position of the hyoid in turn affects the level of the floor of the mouth and the position of the base of the tongue. As the anterior suprahyoid muscles are connected to both the mobile lower jaw and hyoid, their positions are closely linked.

The following model of jaw and hyoid function is based on the results of the anatomical and experimental studies described above.

During Fast Closing, the jaw is elevated by the Temporalis, Masseter and Internal Pterygoid muscles. All three continue to elevate the jaw against the resistance of the food during the slow closing movement of the Power Stroke (Figure 17.6A and B). The backward movement of the hyoid during Fast Closing is produced primarily by the Omohyoid. Contraction of the posterior belly of the Digastric (and presumably the Stylohyoid) produces the slight upward movement sometimes seen. The synchronous low-level activity in the Sternohyoid produces a resultant backward pull on the hyoid by cancelling or reducing the effect of the upward pull of the posterior belly of the Digastric. When the Sternohyoid becomes more active during the Power Stroke, the pull of the posterior belly of the Digastric is further counteracted by the much larger Sternohyoid so that the hyoid

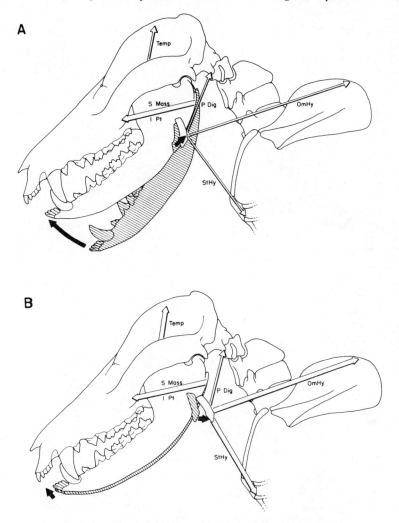

Figure 17.6 Reconstructions from a lateral projection cinefluorographic recording of a typical chewing cycle in the opossum, when feeding on soft food, to show the movements of the lower jaw and hyoid during each stage of the cycle. Tracings of the hyoid and lower jaw at the end of each stage (clear outline) have been superimposed on tracings of their initial position (cross-hatched outlines). The solid black arrows show the direction and amplitude of each movement. The lines of action of the muscles active during each stage are indicated by open arrows. The width of the arrow is used as a guide to the associated level of EMG activity, i.e. high levels of activity are represented by wide arrows, lower levels of activity by narrow arrows. The Superficial Masseter (S Mass) and Internal Pterygoid (I Pt) have closely comparable lines of action, as do the Deep Masseter and bulk of Temporalis (Temp). A, Fast Closing; B, Power Stroke (the teeth did not come into full occlusion in this case); C, Slow Opening; D, Fast Opening. Other abbreviations as in Figure 17.1. Parts C and D are on page 300

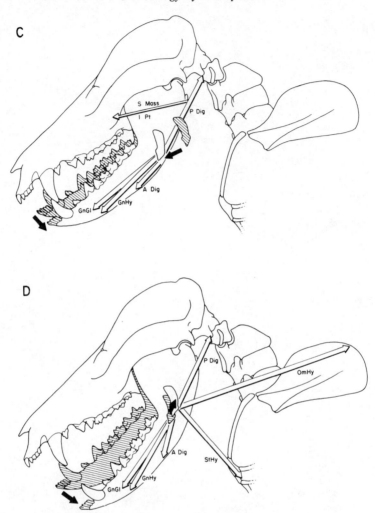

Figure 17.6 (*continued*)

bone is moved markedly downwards as well as backwards (Figure 17.6B). The combination of an extensive upward movement of the lower jaw and a backward and downward movement of the hyoid stretches the anterior suprahyoid muscles. The relative movements of the hyoid and lower jaw during the Power Stroke also alter the position of the tongue, pulling its base downwards and backwards.

During Slow Opening all the movements of the lower jaw, hyoid and tongue are reversed (Figure 17.6C). The jaw opens slightly and the hyoid moves sharply upwards and forwards carrying the base of the tongue with it. This last movement facilitates the protrusive movement of the tongue which occurs during this stage of the cycle and which is associated with the food collection and repositioning functions of that organ (Hiiemae, 1976, 1977). The anterior suprahyoids, which have high and sustained levels of activity during Slow Opening, pull the hyoid forwards. The activity in the posterior belly of Digastric during the earlier part of Slow Opening may transfer some of the pull from the anterior belly of Digastric and Geniohyoid to the cranium and so depress the jaw. When the posterior belly of Digastric ceases to show electrical activity, as happens towards the end of the Slow Opening stage, depression of the lower jaw may cease and the jaws may even close slightly. This hesitation in the jaw opening coincides with a brief pause in the activity in the anterior suprahyoids (Figure 17.5). The weak bursts of EMG

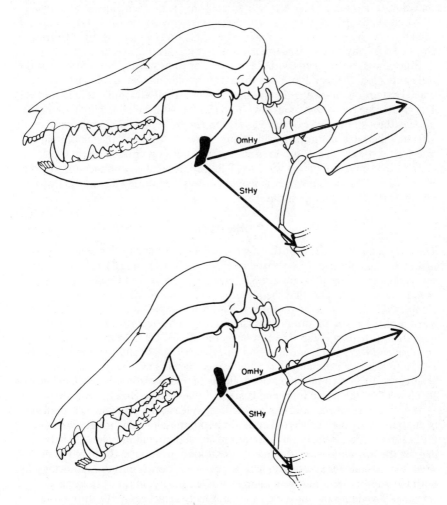

Figure 17.7 To show the effect of cranial extension (A) and flexion (B) on the length of the Omohyoid and Sternohyoid when the hyoid bone and lower jaw are in their normal position at the beginning of the Power Stroke. Both figures are based on tracings from cinefluorographic recordings of normal chewing in the opossum

activity in the Masseter and Internal Pterygoid in Slow Opening restrict mandibular depression.

In Fast Opening (Figure 17.6D) all the hyoid musculature is contracting; the retraction and elevation of the hyoid by the posterior belly of Digastric, Omohyoid and Sternohyoid ensures that the contraction of the anterior suprahyoids moves their mandibular rather than their hyoid attachment, so producing rapid depression of the lower jaw. The orientation of the Geniohyoid, and its larger bulk, suggest that this muscle makes a greater contribution to the production of mandibular depression than either the hyoid slip of the Geniglossus, although that muscle is active at the same time as the Geniohyoid, or the anterior belly of the Digastric. The absence of a discernible effect on jaw opening after severing the anterior belly of the Digastric from its hyoid attachment supports this view.

Since the opossum feeds with its head in positions ranging from fully flexed to fully extended, the length of both the Sternohyoid and Omohyoid also varies, cranial extension increasing the distance between their attachments (Figure 17.7). The greater length of these muscles, as compared with the remaining hyoid musculature, allows this extensive change in length to be accommodated without affecting their action on the hyoid during chewing. Although the Omohyoid is inserted on the scapula, it has been shown that this bone rotates about a point close to that insertion during walking (Jenkins, personal communication, 1975). Were this not the case, the muscle would be stretched during locomotion with a possible effect on hyoid position.

Discussion

The integrated movements of the jaw and hyoid described in this paper are not unique to the opossum. Comparison of Figures 17.2(A) and (B) and 17.8, which shows the equivalent plots for a single chewing cycle in the insectivore *Tenrec ecaudatus*, demonstrates that the patterns are almost identical. Although no other comparable study of hyoid movement has been completed to date, the jaw movement profile has been plotted from cinefluorographic records of chewing in a number of other mammals (the Rat kangaroo: Cook, 1975; several primates: Hiiemae and Kay, 1973, and Kay and Hiiemae, 1974a, b; and the cat: Hiiemae, 1977). In all cases the animal has a four-stage chewing cycle with the same general profile as seen in the opossum and with a long distinct Slow Opening stage. It follows that the hyoid movements are also likely to be closely comparable.

It is clear that in the opossum and tenrec, at least, the generally accepted view of the actions of the suprahyoid and infrahyoid muscles is incorrect. Carlsöö (1956) states that, in man, the function of the suprahyoid muscles '. . . is to depress the mandible when the hyoid is stationary or to raise the hyoid bone when the mandible is stationary, which movement occurs during swallowing'. In both the admittedly primitive mammals studied the hyoid is not stationary in chewing: jaw and hyoid movements are normally achieved by the differential contractions of the adductor and hyoid muscles. This is also likely to be true in man (Crompton *et al.*, 1975). It appears that although slight jaw opening can occur without any contractile activity in the infrahyoid muscles while the hyoid itself moves forward, extensive mandibular depression is always produced by the infrahyoid musculature and the posterior belly of Digastric. The results of this

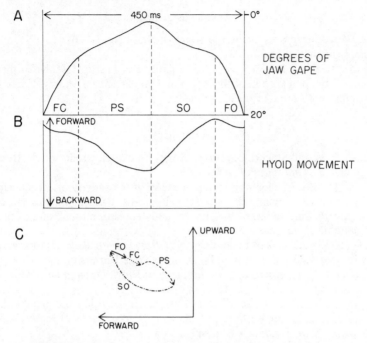

Figure 17.8 Jaw and hyoid movement in the insectivore *Tenrec ecaudatus*, during a single cycle when chewing on soft food. A, The four stages of the jaw movement cycle (compare Figure 17.2A); B, The antero-posterior movement of the hyoid; the normal range of antero-posterior movement of the bone is of the order of 10–15 mm (compare Figure 17.2B); C, The movement path of a point on the hyoid during the same chewing cycle, obtained as for Figure 17.2D

study also indicate that the Digastric in the opossum is not a specialised jaw opening muscle.

Finally, studies such as this are not only essential for the proper understanding of the mechanisms of the jaw apparatus, but also throw some light on the control of mastication in this animal. On the basis of neurophysiological studies, it seems likely that the various modes of feeding found in mammals may be modifications of a basic pattern determined by a 'oscillator' or 'pattern generator' in the central nervous system, the output of which is modified by sensory input from the oral region (Thexton, 1973, 1974, 1976). During the chewing of soft food, the timing and amplitude of the bursts of electrical activity in the muscles shows far less variation than is the case with hard food. This suggests that, in the present study, the motor outflow from the central nervous system owes more to a 'pattern generator' and less to a peripheral (oral) sensory feedback than is the case with hard food. This question is being further examined by the same techniques as in the present paper to examine feeding behaviour on foods of various consistencies (Thexton and Hiiemae, in preparation).

Acknowledgements

This work was made possible by the award of Grant No. DE-02419-08 from the National Institutes of Health.

We wish to thank Brigita Stegers, Al Coleman and Laszlo Meszoly for their assistance in the preparation of this paper and, in particular, Andrew Chester for his assistance with some of the experiments.

References

Basmaijan, J. V. and Stecko, G. (1962). A new bipolar electrode for electromyography. *J. Appl. Physiol.*, **17**, 849.

Carlsöö, S. (1956). An electromyographic study of the activity of certain supra-hyoid muscles (mainly the anterior belly of the digastric muscle) and of the reciprocal innervation of the elevator and depressor musculature of the mandible. *Acta Anat.*, **26**, 81 – 93.

Cook, P. (1975). Aspects of the biology of *Bettongia penicillata,* a rat kangaroo with plagiaulacoid dentition. Ph.D. thesis, Yale University.

Coues, E. (1872). The osteology and myology of *Didelphis virginiana. Mem. Boston Soc. Nat. Hist.*, **2**, 41 – 154.

Crompton, A. W. and Hiiemae, K. M. (1970). Molar occlusion and mandibular movements during occlusion in the American opossum, *Didelphis marsupialis. Zool. J. Linnean Soc.*, **49**, 21 – 47.

Crompton, A. W., Hiiemae, K. M., Thexton, A. J. and Cook, P. (1975). The movement of the hyoid apparatus during chewing. *Nature,* **258**, 69 – 70.

Gardner, A. L. (1973). The systematics of the genus *Didelphis* (Marsupialia: Didelphidae) in North and Middle America. *Spec. Publ. Mus. Tex. Tech. Univ.*, **41**, 3 – 81.

Herring, S. W. and Scapino, R. P. (1974). Physiology of feeding in miniature pigs. *J. Morphol.*, **141**, 427 – 60.

Hiiemae, K. M. (1976). Masticatory movements in primitive mammals. In *Clinical and Physiological Aspects of Mastication* (ed. D. J. Anderson and B. Matthews), John Wright, Bristol (in press).

Hiiemae, K. M. (1977). Mammalian mastication, a review of the activity of the jaw muscles and the movements they produce in chewing. In: *Studies on the Development and Function of Teeth* (ed. P. M. Butler and K. A. Joysey), Academic Press, London.

Hiiemae, K. M. and Ardran, G. M. (1968). A cinefluorographic study of feeding in *Rattus norvegicus. J. Zool. London,* **154**, 139 – 54.

Hiiemae, K. M. and Crompton, A. W. (1971). A cinefluorographic study of feeding in the American opossum, *Didelphis marsupialis.* In: *Dental Morphology and Evolution* (ed. A. A. Dahlberg), University of Chicago Press, pp. 299 – 344.

Hiiemae, K. M. and Jenkins, F. A. Jr (1969). The anatomy and internal architecture of the muscles of mastication in the American opossum, *Didelphis marsupialis. Postilla,* **140**, 1 – 49.

Hiiemae, K. M. and Kay, R. F. (1973). Evolutionary trends in the dynamics of primate mastication. In: *Craniofacial Biology of Primates*, Vol. 3 (ed.

M. R. Zingeser); *Symp. IVth Int. Congr. Primat.*, Karger, Basle.

Hiiemae, K. M. and Thexton, A. J. (1975). Consistency and bite size as regulators of mastication in cats. *IADR Abstracts,* L 375, p. L94.

Kallen, F. C. and Gans, C. (1972). Mastication in the little brown bat, *Myotis lucifugus. J. Morphol.,* **136**, 385 – 420.

Kay, R. F. and Hiiemae, K. M. (1974a). Jaw movement and tooth use in recent and fossil primates. *Am. J. Phys. Anthropol.,* **40**, 227 – 56.

Kay, R. F. and Hiiemae, K. M. (1974b). Mastication in *Galago crassicaudatus,* a cinefluorographic study. In: *Prosimian Biology* (ed. R. D. Martin, G. A. Doyle and A. C. Walker), Duckworth, London.

Luschei, E. S. and Goodwin, G. M. (1974). Patterns of mandibular movement and jaw muscle activity during mastication in the monkey. *J. Neurophysiol.,* **37**, 954 – 66.

Thexton, A. J. (1973). Some aspects of neurophysiology of dental interest. I. Theories of and function. *J. Dent.,* **2**, 49 – 54.

Thexton, A. J. (1974). Oral reflexes and neural oscillators. *J. Dent.,* **2**, 131 – 7.

Thexton, A. J. (1976). To what extent is mastication programmed and independent of peripheral feed-back? In: *Clinical and Physiological Aspects of Mastication* (ed. D. J. Anderson and B. Matthews), John Wright, Bristol (in press).

Thexton, A. J. and Hiiemae, K. M. (1974). Does the jaw opening reflex act as effective protective response? *J. Dent. Res.,* **53**, 1067.

Thexton, A. J. and Hiiemae, K. M. (1975a). Masticatory electromyographic activity as a function of food consistency. *IADR Abstracts,* L378, p. L95.

Thexton, A. J. and Hiiemae, K. M. (1975b). The twitch contraction characteristics of opossum jaw musculature. *Arch. Oral Bio,* **20**, 743 – 8.

Thexton, A. J. and Hiiemae, K. M. Food consistency as a regulator of oral function (submitted for publication).

de Vree, F. and Gans, C. (1973). Masticatory responses of pygmy goats (*Capra hircus*) to different foods. *Am. Zool.,* **13**, 1342 – 3.

Weijs, W. A. and Dantuma, R. (1975). Electromyography and mechanics of mastication in the albino rat. *J. Morphol.,* **146**, 1 – 34.

18 *The penis of dasyurid marsupials*

P. Woolley* and S. J. Webb†

Introduction

Our interest in the penis of dasyurid marsupials began with the observation by
Woolley of an unusual appendage on the penis of one of the rarer species, the
Dibbler, *Antechinus apicalis*. In the course of a study on the reproductive biology
of the Dibbler (Woolley, 1971) two of the three males in the laboratory colony
were seen with the penis everted. Figure 18.1 shows the everted penis of one of
the Dibblers, with the appendage lying above the terminal, free portion of the
penis. This part was flaccid and small by comparison with the appendage. Although it was not known whether or not the penis was fully erect, or the
appearance normal, the male illustrated was able to copulate. This male had
mated on four consecutive days with one female. The penis was first seen everted
two days later. The following day he mated with another female and the penis,
which remained everted after copulation, was photographed. Two days later the
male copulated again with this female. A week after the last mating, the appendage
was shrunken, which suggested that it might contain erectile tissue, but the entire
penis was not fully retracted until three weeks later. The penis of a second male
was seen when he was courting a female before copulating with her. The general
appearance of the penis of this male was the same as that of the first but, in
contrast, the penis was retracted immediately after copulation.

At the time these observations were made a survey of the literature revealed
that three other species of dasyurid marsupials, *Dasyurus viverrinus, D. geoffroii*
and *Myoictis melas,* had an apparently similar appendage. Gerhardt (1904, 1933)
had published an illustration of the penis of *D. viverrinus* showing what he called
a 'dorsaler Anhang' and in the later paper he stated that it also occurred in
Phascologale thorbeckiana (= *Myoictis melas*). van den Broek (1910) described
the appendage in *D. geoffroii* and found it to be an accessory erectile body
derived from the corpora cavernosa. The appendage was not found on the penis
of other species of dasyurid marsupials, including *Thylacinus cynocephalus*

*Pat Woolley is Senior Lecturer in Zoology at La Trobe University, with degrees from the
University of Western Australia and the Australian National University. At the University of
Western Australia Professor Harry Waring did much to foster interest in the study of the
marsupial fauna of Australia. Her main research interests have been in the fields of pharmaco-
logical effects of posterior lobe pituitary hormones, reproductive and nutritional studies on
macropod marsupials and, since 1961, the reproductive biology of dasyurid marsupials.

†Stephen Webb, now a medical student at the University of Papua New Guinea, completed his first degree and a diploma of education at La Trobe University. He has been a
school teacher in Melbourne, and has taught both in school and at a Secondary Teachers
College in Papua New Guinea.

307

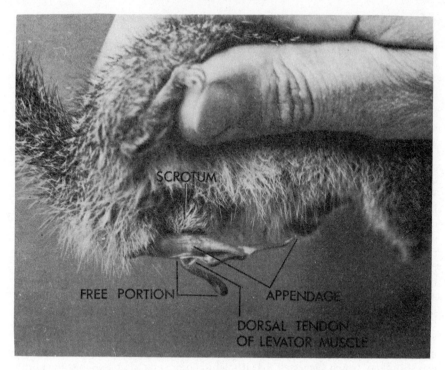

Figure 18.1 The penis of *Antechinus apicalis*.

(Cunningham, 1882), *Phascologale flaviceps* (= *Antechinus flavipes*?) and
Sminthopsis crassicaudata (van den Broek, 1910), *Myrmecobius fasciatus*
(Fordham, 1928) and *Antechinus stuartii* (Woolley, 1966). Moreover, although
the anatomy of the penis of many species has been studied, there has been no
record of an appendage of the type seen in the dasyurids for any other marsupials
or eutherian mammals.

The observation of the appendage on the penis of the Dibbler led to two
studies on the penis of dasyurid marsupials. The first was a study by Webb (1971)
of the anatomy of the penis of *Antechinus apicalis* and *Antechinus stuartii*, a
species which lacks the appendage. The second was a survey of dasyurid marsupials by Woolley to determine which species have the penis appendage, and the
possible significance of this structure in the classification of marsupials of the
family Dasyuridae.

Anatomy of the penis

Introduction

The structure of the penis of marsupials has previously been described in some
detail for a number of species, including the Opossum, *Didelphis virginiana*
(Cowper, 1704; van den Broek, 1910; Retterer and Neuville, 1916); the Potoroo,
Hypsiprymnodon (= Potorous) (Owen, 1839); the Koala, *Phascolarctos cinereus*

(Young, 1879); the Cuscus, *Phalanger maculata* (Desmarest, 1818); the Honey possum, *Tarsipes spenserae* (Rotenburg, 1928); the bandicoot *Perameles obesula* (van den Broek, 1910); as well as the dasyurids *Thylacinus cynocephalus* and *Dasyurus geoffroii* mentioned above. Less detailed information for several other marsupials is available in the papers of Owen and van den Broek.

In the non-erect state the penis is withdrawn into the body and it assumes an S-shaped curve. The terminal, free portion of the penis is surrounded by an invagination of skin which forms the preputial sac. The opening of the preputial sac is directed posteriorly, and it lies ventral to the anus within the shallow cloaca. When erect, the penis is directed forwards and the preputial skin is stretched so that the sac disappears.

Considerable variation is seen in the form of the terminal portion of the penis. In many species the tip is bifid, the depth of the cleft and the form of the apices being variable between species (see Biggers, 1966). In others the tip is undivided, and either tapered or blunt. The urethra may be continued as grooves at the distal end of the penis, and in species with a bifid penis the grooves lie along the inner side of each half. Only in *Perameles (= Macrotis) lagotis* (Owen, 1839) does the urethra also bifurcate and penetrate each half.

The penis consists essentially of a tube containing the terminal portion of the urethra and columns of erectile tissue, the paired corpora cavernosa penis (corpora cavernosa) and the corpus cavernosum urethrae (corpus spongiosum). The corpora cavernosa commence as two widely separated bodies, the crura, that come together at the base of the penis. The variation seen in the arrangement of the corpora cavernosa within the penis has been summarised as follows by van den Broek (1910). In Didelphidae, Dasyuridae and Peramelidae the crura unite medially to form a single mass on the ventro-lateral sides of the urethra. This mass bifurcates distally in species with a bifid penis. In Macropodidae the crura unite but only the left corpus cavernosum runs through the entire penis. In Phascolarctidae the two corpora cavernosa remain entirely separate. The corpora cavernosa have a well-defined, thick tunica albuginea and contain relatively few blood lacunae. The corpus spongiosum commences as two separate bodies, the bulbs, which unite medially, distal to the point of entry of the ducts of the bulbo-urethral glands into the urethra, and surround the urethra. This double bulb of the corpus spongiosum constitutes a basic difference between the penis of marsupials and eutherians. There is no tunica albuginea as an external boundary to the corpus spongiosum and the extent of the development of the blood lacunae is variable. Distally the corpus spongiosum may form a swollen tip to the penis, the corpus cavernosum glandis (glans penis).

The musculature associated with the penis has been described in detail by van den Broek (1910). He considered the major muscles involved in the erection and retraction of the penis to be the ischio-cavernosus and bulbo-cavernosus muscles, encasing the crura and bulbs of the penis, respectively, the paired ventral levator penis muscle, the paired dorsal retractor penis muscle and the cloacal sphincter. The fibres of the retractor muscles are smooth, and those of the other muscles associated with the penis are striated. In some marsupials a suspensory ligament runs from the pubic symphysis to the penis.

The penis of Antechinus stuartii

Material examined. The gross morphology of the penis had previously been studied by Woolley (1966) and the specimens dissected for that study were available for examination. Sections for histological investigation of the internal anatomy were prepared from the penis of each of four adult animals. Two of the four penis specimens were left in the normal retracted position and two were straightened after they were dissected out of the animal. The specimens were fixed in either 70 per cent alcohol or aqueous Bouin's solution and embedded in paraffin wax, and serial transverse sections were cut at 10 μm. All sections of one penis, and up to five series of sections, each series consisting of sections 100 μm apart along the length of the penis, of the other three, were mounted. The sections were stained with either haematoxylin and eosin or Masson's trichrome stain.

Anatomy. The gross anatomy of the penis is illustrated in Figures 18.2 and 18.3 Figure 18.2 shows the penis in the retracted position, and the associated bulbo-

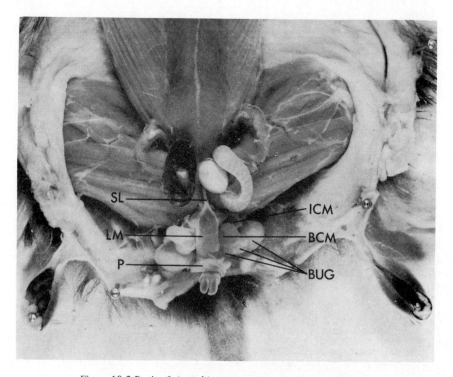

Figure 18.2 Penis of *Antechinus stuartii, in situ.* The ventral portion of the cloacal sphincter has been removed and the testes and epididymides dissected out of the scrotum. SL = suspensory ligament; LM = levator muscle; P = preputial skin; ICM = ischio-cavernosus muscle encasing the crus; BCM = bulbo-cavernosus muscle encasing the bulb (abnormal on right); BUG = bulbo-urethral glands.

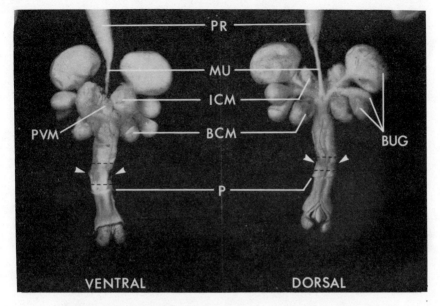

Figure 18.3 Penis of *Antechinus stuartii*, removed from the body
and straightened. The broken lines indicate the approximate levels
of the proximal and distal flexures; the arrow heads, the points of
insertion of the retractor penis muscles. PVM = penis vein muscle;
PR = prostate; MU = membranous urethra; ICM = ischio-cavernosus
muscle; BCM = bulbo-cavernosus muscle; P = preputial skin (site of
attachment to penis); BUG = bulbo-urethral glands.

urethral glands *in situ*. The ventral portion of the cloacal sphincter has been removed. The suspensory ligament can be seen in the mid-line anterior to the penis.
It divides and lies on each side of the penis beside the levator muscle. The tip of the
penis is bifid. The urethral opening is situated at the base of the cleft and the urethra
is continued as two grooves along the inner side of each half of the tip of the penis.

Figure 18.3 shows the ventral and dorsal aspects of the penis removed from the
body and straightened. The site of attachment of the preputial skin to the penis,
the levels at which the flexures in the penis occur and the points of insertion of
the retractor muscles are indicated. On the ventral surface in the mid-line between
the ischio-cavernosus muscles a small bulbous muscle can be seen. This muscle is
inserted onto the tunica albuginea of the crus penis, antero-dorsal to the region of
insertion of the levator penis muscle, on each side. The fibres converge and surround the vein of the penis for a short distance after the vein emerges from the
body of the penis. No previous description of this muscle has been found and in
this study it is called the penis vein muscle.

The internal anatomy of the penis is illustrated in Figure 18.4 in a series of
drawings of transverse sections of the penis at the eight levels shown in the
accompanying outline drawing of the penis. The ventral surface of each section is
shown uppermost because a ventral approach to the penis is made in dissection.

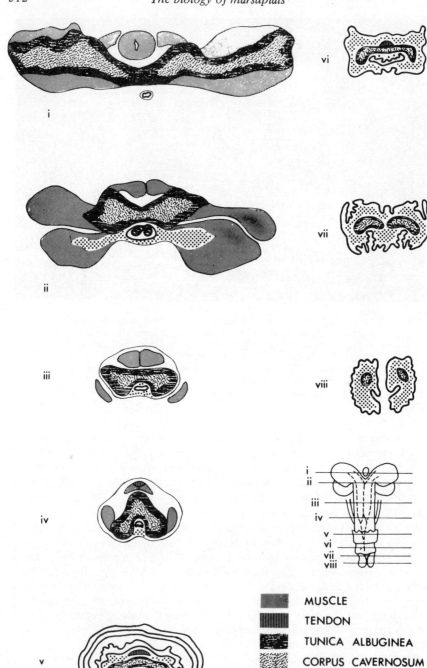

MUSCLE

TENDON

TUNICA ALBUGINEA

CORPUS CAVERNOSUM

CORPUS SPONGIOSUM

EPITHELIUM

A Leitz drawing mirror attached to a microscope was used to project the sections to make outline drawings to the same scale. In the drawings it can be seen that the crura of the corpora cavernosa unite medially (section i) to form a single mass (sections ii – vi) which extends distally on the ventro-lateral sides of the urethra to near the point of bifurcation of the penis, where it also divides (sections vii and viii). The cavernous tissue is surrounded by a thick tunica albuginea. The bulbs of the corpus spongiosum unite medially (section ii), distal to the point of entry of the ducts of the bulbo-urethral glands into the urethra, to form a thin layer of spongy tissue around the urethra. In the terminal free portion of the penis the corpus spongiosum extends around the sides of the corpora cavernosa and forms a swollen glans penis (sections v – viii). The ischio-cavernosus muscles, partially enclosing the crura of the penis, the levator penis muscles, inserting onto the tunica albuginea of the crura, and the penis vein muscle, surrounding the penis vein, can be seen in section i. The levator muscles converge and lie close together ventral to the corpora cavernosa (sections ii – iv). Distally, the tendon of the levator muscles inserts onto the tunica albuginea of the corpora cavernosa (section vi). The bulbo-cavernosus muscles encase the bulbs of the penis (section ii). The retractor penis muscles insert onto the tunica albuginea of the corpora cavernosa at the level of the distal flexure in the penis (section iv).

Apart from the form of the terminal free portion, the anatomy of the penis of *A. stuartii* was found to be basically the same as that described for other dasyurid species that do not have a penis appendage. The location of the previously undescribed penis vein muscle suggests that it may be of importance in the erection of the penis.

The penis of Antechinus apicalis

Material examined. The penis of each of three adult animals was available for study. The penis of one animal was not retracted at the time the animal was killed, but the other two were retracted. The three specimens were fixed in aqueous Bouin's solution and embedded in paraffin wax, and serial transverse sections were cut at 10 μm. Series of sections from each penis were mounted and stained in the same way as the *A. stuartii* material.

Anatomy. The fixed, non-retracted penis specimen is shown in Figure 18.5. The form of the penis is partly obscured by the cloacal sphincter muscle and the rectum. It can be seen that the penis appendage is enclosed in a pocket of preputial skin. The end of the penis is rounded and slightly swollen. On the dorsal surface the urethral grooves form distinct channels from the urethral opening to the tip; these channels can be seen more clearly in the inset photograph of the tip of the penis of another specimen.

Figure 18.4 The internal anatomy of the penis of *Antechinus stuartii*. The position of the levator muscles and tendon is shown in broken lines in the outline drawing of the penis. The ventral surface of each section is uppermost.

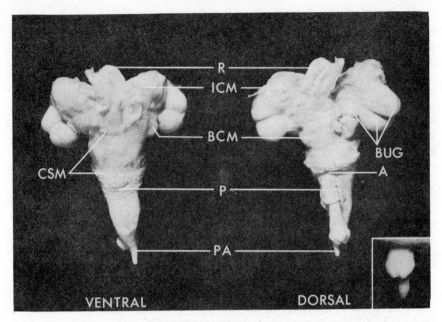

Figure 18.5　Penis of *Antechinus apicalis*, removed from the body
and fixed in the non-retracted position. CSM = cloacal sphincter
muscle; R = rectum; ICM = ischio-cavernosus muscle encasing the
crus; BCM = bulbo-cavernosus muscle encasing the bulb; P = preputial skin; PA = penis appendage; BUG = bulbo-urethral glands;
A = anus.

The internal anatomy of the penis is illustrated in Figure 18.6 in a series of
drawings, prepared as for *A. stuartii*, of transverse sections of the penis at the nine
levels shown in the diagram. Proximal to section i the anatomy of the penis of
A. apicalis was essentially the same as in *A. stuartii* and has not been illustrated.
The penis vein muscle was present. The corpora cavernosa, which unite and lie
ventro-lateral to the urethra (section i) branch distally. A median dorsal branch
lies above the urethra and extends to the tip of the penis (sections ii – ix). Two
lateral branches lie beside the central portion of the united corpora cavernosa and
extend into the penis appendage (sections iii – vii). The central portion extends to
the tip of the penis appendage (sections iv – ix). The corpus spongiosum surrounds the urethra and in the terminal free portion of the penis extends around
the sides of the dorsal branch of the corpus cavernosum (sections v – ix). The
levator penis muscles unite to form a single muscle mass (sections i – iv). The
tendon of the levator muscle bifurcates to form ventral and dorsal tendons. The
ventral tendon is inserted onto the tunica albuginea of the central portion of the

Figure 18.6　The internal anatomy of the penis of *Antechinus apicalis*. The position of the
levator muscles and tendons is shown in broken lines in the outline drawing of the penis. The
ventral surface of each section is uppermost. Key to shading as for Figure 18.4.

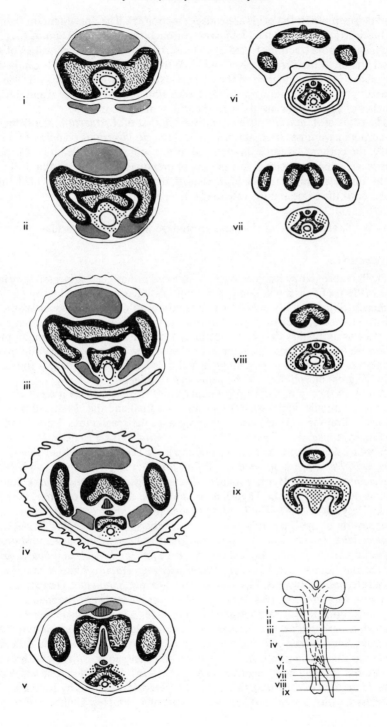

corpora cavernosa in the penis appendage (section vi). The dorsal tendon (which can be clearly seen in Figure 18.1) passes through the central portion of the corpora cavernosa (sections iv and v) and inserts onto the tunica albuginea of the dorsal branch of the corpora cavernosa close to the tip of the penis. It can be seen lying above the dorsal branch of the corpora cavernosa in sections v – viii. The retractor penis muscles insert onto the tunica albuginea of the lateral and dorsal branches of the corpora cavernosa in the region of sections ii – iv.

The penis of *A. apicalis* differs greatly from that of *A. stuartii* in the form of the corpora cavernosa. In *A. apicalis*, as in *Dasyurus viverrinus* (van den Broek, 1910), the corpora cavernosa branch and form a separate erectile body, the penis appendage, which is enclosed in a pocket of preputial skin. The tendon of the levator muscles, which is single in *A. stuartii*, branches in *A. apicalis*. One branch inserts onto the penis and the other onto the penis appendage.

Occurrence of the penis appendage within the dasyuridae

Introduction

Phallic morphology has previously been used, or has been suggested as being potentially useful, as a taxonomic character in studies on the phylogeny of mammals (Slijper, 1938) and of particular groups of mammals for example, rodents (Lidicker, 1968) and bats (Matthews, 1937). It seems possible that a structure as unusual as the penis appendage found in some dasyurids might prove to be a useful taxonomic character. With this in view a survey of dasyurid marsupials was undertaken to determine in which species the appendage is found. Only the gross morphology of the penis was examined.

About 54 species of dasyurid marsupials are recognised at the present time. The precise number varies with the author consulted and the classification followed. Simpson's (1945) classification groups the species into 4 sub-families: the Phascogalinae, with 7 genera (*Phascogale, Antechinus, Planigale, Dasyuroides, Dasycercus, Sminthopsis* and *Antechinomys*); the Dasyurinae, with 4 genera (*Dasyurus, Dasyurinus, Satanellus* and *Dasyurops*); the Thylacininae, with 1 genus (*Thylacinus*); and the Myrmecobiinae, also with 1 genus (*Myrmecobius*). Some authors do not include the Thylacininae, and some the Myrmecobiinae, in the family Dasyuridae. Tate (1947) does not include the Myrmecobiinae in his classification of the Dasyuridae and he groups the species, including *Thylacinus cynocephalus*, into two sub-families: the Phascogalinae, with 7 genera (*Murexia, Thylacinus, Sminthopsis, Antechinomys, Antechinus, Planigale* and *Phascogale*) and the Dasyurinae with 12 genera (*Neophascogale, Parantechinus, Phascolosorex, Pseudantechinus, Myoictis, Dasycercus, Dasyuroides, Satanellus, Dasyurinus, Dasyurus, Dasyurops* and *Sarcophilus*). The most recent list of the species of dasyurids found in Australia (Ride, 1970) includes 37 species in 11 genera (*Dasyurus, Sarcophilus, Phascogale, Dasyuroides, Dasycercus, Antechinus, Planigale, Sminthopsis, Antechinomys, Myrmecobius* and *Thylacinus*). Ride does not group the genera into higher categories. Another genus, *Ningaui*, with two species, has recently been recognised (Archer, 1975). In addition, one species of *Antechinus*, one of *Sminthopsis* and one of *Planigale* not listed by Ride are recognised, giving a total of 42 species in Australia. For New Guinea, Laurie and

Hill (1954), who follow Tate's classification of the Dasyuridae, list 12 species in 8 genera (*Murexia, Sminthopsis, Planigale, Antechinus, Neophascogale, Phascolosorex, Myoictis* and *Satanellus*). The one species of *Sminthopsis* is also found in Australia but the remaining species are endemic to New Guinea. Recently, a new species of *Dasyurus* has been found in New Guinea (Archer, personal communication), bringing the total number of species to 13. Thus, if *Thylacinus* and *Myrmecobius* are included, the total number of species of dasyurid marsupials found in Australia and New Guinea is 54. These species represent the following 16 currently recognised genera: *Dasyurus* (includes *Dasyurops, Dasyurinus* and *Satanellus*), *Sarcophilus, Phascogale, Dasyuroides, Dasycercus, Antechinus, Planigale, Sminthopsis, Antechinomys, Myrmecobius, Thylacinus, Ningaui, Myoictis, Neophascogale, Phascolosorex* and *Murexia.*

Results of the survey

From the direct examination of 41 species, and from records in the literature for another 2 species, it has been determined whether or not the penis appendage is present in 43 of the 54 species, representing 15 of the 16 genera. The results of the survey are shown in Table 18.1. The 16 genera and the numbers of currently recognised Australian and New Guinea species in each are listed in the table. The total number of species in each genus, the number of these examined, and the number in which the appendage is present or absent are shown. All the species in each genus are listed in Table 18.2, together with details of the number and source of the specimens examined and the results of examination of the penis.

Table 18.1 The number of species with and without the penis appendage in each of the 16 genera of dasyurid marsupials. The figures in parentheses are the number of species not listed by Ride (1970) for Australia and by Laurie and Hill (1954) for New Guinea (see text)

| Genus | No. of species | | | | With/without appendage | |
	Australia	New Guinea	Total	Examined		
Dasyurus	4	1 (1)	6	5	5	
Sarcophilus	1		1	1		1
Phascogale	2		2	2		2
Dasyuroides	1		1	1		1
Dasycercus	1		1	1		1
Antechinus	9 (1)	3	13	10	2	8
Planigale	4 (1)	1	6	6		6
Sminthopsis	11 (1)	1*	12	8		8
Antechinomys	2		2	2		2
Myrmecobius	1		1	1		1
Thylacinus	1		1	1		1
Ningaui	(2)		2	2		2
Myoictis		1	1	1	1	
Neophascogale		1	1	1		1
Phascolosorex		2	2	1		1
Murexia		2	2	0		

*Occurs in Australia.

Table 18.2 List of the currently recognised species of dasyurid marsupials* showing the presence or absence of the penis appendage in those species which have been directly examined or for which there are records in the literature. The number and source† of the specimens examined is shown see key on p. 320.

Genus	Australian species	New Guinea species	No. and source of specimens examined	Reference	Penis appendage present (+), absent (−)
Dasyurus	maculatus		2	9029 FWD, 9049 FWD	Gerhardt (1904, 1933)	+
	viverrinus		1	ZMU		+
	geoffroii		1	ZUWA	van den Broek (1910), Archer (1974)	+
	hallucatus				Archer (1974)	+
		albopunctatus sp. nov.	1	10400 PNGM		+
Sarcophilus	harrisii		1	MPL		−
Phascogale	tapoatafa		1	ZLTU		−
	calura		2	M6316 WAM, M13460 WAM		−
Dasyuroides	byrnei		2	ZLTU		−
Dasycercus	cristicauda		4	ZLTU, M9670 WAM		−
Antechinus	flavipes		4	ZLTU	van den Broek (1910)	−
	stuartii		6	ZLTU	Woolley (1966)	−
	bellus		2	M9806 AM, 4592 NTM		−
	godmani					−
	apicalis		4	ZLTU, C7155 NMV	Webb (1971)	+
	rosamondae		1	M6358 WAM		−
	swainsonii		3	ZLTU, ZMU		−
	minimus		2	D426 FWD, D647 FWD		−
	macdonnellensis		4	M7694 AM, M7696 AM, J571 QM, 3260 NTM		+
	bilarni					−
		melanurus	1	M9666 AM		−
		naso	1	M9610 AM		−
		wilhelmina				

Genus	Species	n	Registration numbers	Reference	
Planigale	*masculata*	5	M9196 SAM, M9197 SAM, J21327 QM, J1681 QM, M9946 WAM		—
	ingrami	2	QM, QM		—
	subtilissima	1	M11066 WAM		—
	tenuirostris	1	M8151 AM		—
	gilesi	1	M8410 SAM		—
	novaeguineae	1	M9091 AM		—
Sminthopsis	*murina*	2	M7682 WAM, M9938 WAM		—
	leucopus	2	ZLTU	Archer (1974)	—
	rufigenis (= virginiae)	3	M4056 WAM, J3237 QM, M7445 AM		—
	nitela				
	longicaudata				
	psammophila				
	crassicaudata	1	ZUM	van den Broek (1910)	—
	macroura				
	froggatti	3	M7749 WAM, M5702 WAM, M7693 SAM		—
	granulipes	1	M6062 WAM		—
	hirtipes	1	M10208 WAM		—
	ooldea	1	M8077 WAM		—
Antechinomys	*spenceri*	1	ZLTU		—
	laniger	1	M1806 AM		—
Myrmecobius	*fasciatus*			Fordham (1928)	—
Thylacinus	*cynocephalus*	2	ZANU, 1894 AIA	Cunningham (1882)	—
Ningaui	*ridei*	1	M8081 WAM		—
	timealeyi	1	M5076 WAM		—
Myoictis	*melas*	1	J13/672 QM	Gerhardt (1933)	+
Neophascogale	*lorentzii**	1	M9609 AM		—
Phascolosorex	*doriae*				
	dorsalis				
Murexia	*longicaudata rothschildi*	1	M8902 AM		—

*See note added p. 323.

†AM = Australian Museum; QM = Queensland Museum; NMV = National Museum of Victoria; SAM = South Australian Museum; WAM = Western Australian Museum; NTM = Northern Territory Museum; PNGM = Papua New Guinea Museum; FWD = Fisheries and Wildlife Department, Victoria; AIA = Australian Institute of Anatomy; MPL = Mount Pleasant Laboratories, Tasmania; ZLTU = Zoology, La Trobe University; ZMU = Zoology, Monash University; ZUM = Zoology, University of Melbourne; ZUWA = Zoology, University of Western Australia; ZANU = Zoology, Australian National University

It can be seen from Table 18.1 that the penis appendage has been found in only 3 genera, *Dasyurus, Antechinus* and *Myoictis*. It is present in all species of *Dasyurus* examined and the single species of *Myoictis*. The 2 species of *Antechinus* which have the penis appendage are the Dibbler (*A. apicalis*) and *A. macdonnellensis* (Table 18.2). In so far as it could be determined from gross examination (in some cases of fresh and in others of preserved specimens) and from descriptions, the penis appendage was similar in all species.

Discussion

The classification of the dasyurid marsupials has been based in the past on cranial and dental characters and on the external morphology of the ear, feet and inguinal region (Bensley, 1903; Tate, 1947).

In Simpson's classification of the Dasyuridae 3 of the 8 species (*Antechinus apicalis, A. macdonnellensis* and *Myoictis melas*) in which the penis appendage is found are placed in the sub-family Phascogalinae, while the remaining 5 species (of *Dasyurus*) are placed in the sub-family Dasyurinae. In Tate's classification of the Dasyuridae all species with the appendage have been placed in the sub-family Dasyurinae. It is interesting that Tate erected new genera (*Parantechinus* and *Pseudantechinus*) for the 2 species of *Antechinus* with the appendage, and he transferred these species from the Phascogalinae to the Dasyurinae. This rearrangement was based on Tate's conclusion that the reduction seen in the upper and lower fourth premolars in these species allied them more closely with the Dasyurinae than the Phascogalinae. However, in his suggested phylogeny of the Dasyuridae he places 3 of the species with the appendage on lines separate from the remainder. Figure 18.7 shows the hypothetical phylogenetic tree of the Dasyuridae proposed by Tate (1947), and the species in which the penis appendage is found. Had Tate been aware of the existence of the penis appendage in these species he might well have proposed a different phylogeny.

The occurrence of the penis appendage in *Satanellus, Dasyurinus, Dasyurus* and *Dasyurops* provides additional evidence for the current inclusion of these genera within a single genus, *Dasyurus*. For future studies on relationships within the group the penis appendage may provide a useful additional character for consideration. However, it would be desirable to have more detailed information on the internal anatomy of the penis and its appendage in order to determine if, in fact, the appendage found in the different species is an homologous structure.

Figure 18.7 Hypothetical phylogenetic tree of the Dasyuridae (after Tate, 1947). ● Species with and o species without the penis appendage. The current specific names of those species with the appendage are given in parentheses. *Myoictis wallacei* is now considered to be a sub-species of *M. melas* (Laurie and Hill, 1954).

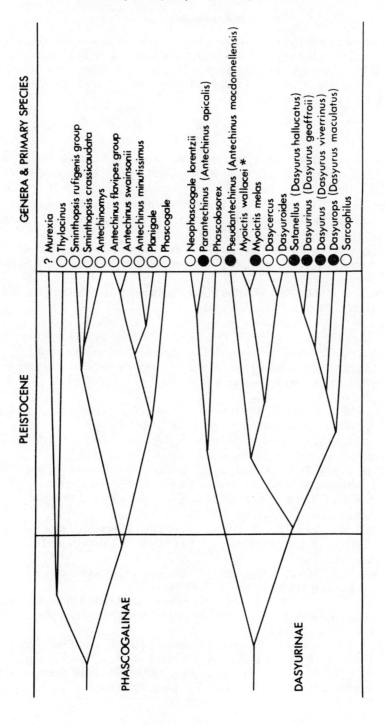

PLEISTOCENE

GENERA & PRIMARY SPECIES

? Murexia
Thylacinus
Sminthopsis rufigenis group
Sminthopsis crassicaudata
Antechinomys
Antechinus flavipes group
Antechinus swainsonii
Antechinus minutissimus
Planigale
Phascogale

Neophascogale lorentzii
Parantechinus (Antechinus apicalis)
Phascolosorex
Pseudantechinus (Antechinus macdonnellensis)
Myoictis wallacei *
Myoictis melas
Dasycercus
Dasyuroides
Satanellus (Dasyurus hallucatus)
Dasyurinus (Dasyurus geoffroii)
Dasyurus (Dasyurus viverrinus)
Dasyurops (Dasyurus maculatus)
Sarcophilus

PHASCOGALINAE

DASYURINAE

Acknowledgements

The authors are grateful to Dr W. D. L. Ride, former Director of the Western Australian Museum, and the Department of Fisheries and Wildlife, Western Australia, for the opportunity to study the rare marsupial *Antechinus apicalis*. We wish to thank all persons who donated specimens, or loaned us specimens in their care, for examination.

References

Archer, M. (1974). Some aspects of reproductive behaviour and the male erectile organs of *Dasyurus geoffroii* and *D. hallucatus*. *Mem. Queensland Museum*, 17, 63 – 7.

Archer, M. (1975). *Ningaui*, a new genus of tiny dasyurids (Marsupialia), and two new species from arid Western Australia, *N. timealeyi* and *N. ridei*. *Mem. Queensland Museum*, 17, 237 - 49.

Bensley, B. A. (1903). On the evolution of the Australian Marsupialia, with remarks on the relationships of the marsupials in general. *Trans. Linnean Soc. London*, 9, 83 - 217.

Biggers, J. D. (1966). Reproduction in male marsupials. *Symp. Zool. Soc. London*, No. 15, 251 - 80.

Cowper, W. (1704). Account of the anatomy of those parts of a male opossum that differ from the female. *Phil. Trans. Roy. Soc.*, 24, 1576 - 90.

Cunningham, D. J. (1882). Report on some points in the anatomy of the Thylacine (*Thylacinus cynocephalus*), Cuscus (*Phalangista maculata*), and Phascogale (*Phascogale calura*). *Rept. Sci. Res. Challenger Zool.*, Vol. 5, part 16.

Fordham, M. G. C. (1928). The anatomy of the urinogenital organs of the male *Myrmecobius fasciatus*. *J. Morphol.*, 46, 563 - 83.

Gerhardt, U. (1904). Morphologische und biologische Studien über die Kopulationsorgane der Säugetiere. *Jena Z. Naturw.*, 39, 43 - 118.

Gerhardt, U. (1933). Kloake and Begattungsorgane. (b) Marsupialier, Beuteltiere. In: *Handbuch der vergleichenden Anatomie der Wirbeltiere* (ed. L. Bolk, E. Göppert, E. Kallius and W. Lubosch). Bd vi. Urban and Schwarzenberg, Berlin, pp. 318 - 21.

Laurie, E. M. O. and Hill, J. E. (1954). *List of Land Mammals of New Guinea, Celebes and Adjacent Islands, 1758 - 1952*, British Mus. (Nat. Hist.), London.

Lidicker, W. Z. (1968). A phylogeny of New Guinea rodent genera based on phallic morphology. *J. Mammal.*, 49, 609 - 43.

Matthews, L. H. (1937). The form of the penis in the British rhinolophid bats, compared with that in some of the vespertilionid bats. *Trans. Zool. Soc. London*, 23, 213 - 23.

Owen, R. (1839). Marsupialia. In: *Cyclopedia of Anatomy and Physiology*, Vol. 3 (ed. R. B. Todd), Sherwood, Gilbert and Piper, London, pp. 257 - 329.

Retterer, E. and Neuville, H. (1916). Due pénis et du clitoris des Saurigues et de leur gland fourchu. *C. R. Seanc. Soc. Biol.*, 79, 760 - 4.

Ride, W. D. L. (1970). *A Guide to the Native Mammals of Australia*, Oxford University Press, Melbourne.

Rotenburg, D. (1928). Notes on the male generative apparatus of *Tarsipes spenserae. J. Proc. Roy. Soc. W. Australia*, **15**, 9 – 14.

Simpson, G. G. (1945). The principles of classification and a classification of mammals. *Bull. Am. Mus. Nat. Hist.*, **85**, 1 – 350.

Slijper, E. J. (1938). Vergleichend anatomische Untersuchungen uber den penis der Säugetiere. *Acta Neerl. Morphol.*, **1**, 375 – 418.

Tate, G. H. H. (1947). Results of the Archbold expeditions. No. 56. On the anatomy and classification of the Dasyuridae (Marsupialia). *Bull. Am. Mus. Nat. Hist.*, **88**, 97 – 156.

van den Broek, A. J. P. (1910). Untersuchungen über den Bau der männlichen Geschlechtsorgane der Beuteltiere. *Morphol. Jahrb.*, **41**, 347 – 436.

Webb, S. J. (1971). A comparative description of the penis of *Antechinus stuartii* and *Antechinus apicalis* (Marsupialia: Dasyuridae). Honours Thesis, La Trobe University, Melbourne.

Woolley, P. (1966). Reproductive biology of *Antechinus stuartii* Macleay (Marsupialia: Dasyuridae). Ph. D. Thesis, Australian National University, Canberra.

Woolley, P. (1971). Observations on the reproductive biology of the Dibbler, *Antechinus apicalis* (Marsupialia: Dasyuridae). *J. Proc. Roy. Soc. W. Australia*, **54**, 99 – 102.

Young, A. H. (1879). On the male generative organs of the Koala (*Phascolarctos cinereus*). *J. Anat. Physiol.*, London, **13**, 305 – 17.

The species listed in Table 18.2 were those recognised at the time the manuscript was prepared. The check list of dasyurid marsupials in Table 2.2 contains several names not included in Table 18.2, and some names used in Table 18.2 are not included in the check list. As explained in the text (p. 316 – 7), the number of species of dasyurid marsupials recognised varies with the author consulted and the classification followed. The equivalent names of the species listed in Table 18.2, but not in the check list, are as follows.

Planigale subtilissima = P. ingrami (subtilissima form, Archer 1976)
Sminthopsis rufigenis, S. nitela = S. virginiae
Sminthopsis froggatti = S. macroura
Antechinomys spenceri = A. laniger

Archer, M. (1976). Revision of the marsupial genus *Planigale* Troughton (Dasyuridae). *Mem. Queensland Museum*, **17**, 341 – 65.

Section 5 Cellular, endocrine and metabolic studies of marsupials

Though early generations of Australians might be said to have shown little scientific interest in their unique flora and fauna, there is now a healthy interest in local species. Expansion of universities, museums and other research institutions during the past two or three decades has produced many new biological research units, and marsupials have come in for their share of attention. Several aspects of their cell biology, endocrinology and physiology are now relatively well studied. Not surprisingly, they have proved in many instances to be mammals first and marsupials second, differing only slightly from eutherians in several aspects of basic physiology.

Reproductive physiology of female marsupials has been particularly well investigated, while that of males has tended to be neglected. In Chapter 24 Dr. Bryan Setchell draws attention to interesting features of male reproductive physiology, and suggests some promising lines of research for future investigations. As might be expected from the wide range of reproductive patterns to be found among marsupials, pituitary function seems to vary considerably from one species to another (Dr. John Hearn, Chapter 20), and may in some respects differ fundamentally from that of eutherians. On the other hand, Dr. Ian Reid's account (Chapter 23) of renal structure and physiology in one Australian species, and his review of work on other species, suggest that marsupial and eutherian kidneys have much in common, and Dr. McDonald (Chapter 21) can identify only trivial differences in adrenocortical function between the two infraclasses. Guiler and Sallis (Chapter 19) draw attention to a curious anomaly in levels of acid phosphatase in the plasma of Tasmanian devils *Sarcophilus harrisii* and Tasmanian tiger cats *Dasyurops maculatus* which may well, as they suggest, prove to have considerable phylogenetic overtones. Dr. Packer's contribution (Chapter 22) describes an investigation of activity levels of quokkas *Setonix brachyurus* in different regimes of temperature and humidity, in a series of experiments relating to observed behaviour and environmental conditions in the field.

19 Unusual plasma acid phosphatase activity in two species of Tasmanian marsupials

J. D. Sallis* and E. R. Guiler†

The unusually high level of acid phosphatase in the plasma of the Tasmanian devil (*Sarcophilus harrisii*) was first highlighted by Parsons *et al.* (1970) during a routine examination of the blood of this animal. It soon became apparent from further investigations (Parsons, Guiler and Heddle, 1971), that the Tasmanian devil, a member of the family Dasyuridae, was indeed unique in respect to its plasma acid phosphatase activity, as no other marsupial from other families which were examined at that time displayed the same characteristics.

Thus members of the Macropodidae, Peramelidae, Vombatidae and Phalangeridae were found to have plasma acid phosphatase values similar to those reported for humans. A comparative scan of other serum enzymes revealed that most fall within the normally accepted range for humans with the exception of lactic dehydrogenase, glutamic-pyruvic aminotransferase and glutamic oxaloacetic aminotransferase. In respect to these latter enzymes, however, such a characteristic elevation would appear to be a normal feature for several non-human species. Giacometti, Berntzen and Bliss (1972), for example, have reported elevated levels of lactic dehydrogenase and serum glutamic-oxaloacetic aminotransferase in the Nine-banded armadillo (*Dasypus novemcinctus*).

Since the initial comparative studies, Parsons and Guiler (1972) have recently had the opportunity to examine samples of blood from one other species, namely the Tiger cat (*Dasyurus maculatus*), another member of the Dasyuridae, which is less common in Tasmania than the Tasmanian devil. It is apparent that this species, along with the Tasmanian devil, shares the distinction of possessing the highest plasma acid phosphatase activity ever recorded for a mammalian species. Table 19.1 illustrates this point by comparing the rates of hydrolysis of the substrate, *p*-nitrophenylphosphate, by the plasma of several mammalian species. Although the Tiger cat would also be a rich source of the enzyme, we have confined most of our investigations to the Tasmanian devil, as Tiger cats are more difficult to trap in sufficient numbers.

*Dr J. D. Sallis is a graduate of the University of Adelaide and has lectured on biochemistry in the University of Tasmania since 1967. He is at present a reader in biochemistry, with research interests including hormonal control of calcium metabolism and the biochemistry of marsupials.

†Dr E. R. Guiler is a graduate of Queen's College, Belfast, and of the University of Tasmania, where he is at present reader in zoology. He has worked on Tasmanian devils since 1966, and is interested in field ecology and its relation to whole-animal biology.

Table 19.1   Rate of hydrolysis of *p*- nitrophenylphosphate by plasma
of some Australian marsupials

Species	μ mol *p*-nitrophenol formed/ min per ml plasma
Potoroo, *Potorous tridactylus*	0.043
Possum, *Trichosurus vulpecula*	0.065
Bandicoot, *Isoodon obesulus*	0.064
Grey kangaroo, *Macropus canguru*	0.094
Wombat, *Phascolomys ursinus*	0.021
Marsupial mouse, *Vombatus swainsonii*	0.030
Tiger cat, *Dasyurops maculatus**	20.14
Tasmanian devil, *Sarcophilus harrisii*	32.50

*Listed as *Dasyurus maculatus* in Kirsch and Calaby (this volume).

The questions arise then as to why these animals possess such unusual acid phosphatase activity and what is the source of the enzyme. In humans, plasma acid phosphatase is a complex mixture of enzymes derived from platelets, erythrocytes and prostatic tissue. Whenever plasma levels of the enzyme are elevated, it is usually indicative of a prostatic carcinoma. The question of origin can sometimes be resolved by measuring activity in the presence of an inhibitor such as tartrate or formaldehyde. For example, activity of acid phosphatase derived from prostate or liver is inhibited by tartrate, whereas formaldehyde completely inhibits erythrocyte acid phosphatase.

In relation to the plasma acid phosphatase of the Tasmanian devil, however, the interpretation of the data is not so simple. Studies using tartrate and formaldehyde have proved inconclusive, as only a partial inhibition can be demonstrated (Parsons *et al*, 1971). Further, values derived from samples of blood from female animals are nearly as high as their male counterparts whether tartrate is present or absent in the assay medium. This suggests, then, that the prostatic tissue is not directly responsible for the major increase in enzyme activity. It seems likely therefore that the plasma acid phosphatase of the Tasmanian devil is of a chemically different nature from that found in other mammals.

A survey of the distribution of acid phosphatase in the Tasmanian devil revealed high levels of activity in most tissues with the highest levels being associated with the reticulo-endothelial system and reproductive organs (Parsons *et al.*, 1971). The nature of the enzyme responsible, however, has yet to be defined. It is clear that many variants of acid phosphatase do exist in mammals but many probably go undetected if they are associated with only a small population of a particular cell type. Once these cell numbers increase, as they may in pathological conditions, then the enzyme activity increases and new variants appear. Perhaps genetic variation or a change in environmental or dietary conditions could also lead to an increase in a particular cell type being present in the Tasmanian devil or Tiger cat.

To study the characteristics of the plasma enzyme, we made an attempt (Sallis, Parsons and Guiler, 1973) to isolate active material by preparative polyacrylamide gel electrophoresis. A band of activity as detected by hydrolysis of the substrate *p*-nitrophenylphosphate was isolated, and this proved to be a single protein band when re-electrophoresed on disc gels. The enzyme does appear to be labile, however, as further bands can be detected on electrophoresis following 24h storage at 4°. We have not yet been able to crystallise the enzyme but the isolated protein has a molecular weight of approximately 85 000 as judged by ultracentrifugation data and gel electrophoresis in the presence of sodium dodecyl sulphate.

For most of our earlier studies the acid phosphatase activity was assayed routinely using *p*-nitrophenylphosphate as the substrate, but it is now apparent that the choice of substrate can be a critical factor in evaluating the role of acid phosphatase in the Tasmanian devil. If the substrate for the reaction is carefully chosen, it is possible to gain information on the origin of the serum enzyme. Babson, Read and Phillips (1959), for example, have described the substrate specificities of the α- and β-naphthylphosphates to determine whether the enzyme arises from erythrocytes or prostatic tissue.

When the rates of hydrolysis of these and other substrates were compared (Table 19.2), it was clear that β-naphthylphosphate in contrast to α-naphthyl phosphate was the preferred substrate for measuring the plasma enzyme activity. Such data would support the concept that erythrocytes and prostatic tissue are not the true origins for the increased plasma levels of enzyme. Of the other substrates, β-glycerophosphate was also readily attacked, a finding that may have important implications relating to metabolic processes in the Tasmanian devil. The optimal activity of the enzyme *in vitro* is expressed at pH 4.8.

Table 19.2 Relative rates of hydrolysis of phosphate esters by purified acid phosphatase (reproduced from Sallis *et al.*, 1973)

Substrate	Product measured	μ mol product formed/ min per ml enzyme
Phenylphosphate	phenol	0.165
p-Nitrophenylphosphate	*p*-nitrophenol	0.212
β-Glycerophosphate	glycerol	0.262
β-Naphthylphosphate	β-naphthol	0.350
α-Naphthylphosphate	α-napththol	0.043

Incubations were made at 37 °C in citrate buffer pH 4.8 with the respective substrate.

Aside from differences in the relative hydrolysis rates of the various substrates by plasma acid phosphatase, complications concerning the choice of substrate arise when attempts are made to detect activity in polyacrylamide gels. As indicated earlier, we noted only a single band of activity following the electrophoresis of plasma on a polyacrylamide gel when *p*-nitrophenylphosphate was used as the substrate and visualisation was effected by lead staining (Figure 19.1a). The mobility of the band corresponded to a β-globulin. However, difficulties arise if the same substrate and visualisation technique are applied to gels that

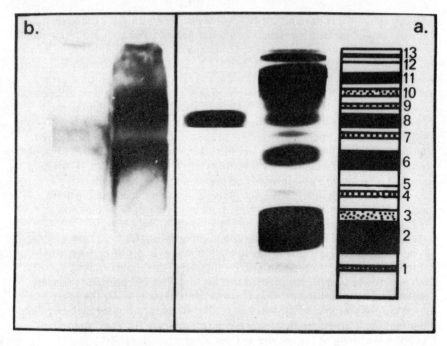

Figure 19.1(a) Polyacrylamide gel electrophoresis of Tasmanian
devil plasma. From left to right: band of acid phosphatase activity
identified by lead staining; protein bands stained with coomassie
brilliant blue; diagrammatic sketch of the electrophoretogram.
Bands 4, 6, 8 and 11 correspond to $\alpha 1$, $\alpha 2$, β and γ globulins,
respectively. (b) Sections of gel after preparative polyacrylamide
gel electrophoresis. The gel on the left has been stained with lead
after incubation with p-nitrophenylphosphate. The gel on the
right has been visualised after incubation with β-naphthylphosphate
by post coupling the liberated β-naphthol with the dye Fast Red B.
Plasma samples were electrophoresed in 5% gels at pH 7.8. (Figure
19.1a reproduced from Sallis *et al.*, 1973).

contain carrier ampholytes. For this reason, we explored the alternative possibility
of detecting acid phosphatase activity in gels using β-naphthylphosphate and
azo coupling with the dye Fast Red B. When plasma is subjected to electrophoresis at pH 7.8 on preparative polyacrylamide gels and subsequently stained in
the presence of β-naphthylphosphate, two distinct zones of activity now appear
positioned above and below the original area previously noted when the gel was
stained in the presence of p-nitrophenylphosphate (Figure 19.1b). We are
currently attempting to characterise these new bands of activity.

Further, β-naphthylphosphate has become the substrate of choice when
screening samples for isoenzymes, as carrier ampholytes are necessary to establish
the pH gradient in this technique. When plasma is subjected to thin-layer gel
isoelectric focusing and subsequently stained with the β-naphthol azo dye system,
at least 15 isoenzymes can be visualised. Figure 19.2 shows a densitometric scan
of a stained gel slab. In addition to numerous minor staining bands, several major

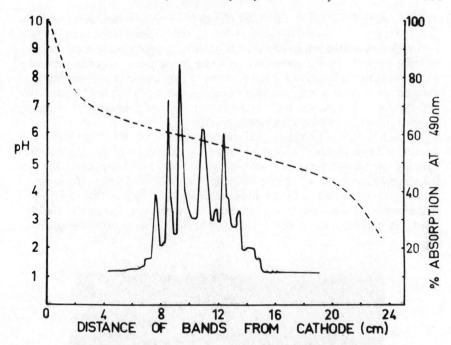

Figure 19.2 Acid phosphatase isoenzymes of Tasmanian devil
plasma separated by thin layer isoelectric focusing in polyacrylamide
gel. The dotted line indicates the final pH gradient produced with a
mixture of carrier ampholytes in the range pH 3 – 8 after electro-
phoresing plasma for 6 h. Gels were subsequently incubated with
β-naphthylphosphate and the liberated β-naphthol was post
coupled to the dye, Fast Red B. The solid line represents a densitometric scan of the gel at 490 nm using a Chromoscan Double Beam
Recording and Integrating Densitometer (Joyce Loebl & Co.).

bands of activity as indicated by the height of the absorption peaks at 490 nm
are clearly discernible. From the pH curve it is evident that the enzymes focus
between pH 4.8 and 6.3. Thin-layer isoelectric focusing is probably the most
powerful tool at present available to separate isoenzymes. Through its use,
together with immunological studies, we hope to gain valuable information
relating to the origin of the plasma acid phosphatase.

The apparent generalised distribution of large quantities of the enzyme(s)
throughout the body tissues of the Tasmanian devil, coupled with the excess
amounts circulating, has made it very difficult to define the tissue of origin. In
animal tissues three of the richest sources are liver, spleen and prostate, and it is
well known also that, within cells, lysosomes are a particularly rich source of
acid phosphatase. As a preliminary to more extensive studies of this aspect, antibodies to the three types of acid phosphatase so far isolated have been prepared
in rabbits. Protein material, recovered after preparative gel electrophoresis of
plasma, has been concentrated using an Amincon B15 miniconcentrator and the
relevant antigen injected into rabbits, three times per week for six weeks.

At the end of this period, the rabbit plasma is normally tested for its antibody titre and it is also subjected to the Ouchterlony gel diffusion technique (1949). Figure 19.3 shows the results of a typical gel diffusion plate, using as antigen the material recovered from a preparative gel which hydrolyses *p*-nitrophenylphosphate. A strong precipitin line forms between antigen and antibody, while a more diffuse precipitin band can be seen between antibody and devil plasma. We have not seen any cross-reactivity between antibody and normal rabbit serum or between the antigen and normal rabbit serum.

Through the use of immunological techniques, then, we have begun a programme of tissue typing of acid phosphatase. Crude extracts of acid phosphatase have been prepared from liver, kidney and intestine by homogenising the tissue in 0.3M acetate buffer pH 4.8 and centrifuging at 12 000g for 15 min. The supernatant was fractionated by the addition of ammonium sulphate, first to 18 per cent saturation and later to a final saturation of 38 per cent. The majority of the activity is retained in the 12 000g precipitate at this concentration of ammonium

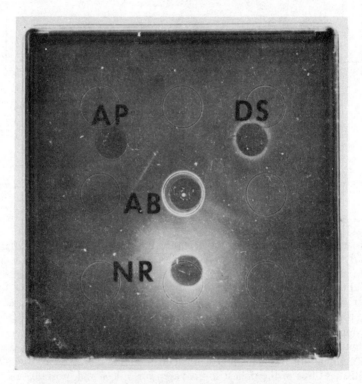

Figure 19.3 Immunodiffusion plate test using antibody (AB) produced by immunising rabbits with acid phosphatase (AP) as antigen. The acid phosphatase was protein recovered from polyacrylamide gel after preparative electrophoresis of 5.0 ml of Tasmanian devil plasma, which hydrolysed *p*-nitrophenylphosphate. NR, normal rabbit plasma; DS, Tasmanian devil plasma.

sulphate. The precipitate dissolved in acetate buffer was dialysed and concentrated to a final volume suitable for immunological studies. Preliminary results indicate that precipitin lines develop when antibody diffuses against all three extracts confirming our previous findings (Parsons *et al.*, 1971) of a widespread distribution of the enzyme. Obviously, a much more detailed investigation is required before any definite conclusions can be drawn regarding the origin of the enzyme.

Although we have virtually no information at this time as to why the Tasmanian devil has such a high plasma level of acid phosphatase, preliminary data from studies with young animals may hold the key to the solution. Tasmanian devils are born during April–May (Southern Hemisphere autumn) and spend until October – November in the pouch or with the female before they assume an independent life. Females do not usually breed until two years of age and animals of 1 – 2 years of age can be regarded as immature. We have noted that pouch-young of three months have very little acid phosphatase activity associated with their plasma and that animals about one year of age have approximately one-half of the maximal activity that one comes to associate with the adult.

These observations suggest that development or induction of the enzyme may occur during the weaning period or, alternatively, there may be a release of suppression of the enzyme's latent activity. Such a developmental pattern for hydrolytic enzymes has already been noted by Raychaudhuri and Desai (1972) in the intestine of developing rats. These workers have suggested that the increased activities of intestinal lysosomal acid hydrolases represent adaptive changes to dietary conditions when the rat changes from the neonatal to post-natal period. Obviously, a long-term study involving the trapping of many Tasmanian devils with pouch-young of differing ages is necessary to clarify this aspect.

Another aspect is whether the enzyme(s) is in fact genetic in origin. Of interest in this respect is the finding that only two species possess this unusual characteristic and that both are true carnivores. Thus it is important to reconsider the classification of these two carnivores in the light of their unusual acid phosphatase activity. Tate (1947), using dentition and foot modifications as criteria, considered that there is a dichotomous pair of branches in the Dasyuridae phylogenetic tree consisting of the Phascogalinae and Dasyurinae. The former group comprise the *Antechinus* group, *Phascogale sensu strictu* and *Thylacinus*, while in the Dasyurinae we find all the Native cats of the genera *Dasyurus, Satanellus, Dasyurinus*, the Tiger cat (*Dasyurops*) and the Tasmanian devil (*Sarcophilus*). On this basis, it might be expected that the Native cat (*Dasyurus*) would also possess high plasma acid phosphatase levels. However, the only species of Native cat so far examined, *Dasyurus quoll*, which is a carnivorous Dasyurid, has low acid phosphatase activity.

Until 1916 the Tiger cat and the Native cat were both considered to be members of the genus *Dasyurus*, having very similar body form, coloration and dentition. Matschie (1916), however, using morphological criteria, suggested that the Tiger cat was sufficiently different to warrant its being placed in a separate genus, *Dasyurops*. Such a conclusion would appear also to be supported by the work of Kirsch and Murray (1969), who showed that the blood of *Dasyurops* had closer serological affinities with that of *Sarcophilus* than with that of *Dasyurus.* Our own findings of a basic biochemical difference between the Native and Tiger cats supports the conclusion of Kirsch and Murray and favours the retention of *Dasyurops* Matschie (1916). We would suggest further that *Dasyurops* and *Sarcophilus* should be placed together on a separate stem of the Dasyuridae.

From the available evidence it would appear that, in general, biochemical parameters of the blood are remarkably uniform throughout the Mammalia. Thus the discovery of a biochemical parameter such as the plasma acid phosphatase activity, which differs so widely from the accepted normal range in the family Marsupialia, may well have considerable phylogenetic overtones.

Acknowledgements

We thank the Tasmanian National Parks and Wildlife Service for their co-operation in the trapping of marsupials around Tasmania. Thanks are also due to Mr. A. McCaw for animal management and Miss J. Short for technical assistance. The programme was supported in part by a grant from the Australian Research Grants Committee.

References

Babson, A. L., Read, P. A. and Phillips, G. E. (1959). The importance of the substrate in the assays of acid phosphatase in serum. *Am. J. Clin. Pathol.*, **32**, 83 – 87.

Giacometti, L., Berntzen, A. K. and Bliss, M. L. (1972). Hematologic parameters of the nine-banded armadillo, *Dasypus novemcinctus. Comp. Biochem. Physiol.*, **43A**, 213 – 7.

Kirsch, J. A. W. and Murray, M. D. (1969). A scheme for the identification of Australasian marsupials. *Aust. J. Zool.*, **17**, 799 – 840.

Matschie, P. (1916). Die verbreitung der Beuteltiere auf Neu Guinea mit einigen Bermerkungen uber ihre Einteilung in Untergattungen. *Mitt. Zool. Mus Berlin*, **8**, 256 – 309.

Ouchterlony, O. (1949). Antigen antibody reactions in gels. *Acta Pathol. Microbiol. Scand.*, **26**, 507 - 15.

Parsons, R. S. and Guiler, E. R. (1972). Observations on the blood of the marsupial Tiger cat, *Dasyurops maculatus* (Kerr) [Dasyuridae] . *Comp. Biochem. Physiol.*, **43A**, 935 – 9.

Parsons, R. S., Guiler, E. R. and Heddle, R. W. L. (1971). Comparative studies on the blood of monotremes and marsupials – II. Electrolyte. Organic constituents, proteins, gas analysis and enzymes. *Comp. Biochem. Physiol.*, **39B**, 209 - 17.

Parsons, R. S., Heddle, R. W. L., Flux, W. G. and Guiler, E. R. (1970). Studies on the blood of the Tasmanian Devil. *Comp. Biochem. Physiol.*, **32**, 345 - 51.

Parsons, R. S., Heddle, R. W. L. and Guiler, E. R. (1971). The distribution of acid phosphatase in the Tasmanian devil, *Sarcophilus harrisii* (Marsupialia; Dasyuridae). *Comp. Biochem. Physiol.*, **39B**, 219 - 26.

Raychaudhuri, C. and Desai, J. D. (1972). Regulation of lysosomal enzymes – III. Dietary induction and repression of intestinal acid hydrolases during development. *Comp. Biochem. Physiol.*, **41B**, 343 - 8.

Sallis, J. D., Parsons, R. S. and Guiler, E. R. (1973). Isolation and some properties of an acid phosphatase from the plasma of the Tasmanian devil, *Sarcophilus harrisii* (Marsupialia: Dasyuridae). *Comp. Biochem. Physiol.*, **44B**, 971 - 80.

Tate, G. H. H. (1947). Results of the Archbold Expeditions No. 56 on the anatomy and classification of the Dasyuridae (Marsupialia). *Bull. Am. Mus. Nat. Hist.*, **88**, 97 - 156.

20 *Pituitary function in marsupial reproduction*

John P. Hearn*

Reproductive cycles of wild animals are strongly linked to environmental events, and often synchronised with environmental cycles in ways which protect and facilitate growth of the offspring. In the higher vertebrates synchrony is effected through the endocrine system, with major points of control in the hypothalamus, gonads and pituitary. The physiological role of the marsupial hypothalamus has been little studied; that of the gonads has been investigated in a range of species and is comparatively better-known. This paper discusses the role of the pituitary, which has recently been investigated in the Tammar wallaby, *Macropus eugenii*. Although the dangers of extrapolating from studies of a single species are recognised, the Tammar wallaby displays most of the important features which distinguish marsupial from eutherian reproduction (for example, short gestation and prolonged lactation) and have allowed the marsupials to achieve evolutionary success no less than that of eutherians.

Patterns of reproduction in macropodid marsupials

All species of macropodids (kangaroos and wallabies) so far studied are monovular and polyoestrous, with cycles varying from 22 to 46 days (Tyndale-Biscoe, Hearn and Renfree, 1974). There is a clearly defined proliferative or pro-oestrous phase of follicular growth and maturation. Ovulation is spontaneous and usually alternate, and the luteal phase which follows affects the endometria of the two quite separate uteri. A rather poorly defined post-luteal phase occurs as the corpus luteum and uteri regress; in an unmated breeding female this phase merges with the next pro-oestrous phase and ovulation. For a review of reproductive patterns in diprotodont marsupials see Sharman, Calaby and Poole (1966).

The Tammar wallaby demonstrates most of the unusual features associated with macropodid reproduction on which the influence of the pituitary gland has been studied. Figure 20.1 shows the reproductive tracts of male and female Tammar wallabies, and Figure 20.2 shows the reproductive cycle of the species. The female is a precise seasonal breeder. In Australia young are usually born in late January or early February, having developed from a quiescent blastocyst

*Dr John Hearn is a graduate of the National University of Ireland and the Australian National University. Born in India and brought up in East Africa, he has worked on enzymes of amphibians and fish and the reproductive endocrinology of wild mammals. He is currently a scientist in the MRC Unit of Reproductive Biology, Edinburgh, studying the endocrinology and immunological control of fertility in primates.

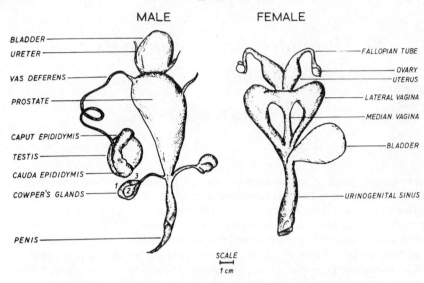

Figure 20.1   Reproductive tracts of male and female Tammar wallabies.

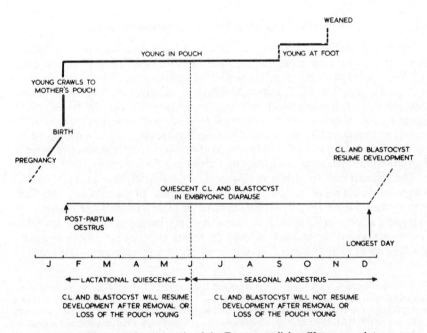

Figure 20.2 The reproductive cycle of the Tammar wallaby. CL = corpus luteum.

carried in the uterus from the previous year. After birth the single young spends up to 18 months in the pouch. During this time the mother is in a state of lactationally induced anoestrus (Berger and Sharman, 1969), which is followed without a break by a seasonal anoestrus between about June and December. A post-partum oestrus occurs between 12 and 24 h after birth of the young, followed by ovulation 12–24 h later. If conception occurs immediately, the embryo develops only to a unilaminar blastocyst with approximately 80 cells (Smith and Sharman, 1969), which remains quiescent and free in the lumen of the uterus for 11 months. The corpus luteum also remains small and inactive through this period of embryonic diapause (Tyndale-Biscoe, 1963).

Loss or removal of pouch young between January and June causes the quiescent corpus luteum and the blastocyst to resume development, and the new young is born about 27 days later (Berger, 1970). Implantation occurs between days 17 and 20 of this 27 day pregnancy, the major part of embryonic growth taking place over the last 10 days. In contrast, if pouch-young are removed between June and December, the corpus luteum and blastocyst remain quiescent and development is resumed in late December, presumably in response to changing photoperiod or other environmental stimuli. Berger (1970) has shown that moving wallabies to the northern hemisphere results in a reversal of breeding season, implicating photoperiod as an important triggering factor. Experiments designed to test whether photoperiod alone is responsible gave inconclusive results, but suggested that pouch-cleaning and other early indications of breeding behaviour might occur without an associated termination of embryonic diapause (Hearn, 1972a); thus photoperiod may not be the only factor involved.

The male Tammar wallaby has received less attention than the female, but there do not appear to be any marked seasonal changes in its reproductive tract. It remains in breeding condition throughout the year, even though the breeding activity of the females is restricted to a period between January and June. Indeed, if there is no loss of pouch-young, the mating activity of the males is restricted to a short spell in late January and early February. Seasonal variation has not been found either in spermatogenesis or in testis size of other seasonally breeding male macropodids (Hughes, 1964; Biggers, 1966), although some other marsupials do show a seasonal testis regression with an interruption of spermatogenesis (Sharman, 1961; Hughes, Thompson and Owen, 1965; Woolley, 1966; Smith, 1969). Such environmental factors as prolonged drought can cause degeneration of the seminiferous tubules in the Red kangaroo (Newsome, 1973).

The pituitary gland in marsupials

The anatomy of the pituitary gland has been examined in several marsupial species. The pars tuberalis in the Quokka (*Setonix brachyurus*) resembles that in monotremes, and its gross anatomy has been suggested as a model for a simple eutherian pituitary type (Hanstrom, 1966). The cytology of the pars distalis of Bennett's wallaby, *Macropus rufogriseus*, was studies by Ortman and Griesbach (1958), and Purves and Sirett (1959) carried out bioassays of the rostral and caudal zones of the pituitary in this species. They found high concentration of thyotrophin and prolactin in the rostral zone and high levels of somatotrophin in the caudal zone, while gonadotrophin and corticotrophin were dispersed generally throughout the gland. Purves and Sirett attributed production of these hormones to similarly distributed cell types.

The pituitary gland of the Tammar wallaby is similar in appearance to that of Bennett's wallaby described by Ortman and Griesbach (1958) – laterally compressed with deep dorsoventral and oval dorsal aspects. The rostral part of the anterior lobe terminates in a rounded, dorsally pointed projection under the median eminence. The posterior lobe is enclosed by the anterior, but separates easily under suction or pressure (Hearn, 1972b).

Exogenous gonadotrophins have been injected into marsupials of several species in attempts to induce ovulation or superovulation, with mixed results which require further investigation; comparable studies involving eutherians have by contrast produced constant results and routine procedures in several species (Robinson, 1951; Moore and Shelton, 1964; Moore, 1970). Tyndale-Biscoe (1963) injected quokkas with gonadotrophins and found that although large cystic follicles were formed, ovulation did not occur. Nelsen and White (1941) obtained young from three out of 30 female Opossums, *Didelphis virginiana*, injected with the gonadotrophic follicle stimulating hormone (FSH) and luteinising hormone (LH), and Smith and Godfrey (1970) had considerable success when they injected female *Sminthopsis crassicaudata* with pregnant mare serum (PMS) and human chorionic gonadotrophin (HCG), inducing ovulation in 20 animals. Possibly these exogenous hormone preparations derived from non-marsupial species were not adequately functional in inducing ovarian effects in all the marsupials studied. More recently Farmer and Papkoff (1974) identified LH and FSH activity in pituitaries from the Red kangaroo, *Macropus rufus*.

The role and function of the pituitary gland in eutherians was initially examined by removal and replacement techniques; more recently radioimmunoassay has been used instead. Over the last 60 years about a dozen species of eutherians, mostly small laboratory or domestic animals but also man and the Rhesus monkey, have been studied in this way (Amoroso and Porter, 1966; Deanesly, 1966) and attention has now focused on the control of positive and negative feedback mechanisms in the hypothalamus. Until recently interpretation of the pituitary's role in the unusual aspects of marsupial reproduction, and particularly in embryonic diapause and pregnancy, was largely guesswork based on the results of eutherian studies. However, when the techniques of hypophysectomy and radioimmunoassay of LH were applied to the Tammar wallaby, the results did not always conform to expectation.

The female Tammar wallaby

Techniques of hypophysectomy (Hearn, 1975a) and radioimmunoassay for gonadotrophin (Hearn, 1972c) were used to examine the role of the pituitary gland during the cycle, pregnancy and embryonic diapause in the Tammar wallaby (Hearn, 1973, 1974). Results from eutherian studies suggested that hypophysectomy before implantation should cause immediate termination of pregnancy, while hypophysectomy after implantation might not. It was further expected that embryonic diapause would be produced by an inactive corpus luteum resulting from insufficient pituitary stimulation. The results obtained with the Tammar wallaby were completely at variance with these hypotheses. Pregnancy continued to term no matter at what stage hypophysectomy was performed, but successful parturition did not occur in absence of the pituitary, and young were retained in the uterus. Hypophysectomy during embryonic diapause, during both the breeding and non-breeding seasons, caused the quiescent corpus luteum and blastocyst to resume development. Pregnancies proceeded to term, but again there was no successful parturition. These results showed clearly that the macropodid pituitary is not essential to activate or maintain the corpus luteum or the blastocyst during pregnancy, but is necessary to ensure successful parturition. This result further suggested that embryonic diapause in the Tammar wallaby is controlled by a tonic pituitary inhibition of the corpus luteum.

Radioimmunoassay of plasma collected from the wallaby during pregnancy and embryonic diapause provided corroborative evidence that there is no marked elevation of gonadotrophin above basal levels at these times. Nor did levels rise after removal of the pouch-young when a surge of gonadotrophin might have been expected to coincide with the resumption of embryonic development.

In other respects hypophysectomy had effects similar to those seen in eutherian mammals. Lactation ceased within two days of the operation, and developing follicles regressed. Ovarian, uterine and vaginal weights declined markedly, exceeding the regression which occurs naturally in non-breeding seasons. This indicates that the pituitary continues to secrete gonadotrophin in normal wallabies during the non-breeding season, a suggestion confirmed by the presence of measurable plasma gonadotrophin throughout the year. However, this species is a precise seasonal breeder, and the mechanism through which its seasonality is controlled appears to be a pituitary-derived inhibition of the corpus luteum. Inhibition may be exerted by secretion of prolactin or oxytocin, or by a hormone as yet undescribed. This must of course in turn be regulated by the hypothalamus, but no work has yet been done to clarify the situation further.

The male Tammar wallaby

The reproductive system of many seasonally breeding eutherian males (for example, the Hedgehog, *Erinaceus europaeus*; Thirteen-lined ground squirrel, *Citellus tridecemlineatus*; Ferret, *Putorius putorius*; and Red deer, *Cervus elaphus*) regresses markedly during non-breeding periods. A similar regression occurs in the platypus (Temple-Smith, 1974) and to some degree in the Brush-tailed possum (Gilmore, 1969). When non-breeding male squirrels and ferrets were hypophysectomised, there were no apparent effects on the reproductive organs, but hypophysectomy during the breeding season caused a rapid regression of the reproductive system to non-breeding condition (Rowlands and Parkes, 1966). Clearly, the effects of hypophysectomy on the male reproductive tract are related to the functional state of the testis. In the wallaby both the testis and the remainder of the reproductive tract regressed at a similar rate at whatever season hypophysectomy was performed (Hearn, 1975b). Radioimmunoassay showed no significant changes in plasma gonadotrophin levels between breeding and non-breeding seasons. These findings indicate that there is no marked seasonal cycle in the male Tammar wallaby, and that the pituitary maintains the reproductive tract in breeding condition throughout the year.

Other macropodids

The macropodids as a group show a wide range of reproductive patterns (Sharman *et al.*, 1966; Tyndale-Biscoe, 1973) although embryonic diapause is found in at least 14 species (Sharman and Berger, 1969), only the Tammar and Bennett's wallaby are known to retain a blastocyst in diapause through a lengthy seasonal anoestrus. There is no precise knowledge of follicular developments during lactational or seasonal anoestrus in marsupials. In the Quokka (*Setonix brachyurus*) on Rottnest Island the uterus regresses during the non-breeding season and any blastocysts in diapause die (Sharman, 1955). A similar regression is seen in the uterus of the Red kangaroo after a lengthy drought (Newsome, 1964). These conditions are unusual, however, as the Quokka on the mainland of Australia (Shield, 1968) and the Red kangaroo normally breed throughout the year; only inadequate diet is believed to be responsible for seasonal breeding on Rottnest Island. Plasma gonadotrophin assays would be needed to establish whether the testicular regression observed during severe drought in the Red kangaroo (Newsome, 1973) results from a fall in pituitary hormonal support for spermatogenesis or from some other effect of the adverse conditions themselves.

Newsome (1964) and Sharman and Clark (1967) showed that ovulation can occur during lactation in the Red kangaroo, and suggested that the corpus luteum may be the only inhibitor of ovulation in this species. A similar situation occurs in the Grey kangaroo, *Macropus giganteus*, where oestrus and ovulation can occur during late lactation (Clark and Poole, 1967), although in this case there is no post-partum oestrus and no corpus luteum is involved. Thus in both Red and Grey kangaroos an inhibitory mechanism independent of lactation must be functioning. This may involve lowered levels of oxytocin or prolactin in the circulation, resulting from the reduced suckling; alternatively there may be an additional inhibiting pituitary factor, linked through neural pathways with size of pouch-young or with the young's initial excursions from the pouch.

If alterations in the secretion of prolactin and oxytocin were the only factors controlling the function of the corpus luteum in the Tammar wallaby, there should be a resumption of embryonic development as the pouch-young mature. This does not occur; thus the Tammar may have an additional hormone in circulation or alternatively, the hypothalamus, controlled by changes in photoperiod, may intervene to prolong the secretion of oxytocin or prolactin independent of lactation.

Macropodids usually resume breeding rapidly when favourable conditions return after drought, either opportunistically (for example, the Red kangaroo) or in accordance with precise seasonal regulation. A pituitary gland which continually secretes gonadotrophin, with a superimposed tonic inhibition of ovulation and of the development of the corpus luteum, would be advantageous in these situations. Clearly, some aspects of pituitary function are different in marsupials and eutherians, and although the marsupial pituitary is closely involved in the control of breeding seasons, its role in relation to the hypothalamus and gonad might vary in different species. The evolutionary success of marsupials in a wide range of difficult environments may well have been assisted by this variability of pituitary function.

References

Amoroso, E. C. and Porter, D. G. (1966). Anterior pituitary function in pregnancy. In: *The Pituitary Gland*, Vol. 2 (ed. G. W. Harris and B. T. Donovan), Butterworths, London, pp. 364 - 412.

Berger, P. J. (1970). The reproductive biology of the tammar wallaby, *Macropus eugenii* Desmarest. Ph.D. Thesis, Tulane University.

Berger, P. J. and Sharman, G. B. (1969). Progesterone induced development of normal blastocysts in the tammar wallaby, *Macropus eugenii, J. Reprod. Fert.*, **20**, 201 - 10.

Biggers, J. D. (1966). Reproduction in male marsupials. *Symp. Zool. Soc. London*, **15**, 251 - 80.

Clark, M. J. and Poole, W. E. (1967). The reproductive system and embryonic diapause in the female grey kangaroo, *Macropus giganteus, Aust. J. Zool.*, **15**, 441 - 59.

Deanesly, R. (1966). The endocrinology of pregnancy and foetal life. In: *Marshall's Physiology of Reproduction*, Vol. 3 (ed. A. S. Parkes), Longmans, London, pp. 891 - 1005.

Farmer, S. W. and Papkoff, H. (1974). Studies on the anterior pituitary of the kangaroo. *Proc. Soc. Biol. Med.*, **145**, 1031 - 46.

Gilmore, D. P. (1969). Seasonal reproductive periodicity in the male Australian Brushtailed possum (*Trichosurus vulpecula*). *J. Zool. London*, **157**, 75 - 98.

Hanstrom, B. (1966). Gross anatomy and evolution of the pituitary. In: *The Pituitary Gland* (ed. G. W. Harris and B. T. Donovan), Butterworths, London.

Hearn, J. P. (1972a). Effect of advanced photoperiod on termination of embryonic diapause in the marsupial *Macropus eugenii* (Macropodidae). *Aust. Mamm.*, **1**, 40 - 2.

Hearn, J. P. (1972b). The pituitary gland and reproduction in the marsupial, *Macropus eugenii.* Ph.D. Thesis, Australian National University, Canberra.

Hearn, J. P. (1972c). The development of a radioimmunoassay for gonadotrophin in the tammar wallaby, *Macropus eugenii. J. Reprod. Fert.*, **28**, 132.

Hearn, J. P. (1973). Pituitary inhibition of pregnancy. *Nature*, **241**, 207 - 8.

Hearn, J. P. (1974). The pituitary gland and implantation in the female tammar wallaby, *Macropus eugenii. J. Reprod. Fert.*, **39**, 235 - 41.

Hearn, J. P. (1975a). Hypophysectomy of the tammar wallaby, *Macropus eugenii*: surgical approach and general effects. *J. Endocrinol.*, **64**, 403 - 16.

Hearn, J. P. (1975b). The role of the pituitary gland in the reproduction of the male tammar wallaby, *Macropus eugenii. J. Reprod. Fert.*, **42**, 399 - 402.

Hughes, R. L. (1964). Sexual development and spermatozoon morphology in the male macropod marsupial *Potorous tridactylus* (Kerr). *Aust. J. Zool.*, **12**, 42 - 51.

Hughes, R. L., Thompson, J. A. and Owen, W. H. (1965). Reproduction in natural populations of the Australian ring tail possum in Victoria. *Aust. J. Zool.*, **13**, 383 - 406.

Moore, N. W. (1970). Fertilisation in ewes treated with progesterone and equine anterior pituitary extract. *J. Endocrinol.*, **46**, 121 - 2.

Moore, N. W. and Shelton, J. N. (1964). Response of the ewe to a horse anterior pituitary extract. *J. Reprod. Fert.*, **7**, 79 - 87.

Nelsen, O. E. and White, E. L. (1941). A method for inducing ovulation in the anoestrous opossum, *Didelphis virginiana. Anat. Rec.*, **81**, 529 - 31.

Newsome, A. E. (1964). Oestrus in the lactating red kangaroo, *Megaleia rufa* (Desmarest). *Aust. J. Zool.*, **12**, 315 - 21.

Newsome, A. E. (1973). Cellular degeneration in the testis of the red kangaroo during hot weather and drought in Central Australia. *J. Reprod. Fert.*, Suppl. 19, 191 - 202.

Ortman, R. and Griesbach, W. E. (1958). The cytology of the pars distalis of the wallaby pituitary. *Aust. J. Exp. Biol.*, **36**, 609 - 18.

Purves, H. D. and Sirett, N. (1959). A study of the hormone contents of the rostral and caudal zones of the pars anterior of the wallaby pituitary. *Aust. J. Exp. Biol.*, **37**, 271 - 8.

Robinson, T. J. (1951). The control of fertility in sheep. II. The augmentation of fertility by gonadotrophin treatment of the ewe in normal breeding season. *J. Agr. Sci.*, **41**, 6 - 14.

Rowlands, I. W. and Parkes, A. S. (1966). Hypophysectomy and the gonadotrophins. In: *Marshall's Physiology of Reproduction*, Vol. 3 (ed. A. S. Parkes), Longmans, London, pp. 26 - 124.

Sharman, G. B. (1955). Studies on marsupial reproduction. 3. Normal and delayed pregnancy in *Setonix brachyurus. Aust. J. Zool.*, **3**, 56 - 70.

Sharman, G. B. (1961). The mitotic chromosomes of marsupials and their bearing on taxonomy and phylogeny. *Aust. J. Zool.*, **9**, 38 - 60.

Sharman, G. B. and Berger, P. J. (1969). Embryonic diapause in marsupials. *Advan. Reprod. Physiol.*, **4**, 211 - 40.

Sharman, G. B., Calaby, J. H. and Poole, W. E. (1966). Patterns of reproduction in female diprotodont marsupials. *Symp. Zool. Soc. London*, **15**, 205 - 32.

Sharman, G. B. and Clark, M. J. (1967). Inhibition of ovulation by the corpus luteum in the red kangaroo, *Megaleia rufa. J. Reprod. Fert.*, **14**, 129 - 37.

Shield, J. (1968). Reproduction of the quokka, *Setonix brachyurus*, in captivity. *J. Zool.*, **155**, 427 - 44.

Smith, M. J. and Sharman, G. B. (1969). Development of dormant blastocysts induced by oestrogen in the ovariectomised marsupial, *Macropus eugenii. Aust. J. Biol. Sci.*, **22**, 171 - 80.

Smith, M. N. and Godfrey, G. K. (1970). Ovulation induced by gonadotrophins in the marsupial, *Sminthopsis crassicaudata* (Gould). *J. Reprod. Fert.*, **22**, 41 - 7.

Smith, R. F. C. (1969). Studies on the marsupial glider, *Schoinabates volans* (Kerr). 1. Reproduction. *Aust. J. Zool.*, **17**, 625 - 36.

Temple-Smith, P. (1974). The reproductive biology of the platypus. Ph.D. Thesis, Australian National University.

Tyndale-Biscoe, C. H. (1963). The role of the corpus luteum in delayed implantation in marsupials. In: *Delayed Implantation* (ed. A. C. Enders), University of Chicago Press.

Tyndale-Biscoe, C. H. (1973). *Life of Marsupials*. Edward Arnold, London.

Tyndale-Biscoe, C. H., Hearn, J. P. and Renfree, M. B. (1974). Control of reproduction in Macropodid marsupials. *J. Endocrinol.*, **63**, 589 - 614.

Woolley, P. (1966). Reproduction in *Antechinus* sp. and other Dasyurid Marsupials. *Symp. Zool. Soc. London*, **15**, 281 - 94.

21 *Adrenocortical functions in marsupials*

I. R. McDonald*

It is now generally accepted that the functions of the mammalian adrenal cortex are intimately related to the metabolic adjustments and other mechanisms of adaptation to environmental stress. As pointed out by Christian (1963), 'it is important not to equate adaptation solely with adrenal function' nor 'to assume that the only function of the adrenals is to enable an organism to meet new and sudden demands'. With this proviso, it is still reasonable to propose that, in the light of the massive amount of information on the functions of the adrenal cortex in the eutherian mammals, comprehension of the role of the adrenocortical secretions in the adaptations of marsupials to their environment would be of value to the management and conservation of this fauna. On the Australian continent the relatively recent invasion of European man has led to severe pressures on habitat and adaptive capacity of marsupials, and evaluation of the effects of these pressures on their adaptive mechanisms is highly desirable.

Some investigations of marsupial physiology have been based on the premise that marsupials represent an early stage in the evolution of 'higher' mammalian forms, and this is implicit in T. H. Huxley's classification of the mammals into the subclasses proto-, meta- and eu-theria (see Tyndale-Biscoe, 1973). As this review will show, there are some peculiarities of adrenocortical function in metatherian mammals — the marsupials — which have not been described in eutherians; but there seems to be no really basic difference between the adrenocortical functions of these two mammalian subclasses. The limited information on prototherian mammals suggests that these too conform to the eutherian pattern (Weiss and McDonald, 1965; McDonald and Augee, 1968; Augee and McDonald, 1973). Differences in detail between the metatheria and the eutheria appear to be no greater than may be found between the several eutherian orders, and even greater differences may be found within or between the several families of the metatheria.

The marsupials on the Australian continent have evolved in isolation for possibly 100 million years (Clemens, 1968) yet exhibit much the same variety of adaptations as the eutherian mammals. It is in this context, of an isolated population of mammals which has evolved, apparently independently, anatomical, physiological and behavioural adaptations so strikingly similar to those of the eutherian mammals, that investigation of their adrenocortical functions is of general interest.

*Dr I. R. McDonald graduated in Medicine and Surgery at the University of Melbourne, and has worked on renal function and comparative endocrinology at the Maudsley Hospital, London, the Australian National University, and the Institute of Animal Physiology, Babraham. Currently a Reader in Physiology at Monash University, his main research interests are in adrenocortical functions of Australian monotremes and marsupials; he is also interested in neuroendocrinology and the physiology of the kidney, exocrine glands and capillaries.

Adrenocortical structure

The anatomy and histology of marsupial adrenal glands has been extensively
reviewed by Bourne (1949), and this has been supplemented briefly by Chester
Jones (1957). In all members of the families investigated, the adrenal glands are
paired, almost spherical or almond-shaped bodies closely associated with the renal
vein on the left side and either the renal vein – vena caval junction or the dorso-
lateral wall of the vena cava on the right side. The cortex is variable in thickness
and surrounds more or less completely a typically 'mammalian' adrenal medulla.

As in the eutherian mammals (Bennett and Kilham, 1940; Harrison, 1951;
Harrison and McDonald, 1966), the arterial blood supply is centripetally arranged
and derived mainly from the aorta via lumbar branches and from the renal arteries.
There is considerable variation in relative preponderance of these sources (personal
observations on specimens from *Macropus giganteus, Macropus rufus, Trichosurus
vulpecula, Vombatus hirsutus, Perameles nasuta, Dasyurus viverrinus, Sarcophilus
harrisii, Antechinus stuartii* and *A. swainsonii* – see also Weiss and McCance, 1974).
The left adrenal gland is usually close to the cranial pole of the kidney and its
venous blood drains via a short venous trunk into the left renal vein a few milli-
metres from its junction with the vena cava (see also Johnston, Davis and Hartroft, 1967). A vein draining the perirenal fat usually enters the cranial pole of the
gland, joining the adrenal vein in the substance of the gland. This gland is easy to
remove surgically and the vascular arrangement is suitable for sampling adrenal
venous blood via the renal vein. The right gland is usually so closely applied to the
wall of the vena cava that its venous blood discharges almost directly into the
lumen of the vena cava itself. As on the left side, a perirenal vein usually enters
the gland near the cranial pole and discharges into the adrenal venous effluent.
Surgical removal of this gland can be extremely difficult, especially when it is
covered by, or embedded in, a lobe of the liver. Accessory adrenal tissue has been
described, in the form of small nodules on the surface of one or other of the two
major glands, or even discrete structures complete with medullary tissue in association with one or other gland (Bourne, 1949).

The histology of the marsupial adrenal cortex conforms, in general, to the basic
eutherian pattern, although in some specimens it is difficult or impossible to define
the classical glomerulosa, fasciculata and reticularis zones. However, the variability
encountered appears to be no greater than that encountered among the different
orders and genera of the eutherian mammals (Bourne, 1949). Much of the published histological material lacks detail and, with a few exceptions, there is no
information about the state of the animal or the circumstances in which the
specimen was obtained. It would seem that a systematic survey of marsupial
adrenal histology, using fresh material obtained from animals in defined circumstances, would be useful. Christian (1963) has also advocated the need for a
detailed, well-illustrated and thorough discussion of the comparative morphology
of the adrenal glands.

The most striking variation from the 'classic' eutherian pattern in marsupial
adrenocortical structure is found in the Brush-tailed possum, *Trichosurus vulpecula.* This was first reported by Bourne (1934) and described more extensively
by Bourne (1949). The cortex of the female, but not the male, has an inner zone
of cellular proliferation which appears to flatten the medulla on one side. This is

extremely small in young virgin females, is more obvious in older, non-pregnant females, particularly during oestrus, and is most conspicuous during pregnancy and lactation. A recent illustration by Vinson *et al.* (1971) indicates that this zone may be quite separate from the rest of the cortex ('definitive' cortex); but more usually there is simply a change in cell type where the special zone becomes distinct from the remainder of the cortex. Bourne (1949) described two types of cell in this zone – a β-cell and a δ-cell. He believed that the β-cells, which are basophilic and resemble reticularis cells, but have little lipid, undergo continuous hyperplasia and metamorphose into the δ-cells, which are larger, stain poorly with an eosinophilic reaction and give the zone its characteristically pale appearance. He noted that, in the adult females, the total mass of the zone remained a constant fraction of the cortical mass during pregnancy and lactation, and that variations encountered in the appearance at different stages of the reproductive cycle were due to variations in the relative proportions of β- and δ-cells. He suggested that there was a constant cycle in which δ-cells degenerated and were replaced by β-cells which then metamorphosed into δ-cells. Another cell type, designated 'λ', with less cytoplasm and containing lipid, is scattered sparsely through the hypertrophied region, but no cyclical changes are described. Chester Jones (1957) compared this zone with the X-zone of small rodents (Howard-Miller, 1927) but conceded, as pointed out by Bourne (1949), that whereas pregnancy made this zone more conspicuous in *Trichosurus*, it abolishes the X-zone in the mouse. No structure analogous to this zone has been found in any other marsupial adrenal gland.

Very few histochemical observations on marsupial adrenal glands have been published. Bourne (1949) described an outer layer of sudanophilic cells in the adrenal cortex of all the dasyurid marsupials which he examined. This has been confirmed for *Antechinus stuartii* by Barnett (1973). Chester Jones (1957) described sudanophilic material in the cortical cells of *Macropus* sp., *Setonix brachyurus* and *Protemnodon (Macropus) eugenii.* Cholesterol was also demonstrated by the Schultz reaction in the cortices of these glands. There were no illustrations and details of distribution were not given. The cortical cells of *Trichosurus vulpecula* contain vesicles which take up Oil-Red 'O' and are distributed fairly uniformly through the 'definitive' cortex of males and females, but absent from the special 'δ'-zone. The vesicles are small in the outer and inner layers of the cortex, but quite large in the middle zone, corresponding to the reticularis of eutherian glands. The distribution of lipid corresponds with that of the enzyme 3β-ol dehydrogenase. In *A. stuartii* 3β-ol dehydrogenase is uniformly distributed through the cortex, whereas the lipid is concentrated in the outer layer (Bradley and McDonald unpublished). Bourne (1949) noted that, in all the marsupial adrenal glands he has tested, there is a high concentration of ascorbic acid in the cortex. Further details are not given.

It would clearly be of interest to have a more comprehensive and systematic array of histochemical observations on marsupial adrenal glands. However, it seems likely that they would confirm Bourne's general conclusion: 'We might expect that such an unusual group of animals would show some variations in the structure of the adrenal cortex from that of the eutherian mammals. But in fact we find surprisingly little difference.' The major exception to this would appear to be the adrenal cortex of the female *Trichosurus vulpecula.*

The adrenal weights of unstressed marsupials are generally similar, in proportion to body weight, to those of most eutherian mammals (Table 21.2). There is considerable variability, and those of the wombats (*Vombatus hirsutus*) and Koala (*Phascolaretos cinereus*) are lower than the others. In histological preparations the medullary tissue usually occupies about 20 per cent of the total cross-sectional area. Weiss and McDonald (1967) measured separately the weights of the cortex and medulla of both adrenal glands from a female Grey kangaroo (*Macropus giganteus*), and found the medulla to be approximately the same weight – 300 and 320 mg (16.5 and 17.6 mg/kg) – on both sides, the difference in weight between right (64.2 mg/kg) and left (76.6 mg/kg) glands being due almost entirely to a difference in weight of the cortical tissue. Vinson *et al.* (1971) recorded the weights of the outer cortex and special 'δ'-zone of female *Trichosurus vulpecula*, but not the total adrenal weight. Using the mean total weights of 108 mg/kg for lactating *Trichosurus* quoted by Weiss and McDonald (1966a), the mean total cortical mass of 85 mg/kg found by Vinson *et al.* (1971) indicates a mean medullary mass of 23 mg/kg – similar to that of the female kangaroo.

In general, sexual dimorphism in adrenal size is evident only in the Macropodidae and Phalangeridae, in which the adrenal/body weight ratio of females is approximately double that of males.

Factors affecting adrenal weight and histology

Few observations have been made on factors which may influence these parameters in marsupials. Reid and McDonald (1968) noted an increase in weight of the remaining adrenal gland of *Trichosurus vulpecula* within 10 days of unilateral adrenalectomy (Table 21.1). and found this to be associated with an increase in the width of the zona fasciculata. Scoggins *et al.* (1970) found that the zona glomerulosa of Eastern grey kangaroos (*M. giganteus*) and Eastern wombats (*V. hirsutus*), shot in alpine regions of Australia, was wider than in animals shot in coastal regions. They related this to evidence for marked renal sodium conservation in kangaroos (but not the wombats) and proposed that hypertrophy of the

Table 21.1 *Trichosurus vulpecula.* Effect of unilateral right adrenalectomy on the weight of the remaining left adrenal gland removed more than 10 days later. Means and S.E.M.s calculated from the raw data of Weiss and McDonald (1966a) and Reid and McDonald (1968). Paired 't' test for estimate of significance

	Right adrenal		Left adrenal	
	Control adrenalectomy (mg/kg ± S.E.M.)		Control adrenalectomy (mg/kg ± S.E.M.)	
Males	27.85 ± 2.24	30.0 ± 2.04	32.40 ± 2.70	45.0 ± 4.8
	(n.s.)		($0.05 > p > 0.01$)	
Females	49.36 ± 5.66	47.50 ± 9.92	77.74 ± 5.18	108.9 ± 3.5
	(n.s.)		($0.05 > p > 0.01$)	

glomerulosa could be an adaptive response to sodium deficiency in the alpine environment. However a similar increase in width of the zona glomerulosa of the alpine wombats was not associated with evidence for renal sodium conservation. Barnett (1973) noted an increase in adrenal weight of male *Antechinus stuartii* following the mating period and just prior to the total male mortality peculiar to some members of this genus (Woolley, 1966, and this volume). This was associated with an increase in size of both the medulla and cortex, and concentration of sud-anophilic material in the glomerulosa, although the increase in size of the cortex was due to an increase in fasciculata and reticularis, which were relatively depleted of lipid.

The total mass of adrenal tissue, expressed as a function of body weight, is frequently used as an indicator of adrenocortical activity. This is not necessarily accurate; Vogt (1955, 1957) found that, in rats and guinea-pigs, oestrogen administration increased adrenal weight, due to hypertrophy of the adrenal cortex, but decreased corticosteroid secretion. The increase in size was shown to be pituitary-dependent, indicating that it was a consequence of increased corticotrophin secretion in response to depressed plasma corticosteroid concentrations. A similar observation has been made on the Brush-tailed possum, *Trichosurus vulpecula*, in which oestrogen administration, sufficient to cause marked hypertrophy of the genital tract, was associated with depressed plasma corticosteroid concentrations and increase in adrenal weight (Khin Aye Than and McDonald, 1975).

Although changes in weight of the adrenal glands are thought to be due mainly to changes in the mass of adrenocortical tissue, hypertrophy of adrenomedullary tissue in stressed rodents and wild ungulates has been reported (see Christian, 1963). In this context, Barnett (1973) has noted increase in the maximum cross-sectional area of the adrenal medulla, associated with increased cortical cross-sectional area and total gland weight of stressed *Antechinus stuartii*.

Secretion of adrenocortical hormones

The first observations on the nature of the adrenocortical secretory products in marsupials were made by Chester Jones *et al.* (1964). They obtained adrenal venous blood from female Brush-tailed opossums, *Trichosurus vulpecula*, anaesthetised with sodium pentobarbitone, by cannulation of the renal vein between ligatures on either side of the adrenal vein (Bush and Ferguson, 1953). They provisionally identified cortisol (pregn-4-ene, 11β, 17α, 21-triol, 3, 20-dione) at a mean concentration of 144 μg/100 ml and testosterone (pregn-4-ene, 17-ol, 3 -one) at a mean concentration of 223 μg/100 ml in the adrenal venous plasma. The testosterone appeared to be conjugated in an unusual way to an unidentified phenolic substance. They stated that the amount of conjugated testosterone secreted varied concomitantly with changes in the special 'δ' zone, but presented no evidence to support this.

Subsequently, Weiss and McDonald (1966a, b, 1967), Weiss and Richards (1970, 1971) made a systematic investigation of the adrenocortical secretory products in representatives of most of the families of Australian marsupial. Their observations were confined to the Δ^4, 3-oxo steroids. They confirmed the identi-fication of cortisol by Chester Jones *et al.* (1964) in the Brush-tailed opossum.

Although their sample included two females with definite special 'δ'-zones, they did not note the presence of testosterone in their extracts. They also found that cortisol was the major Δ^4, 3-oxo steroid present in the adrenal venous blood plasma of the macropods *Macropus giganteus* and *Megaleia rufa* (= *Macropus rufus*), the wombat (*Vombatus hirsutus*), the bandicoot (*Perameles nasuta*), the Koala (*Phascolarctos cinereus*), the Native cat (*Dasyurus viverrinus*) and the Tasmanian devil (*Sarcophilus harrisii*). Johnston *et al.* (1967) also found that cortisol was the major Δ^4, 3-oxo steroid in the adrenal venous blood of the North American opossum, *Didelphis virginiana*; and Illet (1969) identified cortisol as one of the major Δ^4, 3-oxo steroids in the adrenal venous blood of Quokka, *Setonix brachyurus*.

Weiss and McDonald (1964, *et seq.*), attempted to standardise their experiments by collecting adrenal venous blood during ether anaesthesia and constant-rate i.v. infusion of ovine corticotrophin, to obtain an estimate of the maximum stressed secretion rate. This was used as a basis for comparison, both within the order *Marsupialia* and with other mammalian orders. This maximum secretion rate was expressed as a function of adrenal weight, to indicate the secretory capacity of the adrenal glands, and as a function of body weight, to indicate the total adrenal functional capacity in relation to the metabolic requirements of the animal. They found considerable variation, both between and within the several marsupial families (Table 21.2). In general, the maximum cortisol secretion rates under these conditions were within the lower range of, or below, those of the eutherian mammals for which comparable data were available. In these experiments corticotrophin infusion had at the most only minor stimulatory effects on cortisol secretion. In addition to cortisol, corticosterone (pregn-4-ene, 11β, 21-diol, 3, 20-dione) was identified in the adrenal venous plasma of all the marsupials investigated by Weiss and McDonald (1966a, b, 1967) and Weiss and Richards (1970, 1971). The ratio of cortisol to corticosterone varied greatly, from an average of 2 in the Tasmanian devils to more than 100 in the kangaroos. Aldosterone was also identified in the samples from wombats, Red and Grey kangaroos, Native cats and Tasmanian devils. Corticosterone and aldosterone have also been identified in the adrenal venous blood of the North American opossum (*Didelphis virginiana*) the ratio of cortisol to corticosterone being approximately 6 (Johnston *et al.*, 1967). Illet (1969) found corticosterone in the adrenal venous blood of the Quokka (*Setonix brachyurus*) in approximately the same concentration as cortisol.

Varying, and usually minor, amounts of a number of other Δ^4, 3-oxo steroids have been found in the adrenal venous blood of the Australian marsupials. Many of these are known intermediates in the biosynthesis of cortisol and corticosterone. These include Reichstein's compound 'S' (pregn-4-ene, 17α, 21-diol, 3, 20-dione), 17α-hydroxy progesterone (pregn-4-ene, 17α-ol, 3, 20-dione) 11β-hydroxy progesterone (pregn-4-ene, 11β-ol, 3, 20-dione), progesterone (pregn-4-ene, 3, 20-dione) and 21-deoxy cortisol (pregn-4-ene, 11β, 17α-diol, 3, 20-dione), which have all been identified in the several families of Australian marsupials (Weiss and McDonald, 1966b, 1967; Weiss, 1968; Weiss and Richards, 1970, 1971). The secretion of 21-deoxy cortisol by marsupial adrenal glands *in vivo* is unusual, and is thought to occur in eutherian mammals only when there is a defect in the 21-hydroxylase system, as occurs in the adreno-genital syndrome of man (Wieland *et al.*, 1965; Bongiovani *et al.*, 1967). The rate of secretion is usually of the same order as, or less than, that of corticosterone, and considerably

Table 21.2 Major parameters of adrenocortical secretion in eutherian and metatherian mammals. F = cortisol; B = corticosterone. The symbol in parentheses indicates that this steroid is secreted at almost the same rate as the major steroid. The values for each parameter are rounded-off means from those quoted in the various publications cited in the text

	Adrenal weight (mg/kg)	Major corticosteroid	Stressed secretion rate (µg/kg per h)	Stressed plasma concentration (µg/100 ml)
EUTHERIA				
Primates				
Homo sapiens	90–100	F	180	56–60
Carnivora				
Canis familiaris	100	F(B)	75	10–18
Artiodactyla				
Ovis aries	65–110	F	110–320	6–12
Rodentia				
Rattus sp.	100–200	B	115	30
Lagomorpha				
Oryctolagus cuniculus	100–200	B(F)	35 F	50–100
METATHERIA				
Macropodidae				
Macropus giganteus ♂	40	F	22	2–6
♀	107	F	46	
Megaleia rufa ♂	57	–	–	–
(= *Macropus rufus*) ♀	108	F	20	3–4
Phalangeridae				
Trichosurus vulpecula ♂	60	F	12	6
♀	108	F	22	7
Vombatidae				
Vombatus hirsutus ♂	40	F	8	1–2
♀	51	F	18	3
Phascolarctidae				
Phascolarctos cinereus ♂	49	F	13	6
♀	41	F	–	–
Peramelidae				
Perameles nasuta ♂	115	F	50	7
♀	110	–	–	–
Dasyuridae				
Sarcophilus harrisii ♂	171	F	111	3
♀	163	F	190	8
Dasyurus viverrinus ♂	140	F	42	5
♀	–	F	41	7
Antechinus stuartii ♂	168	F(B)*	40	7
Didelphidae				
Didelphis virginiana ♀	300	F	12	–

*Bradley (unpublished observation).

greater than that of Reichstein's compound 'S', which is considered to be the usual immediate precursor of cortisol in eutherian mammals (Hechter and Pincus, 1954; Siiteri, 1971). Another unusual steroid, which has been identified in the adrenal venous plasma of *D. viverrinus* and *S. harrisii*, is 17α, 20 hydroxy-progesterone (pregn-4-ene, 17, 20-diol, 3-one) (Weiss and Richards, 1971). The presence of these latter two steroids suggests a lesser activity of the 21-hydroxylase system in these marsupials than in eutherian mammals, and the large number of 17-hydroxy, 11-deoxy compounds suggest that the 17-hydroxylase system may be relatively more active than the 11-hydroxylase system in the adrenal cortex of marsupials. It would clearly be of interest to have more comparable information on the minor constituents of the adrenal venous blood of the eutherian mammals, collected under similar conditions. The presence of these biosynthetic intermediates in the adrenal venous blood during the stress of ether anaesthesia and corticotrophin infusion does not necessarily indicate that they are normally secreted and exert biological activity, although this has to be considered as a possibility. It is more likely to be a consequence of stimulation of adrenocortical biosynthesis beyond the capacity of the final rate-limiting enzyme system (Weiss and McDonald, 1967).

Finkelstein *et al.* (1967) made use of the observation of 21-deoxy cortisol production by marsupial adrenals to support their hypothesis that 11β-hydroxylation of a 21-deoxy substrate is due to the presence of a 'primitive' or 'immature' enzyme system. This reasoning is an example of the approach of some investigators to marsupial biology, referred to at the beginning of this review.

Corticosteroid biosynthesis

A few investigations have been made on the enzyme systems involved in the adrenocortical synthesis of steroids, using both *in vivo* and *in vitro* methods.

Brownell, Beck and Besch (1967) found that minced adrenal glands from the North American opossum, *Didelphis virginiana*, when incubated with radioactive progesterone substrate, produced radioactive cortisol, corticosterone, aldosterone (pregn-4-ene, 11, 21-diol, 18-ol, 3, 20-dione), 11-deoxy corticosterone, 18-hydroxy, 11-deoxy corticosterone (pregn-4-ene, 18, 21 diol, 3, 20 dione) and 18-hydroxy corticosterone (pregn-4-ene, 11β, 18, 21-triol, 3, 20-dione). They noted slightly higher percentage conversion of progesterone to corticosterone than to cortisol, and unexpectedly high percentage conversion to 11-deoxy corticosterone and aldosterone (3 per cent and 1 per cent, respectively). They suggested that *Didelphis* adrenocortical tissue has a low activity of the 11-β hydroxylase system, which could be the rate-limiting step in corticosteroid production *in vivo*. Although the observations of Johnston *et al.* (1967) do not support this, their methods were such as to permit detection only of those substances they chose to assay. It is conceivable that, as in the Australian marsupials, there may be a high proportion of 11-deoxy steroids in the adrenal venous blood of *Didelphis* during stress and corticotrophin stimulation, which would support their hypothesis. Vinson *et al.* (1971) investigated the steroidal biosynthetic mechanisms in slices of *Trichosurus* adrenal glands with well-defined special 'δ'-zones. They found that both 17-hydroxy and 17-deoxy corticosteroids were produced by incubation with either radioactive pregnenolone (5α-pregnane,

3β-ol, 20-one) or progesterone substrates, and that these conversions occurred in both the 'definitive cortex' and the special 'δ'-zone of the female. They confirmed the earlier, tentative finding of testosterone production by *Trichosurus* adrenal glands which was produced in highest yield by 'δ'-zone tissue, and also identified androstenedione (pregn-4-ene, 3, 17-dione) as a significant *in vitro* product. In this context, Weiss (1968) had assayed the minor Δ^4, 3-oxo steroids of two 'normal' *Trichosurus*, of unstated sex, without noting the presence of either testosterone or androstenedione, but did identify 11β-hydroxy androstenedione (pregn-4-ene, 11β-ol, 3, 17-dione). Vinson *et al.* (1971) concluded that the adrenal cortex of the female *Trichosurus* has acquired an additional physiological role, having 'a remarkable capacity for the formation of the C-19 steroids androstenedione and testosterone, which is unparalleled by adrenal tissue from other species under the same or similar conditions'. They correlate these observations with the finding of unusually large amounts of 'testosterone-like material' in the peripheral blood of female *Trichosurus*. However, they could not find any significant variation in the plasma concentration of this material in relation to the stage of the oestrus cycle and, presumably, the size or state of the special 'δ'-zone. In this context it is curious that the zone contains little lipid and little evidence of 3β-ol dehydrogenase, which are abundant in the outer layers of the cortex. Clearly, the functions of this zone deserve further investigation.

Weiss and McCance (1974) investigated the conversion of 21-deoxy substrates to 17α, 21-hydroxy steroids in the adrenals of *Trichosurus*, both *in vivo* and *in vitro*, and compared their findings with the results of similar experiments of guinea-pig and rabbit adrenals. The sex of the donor animals was not stated. They detected ^{3}H-labelled cortisone in the adrenal venous blood of *Trichosurus* during constant-rate infusion of 21-deoxy cortisone into the arterial supply of the gland, but the conditions of the experiment were not amenable to quantitative interpretation. They also found that *Trichosurus* adrenal gland incubates produced higher yields of labelled cortisol and cortisone from labelled 21-deoxy cortisone than from 11-deoxy cortisol. Much the same results were obtained with rabbit and guinea-pig adrenals incubated under the same conditions. These observations do not suggest that steroidal biosynthesis in *Trichosurus* adrenal glands is in any way uniquely different from that of the eutheria, and the authors suggested instead that 21-deoxycortisol may be a more significant precursor of cortisol in mammalian adrenal tissue than has been formerly realised.

Renfree and Vinson (1974) investigated steroidogenesis by adrenal tissue of the North American opossum, *Didelphis virginiana*. They incubated adrenal tissue from adults, prepared in an unspecified way, with ^{3}H-labelled pregnenolone and obtained low yields of cortisol (1.1 and 1.2 per cent) and testosterone (0.68 and 0.70 per cent). They also incubated adrenal tissue from pouch-young, juveniles and adults in an unspecified way, and measured the absolute rates of production of total corticosteroids (competitive protein binding assay) and testosterone (radioimmunoassay). It is difficult to evaluate the data, because of the lack of experimental detail, but they conclude that testosterone production occurs in *Didelphis* adrenal glands as in *Trichosurus*, and is stimulated by luteinising hormone or follicle stimulating hormone (of unstated origin) but not by synthetic corticotrophin. Luteinising hormone also stimulated corticosteroid production by these incubates. Their yield of cortisol from labelled precursor was so much less than that found by Brownell *et al.* (1967) that the question of methodology must be considered.

Weiss (1974) found that homogenates of *Trichosurus* adrenal glands of unstated sex, incubated with [4,^{14}C]progesterone yielded a large variety of A-ring reduced metabolites and little, if any, of the expected corticosteroids and their normal precursors. Similar incubations, this time with adrenal slices, yielded cortisol, corticosterone, androstenedione and 17α-hydroxy progesterone, the last being in the highest yield. Testosterone was not identified, nor was 11β-hydroxyandrostenedione, and no mention was made of the other quantitatively minor steroids found in adrenal venous blood (Weiss, 1968). A time – yield study indicated that the reductions by the homogenates 'followed the eutherian pattern'. Similarly high ring-A-reductase activity has been found in guinea-pig adrenals (Brown-Grant, Forschelli and Dorfman, 1960). Weiss (1974) also cited unpublished observations that ring-A reduction did not occur in homogenates of kangaroo and wombat adrenal glands. This report illustrates the role of methodology in determining the nature and the yields of biosynthetic products from radioactive substrates in *in vitro* experiments.

At this stage one can only adopt the conservative attitude that *in vitro* studies are useful for demonstrating the presence of enzyme systems in adrenal homogenates or slices, but that caution should be exercised in extrapolating the findings to explain synthesis *in vivo*. The lack of uniformity in both preparative and analytical procedures also makes it difficult to compare the findings made by the different investigators. The evidence so far reported indicates that the marsupial adrenal cortex contains all the enzyme systems necessary for the production of corticosteroids from the same precursors as those believed to be involved in corticosteroid production by eutherian mammals. Confirmation of hypotheses relating to relative activities of enzyme systems, sequence of biosynthetic steps and possible evolutionary significance must await more rigorous experimentation.

Corticosteroids in peripheral blood

Cortisol is the major C-21Δ^4, 3-oxo steroid present in the peripheral blood plasma (or whole blood) of all the marsupials so far examined (Weiss and McDonald, 1966a, b, 1967; Coghlan and Scoggins, 1967; Blair-West *et al.*, 1968; Scoggins *et al.*, 1970; Weiss and Richards, 1970, 1971; Illet, 1969; Barnett, 1973; Khin Aye Than and McDonald, 1973). In addition, a number of other steroids of adrenal origin have been detected. The most important of these is aldosterone, which has been identified and measured in the peripheral blood of kangaroos (*M. giganteus*) and wombats (*V. hirsutus*), at concentrations comparable with those found in eutherian mammals, by Coghlan and Scoggins (1967), Blair-West *et al.* (1968) and Scoggins *et al.* (1970). The latter authors reported that they had detected aldosterone in the peripheral blood of the Tiger cat (*Dasyurus maculatus*), the Pademelon wallaby (*Thylogale billardierii*) and the Red-necked wallaby (*Macropus rufogriseus*) but gave no details. Weiss and McDonald detected a Δ^4, 3-oxosteroid with the chromatographic mobility of either corticosterone or 11-deoxycortisol in the peripheral blood plasma of the kangaroos *M. giganteus* and *M. rufus* and estimated it to be approximately one-tenth the concentration of cortisol. Weiss and Richards detected a similar steroid in the peripheral blood plasma of Koalas

(*P. cinereus*), Native cats (*D. viverrinus*) and Tasmanian devils (*S. harrisii*), and tentatively identified it as corticosterone. Coghlan and Scoggins (1967) identified corticosterone in the peripheral blood of *M. giganteus* and *V. hirsutus*. Illet (1969) identified corticosterone in the peripheral blood plasma of the Quokka (*Setonix brachyurus*). The ratio of cortisol: corticosterone ranged from approximately 2:1 in Quokka and wombats to 10:1 in kangaroos, Native cats and Tasmanian devils. Vinson *et al*. found testosterone in the peripheral blood plasma of female Brush-tailed possums (*T. vulpecula*) and assumed this to be at least partly of adrenal origin. With the present rapidly increasing sensitivity and refinement of analytical techniques, it seems likely that a larger number of quantitatively minor steroids in the peripheral blood of marsupials could be found, and the question of their origin and biological significance would then have to be determined.

Corticosteroid concentrations in peripheral blood

Glucocorticoids

Estimates of peripheral blood and plasma concentrations of cortisol and cortisol-like corticosteroids have been made on a variety of marsupials in differing circumstances. Those made by Weiss and McDonald (1966a, b, 1967), Weiss and Richards (1970, 1971) and Illet (1969) were on large samples of blood obtained at the end of extremely stressful procedures, using analytical techniques with low precision and sensitivity and the findings should be considered to be only approximations. Considering the circumstances of sample collection, the values were low when compared with those of stressed eutherian mammals (Tables 21.2 and 21.3). Coghlan and Scoggins (1967) found even lower cortisol concentrations in the peripheral whole blood of wombats and Grey kangaroos shot in presumably low-stress environments, and their findings have been supplemented by Scoggins *et al*. (1970). The average peripheral blood cortisol concentration of Grey kangaroos shot in coastal regions was higher than in those shot in alpine regions, and was also higher than that of foxes and sheep shot in the same regions. The range of values found in the coastal kangaroos overlapped that found in the highly stressed animals reported by Weiss and McDonald (1967) – 1.55 – 4.4 μg/100 ml and 2.0 – 5.6 μg/100 ml, respectively. In contrast, the blood cortisol concentration of the wombats shot in either region were approximately one-tenth those of the stressed values reported by Weiss and McDonald (1966b). This suggests that the adrenal cortices of wombats have a greater capacity to respond to stress than those of kangaroos. However, as described below, the relationship between plasma cortisol concentration and adrenal cortisol secretion rate in marsupials may vary according to circumstances and it is necessary to be cautious in the interpretation of plasma corticosteroid concentrations.

The advent of the radioligand assay for corticosteroids (Murphy, Engelberg and Pattee, 1963) has enabled reliable estimates of plasma corticosteroid concentrations to be made serially on very small blood samples and also permitted observations to be made on the smaller marsupials. Barnett (1973) measured plasma corticosteroid concentrations (approximately 70 per cent cortisol) at various stages in the annual life cycle of the small dasyurid *Antechinus stuartii*. He found significant variations in both males and females between the ages of 20 and 50

Table 21.3
Peripheral plasma cortisol concentrations (μg/100 ml) of marsupials in various conditions.
'Moderate stress' includes restraint for 5 – 10 min, and light ether anaesthesia.
'Severe stress' indicates laparatomy during surgical anaesthesia. The 'ACTH' values are
the maximum values recorded following ACTH administration

	Unstim.	Moderate stress	Severe stress	ACTH	Other	Source
Trichosurus vulpecula			1.0–2.5			Weiss and McDonald (1966a)
	♂ 0.92 ± 0.48 ♀ 1.01 ± 0.57	3.59±1.04		4.22±1.51	3.4 ± 0.9 (insulin)	Khin Aye Than and McDonald (1973, 1974b)
					6.95 ± 2.12 (epinephrine)	Khin Aye Than (1975)
		♂ 2.7±1.2 ♀ 2.4±0.2		6.0±0.5 6.6±0.7	0.03 ± 0.01 (dexameth azone)	Vinson *et al.* (1973)
Vombatus hirsutus			♂ 1.25			Weiss and McDonald (1966b)
			♀ 3.1			
	C 0.14 ± 0.04 A 0.23 ± 0.23					Scoggins *et al.* (1970)
Phascolartus cinereus			6.3			Weiss and Richards (1970)
Perameles nasuta			7.4			Weiss and McDonald (unpublished)
Macropus giganteus			2.0–5.6			Weiss and McDonald (1967)
	C 2.82 ± 1.2 A 1.39 ± 0.7					Scoggins *et al.* (1970)
Macropus rufus	3.4				4.6 (insulin)	McDonald (unpublished)

Table 21.3 (*continued*)

	Unstim.	Moderate stress	Severe stress	ACTH	Other	Source
Sarcophilus		♂ 2.8 – 3.7				Weiss and
harrisii		♀ 7.5				Richards
Dasyurus		♂ 3.2 – 5.6				(1970)
viverrinus		♀ 7.0				
Antechinus	♂ 0.83	3.48*				Barnett (1973)
stuartii	♀ 3.0	1.40				Bradley
						(unpublished)
	♂ 2.9 ± 1.5	5.2±2.4		7.14±1.33	♂ 6.57 ± 2.27 ♀ 6.32 ± 1.32 (ether)	Bradley *et al.* (1975)

*Post-mating period, just prior to total mortality. C = coastal habitat. A = alpine habitat.

weeks. The mean plasma concentration in males up to the age of 44 weeks (just before mating) was 0.83 $\mu g/100$ ml. This rose to 1.36 $\mu g/100$ ml during the mating period between 44 and 48 weeks and rose further to 3.48 $\mu g/100$ ml between 48 and 50 weeks, just before the total male mortality. In contrast, the plasma corticosteroid concentration of females was consistently higher than that of the males (mean of 3 $\mu g/100$ ml) and fell to a mean of 1 $\mu g/100$ ml after mating. Bradley, McDonald and Lee (1975) confirmed this increase in plasma corticosteroid concentration of the male *A. stuartii* but found higher overall values of 2.9 ± 0.51 SEM $\mu g/100$ ml before mating and 5.2 ± 0.6 $\mu g/100$ ml just prior to death.

Vinson, Tyndale-Biscoe and Bancroft (1973) and Vinson (1974) reported data, apparently from the same experiments, on plasma corticosteroid concentrations in intact Brush-tailed opossums (*Trichosurus vulpecula*). Blood samples were taken either from an ear vein or by cardiac puncture under anaesthesis. The unstimulated cortisol concentrations were 2.7 ± 1.2 (S.D.) $\mu g/100$ ml in males and 2.4 ± 0.2 (S.D.) in females. These were increased to 6.0 ± 0.5 and 6.6 ± 0.7, respectively, 2 h after i.m. injection of 10 I.U. synthetic corticotrophin (approx. 5 I.U./kg). Treatment with 0.2 mg/kg dexamethasone i.m. for 2 days reduced plasma cortisol concentrations to 0.3 - 0.4 $\mu g/100$ ml.

Khin Aye Than and McDonald (1973) implanted catheters into the jugular veins of Brush-tailed opossums to permit blood sampling in the conscious state without restraint. They demonstrated that cortisol contributed more than 90 per cent of the measured corticosteroids and found unstimulated total corticosteroid values of 0.92 ± 0.48 (S.D.) and 1.01 ± 0.57 $\mu g/100$ ml in males and females, respectively. There was a minor but significant diurnal variation, the lower values occurring between 0800 hours and 1100 hours, when the animals were usually asleep, and the higher values between 2000 hours and 2300 hours, when they were awake and active. Serial (15 min) blood sampling revealed no evidence for short-term episodic fluctuations, as found in some eutherians. Ether anaesthesia caused a rapid increase to 3.6 ± 0.8 μg/100 ml within 10 min and the cortisol concentration then declined progressively to reach the control values 6 h later. Constant-rate i.v. infusion of synthetic corticotrophin also caused a rapid increase in plasma cortisol concentration to or above the levels during ether anaesthesia when the injection rate was between 0.12 and 0.4 1.U./kg per hour. Increasing the rate above this did not cause any significant further increase in plasma cortisol concentration. The dose–response curve for synthetic corticotrophin indicated a potency approximately one-twentieth of that found in man (Figure 21.1). As in man, synthetic and porcine corticotrophins were equipotent. Bradshaw and McDonald (unpublished) conducted similar experiments on the Quokka (*Setonix brachyurus*) and found that this macropid marsupial was even less sensitive to synthetic corticotrophin, (Figure 21.1), although, again, synthetic and porcine corticotrophins were equipotent. Bradley (1975) has found that the small dasyurid marsupial *Antechinus stuartii* is surprisingly insensitive to the action of either synthetic or porcine corticotrophins—intramuscular doses of the order of 50 I.U./kg are necessary to raise the plasma total corticosteroid concentration to the levels found in males at the end of the mating period.

Khin Aye Than and McDonald (1974b) found that intravenous insulin injection increased plasma cortisol concentration in *Trichosurus vulpecula*, the degree of increase being proportional to the hypoglycaemia, below 50 per cent of the initial glucose concentration, as in man (Landon, Wynn and James, 1963)

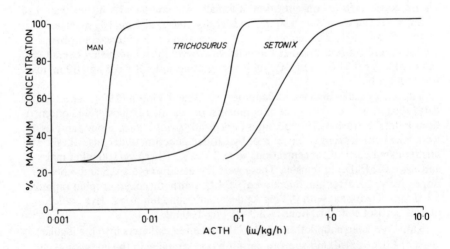

Figure 21.1 Dose–response curves for synthetic (24 amino acid) corticotrophin injection i.v. at constant rates. The values on the ordinate represent the concentrations of plasma cortisol at the end of the 1 h infusion period, expressed as a percentage of the maximum stimulated value (= 100 per cent). (Data from Khin Aye Than and McDonald, 1973; Landon *et al.*, 1963; Bradshaw and McDonald, unpublished.)

(Figure 21.1). Khin Aye Than (1975) also found that intravenous infusion of adrenaline caused a dose-related increase in plasma cortisol concentration, due to an increase in production rate, with little change in metabolic clearance (Figure 21.2). Injection of oestradiol, in doses sufficient to cause hypertrophy of the genital tract, was found to depress plasma cortisol concentration of female *Trichosurus vulpecula* (Khin Aye Than and McDonald, 1975).

In summary, although detailed observations have been made only on *Trichosurus vulpecula*, the evidence so far accumulated indicates that the peripheral blood and plasma concentrations of the major adrenocorticosteroid in marsupials – cortisol – are in the lower part of the range quoted for the various eutherian mammals for which comparable data are available (see Khin Aye Than and McDonald, 1973). Marsupials appear to be considerably less sensitive to the corticosteroidogenic actions of corticotrophins, but plasma cortisol concentrations are affected by stress and adequate corticotrophin stimulation in much the same way as in the eutheria. The marked fall in plasma cortisol concentration of *Trichosurus* following treatment with dexamethasone, and the hypertrophy of the remaining adrenal of this marsupial following unilateral adrenalectomy, suggest that the secretion of adrenocorticotrophic hormone is subject to regulation by the peripheral blood plasma corticosteroid concentration.

Mineralocorticoids

Coghlan and Scoggins (1967) first reported measurements of peripheral blood aldosterone concentration in kangaroos (*M. giganteus*) and Eastern wombats (*V. hirsutus*), their observations being supplemented by Blair-West *et al.* (1968) and Scoggins *et al.* (1970). The blood aldosterone concentration of *M. giganteus* shot in alpine regions, approximately 300 km from the coast, was 27.7 ± 11.5 (S.D.) $\mu g/100$ ml, and 6.6 ± 6.5 $\mu g/100$ ml in those shot near the coast. That of wombats shot either in alpine or coastal districts was 9.0 ± 6.4 $\mu g/100$ ml (the above values have been calculated from the graphically expressed data of Scoggins *et al.*, 1970). Scoggins *et al.* related the high blood aldosterone concentration of the alpine kangaroos to a low salt content of the food consumed and the evidence for renal Na conservation in the low Na concentration of the bladder urine. There was no evidence for renal Na conservation in the wombats: however, as noted previously, they reported an increased in the width of the zona glomerulosa of the adrenals of the alpine wombats, compared with the coastal animals. The high blood aldosterone concentrations of the alpine kangaroos also correlated with high blood renin-like activity, when compared with the coastal animals. Johnston *et al.* (1967) noted increased granulation in the renal juxta-glomerular apparatus of *Didelphis* following Na depletion; and stimulation of aldosterone secretion by a renal extract. Reid and McDonald (1969) found that plasma renin concentration of *Trichosurus vulpecula* was increased in association with Na depletion induced by the natriuretic drug furosimide.

This evidence suggests that peripheral blood aldosterone concentration in marsupials is also subject to the same kind of control as that found in the eutherian mammals. However, there is clearly need for more definitive observations under laboratory conditions to confirm this suggestion.

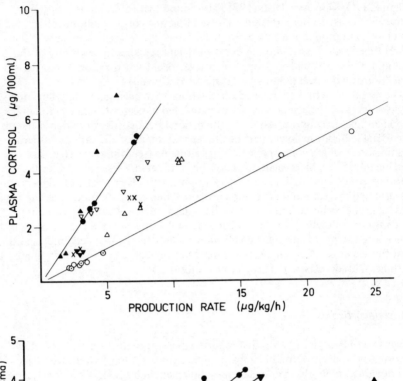

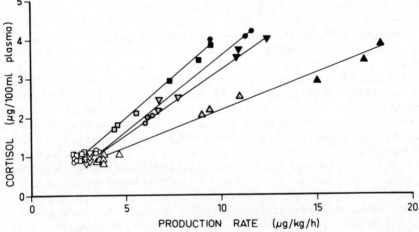

Figure 21.2 Relationship between plasma cortisol concentration and production rate, measured by isotope dilution at equilibrium in conscious *Trichosurus vulpecula*. (a) Control (saline) and ACTH (0.4 i.u./h) infusion. The different symbols represent experiments on individual animals. The line through the symbols (•) shows the relationship at the first experiment on one animal, and the line through the symbols (o,⊙) shows the relationship at two subsequent experiments on the same animal. (From Khin Aye Than and McDonald, 1973, with permission of *J. Endocrinol*). (b) Control saline infusion (open symbols) and adrenaline infusion at 7.2 μg/h (dotted symbols) and 36.0 μg/h (solid symbols). (From Khin Aye Than, 1975, with permission of the author.)

Metabolism and excretion of corticosteroids

Little work has been carried out on this topic. McDonald and Weiss (1967) found that, in *Trichosurus vulpecula*, $(4,^{14}C)$-cortisol disappeared rapidly from the peripheral blood plasma following single-dose intravenous injection with a half-time of only 3 min during the first 10 min, and 10.5 min during the next 30 min. Thereafter, the rate of disappearance of isotopic cortisol was very variable, and significant amounts of radioactivity began to appear in the glucuronide and sulphate conjugated fractions of plasma extracts. The major route of excretion was via the gastrointestinal tract, 30 – 40 per cent of the injected radioactivity being recovered in the bile in both free and conjugated forms in the first 40 min after injection. Only 10 per cent of the injected isotope was recovered in the urine excreted during the first 5 – 8 days after injections, the major amount of this being in a form more polar than tetrahydrocortisol. Analysis of urine samples of *T. vulpecula* revealed the presence of small amounts of free cortisol, cortisone and their tetrahydro derivatives (Weiss and McDonald, 1966a). Conjugated tetrahydrocortisol and tetrahydrocortisone have been tentatively identified in the blood plasma of a female wombat (Weiss and McDonald, 1966b).

Khin Aye Than and McDonald (1973) measured the steady-state metabolic clearance of $[1, 2,^3H]$ cortisol of conscious undisturbed *Trichosurus vulpecula* during constant rate intravenous injection of this isotopic steroid. This was 3.14 ± 0.88 ml/min per kg in males and 4.96 ± 2.66 in females. These values were not significantly altered by intravenous infusion of synthetic corticotrophin at a rate sufficient to treble the cortisol production rate, and are intermediate to those reported for sheep (Paterson and Harrison, 1967) and man (Tait and Burstein, 1964). The rather high variance in metabolic clearance of the opossums, particularly the females, appeared to be related to the experimental conditions. In those animals which had not been handled repeatedly before the experiment, metabolic clearance was consistently lower than in those which were accustomed to handling. As a consequence, although there was a linear relationship between plasma cortisol concentration and cortisol production rate, the slope of this line varied, even in the same individual, from experiment to experiment (Figure 21.2). This suggests that metabolic clearance could be a significant factor determining plasma cortisol concentration in marsupials and that caution should be exercised in the interpretation of variations in plasma cortisol concentration. Khin Aye Than (1975) subsequently found that intravenous injection of adrenaline, at rates sufficient to cause marked increase in cortisol production rate and plasma concentration, had no effect on metabolic clearance (Figure 21.2). Further work is required to determine the factors which may affect metabolic clearance of corticosteroids in marsupials, as this would seem to be essential to the proper evaluation of plasma corticosteroid concentrations in field investigations.

Binding of corticosteroids to plasma proteins

Seal and Doe (1966), using a gel filtration technique, found evidence for a high-affinity corticosteroid binding system in the blood plasma of the North American opossum, *Didelphis virginiana*, with a maximum binding capacity of 8 – 10 μg/100 ml for cortisol and 1 μg/100 ml for corticosterone. Khin Aye Than and McDonald (1975) made a more detailed investigation of corticosteroid binding by the plasma proteins of *Trichosurus vulpecula*. They found that the binding affinity of purified *Trichosurus* albumin for cortisol was much the same as that of sheep albumin, and somewhat less than that of human albumin. The binding of cortisol to whole plasma of *Trichosurus* was much greater than could be accounted for by the presence of albumin alone, and they determined a non-albumin binding affinity constant of approximately 4×10^7 l/mol, which is similar to that quoted for the transcortins of most eutherian mammals, except those for rat and guinea-pig, which are much lower. The concentration of binding sites for this system was only 1.8×10^{-7} mol/l, and of the eutherian data available, only sheep have a lower value (Table 21.4). As in man, approximately 80 per cent of the plasma cortisol of *Trichosurus* is bound to this transcortin-like system, 10 per cent is weakly bound to albumin and 10 per cent is free (Figure 21.3). Using the gel filtration technique of Doe, Fernandez and Seal (1964), the maximum binding capacity for cortisol agreed well with that calculated from the equilibrium dialysis technique used to determine the affinity constants, and that for corticosterone was approximately half the cortisol value. Oestrogen administration had no effect on either the binding affinity or concentration of binding sites of the transcortin-like system of *Trichosurus*.

Table 21.4 Concentration of 'transcortin'-binding sites and association constant for cortisol in different species of mammalian plasma at 36 - 37°C. (From Khin Aye Than and McDonald, 1975, with permission of *J. Endocrinol.*)

Species	ΣP_T (× 10^{-7} mol/l)	K_T (× 10^7 l/mol)	Sources
Human	5.5, 6.1, 8.0	5.2	Mills (1962)
	7.2	3.0	Westphal (1967)
Monkey	9.3	3.0	Westphal (1967)
Rat	11.3	1.0	Westphal (1967)
	–	0.31	Rosner and Hochberg (1972)
Guinea-pig	5.7	0.4	Westphal (1967)
Rabbit	3.4	4.0	Westphal (1967)
Sheep	0.7	8.7	Paterson and Hills (1967)
Possum	1.84 ± 0.62	4.03 ± 0.94	Khin Aye Than and McDonald (1975)

Bradley (1975), using the technique of Doe *et al.* (1964), measured the maximum cortisol binding capacity of male and female *Antechinus stuartii* at different stages in their life cycle. During April the maximum binding capacities of males and females were similar, 7.51 ± 0.53 and 9.41 ± 0.6 μg/100 ml, respectively. However, that of the males declined markedly to only 1.96 ± 0.38 μg/100 ml during and after the mating period in August, just preceding the time of total male mortality (Woolley, 1966; Wood, 1970), while that of the females was still near normal, at 7.11 ± 1.7 (see Lee, Bradley and Braithwaite, this volume). Hence, the increase in free cortisol concentration of the males at this time is even greater than would be expected from measurement of the total concentration alone. Furthermore, administration of corticotrophin to laboratory-held males, captured well before the mating period, caused a similar decrease in plasma cortisol-binding

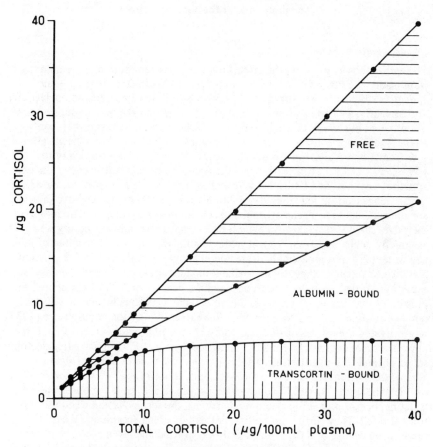

Figure 21.3 Relative preparations of 'free', albumin-bound and 'transcortin'-bound cortisol in the blood plasma of *Trichosurus vulpecula* with increasing plasma total cortisol concentration. Note that the high-affinity binding system saturates at approximately 6 μg/100 ml. The naturally stimulated plasma total cortisol concentration of *Trichosurus* rarely exceeds 10 μg/100 ml. (From Khin Aye Than and McDonald, 1975, with permission of *J. Endocrinol.*)

capacity, whereas there was no change in the saline-injected controls. Administration of testosterone had much the same effect as corticotrophin.

These observations all indicate that marsupials have a transcortin-like high-affinity corticosteroid binding system in their blood plasma. The observations made on the changes in this system during the life cycle of male *A. stuartii* indicates the probable importance of this system in determining the biological effectiveness of the circulating corticosteroids. It would be of great interest to compare the properties of this system in a wider variety of marsupials, particularly in relation to environmental stresses.

Metabolic role of the adrenal cortex

Effects of adrenalectomy

The earliest observations on the effects of bilateral adrenalectomy in marsupials were made by Britton (1931), who found that the North American opossum, *Didelphis virginiana*, may survive for 3 – 33 days following bilateral adrenalectomy. Subsequently, Sylvette and Britton (1936) reported that the average survival time of untreated bilaterally adrenalectomised *Didelphis* was 6 days and that, contrary to the findings in eutherian mammals, the blood serum Na and Cl concentrations rose from 144 ± 6 and 104 ± 5 mM/l, respectively, before adrenalectomy to 153 ± 9 and 115 ± 7 at the time when the signs of adrenal insufficiency (loss of appetite, weakness and lassitude) were well defined. On the other hand, blood 'sugar' concentration fell from 90 ± 10 to 59 ± 8 mg/100 ml. (The above values are all recalculated from the original data.) In animals surviving more than 6 days, their data show that urinary Na excretion 5 – 7 days after adrenalectomy was essentially unchanged from control values in animals maintained on a diet of milk and water (2:1) but moderately decreased in animals maintained on 0.9 per cent NaCl. Unfortunately, they did not make serial measurements of daily urinary Na excretion, so that an early natriuresis may have been missed. They considered the possibility that circulatory changes due to adrenal insufficiency may have interfered with urine flow and Na excretion. Subsequently, Britton and Sylvette (1937) reinvestigated the effects of adrenalectomy in *Didelphis*, and reported that, if the opossums were deprived of food but given access to water, the post-adrenalectomy serum Na and Cl concentrations were reduced at the time of onset of signs of adrenal insufficiency, as in eutherian mammals. If both food and water were withheld the serum Na and Cl concentrations rose as before. If the opossums were lactating, or had post-adrenalectomy diarrhoea, the serum Na and Cl concentrations also fell. They further noted that untreated adrenalectomised *Didelphis* survived an average of 6 days; but if deprived of food and given access to water, they survived for an average of 10 days. Whatever the treatment, the blood 'sugar' and liver glycogen concentrations were always reduced at the time of development of signs of adrenal insufficiency. These observations leave little doubt that, in the absence of supportive treatment, the secretions of the adrenal gland are essential to life in *Didelphis*, and that the symptoms of adrenal insufficiency in this marsupial are similar to those found in eutherian mammals. It seems likely that the initially reported rise in serum Na and Cl concentrations was a consequence of dehydration, and this could account for the reduced urine flow and Na excretion several days after adrenalectomy, when they were receiving only 0.9 per cent NaCl orally.

Anderson (1937) investigated the effects of bilateral adrenalectomy in the Australian Brush-tailed possum, *Trichosurus vulpecula*, and found that they survived only 26 – 40 h after removal of the second gland. Survival time was not altered by intramuscular injection of isotonic saline.

Hartman, Smith and Lewis (1943) reopened the question of the effects of bilateral adrenalectomy in *Didelphis virginiana*, and reported survival times from 5 up to 603 days without supportive treatment. They concluded that: 'In the body economy of the opossum, the adrenal appears to be of less importance than in many other mammals.' This conclusion has led to the frequently quoted assertions that *Didelphis* is unusual among the mammals in being resistant to bilateral adrenalectomy. Fortunately, it has also provided a stimulus to research into the adrenal physiology of other marsupials. The long-term survival of some of Hartman *et al.*'s opossums (1943) can probably be accounted for by the high (2 per cent) salt content of their diet, amounting to 3 g daily, if they ate all that was offered. Given this high salt intake, the progressive rise in serum Na of the adrenalectomised animals can be readily explained, and the equally progressive rise in plasma K suggests that there was a defect in renal K excretion, similar to that found in adrenalectomised eutherian mammals (Table 21.5). A more definitive investigation of the effects of adrenalectomy and the maintenance requirements of adrenalectomised *Didelphis* would be desirable.

Table 21.5 Changes in plasma electrolytes of *Didelphis virginiana* following bilateral adrenalectomy (from Hartman *et al.*, 1943), compared with those of *Macropus rufus* and *Trichosurus vulpecula* when either offered 0.9 per cent NaCl for self-selection (*) or given no supportive treatments (N.T.)

	Days of adrenal insufficiency	Na	Plasma (mmol/l)	K
	Control	141.6 ± 2.07		4.96 ± 0.69
Didelphis	10	147.4 ± 4.45		5.90 ± 0.99
virginiana	29 – 59	147.3 ± 12.15		5.20 ± 0.08
(2% salt in	100 – 224	150.4 ± 10.42		6.60 ± 0.25
diet)	224 – 288	150.2 ± 21.00		7.70 ± 2.99
Macropus	Control	140		4.4
rufus	7	147		6.2
Nacl *				
N.T.	3	129		7.0
Trichosurus	Control	147		4.8
vulpecula	6	145		5.6
NaCl *				
N.T.	1	137		7.7

Buttle, Kirk and Waring (1952) found that the Quokka, *Setonix brachyurus*, survived bilateral adrenalectomy for a maximum period of 48 h, in spite of administration of saline or adrenal cortical extracts. The signs of adrenal insufficiency – anorexia, muscular weakness and lassitude – were the same as those recorded for *Didelphis* and *Trichosurus*, and were associated with a progressive decline in plasma Na concentration and rise in K concentration, as in eutherian mammals. They concluded that *Setonix* was unusually sensitive to the effects of adrenal insufficiency, which were otherwise no different from those of eutherian mammals.

Reid and McDonald (1968) found that bilaterally adrenalectomised *Trichosurus vulpecula* could be maintained in apparently normal condition with daily intramuscular injections of 1 mg/kg cortisol acetate, or smaller doses of cortisol acetate combined with aldosterone diacetate i.m. (0.7 and 0.014 mg/kg, respectively). If these supplements were withheld, the animals lost appetite and became quieter, blood plasma K concentration rose, and Na, Cl and HCO_3 concentration fell, as in eutherian mammals. These changes were associated with an increase in urinary Na excretion and development of an appetite for salt. They were reversed by intra-muscular injection of 1 - 2 mg/kg cortisol acetate or cortisol and aldosterone acetates. Aldosterone alone only reduced plasma K concentration, and did not reverse the changes in Na and Cl concentrations. One possum maintained normal plasma electrolyte concentrations, appetite and activity by self-selection of salt alone for 5 days without steroid supplement. Plasma glucose concentration of adrenally insufficient possums were maintained in the normal way until they stopped eating, when it fell markedly.

McDonald (1974) found that bilaterally adrenalectomised Red kangaroos (*Macropus rufus*) could be maintained indefinitely with daily intramuscular injections of 2 mg/kg cortisol acetate, or 0.5 mg/kg cortisol acetate with 0.05 mg/kg deoxycorticosterone acetate. If they were given access to isotonic NaCl solution, withdrawal of the steroid supplement was followed by a marked increase in salt intake, up to 5 litres of isotonic saline/day. This was sufficient to maintain these animals in apparent good health indefinitely. However, if cortisol alone or salt alone were the only maintenance supplement, plasma K concentration was approximately double the normal value, and plasma Na concentration fluctuated near or above the normal value in the salt-maintained animals. The observations on the salt-maintained Red kangaroos resembled those of Hartman *et al.* (1943) on *Didelphis* (Table 21.5). Withdrawal of steroid supplement without access to salt was followed by a marked but short-lived natriuresis, fall in plasma Na, rise in plasma K and urea concentrations and in the haematocrit. The animals stopped eating and became progressively weaker until they were in a state of collapse 4 - 5 days after the last dose of steroid. They recovered rapidly when given cortisol and aldosterone i.v. followed by access to salt. Withdrawal of salt from a salt-maintained animal was followed by death in 48 h. As with *Trichosurus*, the plasma glucose concentration of adrenally insufficient kangaroos remained within the normal range while they were eating.

All these observations indicate that bilateral adrenalectomy has much the same effects in marsupials as those reported for eutherian mammals.

The only mammal which is able to tolerate bilateral adrenalectomy without any supportive treatment appears to be the prototherian *Tachyglossus aculeatus* (McDonald and Augee, 1968), but long-term survival is only possible in a low-stress environment (Augee and McDonald, 1973).

Metabolic actions of adrenocortical steroids

Mineralocorticoids

The observations of Buttle *et al.* (1952), Reid and McDonald (1968) and McDonald (1974) clearly show that marsupial adrenocortical secretions are concerned with Na conservation and K excretion, and from the evidence summarised in previous sections this mineralocorticoid activity is probably due to aldosterone secretion, although cortisol is also effective in preventing renal Na loss. From the meagre evidence available so far, it seems that in *Trichosurus vulpecula* (Reid and McDonald, 1968) and *Macropus rufus* (McDonald, 1974) mineralocorticoids such as aldosterone and DOC have an important role in maintaining normal extra-cellular K concentrations, and that a combination of both mineralo- and gluco-corticoids is necessary for maintenance of a completely normal extracellular electrolyte pattern.

Glucocorticoids

The early observations of Sylvette and Britton (1936) showed a marked fall in liver glycogen concentration and blood glucose in association with signs of adrenal insufficiency in bilaterally adrenalectomised *Didelphis virginiana*. Anderson (1937) noted an increase in the concentration of reducing substances in plasma and erythrocyte of intact *Trichosurus vulpecula* given injections of adrenocortical extract. Reid and McDonald (1968) found that intramuscular injection of 10 mg/kg cortisol acetate caused a marked rise in blood glucose concentration of *Trichosurus*, but made no detailed observations.

Griffiths, McIntosh and Leckie (1968) investigated the diabetogenic and nitrogen-mobilising actions of cortisone in herbivores, including Red kangaroos. They found that intramuscular doses as high as 14 mg/kg cortisone acetate per day had no effect on the fasting blood glucose or nitrogen balance of intact and alloxan-diabetic Red kangaroos, whereas smaller doses had the expected diabetogenic and

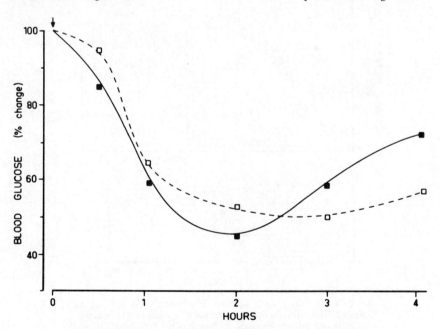

Figure 21.4 Lack of effect of cortisol acetate (10 mg/kg i.m. for 7 days) on sensitivity of *Macropus rufus* to insulin (0.22 i.u./kg i.v.). □ control; ■ cortisol-injected. (Data from Griffiths and McDonald, unpublished.)

nitrogen mobilising effects in sheep and rabbits. The sensitivity of kangaroos to insulin was also unimpaired by cortisone (Figure 21.4). This observation has been confirmed by Griffiths and McDonald (unpublished), using unilaterally and bilaterally
adrenalectomised red kangaroos, and cortisol acetate instead of cortisone acetate.
However, insulin sensitivity in the adrenalectomised kangaroo is enhanced (Figure
21.7). Bradshaw and McDonald (unpublished) found that another macropod
marsupial *Setonix brachyurus* is similarly refractory to the diabetogenic and
nitrogen mobilising actions of cortisol (Figure 21.5).

In contrast to these findings, Khin Aye Than and McDonald (1974a) found
that the Brush-tailed opossum, *Trichosurus vulpecula*, was highly sensitive to the
diabetogenic and nitrogen mobilising actions of cortisol acetate, doses of 1 mg/kg
per day i.m. causing hyperglycaemia, glycosuria, increased nitrogen excretion and
increased fasting liver glycogen concentration. The only eutherian mammal with
comparable sensitivity is the rabbit. Short-term (1 h) intravenous infusion of

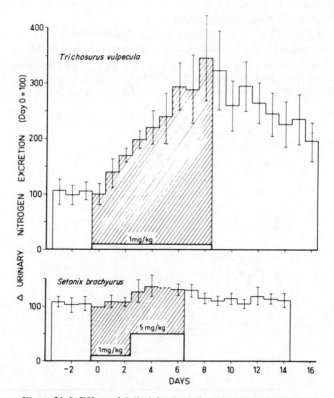

Figure 21.5 Effect of daily injection of cortisol acetate i.m., at the
indicated dose rates, on urinary nitrogen excretion of *Trichosurus
vulpecula* and *Setonix brachyurus*. The data for *Trichosurus* have
been recalculated from those published by Khin Aye Than and
McDonald (1974a) for representation as percentages of the excretion
rates on the first day of cortisol injection. The data for *Setonix* are
from those of Bradshaw and McDonald (unpublished).

cortisol at rates equivalent to the maximum corticotrophin-stimulated secretion rate caused a rapid increase in plasma glucose concentration with no change in plasma urea or α-amino nitrogen concentration. They suggested that the short-term effect of cortisol was to depress peripheral utilisation of glucose, and the long-term effect was to enhance gluconeogenesis through increased tissue nitrogen mobilisation. This was later confirmed by direct measurement of hepatic glucose release and metabolic clearance of glucose in conscious opossums during intra-venous cortisol infusion or at the end of a 5 day course of intramuscular cortisol

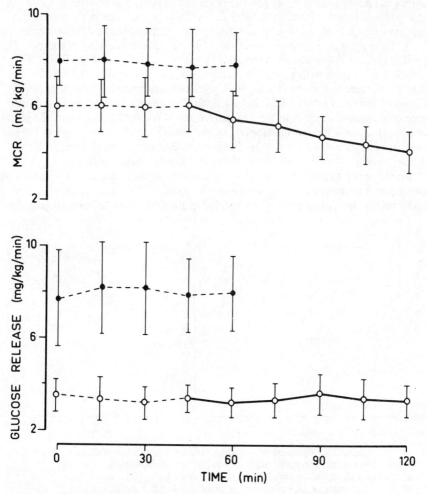

Figure 21.6 *Trichosurus vulpecula*. Short-term and long-term effects of cortisol on metabolic clearance and hepatic release of glucose. 0−−0 i.v. infusion of 0.9 per cent NaCl; 0—0 i.v. infusion of cortisol in 0.9 per cent NaCl at 1 mg/kg per h; •−−• i.v. infusion of 0.9 per cent NaCl at the end of a 5 day course of i.m. cortisol acetate, 1 mg/kg per day. (From McDonald and Khin Aye Than, 1976, with permission of *J. Endocrinol.*)

acetate of cortisol injection (McDonald and Khin Aye Than, 1976). During
short-term infusion, there was no change in hepatic new glucose release, but a
progressive fall in metabolic clearance rate, whereas, at the end of the course of
cortisol acetate injections, hepatic glucose release was approximately doubled,
and MCR moderately increased (Figure 21.6).

Khin Aye Than and McDonald (1974b) also found that the insulin sensitivity
of *Trichosurus vulpecula* was depressed by cortisol acetate (1 mg/kg per day i.m.),
and that there was a linear relationship between the plasma cortisol concentration
within the physiological range and the degree of hypoglycaemic response to a
given dose of insulin (Figures 21.7 and 21.8). This suggests that corticosteroids
are important normal regulators of carbohydrate metabolism in this marsupial.

In the small dasyurid marsupial *Antechinus stuartii* the increase in plasma
cortisol concentration found in the males by Barnett (1973) and Bradley *et al.*
(1975) is associated with a negative nitrogen balance (Woollard, 1971) and a
marked increase in the non-fasting liver glycogen concentration (Barnett, 1973).
However, Barnett (1974) could find no significant change in the non-fasted liver
glycogen concentration or in the concentrations of Na, K and glucose in the plasma
of male *A. stuartii* held in the laboratory and given large (4–40 mg/kg i.m.) doses
of cortisol acetate daily for 4 days. These doses caused a marked increase in the
concentration of hydroxyproline in the dorsal skin of these animals.

Bradley *et al.* (1975) reported that daily intramuscular injection of 1 and 10 mg
cortisol acetate/kg caused a highly significant mortality in male *A. stuartii*, captured before the mating period and isolated in the laboratory, compared with

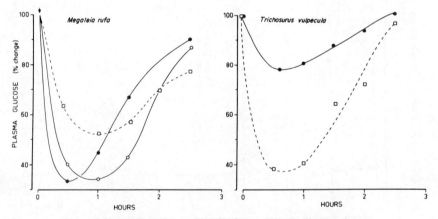

Figure 21.7 Comparison of insulin sensitivity of intact and adrenalectomised *Macropus rufus* with that of intact *Trichosurus vulpecula.*
(a) *M. rufus:* □ normal; 0 adrenalectomised, maintained with 0.9 per
cent saline; ● adrenalectomised, maintained with 2 mg/kg cortisol
acetate i.m./day. (b) *T. vulpecula:* □ normal; ● given 1 mg/kg cortisol
acetate i.m. daily for 5 days. Insulin dose, 0.15 i.u./kg i.v. Note the
lower sensitivity of normal *M. rufus* to insulin, and its enhancement
to approximately the same as that of *T. vulpecula* by adrenalectomy,
with or without cortisol supplementation. (Data from Khin Aye
Than and McDonald, 1974b, and Griffiths and McDonald, unpublished. Note: *Megaleia rufa* = *Macropus rufus.*

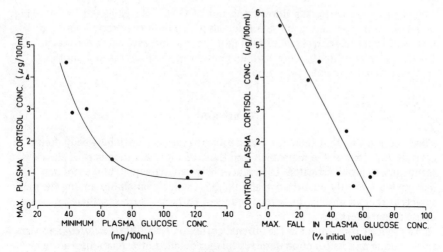

Figure 21.8 (a) Effect of insulin hypoglycaemia on plasma cortisol concentration of *Trichosurus vulpecula*. (b) Relationship between plasma cortisol concentration before insulin injection and sensitivity to insulin, as indicated by its hypoglycaemic effect. The degree of hypoglycaemia is expressed as a percentage of the initial plasma glucose concentration. (From Khin Aye Than and McDonald, 1974b, with permission of *J. Endocrinol.*)

saline injected controls in which there was no mortality. Bradley (1975) correlated this with evidence for increased invasiveness of parasites and micro-organisms, and reduction in the concentrations of presumed IG A, G, and M (see also Lee *et al.*, this volume).

In summary, the data on the metabolic actions of corticosteroids in marsupials are meagre and uncoordinated for comparative purposes. However, it is clear that the actions of glucocorticoids in *Trichosurus* closely resemble the generally accepted actions in eutherian mammals (Cahill, 1971), whereas the macropods appear to be resistant to their diabetogenic and nitrogen mobilising actions. This latter observation is of great interest and fits in with the observation that adrenalectomised Red kangaroos seem to require only salt for normal maintenance. One important adaptation that macropod marsupials appear to have developed in common with eutherian ruminants such as camels (Read, 1925; Schmidt-Neilsen *et al.*, 1957) and sheep (Schmidt-Neilsen *et al.*, 1958) is the ability to conserve nitrogen during periods of nitrogen deprivation and recycle it via their semi-ruminant digestive system (Moir, 1968; Kinnear and Main, 1975). It seems conceivable that the resistance to the nitrogen-mobilising action of glucocorticoids may be another adaptation to prevent nitrogen-wasting in a high-stress environment. It is clearly of interest to learn more about glucocorticoid actions in the macropod marsupials. The observations on *A. stuartii* implicate glucocorticoids in the immune response of these marsupials, and suggest that this glucocorticoid action is an important determinant of their population dynamics, as discussed by Lee *et al.* in this volume.

The mineralo-corticoids aldosterone and DOC have the expected Na-retaining and K-excreting actions in those marsupials which have been tested, and the observations of Scoggins *et al.* (1970) indicate that increased aldosterone secretion is a normal adaptation to a low sodium content of the food taken from the immediate environment.

Conclusion

This survey shows that there is a remarkable overall similarity between adreno-cortical functions of the marsupials and those of the other members of the class Mammalia so far investigated. Differences between marsupials and eutherian mammals are mostly quantitative and there is as much variability among the marsupials as there is among the several orders of eutheria. However, there are some unusual characteristics.

The special zone found in the adrenal cortex of *Trichosurus vulpecula* appears to be unique and its function deserves further detailed investigation. Another character, apparently unique to marsupials, is the insensitivity of the macropod marsupials to the diabetogenic and nitrogen-mobilising actions of the glucocorticoids. This could be an important adaptation related to the success of some species in arid nitrogen-deficient environments, and deserves detailed investigation. This specialisation is not shared by other marsupial families, and discovery of the reasons for the difference between metabolic actions of the glucocorticoids in the several marsupial families could give greater insight into the general topic of the metabolic role of mammalian adrenocortical secretions.

The low sensitivity of those marsupials so far investigated to the corticosteroidogenic action of corticotrophins is another topic which deserves detailed investigation. It is generally assumed that there are no species differences in the biologically active groups of the corticotrophins and, therefore, the adrenocortical receptors of all species should have similar structures. It is difficult to imagine that the low sensitivity of the marsupials could be due to other differences at the tissue level — for example, low density of binding sites — because the relatively enormous doses of exogenous corticotrophin required to raise plasma corticosteroid concentrations to naturally stimulated levels could not possibly be reproduced by the animals' own adenohypophysis. One can only postulate extremely rapid metabolic clearance or inactivation of exogenous hormone, or a more active natural hormone. Present analytical methods are such as to make it reasonably easy to determine the structures of marsupial corticotrophins, and their biological efficacy could be compared in the several families.

So far, the role of the adrenocortical secretions in the natural metabolic adaptations of marsupials to environmental changes has not been rigorously investigated. However, the investigation of the role of the adrenal cortex in the mortality of male *A. stuartii* has revealed very clearly the deleterious effects of increased adrenocortical activity in a natural population. Possibly the most interesting aspect of this work is the marked fall in the high-affinity corticosteroid binding capacity of the plasma in association with the evidence of increased adrenocortical activity in the males just before death. It seems likely that this change in the physical state of the plasma corticosteroids, leading to an increase in the fraction available for biological action, may be an important determinant of adrenocortical function in these marsupials. The possible implications of such changes in association with symptoms of stress in other populations of mammals deserves further investigation.

Acknowledgements

I thank present and former colleagues Dr. S. D. Bradshaw, Dr. M. Griffiths and Dr. M. Weiss, former graduate students Dr M. L. Augee, Dr Khin Aye Than and Dr I. A. Reid, and present graduate student A. J. Bradley, who have all contributed much to the work quoted in this review. The assistance of Miss Diana Harrison and Miss Jill Maplesdon with the drawing of figures and photography, and of Mrs. Lyn Hepburn with typing, is gratefully acknowledged.

References

Anderson, D. (1937). Studies on the opossum (*Trichosurus vulpecula*). II. The effects of splenectomy, adrenalectomy and injection of cortical hormones. *Aust. J. Exp. Biol. Med. Sci.*, **15**, 22 - 32.

Augee, M. L. and McDonald, I. R. (1973). Role of the adrenal cortex in the adaptation of the monotreme *Tachyglossus aculeatus* to low environmental temperature. *J. Endocrinol*, **58**, 513 - 23.

Barnett, J. L. (1973). A stress response in *Antechinus stuartii* (Macleay). *Aust. J. Zool.*, **21**, 502 - 13.

Barnett, J. L. (1974). Changes in the hydroxyproline concentration of the skin of *Antechinus stuartii* with age and hormonal treatment. *Aust. J. Zool.*, **22**, 311 - 8.

Bennett, H. S. and Kilham, L. (1940). The blood vessels of the adrenal cortex of the cat. *Anat. Rec.*,77, 447 - 72.

Blair-West, J., Coghlan, J. P., Denton, D. A., Nelson, J. F., Orchard, E., Scoggins, B. A., Wright, R. D., Myers, K. and Junqueira, C. L. (1968). Physiological, morphological and behavioural adaptation to a sodium-deficient diet by wild native Australian and introduced species of animals. *Nature*, **217**, 922 - 8.

Bongiovani, A. M., Ebehrlein, W. R., Goldman, H. D. and New, M. (1967). Disorders of adrenal steroid biogenesis. *Recent Progr. Hormone Res.*, **23**, 375 - 439.

Bourne, G. H. (1934). Unique structure in the adrenal of the female opossum. *Nature*, **134**, 664 - 5.

Bourne, G. H. (1949). *The Mammalian Adrenal Gland.* Oxford University Press, London.

Bradley, A. J. (1976). Role of the adrenal cortex in the mortality of *Antechinus stuartii. Aust. J. Wildlife Res.* (in press).

Bradley, A. J., McDonald, I. R. and Lee, A. K. (1975). Effect of exogenous cortisol on mortality of a dasyurid marsupial. *J. Endocrinol.*, **66**, 281 - 2.

Britton, S. W. (1931). Observations on adrenalectomy in marsupial, hibernating and higher mammalian types. *Am. J. Physiol.*, **99**, 9 - 14.

Britton, S. W. and Sylvette, H. (1937). Further observations on sodium chloride balance in the adrenalectomised opossum. *Am. J. Physiol.*, **118**, 21 - 5.

Brownell, K. A., Beck, R. R. and Besch, P. K. (1967). Steroid production by the normal opossum (*Didelphis virginiana*) adrenal *in vitro*. *Gen. Comp. Endocrinol.*, 9, 214 – 6.

Brown-Grant, K., Forschelli, E. and Dorfman, R. I. (1960). The Δ^4-hydrogenase system of guinea-pig adrenal gland. *J. Biol. Chem.*, 235, 1317 – 20.

Bush, I. E. and Ferguson, K. H. (1953). Secretion of the adrenal cortex in the sheep. *J. Endocrinol.*, 10, 1 – 8.

Buttle, J. M., Kirk, R. L. and Waring, H. (1952). The effects of complete adrenalectomy on the wallaby *Setonix brachyurus*. *J. Endocrinol.*, 8, 281 – 90.

Cahill, G. F. (1971). Action of adrenal cortical steroids on carbohydrate metabolism. In: *The Human Adrenal Cortex* (ed. N. P. Christy), Harper and Row, New York, pp. 205 – 39.

Chester Jones, I. C. (1957). *The Adrenal Cortex*, Cambridge University Press, London.

Chester Jones, I. C., Vinson, G. P., Jarrett, I. G. and Sharman, G. R. (1964). Steroid components in the adrenal venous blood of *Trichosurus vulpecula*. *J. Endocrinol.*, 30, 149 – 50.

Christian, J. J. (1963). Endocrine adaptive mechanisms and the physiologic regulation of population growth. In: *Physiological Mammalogy*, Vol. 1 (ed. M. V. Meyer and R. G. Von Gelder), Academic Press, New York, pp. 189 – 353.

Clemens, N. A. (1968). Origin and early evolution of marsupials. *Evolution (N.Y.)*, 22, 1 – 18.

Coghlan, J. P. and Scoggins, B. A. (1967). The measurement of aldosterone, cortisol and corticosterone in the blood of the wombat (*Vombatus hirsutus*, Perry) and the kangaroo (*Macropus giganteus*). *J. Endocrinol.*, 13, 218 – 21.

Doe, R. P., Fernandez, R. and Seal, U. S. (1964). Measurement of corticosteroid-binding globulin in man. *J. Clin. Endocrinol. Metab.*, 24, 1029 – 39.

Finkelstein, M., Shoenberger, J., Maschler, I. and Halperin, G. (1967). An attempt to unify concepts of C-21 steroid producing pathways in certain adrenal and ovarian disorders. *Res. Steroids*, III, 233 – 48.

Griffiths, N. McIntosh, D. L. and Leckie, R. M. C. (1968). The effects of cortisone on nitrogen balance and glucose metabolism in diabetic and normal kangaroos, sheep and rabbits. *J. Endocrinol.*, 44, 1 – 12.

Harrison, F. A. and McDonald, I. R. (1966). The arterial supply to the adrenal gland of the sheep. *J. Anat.*, 100, 189 – 202.

Harrison, R. G. (1951). A comparative study of the vascularisation of the adrenal gland in the rabbit, rat and cat. *J. Anat.*, 85, 12 – 23.

Hartman, F. A., Smith, D. E. and Lewis, L. A. (1943). Adrenal functions in the opossum. *Endocrinology*, 32, 340 – 4.

Hechter, O. and Pincus, G. (1954). Genesis of adrenocortical secretion. *Physiol. Rev.*, 34, 459 – 96.

Howard-Miller, E. (1927). A transitory zone in the adrenal cortex which shows age and sex relationships. *Am. J. Anat.*, 40, 251 – 94.

Illet, K. F. (1969). Corticosteroids in the adrenal venous and heart blood of the quokka *Setonix brachyurus* (Marsupialia:macropodidae). *Gen. Comp. Endocrinol.*, 13, 218 – 21.

Johnston, C. I., Davis, J. O. and Hartroft, P. M. (1967). Renin-angiotensin system, adrenal steroid and sodium depletion in a primitive mammal the American opossum. *Endocrinology*, **81**, 633 – 42.

Khin Aye Than (1975). Adrenocortical function in a marsupial. Thesis for Ph. D., Monash University.

Khin Aye Than and McDonald, I. R. (1973). Adrenocortical function in the Australian brush-tailed possum *Trichosurus vulpecula* (Kerr). *J. Endocrinol.*, **58**, 97 – 109.

Khin Aye Than and McDonald, I. R. (1974a). Metabolic effects of cortisol and corticotrophins in the Australian brush-tailed opossum *Trichosurus vulpecula* (Kerr). *J. Endocrinol.*, **63**, 137 – 47.

Khin Aye Than and McDonald, I. R. (1974b). Effect of cortisol on insulin sensitivity of the Marsupial brush-tailed opossum *Trichosurus vulpecula* (Kerr). *J. Endocrinol.*, **63**, 473 – 81.

Khin Aye Than and McDonald, I. R. (1975). Corticosteroid binding by plasma proteins and the effects of oestrogens in the marsupial brush-tailed opossum *Trichosurus vulpecula* (Kerr). *J. Endocrinol.*, **67**, No. 3.

Khin Aye Than and McDonald, I. R. (1976). Effect of cortisol on utilization and hepatic release of glucose in the marsupial brush-tailed opossums *Trichosurus vulpecula*. *J. Endocrinol.*, **68**, 31 – 41.

Kinnear, J. E. and Main, A. R. (1975). The recycling of urea nitrogen by the wild Tammar wallaby (*Macropus eugenii*) – a 'ruminant-like' marsupial. *Comp. Physiol. Biochem.*, **51A**, 793 – 810.

Landon, J., James, V. H. T., Cryer, R. J., Wynn, V. and Frankland, A. M. (1964). Adrenocorticotrophic effects of synthetic polypeptide-B$^{1-24}$ corticotrophin in man. *J. Clin. Endocrinol. Metab.*, **24**, 1206 – 13.

Landon, J., Wynn, V. and James, V. H. T. (1963). The adrenocortical response to insulin-induced hypoglycaemia. *J. Endocrinol.*, **27**, 183 – 92.

McDonald, I. R. (1974). Adrenal insufficiency in the red kangaroo (*Megaleia rufa* Desm). *J. Endocrinol.*, **62**, 689 – 90.

McDonald, I. R. and Augee, M. L. (1968). Effects of bilateral adrenalectomy in the monotreme *Tachyglossus aculeatus*. *Comp. Biochem. Physiol.*, **27**, 669 – 78.

McDonald, I. R. and Khin Aye Than (1976). *J. Endocrinol.*, **68**, 257 – 64.

McDonald, I. R. and Weiss, M. (1967). Turnover and excretion of cortisol in the Australian marsupial *Trichosurus vulpecula*. *Acta. Endocrinol. (Kbh).*, Suppl. 119, 242.

Mills, I. H. (1962). Transport and metabolism of steroids. *Brit. Med. Bull.*, **18**, 127 – 33.

Moir, R. J. (1968). Ruminant digestion and evolution. In: *Handbook of Physiology*. Section 6. *Alimentary Canal*, Vol. 5, American Physiological Society.

Murphy, B. P., Engelberg, W. and Pattee, C. J. (1963). Simple method for the determination of plasma corticoids. *J. Clin. Endocrinol. Metab.*, **23**, 293 – 300.

Paterson, J. Y. F. and Harrison, F. A. (1967). The specific activity of plasma cortisol in sheep during continuous infusion of (1, 2, ^{3}H) cortisol and its relation to the rate of cortisol injection. *J. Endocrinol.*, **37**, 269 – 77.

Paterson, J. Y. F and Hills, F. (1967). The binding of cortisol by ovine plasma proteins. *J. Endocrinol.*, **37**, 261 – 8.

Read, R. E. (1925). Chemical constituents of camel's urine. *J. Biol. Chem.*, **64**, 615 - 7.

Reid, I. A. and McDonald, I. R. (1968). Bilateral adrenalectomy and steroid replacement in the marsupial *Trichosurus vulpecula*. *Comp. Biochem. Physiol.*, **26**, 613 - 25.

Reid, I. A. and McDonald, I. R. (1969). The renin-angiotensin system in a marsupial (*Trichosurus vulpecula*). *J. Endocrinol.*, **44**, 231 - 40.

Renfree, M. B. and Vinson, G. P. (1974). Control of steroidogenesis in adrenal tissue from the developing opossum, *Didelphis marsupialis virginiana*. *J. Endocrinol.*, **63**, 15 - 6P.

Rosner W. and Hochberg, R. (1972). Corticosteroid-binding globulins in the rat: isolation and studies of its influence on cortisol actions *in vivo. Endocrinology*, **91**, 626 - 32.

Schmidt-Neilsen, B., Osaki, H., Murdaugh, H. V. and O'Dell, R. (1958). Renal regulation of urea excretion in sheep. *Am. J. Physiol.*, **194**, 221 - 8.

Schmidt-Neilsen, B., Schmidt-Neilsen, K., Houpt, T. R. and Jarnum, S. A. (1957). Urea excretion in the camel. *Am. J. Physiol.*, **188**, 477 - 84.

Scoggins, B. A., Blair-West, J. R., Coghlan, J. P., Denton, D. A., Myers, K., Nelson, J. F., Orchard, E. and Wright, R. D. (1970). The physiological and morphological response of mammals to changes in their sodium status. In: *Hormones and the Environment* (ed. G. K. Benson and J. G. Phillips), Cambridge University Press, pp. 577 - 600.

Seal, U. S. and Doe, R. P. (1966). Corticosteroid-binding globulin: biochemistry, physiology and phylogeny. In: *Steroid Dynamics*, (ed. G. Pincus, T. Nakao and J. F. Tait), Academic Press, New York, pp. 63 - 90.

Siiteri, P. K. (1971). Qualitative and quantitative aspects of adrenal secretion in steroids. In: *The Human Adrenal Cortex* (ed. N. P. Christy), Harper and Row, New York, pp. 1 - 30.

Sylvette, H. and Britton, S. W. (1936). Carbohydrate and electrolyte changes in the opossum and marmot following adrenalectomy. *Am. J. Physiol.*, **115**, 618 - 26.

Tait, J. F. and Burstein, S. (1964). *In vivo* studies of steroid dynamics in man. In: *The Hormones*, Vol. V (ed. G. Pincus, K. V. Themann and E. B. Astwood), Academic Press, New York, pp. 441 - 557.

Tyndale-Biscoe, C. H. (1973). *Life of Marsupials*, Edward Arnold, London.

Vinson, G. P. (1974). The control of the adrenocortical secretion in the brush-tailed possum *Trichosurus vulpecula*. *Gen. Comp. Endocrinol.*, **22**, 268 - 76.

Vinson, G. P., Phillips, J. G., Chester Jones, I. and Tsang, W. N. (1971). Functional zonation of adrenocortical tissues in the brush possum *Trichosurus vulpecula*. *J. Endocrinol.*, **49**, 131 - 40.

Vinson, G. P., Tyndale-Biscoe, C. H. and Bancroft, B. J. (1973). The effects of trophic hormones on the adrenocortical secretion of the brush-tailed possum *Trichosurus vulpecula*. *J. Endocrinol.*, **55**, XXXV - XXXVI.

Vogt, M. (1955). Inhibition by hexoestrol of adrenocortical secretion in the rat. *J. Physiol.*, **130**, 601 - 14.

Vogt, M. (1957). Effects of hexoestrol and of amphenone β on morphology and function of the rat adrenal cortex. *Yale J. Biol. Med.*, **29**, 469 - 79.

Weiss, M. (1968). 21-Deoxycortisol and other minor steroids in adrenal venous blood of the phalanger (*Trichosurus vulpecula*). *J. Endocrinol.*, **4**, 293 – 4.

Weiss, M. (1974). Steroid formation by the adrenal tissue of the possum (*Trichosurus vulpecula*). *Comp. Biochem. Physiol.*, **50B**, 211 – 3.

Weiss, M. and McCance, I. (1974). 21-hydroxylation of 21-deoxycortisone by adrenal glands of a marsupial and two eutherian species. *Comp. Biochem. Physiol.*, **49B**, 227 – 39.

Weiss, M. and McDonald, I. R. (1965). Corticosteroid secretion in the monotreme *Tachyglossus aculeatus. J. Endocrinol.*, **33**, 203 – 10.

Weiss, M. and McDonald, I. R. (1966a). Corticosteroid secretion in the Australian phalanger (*Trichosurus vulpecula*). *Gen. Comp. Endocrinol.*, 7, 345 – 51.

Weiss, M. and McDonald, I. R. (1966b). Adrenocortical secretion in the wombat *Vombatus hirsutus* (Perry). *J. Endocrinol.*, **35**, 207 – 8.

Weiss, M. and McDonald, I. R. (1967). Corticosteroid secretion in kangaroos *(Macropus cangura major* and *M. (Megaleia) rufa). J. Endocrinol.*, **39**, 251 – 61.

Weiss, M. and Richards, P. G. (1970). Adrenal steroid secretion in the koala *Phascolarctos cinereus. J. Endocrinol.*, **48**, 145 – 6.

Weiss, M. and Richards, P. G. (1971). Adrenal steroid secretion in the Tasmanian Devil (*Sarcophilus harrisii*) and the eastern native cat (*Dasyurus viverrinus*). A comparison of adrenocortical activity of different Australian marsupials. *J. Endocrinol.*, **49**, 263 – 75.

Westphal, V. (1967). Steroid-protein interactions. XIII. Concentration and binding affinities of corticosteroid-binding globulins in sera of man, monkey, rabbit and guinea-pig. *Arch. Biochem. Biophys.*, **118**, 556 – 67.

Wieland, R. G., Maynard, D. E., Riley, T. R. and Hammwi, G. J. (1965). Detection of 21-deoxycortisol in blood from a patient with congenital adrenal hyperplasia. *Metabolism*, **14**, 1276 – 81.

Wood, D. H. (1970). An ecological study of *Antechinus stuartii* (Marsupialia) in a south-east Queensland rain forest. *Aust. J. Zool.*, **18**, 185 – 207.

Woollard, P. (1971). Differential mortality of *Antechinus stuartii* (Macleay): nitrogen balance and somatic changes. *Aust. J. Zool.*, **19**, 347 – 73.

Woolley, P. (1966). Reproduction in *Antechinus* sp. and other dasyurid marsupials. *Symp. Zool. Soc. London*, **15**, 281 – 94.

22 A metabolic study of the Quokka, Setonix brachyurus, in varying regimes of temperature and humidity

Wayne C. Packer*

Introduction and ecological background

An increasing number of studies in recent years have attempted to determine how the individuals in a population survive in harsh environments. Few free-living animals, if any, spend their entire life cycle in constant conditions. Most are exposed to a wide variety of physical and biological factors which fluctuate, either with reasonably predictable periodicity (for example, temperature, light, salinity, etc.) or more irregularly (for example, population density, food supply, rainfall, etc.). Even those factors, such as temperature, which generally follow a predictable pattern (that is, higher during day than night) may vary in intensity from day to day. Short-term adjustments to changes in the physical and biological environment are achieved through physiological and behavioural responses. This paper examines some responses of the Quokka, *Setonix brachyurus*, to varying temperature and humidity regimes under laboratory conditions, relating them to observed behaviour in the field. For further studies of Australian macropods in relation to their environments see Tyndale-Biscoe (1973), Russell (1974), Waring, Moir and Tyndale-Biscoe (1966) and Kaufmann (1974).

Quokkas are small wallabies, weighing 3 to 4 kg, nocturnally active, indigenous to the south-west corner of Western Australia. A number of relict populations persist on the mainland in relatively isolated and undisturbed valleys in the Stirling Ranges and near the south coast. In these situations the Quokka inhabits dense thickets along the valley floor where water seepage during the dry summer maintains fairly luxuriant vegetation. The animals emerge from these diurnal shelter areas at night to graze on the forest floor (Sadleir, 1959; Storr, 1964).

Two island populations also exist. One, which has been little studied, is on Bald Island near Albany on the south coast. The other population is on Rottnest Island, approximately 19 km off the coast from Fremantle. This population has been investigated in many ways since 1948, with particular emphasis on population regulation (especially the occasional crashes (Waring, 1956) and continuing studies of nutritional physiology (Moir, Somers and Waring, 1956; Waring *et al.*, 1966; Brown, 1968)).

*Dr Wayne Packer was educated at Whittier College and Stamford University, USA. Awarded a Fulbright Scholarship to the University of Western Australia, he is now a Senior Lecturer in the Department of Zoology, with research interests in ecology and behaviour of marsupials.

379

Historically, Rottnest has been separated from the mainland since the end of the Pleistocene (Churchill, 1959). The present vegetation, described by Storr, Green and Churchill (1959) consists solely of elements of a coastal complex derived from the adjacent mainland. Closed formations of *Acacia rostellifera* from 10 to 20 feet in height covered much of the island originally and provided shelter for the Quokka, but numerous fires, particularly over the past 40 years, have destroyed most of the tall scrub. Much of the *Acacia* not killed by the fires was unable to regenerate due to overgrazing by the Quokka. Except where reafforestation with exotic species has been attempted, the dominant vegetation has a heath-like appearance and consists of various tussock grasses and small, low shrubs. A number of salt lakes are located on the eastern end of the island near the resort settlement. Fresh-water seepages are exposed in several places around these lakes after evaporation has lowered the lake levels in the summer. These and a couple of fresh-water swamps provide the only naturally occurring potable water during the summer. No fresh water is available on the western half of the island at all during summer except following brief and irregular thunderstorms. The Quokka on Rottnest therefore persists on environment which differs from that of mainland populations, especially in respect of food and water supplies and rest sites offering shade during the day.

While the Quokka occupies the entire island, the total population is effectively subdivided into a large number of sub-populations with relatively little interchange of adults between them. For example, Holsworth (1967) found 15 such groups in the 140 ha comprising the relatively geographically isolated West End area of the island. Within these areas Holsworth (1967) and Storr (1964) noted seasonal changes in the area utilised by the Quokka at night. Nicholls (1971) followed individually marked animals and found that some changed not only their nocturnal feeding site but also their diurnal rest site. These changes he interpreted in terms of the distribution of vegetation with regard to both protection from the weather and seasonal changes in food plants, both from the standpoint of availability (that is, presence or absence) and nutritional preference. Holsworth (1967) applied the term group territory to these sub-populations. However, as Nicholls (1971) points out, these areas correspond to the distributions of vegetation which provides the food and shelter. Since the term 'territory', when used in a behavioural context, normally implies a defended area, the term 'group territory' is misleading because no indication of exclusive occupancy of these areas has been found (Dunnett, 1962; Nicholls 1971; Kitchener, 1972; Packer, personal observation). Such areas clearly represent a group of broadly overlapping home ranges to which there is an individual attachment, as evidenced by the high proportion who return following artificial translocation (Packer, 1963). Movement over long distances sometimes occurs on the island (moves of up to 1.8 km have been recorded; Dunnett, 1962) but these are apparently relatively infrequent, short-term and probably goal oriented (for example, to obtain fresh water during summer).

The behaviour of animals within these sub-population areas varies not only seasonally but from day to day. In the Barker's Swamp area, the animals shelter under dead, felled *Melaleuca pubescens* and other rather open plant cover during the winter, but some change rest sites to an adjacent area during the summer, sheltering beneath both the large sedge *Gahnia trifida* surrounding the swamp and the *Arthrocnemum halocnemoides* which is exposed on the floor of the swamp after the water has mostly evaporated. Movement of animals through gaps in drift fences within the *Melaleuca*, between the *Melaleuca* and the swamp proper, and between these two areas and a portion of the surrounding feeding area, was monitored periodically over one year (Packer, 1965). During even the most severe summers a small pool, about 3 - 4 m in diameter, persists and provides potable water. When the data for all days within each of four 6 week study periods were combined, the pattern of time of activity was clearly bimodal, with peaks near dusk and dawn. Within the *Melaleuca* shelter area there was movement during the day and little at night, while the converse was true for the other two drift fences. Thus a daily cycle of movement, from diurnal shelter to nocturnal feeding sites and return, was documented. Although the overall pattern of movement through the drift fences was bimodal, there was considerable variation from day to day in both the total number of movements recorded and the periods of greatest activity. These differences could to some extent be correlated with the concurrent weather and sometimes with the weather immediately preceding. During the dry part of the year (October to March) high activity was associated with a fall in barometric pressure which directed a flow of warm, dry, easterly air over the island. During the wetter months (April to September) high activity counts were often associated with rain, but direct observation showed the animals sought shelter from sudden, heavy rainfall. This suggests that rain was generally inhibitory to activity and the animals were most active under clearing conditions with rising barometric pressure and cooler temperature. A similar bimodal pattern of activity was observed in captive animals (Packer, 1969).

Working on the same field population subsequently, Kitchener (1970, 1972) demonstrated a dominance hierarchy within the males and observed defence of the rest site by males. Some of the animals moved rest sites on hot summer days from the *Melaleuca* area to under the *Gahnia* or *Arthrocnemum,* where the measured total radiation load was much less. When the metabolic rate, rectal temperature and evaporative water loss of individuals which moved was compared with similar data from those which did not change rest sites, it was found that the 'movers' were those significantly less able to cope with high temperatures — that is, they became hyperthermic at air temperatures above 32 °C.

With this background in mind, it was decided to examine quantitatively the amount and periodicity of activity, and incidentally food and water consumption, in relation to certain temperature and relative humidity regimes in the laboratory, in order to elucidate further, and under controlled conditions, the interaction between these aspects of the Quokka's biology.

Methods

Adult male quokkas were captured on Rottnest Island and maintained in individual pens in the animal yards of the University of Western Australia, feeding in captivity on an enrighed chaff mixture for at least 30 days prior to study. During the experiments, the animals were housed in a running wheel, constructed of plywood and plastics-coated expanded metal, with a diameter of 90 cm and a running track 45 cm wide. The wheel was illuminated during the 'light' phase of each day by three 150 W incandescent bulbs mounted in a reflector which directed most of the light (1270 $1/m^2$) onto the wheel. These were controlled automatically through a dimming switch which simulated, over a half-hour period, dawn (beginning at 0600 hours) and dusk (beginning at 1800 hours). Two 6 V flashlight bulbs on the sides of the reflector provided a dim light (subjectively, roughly similar to full-moon light) during the 'dark' phase of each day.

The experiments were run in a lightproof, air-conditioned room where both temperature and relative humidity could be controlled, as detailed below. Food (the enriched chaff mixture to which they had become accustomed in the yards) and water were provided *ad lib.* in dishes mounted outside the wheel.

The following parameters were recorded either automatically or manually: (1) Activity – number of revolutions per hour was recorded on a bank of 24 telephone-type counters, with a graphic display on an Esterline-Angus Event Recorder; the number of minutes per hour spent running was estimated from this display, and the mean rate per minute derived arithmetically. (2) Food consumption – a measured quantity of food was provided daily and the excess saved and dried at 105 °C to determine the daily dry matter intake (D.M.I.). (3) Water consumption – a measured amount was provided daily and the excess remaining, allowing for evaporation, was recorded. The room was entered at varying times between 1000 hours and 1600 hours to remove excess food and water and provide fresh. Data for both food and water consumption are adjusted in the results to a standard size of animal, using the expressions ml water/day per kg bodyweight$^{0.80}$ (Richmond, Langham and Trujillo, 1962) and g D.M.I./day per kg bodyweight$^{0.75}$ (Kleiber, 1961).

The experimental procedure involved the use of four different combinations of temperature and humidity (treatments shown in Table 22.1). The air-conditioning system was unreliable, particularly with regard to humidity. The relative humidities used, therefore, represent the maximum and minimum that the system could provide at each of the selected temperatures rather than a considered choice by the investigator, and the data were analysed on the basis of whether the temperature and relative humidity were comparatively high or low (for example, in Table 22.1 and subsequently LL indicates low temperature, about 20°C, and low relative humidity, about 50 per cent). The temperatures used are near the extremes of the thermal neutral zone for *Setonix* (Jones, 1970).

Four animals were used. Each was placed in the wheel individually and given approximately 24 h to adjust to the small area before the wheel was allowed to revolve. All of the animals learned without specific training to run and stop the wheel at will within one night of first experiencing it. Each experiment lasted 28

Table 22.1 Actual temperature and relative humidity employed for each experimental treatment

Treatment	Temp. (°C)	Rel. Hum. (%)
1 (LL)	20 ± 2	50 ± 5
2 (LH)	20 ± 2	66 ± 5
3 (HL)	30 ± 2	40 ± 5
4 (HH)	30 ± 2	67 ± 5

days, following a 7 day accommodation period, with each treatment serving as the accommodation and base condition for one animal. After a 4 day base period the treatment condition was then changed in a random manner to one of the other three for a further period of 4 days and then the animal was exposed to the base treatment again for 4 days. This pattern continued until the animal experienced all three of the non-base treatments (for 4 days each) alternating with identical periods of the base treatment (for a total of 16 days). Treatment changes were instituted at 1200 hours. Since considerable variability was expected between animals, this experimental design, while perhaps not ideal, allowed each animal to serve as its own control and permitted an orthogonal analysis of the data. The period of 4 days was selected because this is a little over one day longer than the period of the average rate of passage of food through the digestive tract (Calaby, 1958).

Several additional animals were studied to look at, in a different manner, the relationship between activity and D.M.I.; the procedures are detailed in the Results section.

In order to examine the interaction between these variables and the treatments, the data were examined by a regression approach to the analysis of variance in which the temperature, humidity and the interaction between temperature and humidity were taken out as factorial effects from a 2 × 2 factorial. Because the primary interest was to look for differences in behaviour under certain temperature and humidity regimes, and because these factors were not absolutely constant (see Table 22.1), numerical values of one (1) and two (2) were assigned to the low and high regimes, respectively, of both temperature and humidity.

Results

The data were initially analysed by single-factor analysis of variance. Table 22.2 shows the mean value (± standard error of the mean) for each animal for the five

Table 22.2 Mean values of measured behaviour parameters for each of the four *Setonix* studied and the between-animal variance ratio (V.R.) from a single factor analysis of variance for each measurement (level of statistical significance here and subsequently:* = 0.05, ** = 0.01, *** = 0.001)

	Mean value (± S.E.M.) − Animals				V.R. (3108 d.f.)	Stat. sign.
Measurement	3014	3089	3086	3085		
Revolution per day	881.2 ± 147.9	401.9 ± 38.1	808.9 ± 126.9	810.6 ± 133.1	3.34	*
Minutes run per day	130.2 ± 16.6	71.6 ± 6.4	142.3 ± 20.0	180.7 ± 24.8	6.16	***
Mean revolutions per minute	6.52 ± 0.19	5.49 ± 0.14	5.25 ± 0.19	4.07 ± 0.18	31.89	***
Food intake (g/kgW $^{0.75}$/day)	29.39 ± 0.57	31.09 ± 0.69	46.19 ± 1.12	36.10 ± 0.88	80.61	***
Water intake (ml/kgW $^{0.8}$/day)	96.30 ± 5.58	71.84 ± 2.30	100.39 ± 5.12	103.53 ± 4.32	10.23	***

Table 22.3 Mean values of measured behaviour parameters for each of the four treatments on *Setonix* and the between-treatments variance ratio from a single factor analysis of variance for each measurement

Measurement	Mean value (± S.E.M.) – Treatment				V.R. (3108 d.f.)	Stat. Sign.
	1 (LL)	2 (LH)	3 (HL)	4 (HH)		
Revolutions per day	915.0 ± 163.3	1093.1 ± 128.5	477.9 ± 69.2	418.9 ± 49.99	8.62	***
Minutes run per day	150.3 ± 21.9	195.1 ± 23.0	83.3 ± 11.1	96.1 ± 11.1	8.45	***
Mean revolutions per minute	5.93 ± 0.22	5.51 ± 0.17	5.58 ± 0.22	4.31 ± 0.23	10.86	***
Food intake $(g/kgW^{0.75}/day)$	36.45 ± 1.32	42.47 ± 1.61	30.01 ± 1.02	33.84 ± 0.85	17.92	***
Water intake $(ml/kgW^{0.8}/day)$	74.93 ± 2.26	84.22 ± 3.94	101.56 ± 5.85	111.48 ± 4.40	14.69	***

Table 22.4 Summary of the regression approach to analysis of variance showing the variance ratio for various behavioural measurements compared with environment parameters in *Setonix*

Measurement (Y)	Environment parameter (X)	V.R. by each (X)	Degrees of freedom	Stat. sign.	R^2	F-value of regression (3108 d.f.)
(1) Revolutions per day	Temp.	24.60	1110	***	0.193	6.41
	Humid.	0.28	1109	N.S.		***
	$T \times H$	1.11	1108	N.S.		
(2) Minutes run per day	Temp.	21.61	1110	***	0.190	6.28
	Humid.	2.65	1109	N.S.		***
	$T \times H$	0.82	1108	N.S.		
(3) Mean revolutions per minute	Temp.	11.28	1110	**	0.232	8.07
	Humid.	15.26	1109	***		**
	$T \times H$	3.85	1108	*		
(4) Food intake $(g/kgW^{0.75}/day)$	Temp.	33.49	1110	***	0.343	13.946
	Humid.	16.87	1109	***		***
	$T \times H$	1.07	1108	N.S.		
(5) Water intake $(ml/kgW^{0.8}/day)$	Temp.	37.90	1110	***	0.289	10.89
	Humid.	5.06	1109	*		***
	$T \times H$	0.01	1108	N.S.		

variables examined (that is, number of revolutions per day, minutes spent running per day, the mean rate of running per minute, food intake (D.M.I.) and water intake) and the variance ratio for each. Table 22.3 shows the mean value (± SEM) and the variance ratio for these variables when compared by treatment (see Table 22.1).

The results of the regression analysis of variance are summarised in Table 22.4 and the regression equations for each of the measurements is shown below (equations 22.1 – 22.5).

Revolutions per day:
$$Y = -200.0X_1 + 415.1X_2 - 237.1X_3 + 936.9 \qquad (22.1)$$
Minutes spent running per day:
$$Y = -34.9X_1 + 76.89X_2 - 32.0X_3 + 140.4 \qquad (22.2)$$
Mean revolutions per minute:
$$Y = 0.49X_1 + 0.41X_2 - 0.84X_3 + 5.88 \qquad (22.3)$$
Food intake ($g/kW^{0.75}$ per day):
$$Y = -3.84X_1 + 8.98X_2 - 2.58X_3 + 33.86 \qquad (22.4)$$
Water consumption ($ml/kW^{0.8}$ per day):
$$Y = 25.76X_1 + 8.52X_2 + 0.75X_3 + 39.89 \qquad (22.5)$$

where X_1 = temperature, X_2 = relative humidity and X_3 = temperature × humidity interaction.

Regression equations for D.M.I. on number of revolutions and water consumption on D.M.I. were calculated and are as follows:

D.M.I. ($g/kgW^{0.75}/day$):
$$Y = 0.0039X_1 + 32.88 \ (r = 0.3201, **) \qquad (22.6)$$

where X_1 = number of wheel revolutions (see also the caption to Table 22.2).

Water consumption ($ml/kgW^{0.8}$ per day):
treatment 1 $Y = 1.059X_1 + 36.336 \ (r = 0.6185, ***)$
treatment 2 $Y = 1.295X_1 + 29.207 \ (r = 0.5310, **)$
treatment 3 $Y = 2.623X_1 + 22.736 \ (r = 0.4591, *)$
treatment 4 $Y = 3.673X_1 - 12.804 \ (r = 0.7124, ***)$

$$(22.7)$$

where X_1 = D.M.I. ($g/kgW^{0.75}$ per day)

The relationship between activity and D.M.I. was pursued in another series of experiments. Two animals, with pretreatment similar to the above, were tested at 20°C with low relative humidity and two at 30°C with low relative humidity. In these cases the experimental treatment involved alternately permitting the animal to run in the wheel for four days and locking the wheel to prevent its revolving for four days. The numbers of revolutions run under the two temperature regimes were not different from that expected as a result of the earlier work (see Table 22.3). When the D.M.I. and water intake for each animal were compared under the two conditions (that is, wheel free versus wheel not free) using Student's 't'-test, all of the values of 't' were small and none even approached the 5 per cent level of probability. One other animal was tested. It was acclimated in the locked wheel for 30 days at 30°C and low relative humidity, and then food and water intake were recorded for 10 days followed by another 10 days with the wheel free to revolve. The mean D.M.I.s, 22.27g and 21.74g, respectively, (t = 0.1487 with 18 d.f.) were not different.

The general pattern of activity in the wheel (that is, the time when wheel-running occurred, and the duration) was variable between animals; activity occurred mostly during the dark phase and usually exhibited peaks near dusk

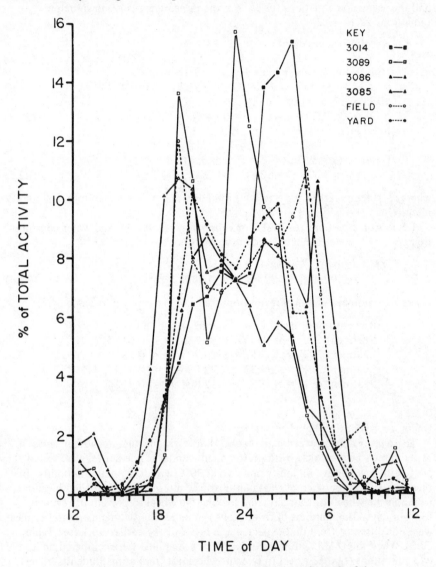

Figure 22.1 The total number of revolutions per hour for each of the four quokkas, transformed and plotted as a percentage of the total for each animal per hour. For comparison, activity curves derived from captive (yard) and free-living (field) animals are included.

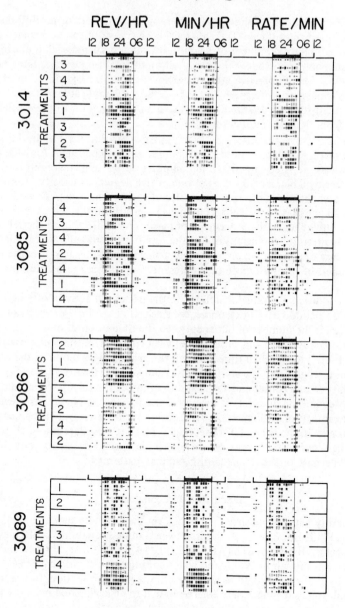

Figure 22.2 Pattern of activity of quokkas in a wheel, showing the time and relative amount per hour of number of wheel revolutions, minutes spent running and mean rate of running. Blanks = no activity: the five symbols of increasing density indicate stages of increasing activity, each symbol representing 20 per cent of the values above zero

and/or drawn. Figure 22.1 shows the pattern in total number of revolutions per hour for each of the animals. The data have been transformed and graphed as a percentage of the total number of revolutions which occurred each hour, in order to facilitate comparison between grossly different total frequencies. Superimposed on these data, for comparison, are curves showing the time of utilisation of a food point by a small group of captive but non-caged quokkas (yard) (Packer, 1969) and movement through drift fences by free-living quokkas on Rottnest (field) (adapted from figure 2-1V, Packer, 1965).

Figure 22.2 shows the pattern in the number of revolutions, number of minutes and mean revolutions per minute for each hour of the experiment for each animal. The figure was prepared directly from computer output and each hour is repre-sented either by a blank (= 0 value) or one of five degrees of density of print. Values greater than zero for each parameter of each animal were assigned symbols such that 20 per cent of the values fall in each density. In this way, the figure shows not only when activity occurred but also some indication of the relative amount. Arbitrarily, revolutions of less than six in an hour and rate calculations based on only one minute in an hour were assigned the value zero because very small numbers of revolutions do not represent in most cases actual running in the wheel.

Certain additional parameters related to the pattern of activity often assessed in studies on periodicity (for example, see Aschoff *et al.*, 1971) were examined. These include: the number of hours between the time of onset of activity (T-onset) on two consecutive days and similarly between consecutive times of cessation of activity (T-off); the lapse in time between the beginning of action of the dimmer control on the lights and the start (at dusk) and stop (at dawn) of the active phase; the actual time of T-onset and T-off and the time of the calculated middle of the active phase; and the duration of the active and non-active phases. As is evident from an examination of Figure 22.2, there are some differences between animals — for example, in the time of T-onset and T-off, but consideration of these is beyond the scope of this paper. No differences between treatments could be identified because of the strong 'within-animal' patterns.

Discussion and conclusion

The variability between animals (Table 22.2) in their activity measurements is not surprising. While not usually stressed, between-animal variability is often evident in biological rhythms research (Aschoff *et al.*, 1971). In theory, however, differ-ences between animals in D.M.I. and water consumption should have been removed by the transformations described in the methods section. In normal nutritional studies it is necessary for the animal to be acclimated not only to temperature, etc., but also to the cage in which the work is to be carried out until a constant weight is achieved (for example, Brown, 1968). The acclimation period was short in these experiments and the treatment changes sudden. These factors combined probably account for some of the variability in water consumption and D.M.I. between animals.

Despite the high variability between animals, the treatment effects reflected in Tables 22.3 and 22.4 and equations 22.1 – 22.5 are unmistakable. There is clearly a negative correlation between temperature and both the number of revolutions of the wheel and the time spent running per day, while humidity has no significant influence. In each case approximately 19 per cent of the variation in the data is accounted for by the regression. Both temperature and humidity appear to be positively correlated with the rate of running, although the regression coefficients of equation 22.3 are quite small. To a greater extent than in the other measures of activity, the negative but statistically significant interaction term (that is, X_3 in equation 22.3) influences the trend seen across Table 22.3 for this measurement. It is uncertain to what extent these arithmetically derived rate figures accurately reflect the actual rate of running. The rates of four to six per minute, shown on Tables 22.2 and 22.3, effectively represent the rate of slow walking. On several occasions, rates of 30 – 40 revolutions per minute by undisturbed animals were noted when watching the recording equipment. These bursts lasted only 10 – 20s and could not be distinguished from much slower rates on the event recorder which advanced at 3 in/h.

The strong influence of humidity as well as temperature on D.M.I. is clearly shown in Table 22.3, where the values are greater at high relative humidity than at low humidities for a given temperature, while the overall trend is toward lower D.M.I. with increasing temperature (see equation 22.4). The simplest explanation of this effect is that, with high relative humidity, evaporative water cooling would be less effective and the animals would require extra D.M.I. (that is, energy) to handle this. A technical problem may also contribute. The food provided daily normally contained about 5 per cent by weight of water. Some of the additives to the chaff mixture were hygroscopic and after 24 h, particularly at high relative humidity, the unconsumed food contained up to 15 per cent by weight of water. Replicate samples were taken to determine water content before calculating the D.M.I., but if the averages calculated for the high humidity treatments underestimated the water content, then the D.M.I. would be correspondingly overestimated. Since it was impossible to monitor the water content of each mouthful of food, this remains unresolved.

Water consumption increased both with increasing temperature and increased humidity (see equation 22.5), although it is clear from the respective variance ratios shown in Table 22.4 that temperature had the greater effect. Overall, 28.9 per cent of the variability was accounted for by the regression. The consumption figures shown in Tables 22.2 and 22.3 underestimate actual water intake because the water in the food, as mentioned above, was not included.

D.M.I. does not appear to depend upon the number of revolutions of the wheel. The regression coefficient is very small (see equation 22.6) and, although the correlation coefficient ($r = 0.3201$) indicates a probability of less than 1 per cent, a 't'-test shows that the slope is not different from zero ($t = 0.0156$). No correlation was found in the supplementary experiments either. Since more energy should be required to run in the wheel than just for sitting, the most logical explanation for this apparent anomaly may lie in the large variations in the daily D.M.I. Bakker (1968 and personal communication) suggests that in the wallaby *Macropus eugenii* it is necessary to limit D.M.I. to the average daily maintenance requirement rather than providing food *ad lib*. in order to undertake meaningful nutritional studies.

D.M.I. and water consumption are not correlated overall ($r = 0.158$) because each of the parameters responds differently to the treatments as described above.

Within treatments, however, there is a significant positive correlation between these two. Temperature evidently exerts the stronger influence because the slope for the regressions for treatments 1 and 2 (about 20 °C) and for 3 and 4 (about 30 °C) are not significantly different from each other but 1 and 2 are different from 3 and 4.

It is clear that with a constant light cycle involving 12 hours of light and 12 hours of darkness, changes in temperature and humidity strongly and inversely influence the amplitude of most of the parameters of the activity cycle as studied here. If temperature or humidity have any other influence on the activity rhythm, they are too subtle to be distinguished by the techniques employed in view of the large variation in most of the measured data. This is not surprising if some aspect of the light cycle itself acted as the *zeitgeber* to entrain the general pattern of activity. Certainly, changing the duration of the L:D phases or the intensity of illumination can produce different patterns of activity (Packer, unpublished data).

The response of the Quokka in an activity wheel does not fully mimic the response of field animals. The basically nocturnal pattern of activity with peaks near dusk and/or dawn is found here as well as all other situations tested (Packer, 1965, 1969). However, in the field high activity was associated, during the dry months, with higher temperature and not lower as found in these experiments. Since field temperatures were uncontrolled, and often varied greatly from day to day as well as seasonally to extremes outside the thermoneutral zone of the Quokka, the differences in behaviour are not surprising.

Acknowledgments

The work was financed by a research grant from the University of Western Australia. Dr N. Goodchild, Biometrics Unit, Institute of Agriculture, University of Western Australia, and Mr N. Campbell, Division of Mathematics and Statistics, Commonwealth Scientific and Industrial Research Organisation, advised on statistical problems. Professor H. Waring, Department of Zoology, University of Western Australia, read a draft of the manuscript and made valuable suggestions.

References

Aschoff, J., Gerecke, U., Kureck, A., Pohl, H., Rieger, P., Saint-Paul, U. V. and Wever, R. (1971). Interdependent parameters of circadian activity rhythms in birds and man. In: *Biochronometry* (ed. M. Menaker), Nat. Acad. Sci., Washington DC, pp. 3 – 29.

Bakker, H. R. (1968). Ecology of the tammar, *Macropus eugenii.* B.Sc. (Hons.) Thesis, University of Western Australia.

Brown, G. D. (1968). The nitrogen and energy requirements of the euro (*Macropus robustus*) and other species of macropod marsupials. *Proc. Ecol. Soc. Australia*, 3, 106 – 12.

Calaby, J. H. (1958). Studies on marsupial nutrition. II. The rate of passage of food and digestibility of crude fibre and protein by the quokka, *Setonix brachyurus* (Quoy and Gaimard). *Aust. J. Biol. Sci.*, 11, 571 – 80.

Churchill, D. M. (1959). Late Quaternary eustatic changes in the Swan River District. *J. Roy. Soc. W. Australia*, 42, 53 – 5.

Dunnet, G. M. (1962). A population study of the quokka, *Setonix brachyurus* (Quoy and Gaimard) (Marsupialia) II. Habitat, movements, breeding and growth. *C.S.I.R.O. Wildlife Res.*, 7, 13 - 32.

Holsworth, W. N. (1967). Population dynamics of the quokka, *Setonix brachyurus*, on the West End of Rottnest I., Western Australia I. Habitat and distribution of the quokka. *Aust. J. Zool.*, 15, 29 - 46.

Jones, A. C. (1970). Energy metabolism and development of homeothermy in the quokka (*Setonix brachyurus*). Ph.D. Thesis, University of Western Australia.

Kaufmann J. (1974). Social ethology of the whiptail wallaby, *Wallabia parryi*, in north-east New South Wales. *Animal Behav.*, 221, 281 - 369.

Kitchener, D. J. (1970). Aspects of the response of the quokka to environmental stress. Ph.D. Thesis, University of Western Australia.

Kitchener, D. J. (1972). The importance of shelter to the quokka, *Setonix brachyurus* (Marsupialia), on Rottnest Island. *Aust. J. Zool.*, 20, 281 - 99.

Kleiber, M. (1961). *The Fire of Life*, Wiley, New York.

Moir, R. J., Somers, M. and Waring, H. (1956). Studies on Marsupial nutrition I. Ruminant-like digestion in a herbivorous marsupial (*Setonix brachyurus* Quoy and Gaimard). *Aust. J. Biol. Sci.*, 9, 293 - 304

Nicholls, D. G. (1971). Daily and seasonal movements of the quokka, *Setonix brachyurus* (Marsupialia), on Rottnest Island. *Aust. J. Zool.*, 19, 215 - 26.

Packer, W. C. (1963). Homing behaviour in the quokka, *Setonix brachyurus*, (Quoy and Gaimard) (Marsupialia). *J. Roy. Soc. W. Australia*, 46, 28 - 32.

Packer, W. C. (1965). Environmental influences on daily and seasonal activity in *Setonix brachyurus* (Quoy and Gaimard) (Marsupialia). *Animal Behav.*, 13, 270 - 83.

Packer, W. C. (1969). Observations on the behavior of the marsupial *Setonix brachyurus* (Quoy and Gaimard) in an enclosure. *J. Mammal.*, 50, 8 - 20.

Richmond, C. R., Langham, W. H. and Trujillo, T. T. (1962). Comparative metabolism of tritiated water by mammals. *J. Cell. Comp. Physiol.*, 59, 45 - 53.

Russell, E. M. (1974). Recent ecological studies on Australian marsupials. *Aust. Mammal.* 1, 189 - 211.

Sadleir, R. M. (1959). Comparative aspects of the ecology and physiology of the Rottnest and Byford populations of the quokka, *Setonix brachyurus* (Quoy and Gaimard). B.Sc. (Hons.) Thesis, University of Western Australia.

Storr, G. M. (1964). The environment of the quokka (*Setonix brachyurus*) in the Darling Range, Western Australia. *J. Roy. Soc. W. Australia*, 47, 1 - 2.

Storr, G. M., Green, J. W. and Churchill, D. M. (1959). The vegetation of Rottnest Island. In: Rottnest Island: The Rottnest Biological Station and Recent Scientific Research (ed. E. P. Hodgkin and K. Sheard), *J. Roy. Soc. W. Australia*, 42, 65 - 95.

Tyndale-Biscoe, C.H. (1973). *Life of Marsupials*, Edward Arnold, London.

Waring, H. (1956). Marsupial studies in Western Australia. *Aust. J. Sci.*, 18, 66 - 73.

Waring, H., Moir, R. J. and Tyndale-Biscoe, C. H. (1966). Comparative physiology of marsupials. *Adv. Comp. Physiol. Biochem.*, 2, 237 - 376.

23 Some aspects of renal physiology in the Brush-tailed possum, Trichosurus vulpecula

Ian A. Reid*

Introduction

In the vertebrates the kidneys play a vital role in the regulation of the volume and composition of the extracellular fluid. To control this function a variety of mechanisms, both neural and endocrine, have evolved. These mechanisms have been studied extensively in eutherians but few detailed comparative investigations have been made. This is particularly true in the case of the marsupials. Until quite recently, little information was available on kidney structure and function and, in general, the data concerning the role of the adrenal cortex and the renin - angiotensin system in the regulation of fluid and electrolyte balance was very limited. It therefore seemed desirable to study renal physiology in the marsupials and to investigate in detail some of the mechanisms responsible for the regulation of renal function.

The purpose of this chapter is to describe the results of such an investigation in the Brush-tailed possum, *Trichosurus vulpecula*. This species was chosen because it is readily available and easily maintained in captivity. The studies were performed in the Department of Physiology at Monash University, Melbourne, Australia in association with Dr I. R. McDonald.

Three general areas are considered: renal structure and function; the role of the adrenal cortex; the renin - angiotensin system. Where appropriate, the information that is available in other marsupial species is also reviewed.

Renal structure

The structure of the kidney in *Trichosurus* resembles that described in several other marsupial species (Sperber, 1944; Bentley, 1955). There is a well-defined cortex and medulla and the arciform arteries and veins are clearly visible at the junction of these two areas (Figure 23.1a). The medulla is divided into an inner and outer zone and there is a single low papilla projecting into the pelvis. There are two types of nephron: cortical, with short loops of Henle which turn in the

*Dr Ian Reid is Adjunct Assistant Professor of Physiology, Department of Physiology, University of California, San Francisco, USA. His research interests include regulation of fluid and electrolyte balance, with particular emphasis on the role of the renin angiotensin system; and comparative aspects of renal physiology.

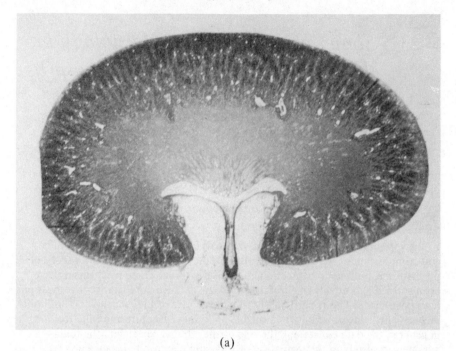

(a)

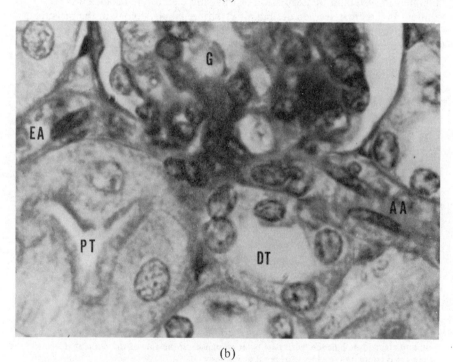

(b)

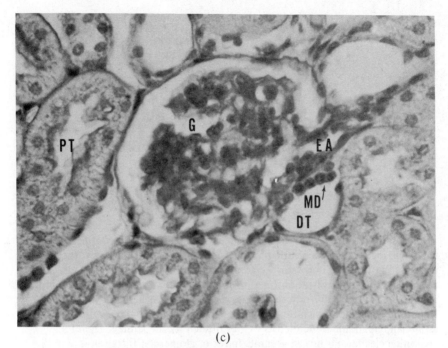

(c)

Figure 23.1(a). Mid-sagittal section of a possum kidney. Stain:
Mallory's tri-chrome. (b) and (c) Sections of renal cortex
showing details of the glomerulus (G) the afferent (A A) and
efferent (E A) arterioles and the proximal (P T) and distal
(D T) tubules. M D = macula densa. Stain: PAS and hematoxylin.
(a) and (b) from Reid and McDonald (1968a).

cortex; and juxtamedullary, with long loops of Henle that descend for a variable
distance into the medulla – in some cases, almost to the pelvis. The loops of
Henle have a thin segment. Vasae rectae are visible as discrete bundles of straight
vessels which descend deeply into the medulla in a course parallel to the loops of
Henle. Glomerular structure in *Trichosurus* closely resembles that in eutherians
(Pak Poy, 1957; Figure 23.1, b, c). In addition, a well-defined juxtaglomerular
apparatus is present. This includes the juxtaglomerular cells, which are modified
smooth muscle cells in the media of the afferent arteriole, and the macula densa,
an area of modified renal tubular epithelium at the point where the distal tubule
contacts the glomerular pole. Typical arrangements are shown in Figure 23.1, b, c.

Renal function

Glomerular filtration rate and renal plasma flow

In order to estimate glomerular filtration rate and renal blood flow in *Trichosurus*,
the renal clearances of inulin and para-aminohippuric acid were measured. Creatinine clearance, which is a good index of glomerular filtration rate in several
eutherian species, was also measured. The results are summarised in Table 23.1.

Table 23.1 Renal clearances in the possum. Each value represents the mean ± one standard deviation. PAH = para-aminohippuric acid. Data from Reid and McDonald (1968a)

Inulin clearance (ml/min per kg)	2.5 ± 0.3
Creatinine clearance (ml/min per kg)	3.3 ± 0.7
PAH clearance (ml/min per kg)	13.0 ± 2.1
PAH extraction ratio (%)	89.0 ± 4.0
Filtration fraction	0.20 ± 0.03

Expressed in terms of body weight, the clearance values for *Trichosurus* fit into the range recorded in four eutherian species (Table 23.2). If, however, they are expressed in terms of body surface area, they fall at the end of a wide eutherian range. The values for filtration fraction and extraction ratio of para-aminohippuric acid differ only slightly from the figures recorded for eutherians (Smith, 1951). Creatinine clearance is significantly higher than inulin clearance, indicating that, as in man, creatinine is secreted as well as filtered by the kidney. Therefore creatinine clearance is not an accurate index of glomerular filtration rate in this species.

Table 23.2 Comparison of inulin and para-aminohippuric acid (PAH) clearances in the possum and four eutherian species. Data from Reid and McDonald (1968a) and Smith (1951)

	kg^{-1} body wt		m^{-2}	
	C_{inulin} (ml/min)	C_{PAH} (ml/min)	C_{inulin} (ml/min)	C_{PAH} (ml/min)
Possum	2.5	13.0	20.6	106
Rat	6.0	22.0	40.4	146
Rabbit	3.1	18.2	50.3	296
Dog	4.3	13.5	84.4	266
Man	2.0	10.0	68.2	347

Stop flow analysis

The technique of stop flow analysis has been employed to localise the sites at which various substances are transported in the eutherian kidney. The details of the technique have been described elsewhere (Malvin, Wilde and Sullivan, 1958). In *Trichosurus* the concentration patterns developed for sodium, potassium, glucose, inulin and para-aminohippuric acid during stop flow are similar to those reported for the dog (Malvin *et al.*, 1958; Wilde and Malvin, 1958) and man (Jahnecke *et al.*, 1966). Taken together with the clearance data described above, these observations indicate that glomerular and tubular function in this marsupial closely resembles that in eutherians.

Renal concentrating capacity

The presence of long loops of Henle in the possum kidney suggests that this marsupial possesses the ability to form a concentrated urine. To test this, possums were deprived of food and water for three days. During this period, mean urinary osmolality increased from 184 ± 18 to 1061 ± 151 mOsm/kg H_2O (Table 23.3). The maximum urinary osmolality recorded was 1504 mOsm/kg H_2O. The corresponding plasma osmolality was 316 mOsm/kg H_2O, giving a urine-to-plasma osmolality ratio of 4.8. It is likely that a higher figure would have been obtained if the animals had been deprived of food and water for a longer period of time, or allowed to eat dry food during the period of water deprivation.

Thus it is clear that *Trichosurus* is able to produce a concentrated urine. Other marsupials, including *Setonyx* (Bentley, 1955), *Macropus* (Ealey, Bentley and Main, 1965) and *Dasycercus* (Schmidt-Nielsen and Newsome, 1962), are also able to produce concentrated urine.

Table 23.3 Changes in urine volume, urine osmolality and plasma osmolality in possums during three days without food and water. Each value represents the mean ± one standard error; ranges are shown in parenthesis. Data from Reid and McDonald (1968a)

	Control	Day 1	Day 2	Day 3
Urine volume (ml/24 h.)	170 ± 9 (127–213)	77 ± 14 (58–118)	27 ± 8 (8–45)	23 ± 8 (8–42)
Urinary osmolality (mOsm/kg H_2O)	184 ± 18 (112–314)	309 ± 40 (220–388)	895 ± 141 (704–1170)	1061 ± 151 (844–1504)
Plasma osmolality (mOsm/kg H_2O)	303 ± 5 (288–316)	309 ± 4 (298–319)	314 ± 5 (300–320)	316 ± 7 (308–330)

Role of the adrenal cortex

The steroids of the eutherian adrenal cortex play a vital role in the control of fluid and electrolyte balance, primarily by regulating the renal excretion of electrolytes and water. In this regard, the steroids with mineralocorticoid activity which increase sodium reabsorption in the distal tubule are the most significant. Aldosterone is the most potent of these, but deoxycorticosterone and corticosterone also have some activity. Glucocorticoids such as cortisol have some mineralocorticoid activity, but also have other important renal actions, including effects on glomerular filtration rate and water excretion. This section is concerned with the role of the adrenal cortex in the regulation of fluid and electrolyte balance in marsupials.

Structure of the adrenal cortex

The structure of the marsupial adrenal cortex has been reviewed by Chester Jones (1957). As in eutherians, the adrenal consists of two endocrine glands: an inner medulla and an outer cortex. The cortex is further differentiated into three zones, the zona glomerulosa, the zona fasciculata and the zona reticularis. The relative widths of these zones have been reported to vary under certain conditions. For example, hypertrophy of the remaining adrenal occurs following unilateral adrenalectomy in *Trichosurus* (Anderson, 1937; Reid and McDonald, 1968b); this is associated with an increased width of the zona fasciculata (Reid and McDonald, 1968b). In eutherians compensatory adrenal hypertrophy appears to be due to increased ACTH secretion. Hypertrophy of the possum adrenal, therefore, suggests that the anterior pituitary participates in the regulation of adreno-cortical activity in marsupials. This conclusion is supported by reports that adrenocortical secretion in marsupials is increased by the administration of ACTH (Weiss and McDonald, 1966a; Weiss and McDonald, 1967; Johnston, Davis and Hartroft, 1967) and is decreased following hypophysectomy (Johnston *et al.*, 1967).

Hypertrophy of the zona glomerulosa has been observed during sodium deficiency in *Didelphis* (Johnston *et al.*, 1967) and in the wombat and kangaroo (Blair-West *et al.*, 1968). Hypertrophy of the zona glomerulosa also occurs during sodium deficiency in eutherians and is apparently mediated via the renin – angiotensin system. Evidence for participation of the renin – angiotensin system in the control of adrenocortical secretion in marsupials is reviewed later in this chapter.

Effects of adrenalectomy and steroid replacement

A classical approach to evaluating the role of an endocrine gland is to study the effects of removing the gland. Removal of the adrenals in eutherians causes wide-spread changes, several of which involve the kidney, and which are fatal unless appropriate treatment is instituted.

The effects of bilateral adrenalectomy have been studied in three marsupials – *Didelphis* (Britton, 1931; Sylvette and Britton, 1936; Britton and Sylvette, 1937; Hartman, Smith and Lewis, 1943), *Setonyx* (Buttle, Kirk and Waring, 1952) and *Trichosurus* (Anderson, 1937; Reid and McDonald, 1968b). *Didelphis* appears to tolerate adrenalectomy with little disturbance to body function and survives for long periods without supportive treatment. Plasma sodium and potassium concentrations and blood glucose concentration remain within the normal range, unless the animals are subjected to some form of stimulation or trauma.

The findings in *Didelphis* raised the possibility that marsupials are less depen-dent on adrenocortical secretion than eutherians and led Buttle *et al.* (1952) to study the effects of adrenalectomy in another marsupial, *Setonyx*. In marked contrast to *Didelphis*, it was found that *Setonyx* died within two days of bilateral adrenalectomy exhibiting the signs of adrenal insufficiency observed in eutherians, including falls in plasma sodium and blood glucose concentrations and a rise in plasma potassium concentration. It is of interest that treatment with adrenal cortical extract, deoxycorticosterone acetate or 1 per cent saline prolonged survival time only slightly.

In view of these differing effects of adrenalectomy in *Didelphis* and *Setonyx*, and the failure of steroid treatment to prevent death in *Setonyx*, it seemed desirable to investigate the effects of adrenalectomy and steroid replacement in greater detail in another species. Anderson (1937) had reported that adrenalectomy is fatal in *Trichosurus*, death occurring 26 – 40h following removal of the second adrenal. However, blood or urine samples for analysis were not collected and the effects of steroid administration in adrenalectomized animals were not investigated. To obtain such information, Reid and McDonald (1968b) studied the effects of bilateral adrenalectomy on blood composition and urinary electrolyte and water excretion in *Trichosurus*. The effects of a variety of adrenocortical steroids on these parameters were also investigated.

In agreement with Anderson's (1937) report, it was found that adrenalectomy was fatal unless some form of treatment was instituted. Adrenal insufficiency was accompanied by blood and urinary changes characteristic of adrenal insufficiency in eutherians. The results of a representative study are shown in Table 23.4. There were progressive decreases in plasma sodium, chloride and bicarbonate concentrations and increases in plasma potassium and urea concentrations. Blood glucose concentration and pH also decreased. These changes were associated with a marked decrease in urine volume and increases in urinary sodium, potassium and urea concentrations. Urinary sodium excretion decreased, but there were no consistent changes in potassium or urea excretion rates.

Three adrenal corticosteroids – cortisol, aldosterone and deoxycorticosterone acetate (DOCA) – were effective in prolonging the survival time of adrenalectomised possums and in preventing or correcting the plasma electrolyte disturbances of adrenal insufficiency (Reid and McDonald, 1968b). Using these steroids, six animals were maintained for up to 59 days following removal of the second gland. Cortisol was effective by itself in maintaining plasma electrolytes within the normal range. The maintenance dose required was of the order expected from published data on cortisol secretion rate in this species (Weiss and McDonald, 1966a).

Aldosterone by itself decreased plasma potassium concentration, but did not prevent the fall in plasma sodium concentration normally associated with adrenal insufficiency. This may indicate that marsupials, as well as eutherians, develop a deficit in water excretion following adrenalectomy which may be corrected by glucocorticoids but not mineralocorticoids. In some animals it appeared that if cortisol and aldosterone were given together, the dose of cortisol required for maintenance was reduced.

The effect of deoxycorticosterone was investigated in two adrenalectomised possums using DOCA pellets which were implanted subcutaneously. Administered in this fashion, DOCA maintained the two animals in apparently normal condition for 31 and 47 days.

During treatment with these steroids, decreases in urinary sodium excretion and the urinary sodium to potassium ratio were observed.

Table 23.4 Blood and urinary changes in an adrenalectomised possum following withdrawal of steroid replacement. Data from Reid and McDonald (1968b)

		Control	Day 1	Day 2	Day 3
Plasma	Sodium concentration (mEq/l)	150	140	127	117
	Potassium concentration (mEq/l)	5.0	6.0	6.7	10.0
	Chloride concentration (mEq/l)	102	97	85	83
	Bicarbonate concentration (mEq/l)	30	23	22	14
	Urea concentration (mg%)	20	38	42	120
Blood	Glucose concentration (mg%)	170	139	138	112
	pH	7.42	7.37	7.42	7.23
Urine	Volume (ml/24 h)	240	145	97	43
	Sodium concentration (mEq/l)	21	46	176	272
	Sodium excretion (mEq/24 h)	5.0	6.7	17.1	11.8
	Potassium concentration (mEq/l)	23	48	59	180
	Potassium excretion (mEq/24 h)	5.5	7.0	5.7	7.8
	Urea concentration (mg%)	280	860	760	2060
	Urea excretion (mg/24 h)	67	125	74	89

When adrenalectomised possums were given a choice of tap-water or salt water they selected the salt solution. Ingestion of saline delayed the onset of adrenal insufficiency but was usually not sufficient to prevent it.

Taken together, the results suggest no basic differences between the effects of adrenalectomy in *Trichosurus* and in eutherians. This also appears to be the case in *Setonyx*, but it would seem worthwhile to reinvestigate the effectiveness of steroid replacement in this species using the normally secreted steroids for supportive treatment. The lack of major effects of adrenalectomy in *Didelphis* is difficult to explain, particularly in view of a report that adrenal steroid metabolism and utilisation in this marsupial are similar to those in eutherians (Beck, Brownell and Besch, 1969).

Aldosterone secretion

The marked changes in plasma electrolyte concentrations and urinary water and electrolyte excretion associated with adrenal insufficiency in marsupials closely resemble the changes which result from loss of mineralocorticoid secretion in eutherians. The most potent naturally occurring mineralocorticoid is aldosterone.

This steroid has now been identified and measured in the peripheral blood of a variety of marsupials, including the wombat and kangaroo (Weiss and McDonald, 1966b, 1967; Coghlan and Scoggins, 1967; Blair-West *et al.*, 1968). In addition, aldosterone has been detected in adrenal venous blood of *Didelphis* (Johnston *et al.*, 1967; Beck *et al.*, 1969) and *Didelphis* adrenals produce aldosterone when incubated *in vitro* (Brownell, Beck and Besch, 1967; Beck *et al.*, 1969).

In eutherians aldosterone is secreted by the zona glomerulosa. Indirect evidence suggests that this is also true in marsupials. For example, the elevated peripheral blood aldosterone levels that occur during sodium deficiency in kangaroos and wombats are associated with an increased width of the zona glomerulosa (Blair-West *et al.*, 1968).

Thus it is clear that marsupials do secrete aldosterone and it seems likely that the zona glomerulosa is responsible, at least in part, for this function.

The control of aldosterone secretion in eutherians is complex and involves the interaction of a variety of factors, including ACTH, plasma sodium and potassium concentrations and the renin – angiotensin system (for review, see Reid and Ganong, 1974). It appears that the most important of these is the renin – angiotensin system. The remainder of this chapter is therefore concerned with the description of this system in marsupials. The evidence that this system participates in the regulation of fluid and electrolyte balance in the marsupials is also reviewed.

The renin – angiotensin system

Eutherians

The renin – angiotensin system has been studied most extensively in eutherians, particularly the dog, rat and man. The system is summarised in Figure 23.2. Renin

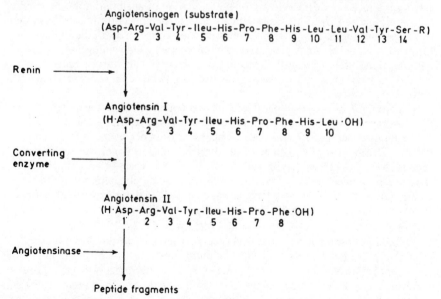

Figure 23.2 The renin – angiotensin system. Reproduced from Reid and Ganong (1974).

is a proteolytic enzyme synthesised and secreted by the juxtaglomerular cells of
the kidney. In the circulation renin acts on its substrate angiotensinogen, a
glycoprotein which occurs in the α_2-globulin fraction of the plasma and which is
produced by the liver. The product of this reaction is angiotensin I, a decapeptide
which is biologically inactive. Angiotensin I is rapidly converted to the octapep-
tide angiotensin II; this conversion occurs primarily in the pulmonary circulation
but also in several other areas, including the kidney. Angiotensin II is inactivated
in the circulation by a group of enzymes collectively referred to as angiotensinase.

Angiotensin II is the active component of the system and has several important
biological actions. It is a potent stimulator of aldosterone secretion, thus producing sodium and water retention and increased potassium excretion. It is also a
very potent constrictor of vascular smooth muscle and increases arterial blood
pressure. Recently angiotensin II has been shown to have marked dipsogenic
activity in several eutherian species (Epstein, Fitzsimons and Rolls, 1970).
Through these actions of angiotensin, the renin – angiotensin system plays a key
role in the regulation of fluid and electrolyte balance and blood pressure.

Marsupials

Several components of the renin – angiotensin system have been identified and
measured in marsupials

Renin. The enzyme kinetic methods of Skinner (1967) were employed by Reid
and McDonald (1969a) to measure plasma and renal renin levels in *Trichosurus.*
Both plasma renin 'activity' and 'concentration' were determined. Plasma renin
activity is a measure of the velocity of the formation of angiotensin during incubation of plasma at $37°C$. This measurement reflects the concentrations of both renin
and angiotensinogen. In the assay for plasma renin concentration, the velocity of
formation of angiotensin is measured when plasma, in which endogenous angiotensinogen has been destroyed, is incubated with a fixed amount of exogenous
(sheep) angiotensinogen. This measurement thus reflects renin concentration
only.

In a group of 18 possums, peripheral plasma renin concentration varied over
the wide range of 10 – 855 units/ml, while plasma renin activity varied from
1.0 – 3.6 ng/ml per hour. There was no apparent difference between males and
females. The renin appeared to be of renal origin, since its concentration in renal
venous blood was higher than in peripheral blood and because circulating levels
were usually reduced to very low values following bilateral nephrectomy. In
addition, it was present in very large amounts in the renal cortex, the concentration in three possums being 13 500, 33 200 and 35 600 units/g.

Using similar techniques, Simpson and Blair-West (1971, 1972) confirmed
these findings in seven other marsupial species. They also observed that plasma
renin concentration values obtained using sheep substrate were higher than plasma
renin concentration values obtained using exogenous marsupial substrate. They
therefore concluded that the affinity of marsupial renin for sheep angiotensinogen
is greater than for marsupial angiotensinogen. Thus the use of sheep substrate
provides a sensitive means of measuring marsupial renin.

In general, marsupial renins closely resemble eutherian renins. They are stable at low pH, heat-labile and non-dialysable (Reid and McDonald, 1969a; Johnston *et al.*, 1967). Reid and McDonald (1969b) estimated the molecular weight of renal renin from *Trichosurus* using Sephadex G-75 chromatography. These preliminary measurements indicated a molecular weight of 53 000, which is slightly higher than the value of 40 000 reported for human renin (Tubash and Peart, 1966).

Reid and McDonald (1969b) also measured the rate of disappearance of renin from the peripheral blood of *Trichosurus* following either bilateral nephrectomy or a single injection of homologous renin in nephrectomised animals. In both cases the disappearance curves could be resolved into two exponential components suggesting a simple two-pool system of renin distribution. The metabolic clearance rate of renin was also calculated and was of the same order as in the dog (Schneider, Rostorfer and Nash, 1968).

Angiotensinogen. Plasma angiotensinogen concentration is measured by incubating plasma with an excess of renin and is expressed as the amount of angiotensin released under these conditions. Plasma angiotensinogen concentration has been measured in the wombat and in two species of kangaroo and is in the same general range as in the sheep, rat and rabbit (Simpson and Blair-West, 1971). Plasma angiotensinogen concentration increases following nephrectomy in eutherians and this also appears to be the case in marsupials (Simpson and Blair-West, 1971). Apart from this, no information is available on the regulation or properties of angiotensinogen in marsupials.

Angiotensin. The material formed during incubation of marsupial renin with marsupial substrate is peptide in nature, but beyond this its chemical structure is unknown. However, the end-product of the reaction in eutherians is angiotensin I (Hollemans, Van der Meer and Kloosterziel, 1969) and it is likely that this is also true in marsupials. Its pressor activity in the rat bioassay for angiotensin closely resembles that of synthetic angiotensin II and this presumably reflects rapid conversion to angiotensin the assay rat. Best *et al.* (1974) reported that kangaroo angiotensin I is not reactive in a radioimmunoassay for angiotensin I and therefore suggested that the amino acid sequence of kangaroo angiotensin is different from that of other angiotensins. Circulating angiotensin levels have not been measured in marsupials.

Factors affecting renin secretion

In eutherians a variety of mechanisms — both intrarenal and extrarenal — participate in the control of renin secretion (for review, see Reid, 1973). These include a baroreceptor mechanism, probably located in the afferent arteriole, which signals an increase in renin secretion when renal perfusion pressure falls. The macula densa (see Figure 23.1c) also plays a role and apparently mediates the increase in renin secretion which is observed when the amount of sodium delivered to the distal tubule is decreased. In addition to these intrarenal mechanisms, the sympathetic nervous system also plays a significant role. The renin-secreting juxta-glomerular cells are sympathetically innervated and there is a direct relationship between sympathetic activity and the rate of renin secretion. Finally, circulating factors also play a role and it is now well established that elevation of the circulating levels of vasopressin, angiotensin II and potassium leads to suppression of renin secretion.

These diverse mechanisms mediate, to varying degrees, the alterations in renin secretion observed in a variety of physiological and experimental situations. Probably the most-studied and best-known of these are sodium depletion and haemorrhage, situations in which renin secretion is increased, leading to increased circulating angiotensin levels. As mentioned above, this results in increased aldosterone secretion, sodium and water retention and increased blood pressure. Thus the renin – angiotensin system functions as a truly homeostatic control system.

Information is now available on the effects of haemorrhage and sodium deficiency on renin secretion in marsupials.

Haemorrhage. Reid and McDonald (1969a) studied the effect of haemorrhage on plasma renin levels in *Trichosurus*. The results are shown in Figure 23.3. In

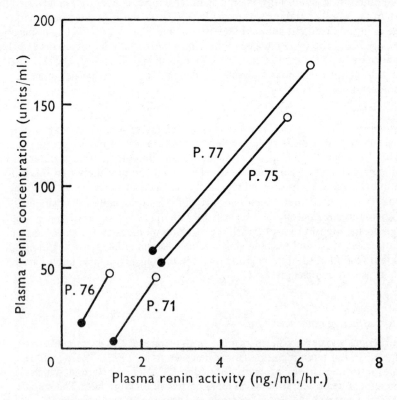

Figure 23.3 Plasma renin activity and concentration before (•) and following (o) haemorrhage in four possums. Reproduced from Reid and McDonald (1969a).

each of four animals hæmorrhage produced marked increases in plasma renin activity and concentration.

Sodium deficiency. The effects of experimental or naturally occurring sodium deficiency have been studied in a variety of marsupial species. In *Trichosurus* administration of mercurial (mersalyl) or non-mercurial (frusemide) diuretics increases plasma renin levels (Reid and McDonald, 1969a). The effects of frusemide are summarised in Figure 23.4. Administration of this diuretic increased urinary sodium excretion and decreased plasma concentration; these changes were associated with marked increases in plasma renin activity and concentration.

A less direct but reliable index of the rate of renin secretion is the degree of granulation of the juxtaglomerular apparatus. This can be readily determined by light microscopy using Bowie's stain (Pitcock and Hartroft, 1958) and is increased

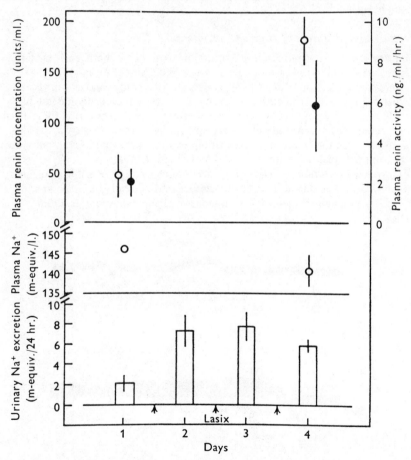

Figure 23.4 The effect of frusemide (Lasix) on plasma sodium concentration, urinary sodium excretion and plasma renin activity (•) and concentration (o) in possums. Reproduced from Reid and McDonald (1969a).

in situations associated with increased renin secretion. In both *Trichosurus* (Reid and McDonald, 1969a; Figure 23.6) and *Didelphis* (Johnson *et al.*, 1967) sodium deficiency is associated with a striking increase in juxtaglomerular granulation. Consistent with these observations, Blair-West *et al.* (1968) reported 'hypertrophy of the juxtaglomerular region' in sodium-deficient wombats and kangaroos. Thus both direct and indirect evidence indicates that renin secretion in marsupials is increased during sodium deficiency.

Conversely, increased sodium intake is associated with a decrease in renin secretion. Reid and McDonald (1969a) reported that when possums with high renin levels were offered a choice of fresh water or 0.9 per cent NaCl solution, they selected the salt solution. Ingestion of saline under these conditions was associated with a decrease in plasma renin levels. Taken together, these data indicate that renin secretion in marsupials varies inversely with the state of sodium balance.

Effects of administration of renin and angiotensin

Blood pressure. Intravenous administration of renin or synthetic angiotensin increases arterial blood pressure in *Trichosurus* (Reid and McDonald, 1969a) and *Didelphis* (Johnston *et al.*, 1967). The results of a study in *Trichosurus* are shown in Figure 23.5. As in eutherians, renin produces a prolonged increase in arterial pressure; by comparison, the pressor effect of angiotensin II is short-lived. Human renin does not increase blood pressure in *Trichosurus* (Figure 23.5); this is consistent with the results of *in vitro* studies which show that human renin does not act on marsupial substrate (Reid and McDonald, 1969a).

Aldosterone secretion. In *Didelphis,* intravenous infusion of renin increases the secretion rate of aldosterone, corticosterone and cortisol (Johnston *et al.*, 1967). This stimulation occurs in the absence of changes in urinary sodium

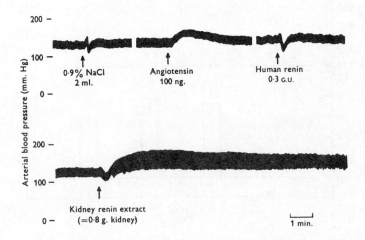

Figure 23.5 The effects of possum renin, synthetic angiotensin II and human renin on arterial blood pressure in an anaesthetised possum. Reproduced from Reid and McDonald (1969a).

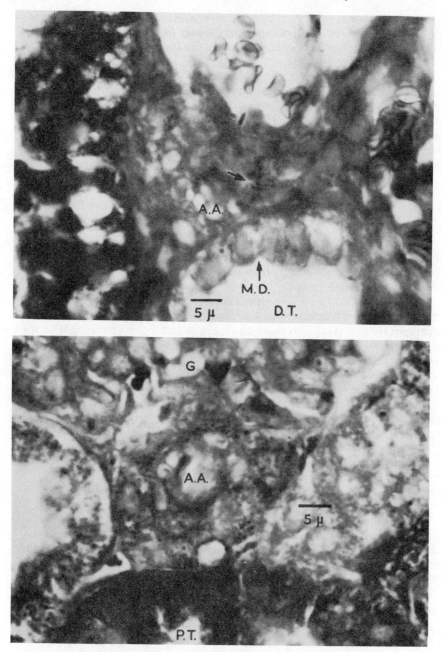

Figure 23.6 Bowie-stained section of a kidney from a normal possum (upper) and a sodium-deficient possum (lower). Note the increase in the degree of the granulation of the juxtaglomerular cells associated with sodium deficiency AA = afferent arteriole; DT = distal tubule; G = glomerulus; MD = macula densa; PT = proximal tubule. Reproduced from Reid and McDonald (1969a).

excretion and plasma sodium and potassium concentrations and is independent of the pituitary, since it occurs in hypophysectomised animals (although administration of ACTH does increase adrenocortical secretion).

From the above, it is clear that there is a renin – angiotensin system in marsupials. It is also clear that administration of renin or angiotensin increases blood pressure and aldosterone secretion and that aldosterone decreases urinary sodium excretion. Taken together with the observations that renin secretion and juxtaglomerular granulation are markedly affected by changes in sodium balance or haemorrhage, the data indicate that in the marsupials, the renin – angiotensin system is an important homeostatic mechanism concerned with the regulation of extracellular fluid volume and blood pressure.

Conclusions

(1) In marsupials, the structure of the kidney is similar to that in eutherians. Renal clearances and stop flow patterns in *Trichosurus* resemble those of eutherians, indicating a close similarity between glomerular and tubular function in these two mammalian groups.

(2) An important role of the adrenal cortex in the regulation of fluid and electrolyte balance is indicated by the effects of bilateral adrenalectomy in two marsupial species. These effects, which closely resemble those resulting from the loss of mineralocorticoid activity in eutherians, can be prevented or corrected in *Trichosurus* by the administration of adrenocortical steroids.

(3) A renin – angiotensin system is present in the marsupials. Renin secretion varies in response to alterations in sodium balance and blood volume and administration of renin or angiotensin stimulates adrenocortical secretion and increases blood pressure. It therefore appears that the renin – angiotensin system is an important homeostatic control mechanism in marsupials.

(4) Thus at least some of the mechanisms which control extracellular fluid volume and blood pressure in the marsupials do not appear to differ significantly from those in eutherians.

Acknowledgements

I thank the following for permission to include figures wich previously appeared in their publications: Pergamon Press and the editors of *Comparative Biochemistry and Physiology* (Figures 23.1a, b); Medical and Technical Publishing Co. Ltd, Butterworths and the editors of *MTP International Review of Science, Physiology Series I Vol. 5 Endocrine Physiology* (Figure 23.2); the editors of *Journal of Endocrinology* (Figures 23.3 – 23.6).

References

Anderson, D. (1937). Studies on the opossum (*Trichosurus vulpecula*) II. The effect of splenectomy, adrenalectomy and injections of cortical hormone. *Aust. J. Exp. Biol. Med. Sci.*, **15**, 24 – 32.

Beck, R. R., Brownell, K. A. and Besch, P. K. (1969). A further study of adrenal function in the opossum (*Didelphis virginiana*). *Gen. Comp. Endocrinol.* **13**, 165 - 72.

Bentley, P. J. (1955). Some aspects of the water metabolism of an Australian marsupial *Setonyx brachyurus. J. Physiol.*, **127**, 1 - 10.

Best, J. B., Blair-West, J. R., Coghlan, J. P., Cran, E. J., Fernley, R. T. and Simpson, P. A. (1974). A novel sequence in kangaroo angiotensin I. *Clin. Exp. Pharmacol. Physiol.*, **1**, 171 - 4.

Blair-West, J. Coghlan, J. P., Denton, D. A., Nelson, J. F., Orchard, E., Scoggins, B. A., Wright, R. D., Myers, K. and Junqueira, C. L. (1968). Physiological, morphological and behavioural adaptation to a sodium deficient diet by wild native Australian and introduced species of animals. *Nature*, **217**, 922 - 8.

Britton, S. W. (1931). Observations on adrenalectomy in marsupial, hibernating and higher mammalian types. *Am. J. Physiol.*, **99**, 9 - 14.

Britton, S. W. and Sylvette, H. (1937). Further observations on sodium chloride balance in the adrenalectomised opossum. *Am. J. Physiol.*, **118**, 21 - 5.

Brownell, K. A., Beck, R. R. and Besch, P. K. (1967). Steroid production by the normal opossum (*Didelphis virginiana*) adrenal *in vitro. Gen. Comp. Endocrinol.*, **9**, 214 - 6.

Buttle, J. M., Kirk, R. L. and Waring, H. (1952). The effect of complete adrenalectomy on the wallaby (*Setonyx brachyurus*). *J. Endocrinol.*, **8**, 281 - 90.

Chester Jones, I. (1957). *The Adrenal Cortex*, Cambridge University Press, London.

Coghlan, J. P. and Scoggins, B. A. (1967). The measurement of aldosterone, cortisol and corticosterone in the blood of the wombat (*Vombatus hirsutus* Perry) and the kangaroo (*Macropus giganteus*). *J. Endocrinol.*, **39**, 445 - 8.

Ealey, E. H. M., Bentley, P. J. and Main, A. R. (1965). Studies on water metabolism of the hill kangaroo, *Macropus robustus* (Gould) in Northwest Australia. *Ecology*, **46**, 473 - 9.

Epstein, A. N., Fitzsimons, J. T. and Rolls, B. J. (1970). Drinking induced by injection of angiotensin into the brain of the rat. *J. Physiol.*, **210**, 457 - 74.

Hartman, F. A., Smith, D. E. and Lewis, L. A. (1943). Adrenal functions in the opossum. *Endocrinology*, **32**, 340 - 4.

Hollemans, H. J. G., Van der Meer, J. and Kloosterziel, W. (1969). Identification of the incubation product of Boucher's renin activity assay, by means of radio-immunoassays for angiotensin I and angiotensin II, and a converting enzyme preparation from lung tissue. *Clinica Chim. Acta*, **23**, 7 - 15.

Jahnecke, J., Sokeland, J., Schmidt, A. W. and Kruck, F. (1966). Stop flow analysis in man. *Metabolism*, **15**, 1076 - 83.

Johnston, C. I., Davis, J. O. and Hartroft, P. M. (1967). Renin - angiotensin system, adrenal steroids and sodium depletion in a primitive mammal, the American opossum. *Endocrinology*, **81**, 633 - 42.

Malvin, R. L., Wilde, W. S. and Sullivan, L. P. (1958). Localisation of nephron transport by stop flow analysis. *Am. J. Physiol.*, **194**, 135 - 42.

Pak Poy, R. K. F. (1957). Electron microscopy of the marsupial renal glomerulus. *Aust. J. Exp. Biol.*, **35**, 437 - 48.

Pitcock, J. A. and Hartroft, P. M. (1958). The juxtaglomerular cells in man and their relationship to the level of plasma sodium and to the zona glomerulosa of the adrenal cortex. *Am. J. Pathol.*, **34**, 863 – 73.

Reid, I. A. (1973). Physiological mechanisms involved in the control of renin secretion. In: *Excerpta Medica International Congress Series* **302**, pp. 300 – 11.

Reid, I. A. and Ganong, W. F. (1974). The hormonal control of sodium excretion. In: *Endocrine Physiology* (ed. S. M. McCann), MTP and Butterworths, London, pp. 205 – 37.

Reid, I. A. and McDonald, I. R. (1968a). Renal function in the marsupial *Trichosurus vulpecula. Comp. Biochem. Physiol.*, **25**, 1071 – 9.

Reid, I. A. and McDonald, I. R. (1968b). Bilateral adrenalectomy and steroid replacement in the marsupial *Trichosurus vulpecula. Comp. Biochem. Physiol.*, **26**, 613 – 25.

Reid, I. A. and McDonald, I. R. (1969a). The renin – angiotensin system in a marsupial *Trichosurus vulpecula. J. Endocrinol.*, **44**, 231 – 40.

Reid, I. A. and McDonald, I. R. (1969b). The clearance of renin from the peripheral plasma of the marsupial *Trichosurus vulpecula. J. Endocrinol.*, **45**, 121 – 9.

Schmidt-Nielsen, K. and Newsome, A. E. (1962). Water balance in the mulgara (*Dasycercus cristicauda*), a carnivorous desert marsupial. *Aust. J. Biol. Sci.*, **15**, 683 – 9.

Schneider, E. G., Rostorfer, H. H. and Nash, F. D. (1968). Distribution volume and metabolic clearance rate of renin in anaesthetised dogs. *Am. J. Physiol.*, **215**, 1115 – 22.

Simpson, P. A. and Blair-West, J. R. (1971). Renin levels in the kangaroo, the wombat and other marsupial species. *J. Endocrinol.*, **51**, 79 – 90.

Simpson, P. A. and Blair-West, J. R. (1972). Estimation of marsupial renin using marsupial renin substrate. *J. Endocrinol.*, **53**, 125 – 30.

Skinner, S. L. (1967). Improved assay methods for renin 'concentration' and 'activity' in human plasma. *Circulation Res.*, **20**, 391 – 402.

Smith, H. W. (1951). *The Kidney*, Oxford University Press, New York.

Sperber, I. (1944). Studies on the mammalian kidney. *Zool. Bidr. Upps.*, **22**, 249 – 432.

Sylvette, H. and Britton, S. W. (1936). Carbohydrate and electrolyte changes in the opossum and marmot following adrenalectomy. *Am. J. Physiol.*, **115**, 618 – 26.

Tubash, G. D. and Peart, W. S. (1966). Purification of human renin. *Biochem. Biophys. Acta*, **122**, 289 – 97.

Weiss, M. and McDonald, I. R. (1966a). Corticosteroid secretion in the Australian phalanger (*Trichosurus vulpecula*). *Gen. Comp. Endocrinol.*, **7**, 345 – 51.

Weiss, M. and McDonald, I. R. (1966b). Adrenocortical secretion in the wombat, *Vombatus hirsutus* Perry. *J. Endocrinol.*, **35**, 207 – 8.

Weiss, M. and McDonald, I. R. (1967). Corticosteroid secretion in kangaroos (*Macropus canguru major* and *M. (Megaleia) rufus*). *J. Endocrinol.*, **39**, 251 – 61.

Wilde, W. S. and Malvin, R. L. (1958). Graphical placement of transport segments along the nephron from urine concentration patterns developed with stop flow technique. *Am. J. Physiol.*, **195**, 153 – 60.

24　Reproduction in male marsupials

B. P. Setchell*

Introduction

For as long as marsupials have been known, the female has always excited more scientific interest than the male, and many of the early accounts of marsupials concerned themselves only with the female (see Hartmann, 1952). Nevertheless, the male marsupial has many features of interest and there is an extensive literature dating back to Cowper (1704). In this review I have endeavoured to describe aspects of reproductive physiology of the male marsupial where some advances have been made recently or where the situation in the marsupial is of particular interest. Structure of the mature spermatozoa, anatomy of the penis and seasonal patterns of reproduction have not been included.

The testis

The scrotum and descent of the testis

The testes of most marsupials are found in a prominent pre-penial scrotum. The exception is the Marsupial mole (*Notoryctes typhlops*), whose testes are abdominal (Stirling, 1891; Sweet, 1897) but in the wombat the testes are scrotal only during the breeding season, returning to the inguinal region at other times (Sharman, 1959a, 1961). In the Tasmanian wolf (*Thylacinus cynocephalus*) the scrotum is situated within a 'pouch' (Beddard, 1891; Pocock, 1926; Boardman, 1945) which Pocock suggested would prevent the testes swinging about too violently while the animal is in swift pursuit of its prey.

As in many eutherian mammals, the testis migrates to the scrotum from a site near the kidney. The scrotum normally appears after birth (Table 24.1). In *Didelphis* the testes move out of the abdominal cavity to an inguinal position at about day 31, reach the scrotum at day 52 and enter it at day 74, and descent is complete by day 77 (Finkel, 1945). The descent of the testis is complete by 91 days in *Wallabia parma* (Maynes, 1973). It has been suggested that the scrotum and pouch arise from the same origin (McCrady, 1938; Burns, 1939b) and that in immature or castrated adult male *Trichosurus vulpecula* the scrotum can be transformed into a sort of pouch by the injection of oestrogens (Bolliger, 1943, 1944, 1954).

*Dr B. P. Setchell is a graduate of the universities of Sydney and Cambridge. Formerly a Veterinary Research Officer with the New South Wales Department of Agriculture and a Research Scientist with the C.S.I.R.O. division of Animal Physiology, he is now a Senior Principal Research Officer in the Biochemistry Department of the Agricultural Research Council Institute of Animal Physiology, Babraham. His main research interests are in the physiology, biochemistry and endocrinology of the testis, oxidative metabolism of organs *in vivo* and comparative aspects of male reproduction.

411

The biology of marsupials

Table 24.1 Time of the first appearance of a scrotum in various marsupials

Species	Age in days after birth	Reference
Sarcophilus harrisii	0	Guiler (1970)
Macropus parma	7	Maynes (1973)
Macropus giganteus	8	} Poole (1973)
Macropus fulginosus	8	
Didelphis virginiana	10	McCrady (1938); Burns (1939a, b)
Bettongia lesueur	13(8–16)	Tyndale-Biscoe (1968)
Schoinobates volans	13*	Bancroft (1973)
Didelphis aurita	20†	Kaiser (1931)
Trichosurus vulpecula	14–21	Broom (1898); Buchanan and Fraser (1919); Fraser (1919); Bolliger and Carrodus (1939; 1940)
Dasyurus viverrinnus	15–19	Hill and Hill (1955)
Macropus rufus	17–20	Sharman and Pilton (1964); Sharman, Frith and Calaby (1964).
Antechinus flaviceps	25	Marlow (1961)

*Head length of 11 mm; age estimated from figure 7 of Lyne and Verhagen (1957) for *T. vulpecula*, which is about the same size as *S. volans* as an adult.

† Crown to rump length 35 mm; age estimated from Moore (1943).

Castration of male pouch-young *Didelphis virginiana* at the age of 20 days has little effect on the development of the Wolffian duct or on the regression of the Mullerian duct (Moore, 1941, 1943, 1950; Rubin, 1943). The development of the scrotum is also unaffected by treatment with testosterone or oestradiol from day 3 to 20, 40 or 50, although testosterone treatment retarded testibular descent and caused enlargement of parts of the Wolffian duct system and of the uterine and tubal parts of the Mullerian duct system (the vaginal part was lacking as in normal males) (Burns, 1939a, b). This is rather surprising in view of the fact that castration of eutherian embryos caused regression of the Wolffian ducts, but persistence of the Mullerian ducts and treatment with testosterone stimulated the male structures without causing regression of the female ducts (Jost, 1953; Jost *et al.*, 1973). Jost (1953) says that 'the conclusions drawn from the experiments on the opossum are in opposition to those resulting from the experimentation on placental mammals'; but the earliest castrations in the opossums may have been done after differentiation had begun. Removal of the gonads should be attempted in younger marsupials to settle this most important point.

Anatomy of the scrotum and its tunicae. The scrotal skin in an adult marsupial is thin and covered with finer and sparser hair than the rest of the body. In the subcutaneous tissue there is usually some smooth muscle, analogous to the dartos muscle. The tunicas vaginalis which surrounds each testis is darkly pigmented in many American marsupials (Finkel, 1945; Biggers, 1964), in *T. vulpecula* (Bolliger and Carrodus, 1939), in *Macropus canguru* (Heddle and Guiler, 1970), *M. eugenii* and *M. rufogriseus*. It has been suggested that this acts as a 'black body radiator' (Biggers, 1964) but this does not seem likely, as the tunica of the Tasmanian devil

(Guiler and Heddle, 1970) and of the Red kangaroo (*M. rufus*), an inhabitant of some of the hottest parts of Australia, is not pigmented. The pigmentation in *Didelphis* is increased by androgen treatment but also, curiously, by castration (Finkel, 1945).

Temperature of the testis and the effects of heat. From an extensive series of measurements of testicular and body temperature in conscious and lightly anaethetised marsupials, there is no doubt that testicular temperature is maintained at least several degrees cooler than body temperature (Table 24.2).

Table 24.2. The temperature of the testis and of the body of some marsupials (unpublished data; and data from Carrick, 1969, 1972; Costa, Gardinho and Cardosa, 1975; Setchell and Thorburn, 1969; Setchell and Waites, 1969; and Waites, 1970).

Species	Conditions of measurement	Temperature (°C)		
		Body	Testis	Difference
Macropus eugenii	Conscious	36.7	32.3	4.4
	Anaesthetised	36.5	31.1	5.4
	Anaesthetised, scrotal testis	35.5	31.5	4.0
	cryptorchid testis		35.4	0.1
M. rufogriseus	Anaesthetised	37.7	35.2	2.5
M. robustus	Conscious, air temperature 32°C	38 – 39	33 – 34	5
Megaleia rufa	Anaesthetised	36.0	31.0	5.0
Trichosurus vulpecula	Conscious, ambient temp. 21°C	36.1	33.0	3.1
	Conscious, ambient temp. 40°C:			
	scrotal licking prevented	39.8	34.2	5.6
	after scrotal licking		25.9	13.9
	Conscious: scrotal testis	35.4	29.1	6.3
	cryptorchid testis		35.4	0
	Anaesthetised	35.0	32.2	2.8
	Anaesthetised	33.9	31.6	2.1
Isoodon obesulus	Anaesthetised	29.4	24.5	4.9
Sarcophilus harrisii	Anaesthetised (1)	33.3	27.7	5.6
	(2)	35.5	35.4	0.1
Didelphis albiventris	Anaesthetised	31.5	30.1	1.4
Eutherian with scrotal testis	Conscious	36.5 – 39.8	30.1 – 36.7	2.2 – 6.3
	Anaesthetised	35.5 – 39.9	28.5 – 37.9	2.0 – 8.5

The statement by Guiler and Heddle (1970) that testis temperature is the same as body temperature in the Tasmanian devil and two other marsupials is not substantiated by the data they themselves present, except for one devil (see Table 24.2) and is certainly not in agreement with out own much more extensive data.

The testis is cooled by countercurrent exchange of heat in the spermatic cord (see next section). The primary loss of heat is from the scrotal skin, aided by the evaporation of sweat. Licking the scrotum seems to be an important additional way of cooling the testes in some marsupials (Carrick, 1969).

It was suggested by Sadleir (1965) that male Red kangaroos and Euros remain fertile at temperatures which cause infertility in rams in the same environment, and Frith and Sharman (1964) and Newsome (1965) found that sperm were produced in marsupials all the year despite exposure to high summer temperatures.

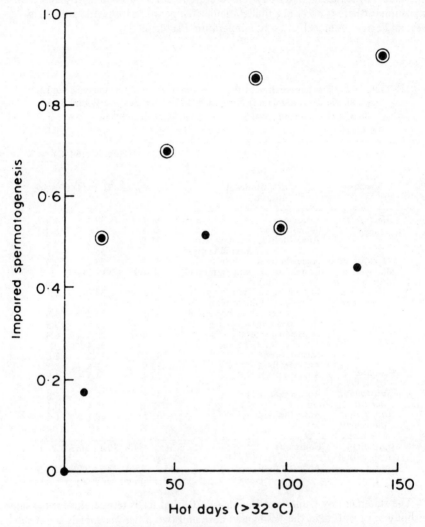

Figure 24.1 The proportion of Red kangaroos with impaired spermatogenesis related to the number of previous days during the summer when maximum ambient temperature exceeded 32°C. •Adequate feed available; ⊙ drought conditions. At least 15 animals per point. Reproduced, with permission of the *Journal of Reproduction and Fertility*, from Newsome (1973).

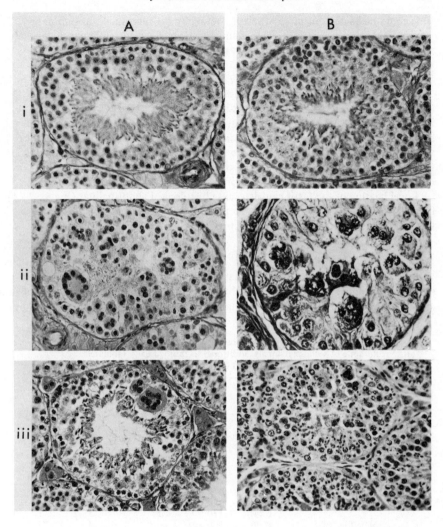

Figure 24.2 Sections through seminiferous tubules from the testes
of (A) Tammar wallabies (*Macropus eugenii*) and (B) Brush-tailed
possums (*Trichosurus vulpecula*). (i) Normal testes at stage 4 of the
cycle of the seminiferous epithlium, showing spermatocytes in
meiosis with developing spermatozoa near the lumina and leptotene
spermatocytes near the tubular wall. (ii) Testes which had been
made cryptorchid 7 days previously, showing giant cells and the
absence of a lumen. (iii) Testes which had been heated 24 h
previously, to 40°C for 3 h, showing giant cells, pycnosis of some
nuclei, and in the possum, absence of a lumen. [A(i), B(i) and
A(ii): Feulgen-alcian blue; B(ii): Verhoff's haematoxylin, reproduced
by courtesy of F. N. Carrick from Carrick (1969); A(iii) and B(iii):
haematoxylin and eosin.]

This could be because of more efficient cooling by the marsupial scrotum or because of a lower sensitivity of the marsupial testis to the effects of heat. Later studies showed that spermatogenesis was impaired in Red kangaroos during hot weather. The degree of impairment was related to the heat load and was exacerbated by poor nutrition (Newsome, 1973; Figure 24.1)

It is quite clear from histological studies that the testis of Tammar wallabies (Setchell and Thorburn, 1970) and Brush-tailed possums (Carrick, 1969) degenerate when returned to the abdominal cavity (Figure 24.2). The testes of an intersex Tammar, Euro and Brush-tailed possum were cryptorchid and had no meiotic stages, although numerous mitoses were seen (Sharman *et al.*, 1970). However, it is not possible to say whether the state of the testis was due to its failure to descend into the scrotum or to the chromosomal abnormality.

Local heating of the testis can also cause histological abnormalities (Figure 24.2) but insufficient information is available to say whether the marsupial testes are inherently less sensitive than eutherian testes to thermal stress.

Spermatic cord, structure and function. The neck of the marsupial scrotum contains the two spermatic cords but it is very narrow, much more obviously so than in eutherian mammals. Each cord consists of the testicular arteries, veins and lymphatics and the ductus deferens. In the larger marsupials, these are virtually surrounded by a

Table 24.3 Details on the number of vessels in the testicular rete mirabile of some marsupials, additional to those already tabulated by Heddle and Guiler (1970)

Species	No. of arteries	No. of veins	No. of lymphatics	Reference
Didelphis marsupialis	50	35	2 or 3	1,4
D. albiventris (azarae)	28		2 or 3	1,3
Perameles nasuta	27	31	3	5
Trichosurus vulpecula {	49	56		2
	43	39	3	5
Macropus rufus	95	74	4	5
M. eugenii {	154	50		4
	240	170	5	5
M. rufogriseus {	96	100		4
	97	122	5	5
M. robustus	120	100	4	5

1. Azzali and DiDio (1965). 2. Barnett and Brazenor (1958).
3. Costa, Godinho and Cardosa (1975). 4. Harrison (1948).
5. Setchell (1970a) and unpublished.

Figure 24.3 Cross-section through the spermatic cord of a Tammar wallaby, *Macropus eugenii*. The arteries are the smaller, thicker-walled vessels. Note the large peripheral lymphatics (L), the separate group of vessels (E) running to the epididymis and the ductus deferens (D) and its associated vessels. (Reprinted with permission of Pergamon Press from Setchell and Thorburn, 1969.)

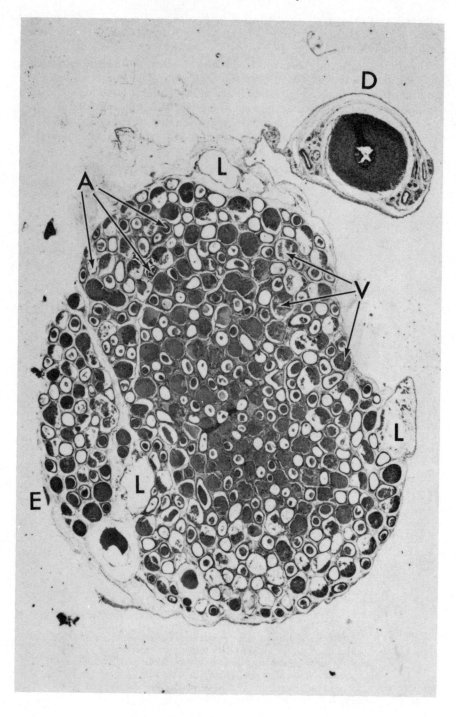

well developed cremaster muscle, which is quite capable of pulling the testes up into contact with the abdominal wall which is the usual position when the animal is moving about (Home, 1795).

The arteries and veins in the cord are parallel branches of the testicular artery and vein, and lie intermingled with one another and the lymphatic vessels. van den Broek (1910) described the arrangement in the 'Funiculus spermaticus' of marsupials as follows: 'The great number of blood vessels, principally extremely fine arteries, are revealed by a transverse section. Towards the periphery are situated the larger veins'. He thought, however, that the structure was similar

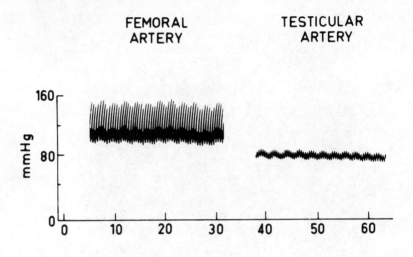

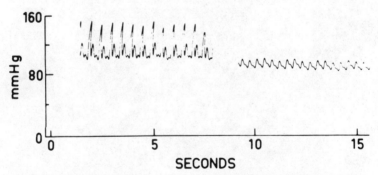

Figure 24.4 Recordings of blood pressure in the femoral artery and testicular artery on the surface of the testis of an anaethetised Tammar wallaby (*Macropus eugenii*), showing the large reduction in pulse pressure with only a small change in mean blood pressure. Reproduced with permission of the *Journal of Reproduction and Fertility,* from Setchell and Waites (1969).

to that in other mammals. More than 200 branches of the artery and a similar number of veins have been demonstrated in some species of marsupials, although the number may be much smaller, and there is only a single artery and vein in *Notoryctes typhlops* (Table 24.3) (Harrison, 1948, 1949, 1951; Barnett and Brazenor, 1958; Setchell, 1970a; Heddle and Guiler, 1970; Guiler and Heddle, 1970). Guiler and Heddle (1970) also state that lymphatics are found in the spermatic cord of the Tasmanian devil but not that of other marsupials, but this is clearly not correct (see Figure 24.3 and Table 24.3; also Azzali and DiDio, 1965; figure 2 of Setchell and Thorburn, 1970; figure 2c of Setchell, 1970a).

We cannot yet say what is the main function of this vascular rete, but two functions have been demonstrated and two others suggested. First, the pulse is eliminated from the arterial blood, with only a small reduction in mean blood pressure, during the passage of the arterial blood through the rete (Setchell and Waites, 1969; Figure 24.4). A similar situation exists in the eutherian mammals with scrotal testes (Waites and Moule, 1960) but no convincing explanation has been advanced for this reduction in pulse pressure. In some other organs a pulsatile arterial supply appears to be necessary for normal function (Wilkens, Regelson and Hoffmeister, 1962).

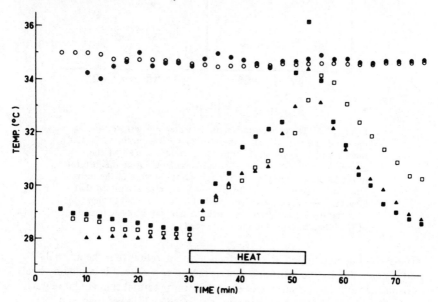

Figure 24.5 Temperature in the rectum (0), testis (□), a testicular artery (■), and vein (▲) and an internal spermatic artery at the inguinal ring (●) during local heating of the testis of an anaesthetised Tammar wallaby (*Macropus eugenii*). Note that the temperature in the vein at the top of the spermatic cord is similar to body temperature, but the temperatures in both the artery and vein below the cord are similar to that in the testis, and change in a similar fashion during heating. This is strong evidence for the vessels in the spermatic cord acting as a countercurrent heat exchanger. Reproduced, with permission of the *Journal of Reproduction and Fertility*, from Setchell and Waites (1969).

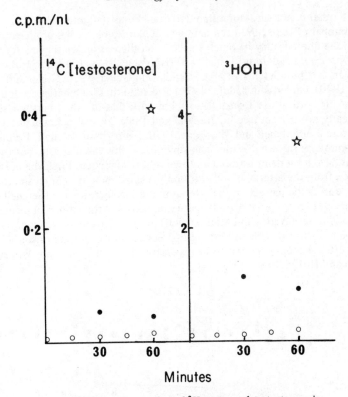

Figure 24.6 The concentration of ^{3}H-water and testosterone in
the femoral artery (A, o), in the testicular artery on the surface of
the testis (TA, ●) and in the internal spermatic vein above the
spermatic cord (ISV, ☆) of a Bennett's wallaby during the infusion
of the labelled compounds into a vein on the surface of the testis.
Using the formula (TA − A)/(ISV − A), it can be calculated that
about 21 per cent of the ^{3}H-water but only 7 per cent of the
testosterone is transferred from vein to artery during one passage
of the blood through the spermatic cord.

The second demonstrated effect of the spermatic rete is that the arterial
blood is precooled by the returning venous blood, which itself is cooled by heat
loss from the veins on the surface of the testis to the scrotal skin (Figure 24.5)
Thus the testis is kept at a uniform temperature which is lower than body
temperature (Setchell and Thorburn, 1969; Setchell and Waites, 1969; Waites,
1970).

It has also been suggested that substances may be transferred from the testi-

Figure 24.7 The vascular supply of the testis of a Tammar wallaby *(Macropus eugenii)*
shown (A) from the epididymal margin, (B) from the hilus and (C) from the free margin.
a: testicular artery; v: testicular veins, ed: efferent ducts; h: head of epididymis; t: tail
of epididymis. Reproduced by permission of Academic Press, from Setchell (1970a).

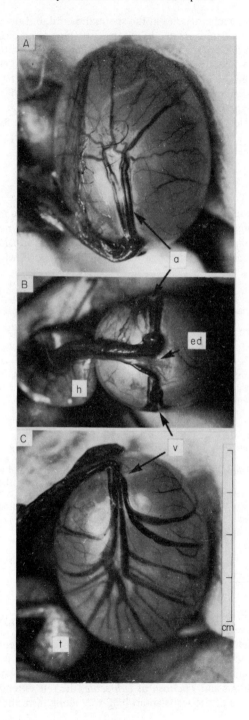

cular veins to the testicular arteries in the spermatic cord, but no differences in composition could be demonstrated between aortic blood and blood from the testicular artery on the surface of the testis (Setchell and Hinks, 1969). However, experiments with radioactive tracers have indicated that some exchange can occur from venous to arterial blood (Figure 24.6; Jacks and Setchell, 1973) but the quantitative significance of this is probably minimal (see Setchell, 1973). The second function suggested for the venous and arterial retia is that of a venous pump (Guiler and Heddle, 1970), but no experimental evidence for this has been presented.

Spermatogenesis

Anatomy of the seminiferous tubules, rete testis and efferent ducts. In contrast to most eutherian testes, which are prolate spheroids, the testes of most marsupials are ellipsoidal in shape. Therefore the appropriate formula for calculating testicular weight from its linear measurement is: testis weight = $\frac{1}{6} \pi \, abc$, where a, b and c are the three diameters.

Table 24.4 Diameter of the seminiferous tubules in the testes of some marsupials

Species	Diameter (μm)	Reference
Didelphis virginiana	230 – 250	Moore and Morgan (1943)
Antechinomys spenceri	490	
Sminthopsis crassicaudata	510	
S. leucopus	430	
Antechinus stuartii	360 – 430	Woolley (1975)
Dasyuroides bymei	400 – 440	
Sarcophilus harrisii	360	
Phalangista	250	Benda (1906)
Trichosurus vulpecula	226 – 288	Smith, Brown and Frith (1969)
	250 ± 7(10)	Setchell (unpublished)
Perameles	390	Benda (1906)
P. nasuta	355 ± 8(15)	Setchell (unpublished)
Macropus rufus	200 – 250	Sharman and Calaby (1964)
Macropus giganteus } *M. fulginosus* }	180 – 320	Poole (1973)
M. eugenii	254 ± 6(10)	Setchell (unpublished)
M. rufogriseus	270 ± 6(10)	

The dimension through the hilus is the longest, that from the mesorchium the least, and the transverse diameter is intermediate (see Figure 24.7).

The rete testis has begun to develop at birth, but a sex difference does not appear until between 12 and 14 days (Fraser, 1919; Burns 1941). The situation in the adult is also unusual, not being centrally or marginally located as in eutheria. In macropods and *Trichosurus vulpecula,* the rete extends like a horse-shoe on either surface of the testis about half-way along from the hilus. The rete lies under the veins on one surface and under the arteries on the other (Figure 24.8; Setchell, 1970a). The seminiferous tubules extend from one arm of the rete to the other (de Burlet, 1921) but are extensively convoluted along their length. The seminiferous tubules are joined to the rete by the tubuli recti, which, as in the eutheria, are short straight tubules continuous with the seminiferous

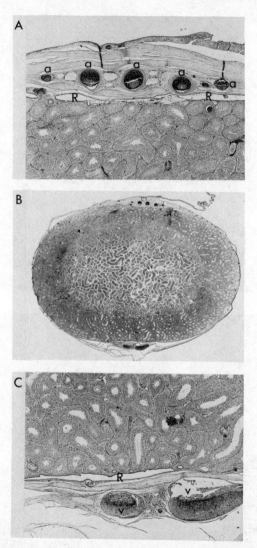

Figure 24.8 (B) Equatorial section through the testis of a Brush-tailed possum, *Trichosurus vulpecula,* with enlarged detail of (A) the epididymal and (C) the free margins. In (A) can be seen the branches of the testicular artery (a) with the rete (R) immediately underneath; in (C) the other arm of the rete (R) lies immediately underneath the testicular veins (v).

tubules, but lined only with cells which look like Sertoli cells. The two parts of the rete join together near the hilus and are there joined to the efferent ducts, which carry the spermatozoa to the epididymis. These ducts are comparatively

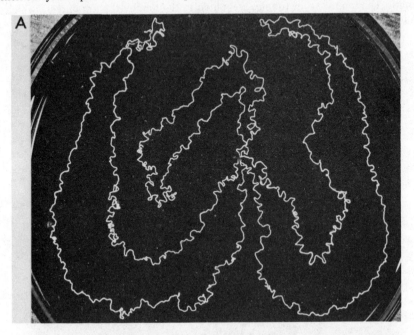

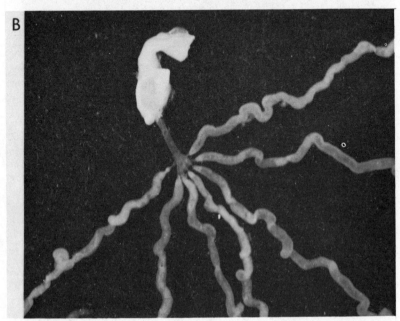

long in marsupials and the epididymis can be separated much further from the testis than in most eutherians. It was originally suggested that there was only one duct in *Didelphis* (Fraser, 1919; Chase, 1939), but later observations did not support this suggestion (Ladman, 1967); certainly in the Australian macropod marsupials examined so far there are about 12 ducts.

In Dasyurids there is only efferent duct and an insignificant rete situated at the hilus (Fraser, 1919; Woolley 1975). The seminiferous tubules in these animals are considerably larger than those of other marsupials or of eutheria (Table 24.4), and in some Dasyurids there are as few as one or two seminiferous tubules which are usually single two-ended loops with two openings into the rete (Figure 24.9). Frequently the tubules are branched or interconnected and then have more than two openings into the rete (Woolley, 1975).

Structure of the germinal epithelium. The epithelium of the seminiferous tubules in marsupials has essentially the same cell types as those found in eutherian mammals, namely the various germinal cells (that is, spermatogonia, spermatocytes and spermatids) and the nutritive or supporting Sertoli cells. There are a series of mitotic divisions by the diploid spermatogonia, then the cells become primary spermato - cytes and enter the long meiotic prophase, which culminates in the meiotic divisions and the production of the haploid spermatids. These cells then become transformed from ordinary-looking round cells into the intricately formed spermatozoa which are finally liberated into the lumen of the tubule.

The general arrangement of the cells is also like that in eutheria. The Sertoli cells extend from the wall of the tubule to its lumen. The spermatogonia are found between the wall of the tubule and the Sertoli cells, while the other germinal cells are either sandwiched between the cytoplasm of adjacent pairs of Sertoli cells, or are completely or partially embedded in the cytoplasm of individual Sertoli cells (Sapsford, Rae and Cleland, 1967, 1969a).

The testis of *Wallabia bicolor* and *Vombatus ursinus* contain a testis-specific isozyme of lactate dihydrogenase (LDH-X), similar to that found in the testes of eutheria (Holmes, 1972; Holmes, Cooper and Vanderbilt, 1973; Baldwin and Temple-Smith, 1973).

Spermatogonia. It is not known whether the numbers of spermatogonia, the diploid germinal cells, are maintained in marsupials by asynchronised divisions of a reserve cell population (Huckins, 1971) or by reversion of some of the daughter cells of one of the synchronised divisions (Clermont, 1972). From the numbers of labelled cells per tubular cross-section at various stages of the cycle it has been suggested that in *M. rufogriseus* there are at least four synchronised spermatogonial

Figure 24.9 (A) Teased out testis of *Dasyuroides byrnei*, showing the three complete seminiferous tubules which formed the testis. Each tubule was a single loop opening at both ends into the rete testis. (B) Connections of the four seminiferous tubules with the rete testis in another *D. byrnei* (one of the connections was broken and cannot be seen). Note the simple rete, the single efferent duct and the narrow tubuli recti. Reproduced, by permission of the *Journal of Reproduction and Fertility*, from Woolley (1975).

divisions (Setchell and Carrick, 1973), but further work is needed on this point.

The spermatogonia of *Potorous tridactylus* appear to be highly sensitive to the damaging effect of X-irradiation, as they are in eutheria (Sharman, 1959b).

Spermatocytes and meiosis. The meiotic division in marsupials was extensively investigated about 50 years ago because the small numbers of chromosomes enabled the various features to be more clearly studied, and in the course of these studies, the existence of the Y-chromosome in mammals was first conclusively demonstrated in *Didelphis* (Painter, 1922). The various steps to meiosis – leptotene, pachytene, zygotene and diplotene – appear to be very similar in marsupials and eutherians (Jordan, 1911; Duesberg, 1920; Agar, 1923; Greenwood, 1923).

Spermiogenesis. The transformation of a spermatid into a spermatozoa is one of the most complex cell transformations occurring in animals and it is therefore not surprising that it has been the subject of considerable interest among light and electron microscopists. The process is probably even more complex in marsupials because of the unusual form of the heads of their spermatozoa (Figure 24.10). In the testicular spermatozoa released from the germinal epithelium the head is set at an acute angle to the tail but the sperm heads develop with their long axes at right angles to the tail and therefore must rotate about the point of attachment to assume their final position.

The earlier microscopists described the various stages of the transformation as seen by the light microscope (Fürst, 1880; von Bardeleben, 1896; Benda, 1906; Binder, 1927), and more recently the results of a most thorough and detailed study with the electron microscope in the bandicoot has been reported (Sapsford *et al.*, 1967, 1969a, 1969b, 1970; Sapsford and Rae, 1969). These last authors divide the process into six stages: (1) early spermatid, (2) nuclear protrusion, (3) nuclear flattening and condensation, (4) nuclear rotation, (5) early post-rotation and (6) late post-rotation.

The first changes which constitute the 'early spermatid' stage are not remarkably different from those in eutheria. In this stage the middle piece and principal piece of the sperm tail begin to develop from the longitudinal centriole while this lies near the Golgi complex; the latter forms the acrosomal vacuole, which then attaches to the nucleus to become the acrosome. In the marsupials, however, the condensation of the acrosome occurs comparatively late, and its insertion changes as the spermatozoa develops (Cleland, 1956). At about the time of the attachment of the aerosomal vacuole to the nucleus the two centrioles migrate to the opposite pole of the nucleus. There the development of the filaments of the tail continues and at the same time a junctional complex is formed which joins the tail to the nucleus (Sapsford *et al.*, 1967; Phillips, 1970).

In the second stage, 'nuclear protrusion', the perinuclear spermatid cytoplasm is moved towards the tubular lumen away from the spermatid nucleus. The beginning of this stage coincides with the shedding of the previous generation of spermatozoa and the obvious encirclement of the new spermatids by Sertoli cell cytoplasm. Thus the changes in the shape of the spermatid and in the orientation

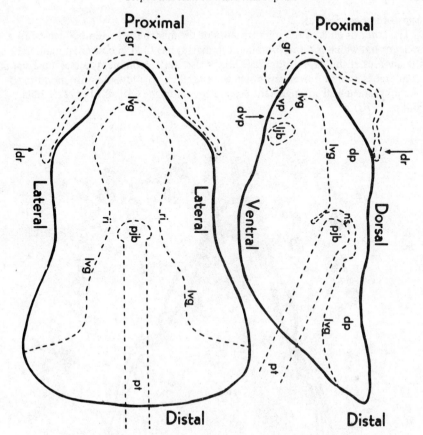

Figure 24.10 Diagrammatic representation of the outlines of the spermatid nucleus of the bandicoot (*Perumeles nasuta*) viewed from the lateral and dorsal aspects, at the time of the formation of the early mitochondrial sheath of the middle piece. Note how the tail is attached in a groove on the ventral surface of the nucleus. dp: dorsal plate of nucleus; dr: distal limit of rostrum; dvp: distal extremity of ventral plate; gr: acrosome; ljb; lateral junctional body; lvg: lateral limit of ventral groove, ns: nuclear socket; pjb: proximal junctional body; pt: proximal part of tail; ri: dorso-ventral ridge of ventral nuclear groove; vp: ventral plate of nucleus. Reproduced, with permission of the *Australian Journal of Zoology,* from Sapsford *et al.* (1969a).

of nucleus could either be inherent in the spermatids or be produced by the Sertoli cells.

The development of the manchette, that strange evanescent structure, which occurs at this stage is also essentially similar in marsupials and eutherians (Sapsford *et al.* 1969; Phillips, 1970), except that in the bandicoot, soon after the manchette begins to develop, it is found immediately beneath the plasma membrane of the spermatid. Here it appears to have an important function in moulding the shape of the anterior extremity of the spermatid, and also possibly of shaping the nucleus

(Sapsford *et al.*, 1969a).

The anterior ends of the microtubules of the manchette are embedded in a nuclear ring, which is more prominent in marsupials than in eutherian mammals. The annulus at the base of the flagellum is also quite prominent in marsupials and it undergoes a rapid movement down the flagellum to allow the mitochondria to take up their helical arrangement around the midpiece (Sapsford *et al.*, 1969a, Phillips 1970).

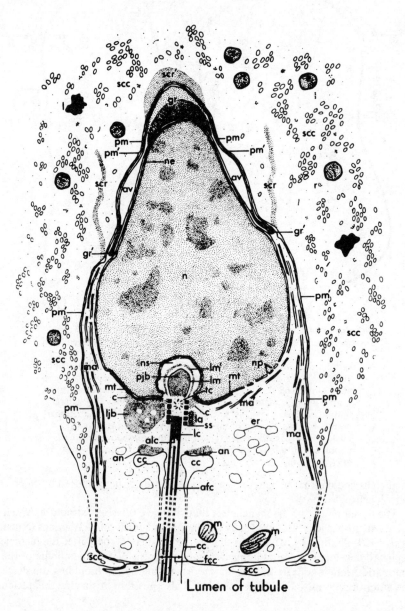

Lumen of tubule

Also at this stage in the bandicoot, there is obvious thickening of the nuclear membrane near the attachment of the tail. Development of this thickening continues during the next stage of nuclear flattening and condensation, and may be associated with the disposal of redundant nuclear envelope, because by the end of the reduction in nuclear volume, the thickening has disappeared. The appearance of the nucleus at the end of the stages is shown in Figure 24.11.

In the stage of nuclear flattening and condensation, the nucleus is flattened in a plane at right angles to the axis of the tail. This contrasts with the situation in eutherian mammals in which the flattening occurs in a plane parallel to the long axis of the tail.

In the next stage, the nucleus is rotated about its point of attachment to the tail (Figure 24.5) and again this may be achieved by the efforts of the Sertoli cells. The spermatid outlines, which are smooth and regular during the stage of nuclear flattening become extremely irregular due to the projection into these cells of large masses of Sertoli cell cytoplasm. These incursions are found mainly below the level of the distal border of the manchette, in the proximal part of the spermatid. Similar invasions by Sertoli cell cytoplasm can be seen in eutherian mammals, but they are much more extensive and pronounced in the bandicoot (Sapsford *et al.*, 1969a).

Once nuclear rotation is complete, the invasion of the spermatid by the cytoplasm becomes less marked. Then, at the beginning of the late post-rotational stage, a second wave of invasion of the spermatid by tongues of Sertoli cell cytoplasm begin, but these are of a different type. They are less massive than before and are found as far distal as that part of the spermatid cytoplasm surrounding the intraspermatid part of the principle piece. These latter Sertoli cell invasions often taken place along the border of the mitochondrial sheath of the middle piece, or along the principle piece, and in other cases they follow the cytoplasmic canal.

Once this second invasion is complete, the spermatid nucleus is moved towards the lumen of the tubule through its own cytoplasm. Again the motive force appears to come from the Sertoli cell, through a specialised tongue of Sertoli cell cytoplasm which is applied to the aerosome-capped region of the nucleus. Surplus spermatid cytoplasm is retained by the invading Sertoli cell branches as the residual body (Sapsford and Rae, 1969).

Figure 24.11 Diagrammatic representation of the structure of the spermatid of the bandicoot at the nuclear protrusion stage. afc: axial filament complex; alc: axial laminated complex; an: annulus; av: electron-lucent part of the acrosomal vacuole; c: cementation; cc: cytoplasmic canal; er: endoplasmic reticulum; fcc: cytoplasm surrounding axial filament complex; gr: granular component of acrosome at its apex; gr': granular component of acrosome at its base; l: lipid; la: lateral appendage; lc: longitudinal centriole; ljb: lateral junctional body; lm: lamina of nuclear socket; lm': additional lamina of moderate electron density in the interspace between the lm and the walls of the socket; m: mitochondria; ma: manchette; ma': offshoot from the inner surface of ma extending along the basal part of the nucleus to terminate in the neck region; mt: membrane thickening; n: nucleus; ne: nuclear envelope; np: nuclear pore; ns: nuclear socket; pjb: proximal junctional body; pm: plasma membrane of spermatid; pm': plasma membrane of Sertoli cell; r: ribosome; scc: Sertoli cell cytoplasm; scr: Sertoli cell reaction or Sertoli cell spurs; scr': second aggregation of granular material in Sertoli cell cytoplasm surrounding the apex of the acrosome-covered nucleus; ss: segmented sheath; tc: transverse centriole. Reproduced, with permission of the *Australian Journal of Zoology*, from Sapsford *et al.* (1969a).

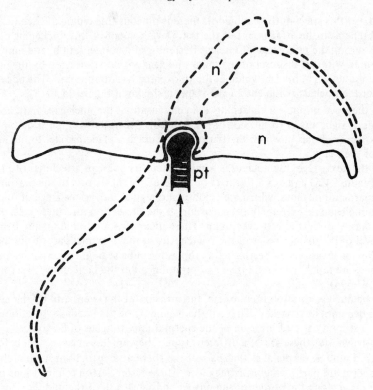

Figure 24.12 The rotation of the bandicoot spermatid nucleus
from its position at the end of nuclear flattening (n) to that at the
end of the rotational stage (n′). The axis about which rotation takes
place is located at the junction of the nucleus with the proximal
part of the tail (pt) and is perpendicular to the plane of the diagram
and thus perpendicular to the long axis of the tail (arrow). Reproduced, with permission of the *Australian Journal of Zoology*, from
Sapsford *et al*. (1969a).

When the spermatozoa are shed into the lumen, the residual bodies move within
the Sertoli cell cytoplasm towards the wall of the tubule. They become more dense
and gradually break up and disappear (Sapsford *et al*., 1969b).

A 'lipid cycle' similar to that described by Lacy (1962, 1967) in the rat occurs
in the bandicoot but in the latter species the increase in lipid occurs before there
is any decrease in size of the residual body. Therefore the 'lipid cycle' does not
appear to be due to recycling of lipid from the retained spermatid cytoplasm but
to some metabolic change within the Sertoli cells (Sapsford *et al*., 1969b).

As well as these changes just described, the bandicoot Sertoli cells show other
peculiar alterations during spermatid maturation. By the stage of nuclear protrusion, a reaction has occurred in the Sertoli cells in the form of an aggregation
of granular material near the apex of the acrosome. The second reaction takes the
form of a hollow cylinder which, when sectioned longitudinally, appears as two
slightly wavy dense lines, one on each side of the acrosomal vacuole (Figure 24.11).

Other marsupials seem to show similar reactions but these are not seen in eutheria (Sapsford *et al.*, 1969a).

There are also peculiar changes in the endoplasmic reticulum (ER) of the Sertoli cells. In the early post-rotational phase all parts of the Sertoli cell cytoplasm show vesicular ER, smooth in type. After the liberation of the spermatozoa, as the next generation of spermatids reach the nuclear flattening stage, the Sertoli cell cytoplasm near them shows a branched lamelliform type of ER, and during nuclear rotation there are, as well, some isolated aggregations of smooth vesicles. The Sertoli cell cytoplasm near the wall of the tubule, between the spermatogonia and spermatocytes, is vesicular during the post-rotational stages, but becomes branched lamelliform ER, with many dense vesicles during nuclear protrusion, flattening and rotation (Sapsford *et al.*, 1969a). These changes are presumably related to cyclical changes in Sertoli cell metabolism already suggested by the changes in Sertoli cell lipid but they have not been reported in other marsupial species.

In *Didelphis* the mitochondria in the spermatocytes and early spermatids are either granules or small rods which give a strong histochemical reaction for protein and a week reaction for lipids. Mitochondria during the later stages of spermiogenesis appear as large spheres which give a strong histochemical reaction for lipids, especially phospholipids (Guraya, 1971).

Kinetics of spermatogenesis. It was recognised long ago that the cells of the seminiferous epithelium of marsupials were always found in definite groupings or associations. Von Korff (1901) described four stages; Benda (1906) subdivided von Korff's stage II into three stages to give a total of six; and Binder (1927) suggested ten stages, based on the scheme of Regaud (1901) for eutherian mammals. None of these authors reported the frequency with which the various stages occurred, but recently we (Setchell and Carrick, 1973) have suggested that marsupial spermatogenesis could be divided on the basis of cellular associations into eight stages, corresponding to those described by Curtis (1918), Roosen-Runge and Giesel (1950) and Courot, Hochereau de Reviers and Ortavant (1970) for eutherian mammals.

The stages are illustrated in Figures 24.2, 24.13 and 24.14. Stage VIII corresponds to the 'early spermatid' stage of Sapsford *et al.* (1967, 1969a); Stages I and II to the 'nuclear protrusion and nuclear flattening' stages; Stage III to 'nuclear rotation and early post-rotational' stages; and stages IV – VIII to the 'late post-rotational' stage. Similar stages have been identified in the testis of *Isoodon macrourus* (Mason and Blackshaw, 1973).

The frequency with which the various stages occurred was reasonably consistent for individuals of the species *M. eugenii, M. rufogriseus, T. vulpecula* and *P. nasuta,* but in *M. eugenii*, like the rat, mouse and boar, the post-meiotic stages 5 and 8 are more common than the pre-meiotic stages 1 to 3, while in the other species the reverse is true, as it is in the ram, bull and rabbit.

Furthermore, by injecting tritiated thymidine either into a testicular artery or directly into a testis and removing the testis at various intervals afterwards, we also showed that the length of the spermatogenic cycle was about 17 days in *M. rufogriseus* and about 16 days in *M. eugenii* and *T. vulpecula* (Setchell and

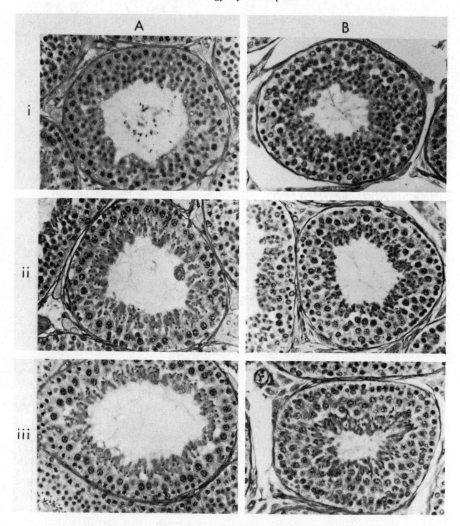

Figure 24.13 The premeiotic stages of the spermatogenic cycle in (A) a Tammar wallaby (*Macropus eugenii*) and (B) a Brush-tailed possum (*Trichosurus vulpecula*). (i) Stage 1: the lumen of the tubule is lined with round spermatids; outside these are pachytene spermatocytes and a new generation of preleptotene spermatocytes; these are the last cells to take up ^{3}H-thymidine; all sections contain Sertoli cells and spermatogonia near the tubule wall. (ii) Stage 2: the spermatid nuclei have condensed and the pachytene spermatocytes outside them have entered diplotene, with large nuclei and prominent chromosomes; the outer spermatocytes still take up ^{3}H-thymidine early in this stage so the transition from preleptotene to leptotene must occur during this time. (iii) Stage 3: the nuclei of the spermatids are now elongated and bifid and their cytoplasm is eccentrically placed; the inner spermatocytes are still diplotene but the outer spermatocytes have entered zygotene.

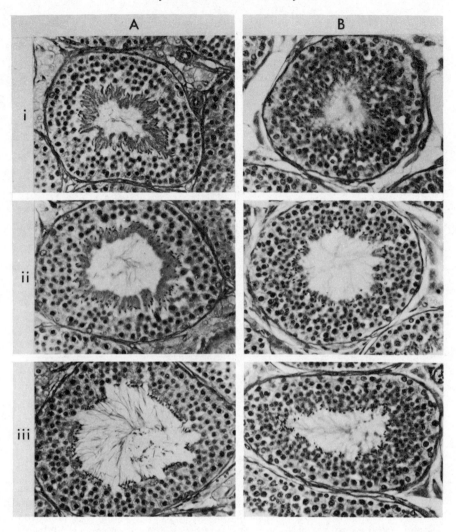

Figure 24.14 The post-meiotic stages of the spermatogenic cycle in
(A) wallaby and (B) possum testes, as in Figure 24.13. (i) Stages 5
and 6: these two stages are very difficult to separate as the distinction
depends on the granularity of the spermatid nucleus; the old spermatids
are grouped in bundles with their nuclei arranged radially; the new spermatids lie outside them and then the pachytene spermatocytes. (II)
Stage 7: this stage is similar to the previous two but the spermatozoa are
evenly arranged around the luminal edge of the epithelium, instead of
being grouped in bundles. (iii) Stage 8: in this stage, the nuclei of the
spermatozoa are set parallel to the wall of the tubule, not radially as in
earlier stages; the other cell types are unchanged.

Carrick, 1973). These values lie towards the upper end of the reported range for eutherian mammals, which extends from 8 days in the boar (Ortavant, Orgebin and Singh, 1962; Swierstra, 1968) to 16 days for man (Heller and Clermont, 1964). Values can also be calculated for the meiotic prophase, spermiogenesis and the total time from the beginning of the meiotic prophase to the liberation of the spermatozoa from the epithelium (see Table 24.5) and these values are again similar to those reported for the few eutherian mammals so far studied (see Courot *et al.*, 1970; Table 24.5).

Table 24.5 The duration in days of various aspects of spermatogenesis in some marsupials

Species	*Macropus eugenii*	*M. rufogriseus*	*Trichosurus vulpecula*	Eutheria (range)
Spermatogenic cycle	16	17	15	8 – 16
Meiotic prophase	21	24	21	12 – 23
Spermiogenesis	25	25	22	13 – 23
Preleptotene to release of spermatozoa	48	51	45	24 – 48
Preleptotene to first labelled spermatozoa in urine	61		56	
Minimum epididymal transit	13		11	7 – 25

Marsupial data from Setchell and Carrick (1973); eutherian data from Courot *et al.* (1970), Hamilton (1972).

In some Dasyurid marsupials spermatogenesis is very short-lived. In *Antechinus stuartii* the tubules of 5 month old animals contain only Sertoli cells and spermatogonia but by 7 months of age spermatocytes are also present. At 9 months there are Sertoli cells and all types of germinal cells except spermatogonia, and by 10 months the spermatocytes have also vanished (Woolley, 1975).

It is interesting also that in the wombat *Lasiorhinus latifrons* spermatogonial mitoses are abundant before the breeding season when the testes are beginning to descend into the scrotum, and absent at later times in the season when the later stages of spermatogenesis are abundant (Sharman, 1961).

Secretion of fluid by the testis. The spermatozoa leave the testis as a dilute suspension in a fluid secreted by the testis (see Setchell, 1970a, 1974). This 'rete testis fluid' has been collected from Tammar wallabies by cannulating the efferent ducts between the testis and the epididymis (Setchell, 1970b). Fluid flowed from the catheter at the rate of about 0.25 ml/h or 0.0125 mg/g testes per hour – that is, at a rate intermediate between that in rams and rats (see Setchell, 1974). It contained about 8×10^7 spermatozoa/ml and; apart from containing no detectable inositol, the wallaby fluid resembled rete testis fluid from eutherian mammals in containing slightly less sodium, more potassium and chloride than blood plasma, practically no glucose or protein and comparatively high concentrations of certain amino acids (Setchell, 1970b).

Vascular anatomy inside the testis. At the bottom of the spermatic cord, the arterial branches reunite to form between two and five vessels which then, in the macropods, run along the epididymal margin of the testis, giving out rib-like branches which then supply the parenchyma. The parenchymal veins drain in a similar fashion to the free margin of the testis to form several main venous trunks which run along the testis to the hilus. There the veins redivide to form the venous branches of the spermatic rete (Figure 24.7). This situation does not exist in the Didephids, where the arteries penetrate directly into the testis at the hilus (Godinho, personal communication).

Blood and lymph flow. No really satisfactory measurements of testicular blood flow under physiological conditions have been made in marsupials. Some measurements have been made in several anaethetised Australian marsupials by two techniques. Carrick and Cox (1973), in a study concerned with the secretion rate of testosterone, measured testicular blood flow by cannulating the internal spermatic vein and collecting the entire venous drainage. This technique requires

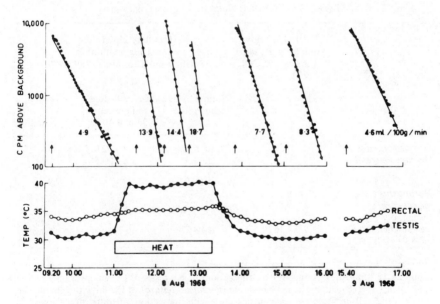

Figure 24.15 The effect of testicular temperature on the clearance of [8 5]Kr from the testis of an anaesthetised Brush-tailed possum (*Trichosurus vulpecula*) after a series of intratesticular injections (arrows) of the gas dissolved in saline (0.05 ml). The figures on the upper panel of the graph are the values for blood flow calculated from the half-time of clearance (see Setchell, 1970a) and the lower panel shows rectal and testicular temperatures and the time of heating. (See Setchell and Thorburn, 1969, for further details.)

The biology of marsupials

fairly extensive surgery, and this, with the alterations in venous pressure inevitable with this technique, introduce serious uncertainties.

The other technique used (Setchell and Thorburn, 1969, 1970) involves measurements of the rate of clearance of radioactive krypton injected into the testis (Figure 24.15). While this technique can be used in lightly anaesthetised animals with no surgical intervention, it suffers from the known fault that the marker has been shown to cross in appreciable amounts, presumably by diffusion down the concentration gradient from testicular venous to testicular arterial blood in rats (Einer-Jensen, 1974), thus leading to an underestimate of its clearance from the testis and consequently of blood flow.

With these limitations in mind, it can be stated that testicular blood flow in those marsupials studied is of the same order as in eutherian mammals, and that the two techniques give results which are in reasonable agreement, considering the different conditions under which the measurements were made (see Table 24.6).

One strange feature of testicular blood flow in the marsupials studied was that flow increased when testicular temperature was raised by externally heating the testis (Setchell and Thorburn, 1969) but not when the testis was made surgically cryptorchid (Setchell and Thorburn, 1970; see Table 24.6). No measurements of testicular lymph flow appear to have been made in marsupials.

Table 24.6 Testicular blood flow in some marsupials

Species	Total blood flow (ml/100g per min) (Carrick and Cox, 1973)	Capillary blood flow (ml/100g per min) (Setchell and Thorburn, 1969, 1970)		
		Normal testis temperature	Testes heated to between 36 and 40°C	Testis Cryptorchid
Trichosurus vulpecula	3.5	5.1	6.2 – 22.0	–
Macropus eugenii	–	{ 9.3	15.5 – 20.5	–
		{ 10.0	–	9.9
M. robustus	9.7	–	–	–
Megaleia rufa	9.5	3.5	5 – 8.5	–
Eutheria (see Setchell, 1970a)	7.7 – 30	8.4 – 27.7	no change	no change

Figure 24.16 (opposite) (A and B) Paraffin sections of the testes of (A) an opossum(*Didelphis marsupialis*) and (B) a Long-nosed bandicoot (*Perameles nasuta*) which show the abundance of interstitial cells in these species; contrast these with the sections of testes of wallabies and Brush-tailed possums (Figures 24.2, 24.13 and 24.14) which have much less interstitial tissue; note also the thick walled blood vessels in the interstitial tissue [(A) kindly supplied by Professor D. W. Fawcett and (B) stained with Feulgen – alcian blue]. (C and D) Electron micrographs of the interstitial tissue of the opossum, showing the abundant Leydig cells, the centrally placed lymphatic vessel and in (C) the thick walled blood vessels which are unlike those found in eutherian mammals [(C) reproduced, with permission of *Biology of Reproduction,* from Fawcett, Neaves and Flores, 1974; (D) kindly supplied by Professor D. W. Fawcett].

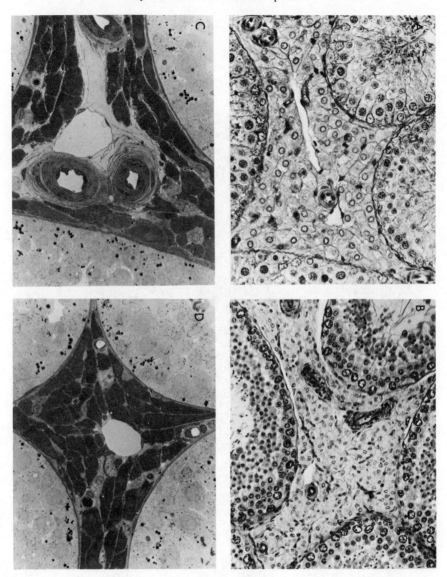

Endocrinology of the testis

Structure of the interstitial tissue and Leydig cells. The tissue between the seminiferous tubules of marsupials shows the same basic structure and the same variability between species as in eutherian mammals. In all animals there are the blood

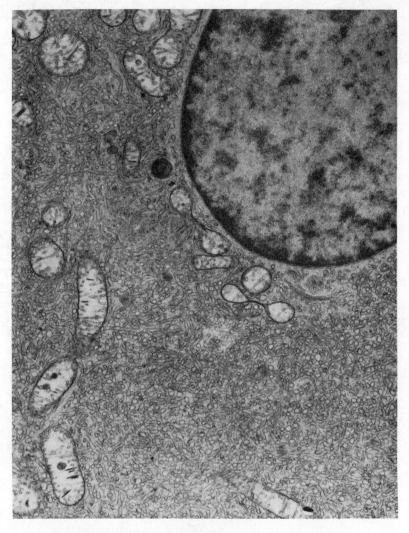

Figure 24.17 An electron micrograph of a Leydig cell from the
testis of an opossum, *Didelphis virginiana*, showing part of the
nucleus, some mitochondria and the prominent agranular endoplasmic reticulum which practically fills the cytoplasm of these cells.
(Kindly supplied by Professor D. W. Fawcett, Department of
Anatomy, Harvard Medical School, Boston.)

vessels and lymphatics, and associated with them are varying numbers of Leydig cells, which are assumed to produce the androgens.

In *Didelphis*, as in animals like the pig, there are very abundant Leydig cells, with very small lymphatic spaces (Duesberg, 1920; Christensen and Fawcett, 1961; Fawcett, Neaves and Flores, 1974; Figure 24.16). In the Australian macropods examined and *T. vulpecula* the Leydig cells are much less abundant (Green, 1963; see also Figures 24.2, 24.13 and 24.14), the appearance of sections being much more like sections of ram or bull testis. The testis of the bandicoot contains almost as high a proportion of Leydig cells as that of *Didelphis* (Figure 24.2).

The ultrastructure of the Leydig cells has also been studied in *Didelphis* by Christensen and Fawcett (1961). They found that these cells, like Leydig cells in eutherian mammals, contained an exceedingly abundant agranular endoplasmic reticulum (Figure 24.17), generally in the form of a meshwork of interconnected tubules about 300 to 450 Å in diameter. Occasionally there were flattened, fenestrated cisternae. The Leydig cells varied considerably in their cytoplasmic density, the majority being quite light in appearance, while some appeared extremely dense; these latter cells also show a more irregular cell surface, with numerous small pseudopodia.

Effects of hormones on the testis. While there is some disagreement about the exact modes of action (for recent reviews, see Setchell and Hinton, 1973. Steinberger, 1971; Steinberger and Steinberger, 1974), there is general agreement that the testis of eutherian mammals require both FSH and LH (ICSH) for the establishment of spermatogenesis, but possibly only LH or testosterone for its maintenance. All that has so far been established in marsupials is that, in the Tammar, hypophysectomy leads to rapid atrophy of the testis and decrease in the diameter of the seminiferous tubules. Eventually out Sertoli cells and few peripherally scattered spermatogonia and primary spermatocytes remained (Hearn, 1975). Further studies will depend on the preparation of pure marsupial gonadotrophins.

The testis of 70 day old pouch-young *Didelphis* responds to injections of pregnant mare serum gonadotrophin by increasing its androgen production, but younger animals were unresponsive as judged by prostate response (Moore and Morgan, 1943). However, when Cowper's gland was used as the criterion, the testis of the youngest animals studied (17 days) were already responsive. This may be due to a small response in androgen secretion in the testes of the younger animals which was sufficient to stimulate the Cowper's glands but not the prostate (Rubin, 1944).

Hormone production by the testis. It was originally suggested that similar concentrations of testosterone are found in the blood of male and female *T. vulpecula* (Vinson and Phillips, 1969), implying a non-testicular origin for this steroid. However, more recent investigations (Carrick and Cox, 1973) have shown conclusively that testosterone is secreted by the testis into the blood in several Australian marsupials at a rate comparable with that for eutherian species (Table 24.7). However,

Table 24.7   Testosterone production *in vivo* by the testes of some Australian marsupials

Species	Body wt (kg)	Wt of both testes (g)	Testosterone concentration in plasma from the internal spermatic vein (ng/ml)	Testosterone secretion rate (ng/min per g testes)	(mg/day)
Perameles nasuta	1	4	13	0.1	0.001
Trichosurus vulpecula	3	8	100	2.3	0.025
Pseudocheirus peregrinus	1.5	8	1	–	–
Macropus robustus	30	40	98	6.8	0.4
Megaleia rufa	50	40	83	2.6	0.15
Ovis aries	70	500	60	6.0	4
Rattus norvegicus	0.5	3	40 – 110	9 – 25	0.05 – 0.13
Homo sapiens	70	40	270	135	8

Marsupial data from Carrick and Cox (1973); sheep values from Setchell, Waites and Lindner (1965); rat plasma testosterone concentrations from Suzuki and Eto (1962), Bardin and Peterson (1967) de Jongh, Hey and Van Der Molen (1973); testicular blood flow from Waites, Setchell and Quinlan (1973); human values from Lipsett (1970).

no testosterone could be found in the blood from the internal spermatic vein of the Ring-tailed possum (*Pseudocheirus peregrinus*) and comparatively low concentrations were found in the bandicoot *P. nasuta,* so some further work is clearly indicated as other steroids may be involved.

Slices of the testes of *Didelphis marsupialis* incubated *in vitro* for 3 h incorporated appreciable amounts of $[1,^{14}C]$ acetate into both androstenedione and testosterone, and this was stimulated by LH, either added *in vitro* or injected into the animal for 3 days before removing the testis. $[^{3}H]$ cholesterol and $[^{3}H]$ pregnenolone were also converted to labelled androstenedione and testosterone, and androstenedione and testosterone were present in peripheral blood of male opossums (Cook *et al.*, 1974).

The epididymis

Very little attention has been directed towards the marsupial epididymis. It appears to be grossly similar to the epididymis of eutherians, but is not as closely attached to the testis. Histologically, its appearance is also quite similar and, from the sperm density, there appear to be areas of fluid resorption, secretion and sperm storage as in eutherian mammals (Morehead and Setchell, 1976).

The spermatozoa take at least 11 – 13 days to pass through the epididymis in *M. eugenii* and *T. vulpecula* (Setchell and Carrick, 1973). During this time their appearance changes. The bell-shaped structure behind the head, which is analogous to the kinoplasmic droplet, disappears and the head is held in line with the tail, not at an angle as it is when the sperm leave the testis (Jordan, 1911; Setchell, 1970b). This is presumably achieved by the complete dislocation of the proximal extremity of the neck from the nuclear socket, which is said to occur while the spermatozoa are in the epididymis. When this happens, the neck and proximal part of the mid-piece enclosed within the ventral nuclear groove, move forwards towards the rostrum of the spermatozoal head (Sapsford *et al.*, 1969a).

Spermatozoa are found regularly in the urine of marsupials (Bolliger and Carrodus 1938; Bolliger 1942; Setchell and Carrick 1973), so presumably they pass along the ductus deferens into the urethra even when the animal is not sexually active.

Accessory glands

Anatomy

The only accessory sex glands marsupials possess are the prostate and Cowper's glands. There are no ampullary glands, seminal vesicles or coagulating glands (Cowper, 1704; Cuvier, 1805; Owen, 1839-47; 1868; Young, 1879; Oudemans, 1892; Disselhorst, 1904; van den Broek, 1910, 1933 and Figure 24.18). The prostate varies in size in different species and with the stage of sexual activity of the animal (Gilmore, 1969) but it can be proportionately larger than in eutherian mammals. Indeed, it has been described in *T. vulpecula* as 'the biggest organ in the abdominal cavity excepting the liver' (Bolliger, 1946). Rodger and Hughes (1973) divided Australian marsupials into two groups, those with a 'carrot-shaped prostate' and those with a 'heart-shaped prostate' (Figure 24.19); their results, those of earlier workers and some results with American marsupials are summarised in Table 24.8. The glands could be divided into three segments, anterior, central and posterior, in some species, and into ventral and dorsal in others (Figure 24.19) (Chase, 1939; Carrodus and Bolliger, 1939; Rodger and Hughes, 1973).

The Cowper's glands are between one and three pairs of glands lying on either side of the urethra. They were originally described in *Didelphis* by Cowper himself, in 1704, soon after his description of the glands in the human (Cowper, 1699), and other authors have described them in a number of other marsupials (see Table 24.8).

After castration (Chase, 1939; Bolliger and Tow, 1946) or hypophysectomy (Hearn, 1975), the size of the accessory glands is very much reduced and the size can be increased by injections of androgens (Chase, 1939).

Androgens (Burns, 1939a, b; Moore, 1941, Rubin, 1944) and gonadotrophic hormones (Moore and Morgan, 1943; Rubin, 1944) also cause precocious stimulation of growth of the accessory glands in male pouch young of Didelphis, although development between the ages of 22 and 100 days appears to continue normally after castration. Early differentiation (up to 39 days), as opposed to growth, of Cowper's glands appears to be inhibited by treatment with androgens (Rubin, 1944). Androgens also stimulate development of the Mullerian duct system as well as that of the Wolffian duct, in both female and male pouch young (Burns, 1939a, b).

The biology of marsupials

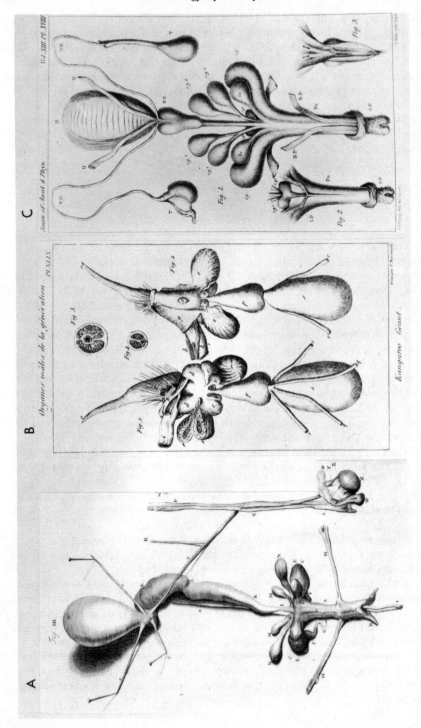

Figure 24.18 The accessory glands of some marsupials. (A) the opossum (*Didelphis marsupialis*) from Cowper (1704). (B) the kangaroo (probably *Macropus giganteus*) from Cuvier (1805). (C) the koala (*Phascolarctos cinereus*) from Young (1879). Note the carrot shaped prostate in (A) and (B) and the heart-shaped prostate in (C), and the three pairs of Cowpers glands in all species shown.

(A) Cowper's key reads as follows: 'The backside of the genitals of the male opossum. A the body of the *Penis*; B its glans; CC The *Bulbi* of the *Corpora Cavernosa Penis* covered with their muscles; DD the *Corpora Cavernosa Penis*; EE the two distinct bulbs of the cavernous body of *Urethra,* inclosed with their particular muscles; FFG parts of the muscles exprest on the fore part of the penis; HH the other pairs of muscles springing from the *Rectum* and inserted to the sides of the *Corpora Cavernosa Penis*; IKL the *Urethra* covered with the prostate KLK; MN the two mucous bags on each side [now called Cowper's glands]; O the bladder of urine; P the *Musculus Cremaster*; Q the *Tunica Vaginalis* opened; R *Vasa Praeparantia* cut from the great trunks; SS the *Vas Deferens* on each side; WXYZ the left testicle as it appeared on opening the *Tunica Vaginalis*; W its *Epididymis*; X the body of the testicle; Y the spermatic vein and artery, as they pass to and from the testicle; Z the excretory duct of the testicle, which could be distinctly seen arising from the *testes* and marching to the *Epididymis*; W, where it is folded up and constitutes that body, whence it is continued to the bladder of urine and called *Vas Deferens* SS; ee parts of the urethra; **a probe inserted into part of the *Urethra.'*

(B) Cuvier's key can be translated as follows: "Male organs of reproduction in the giant kangaroo, in Fig. 1 from the dorsal aspect, in Fig. 2 from the ventral aspect. S Urinary bladder; p (sic) the ureters; V the deferent ducts; p the first part of the urethral canal, enveloped by the prostate; abc the three Cowper's glands on each side; dd the two parts of the urethral bulb encased in their muscles; ee the two parts of the corpus cavernosus, encased in their muscles – the one on the right side (Fig. 1) has been split along its length to show it in section with the enveloping muscle; g a muscle which serves as the anal sphincter, it is folded back in Fig. 1; k (in Fig. 1) a part of the levator ani muscle; hh (in Fig. 2) the retractor muscles of the penis; 1 the anus; o the exterior orifice of the urethra; m (in Fig. 2) part of the suspensory ligament of the penis. [Some of the letters appear incorrect and several others in the figure e.g. 'f', 'i', and 'u' are not listed in the legend] . Fig 3: section through the penis just after the union of the branches of the corpora cavernosa: "a" section through the urethral canal, situated almost at the centre of the corpus cavernosum in a canal which this body forms for it; Fig. 4 section through the same organ a short distance from its end; "a" indicates as before the section through the urethral canal, which has approached the lowest edge of the corpus cavernosus, and is on the point of emerging from it."

(C) Young's key reads as follows: 'Fig. 1: male generative organs of the Koala, rectal surface. B bladder; UU ureters; TT testicles (covering removed from left to show the epididymis); V.D. vasa deferentia; P.G. prostate gland; G.P. glans penis, prepuce retracted; R.P. retractores penis; Pe penis; c.p. crura penis, with erectores partially reflected; b.s. and b.s.[1] the two separate bulbs of the corpus spongiosum; c.g.[1], c.g.[2] and c.g.[3] Cowper's glands (three pairs). Fig. 2: pubic aspect of the penis; L.P. levator penis; other letters as in Fig. 1. Fig. 3: intra-pelvic portion of the urethra laid open to show the verumontanum and the utricular depression; The two bristles are placed in the openings of the ureters.'

Table 24.8 The anatomy of the accessory glands of male marsupials
(based mainly on Rodger and Hughes, 1973)

Family	Species	Prostate carrot- or heart-shaped	Division (see figure 24.19)	Cowpers gland No. of pairs	Reference
Didelphidae	*Didelphis virginiana*	C	ACP	3	3,5,16,22
	D. azarae	C	ACP	2	14
	D. cancrivora	C		3	16
Dasyuridae	*Dasyurus viverrinus*	C		3	16
	Sarcophilus harrisii	C		2	13
	Antechinus stuartii	C	$AP_1 P_2$	3	19,25
	Sminthopsis Crassicaudata	C	$AP_1 P_2$	2	19
	Myrmecobius fasciatus	C		3	8
Thylacinidae	*Thylacinus cynocephalus*	C		2	5
Notoryctidae	*Notoryctes typhlops*	C			24
Peramelidae	*Perameles gunnii*	H		2	16
	P. nasuta	H	DV	2	19
	Isoodon macrourus	H	DV	2	19
Caenolestidae	*Caenolestes obscurus*	H		3	15
Vombatidae	*Vombatus ursinus*	absent?		3	1,13
Phascolarctidae	*Phascolarctos cinereus*	H	longitudinal (see Figure 24.18)	3	26
Phalangeridae	*Phalanger maculatus*	C		1	5,7
	Trichosurus vulpecula	C	ACP	2	7,10,12, 13,19
Petauridea	*Pseudocheirus peregrinus*	C	$AP_1 P_2$	1	19
	Schoinobates volans			2	23
Macropodidae	*Potorous tridactylus*	C	ACP	2	9,19
	Aepyprymnus rufescens	C		3	7
	Bettongia sp.	C		3	17
	Macropus giganteus	C	ACP	3	6,1,20
	M. eugenii	C	ACP	3	19,17
	M. rufogriseus	C		3	16,18
	Megaleia rufa	C	ACP	3	16,19
	Petrogale penicillata			3	16
Tarsipedidae	*Tarsipes spencerae*	C	AP	1	21

1. Adams (1847–9). 2. Carrodus and Bolliger (1939). 3. Chase (1939). 4. Cowper (1704). 5. Cunningham (1882). 6. Cuvier (1805). 7. Disselhorst (1904). 8. Fordham (1928). 9. Fraser (1919). 10. Gilmore (1969). 11. Home (1808). 12. Howarth (1950). 13. Mackenzie and Owen (1919). 14. Martin and Jones (1834). 15. Osgood (1921). 16. Oudemans (1892). 17. Owen (1839–47, and 1868). 18. Pavaux (1962). 19. Rodger and Hughes (1973). 20. Rodger and White (1974a, b). 21. Rotenberg (1929). 22. Rubin (1943, 1944). 23. Smith (1969). 24. Sweet (1897). 25. Woolley (1964). 26. Young (1879).

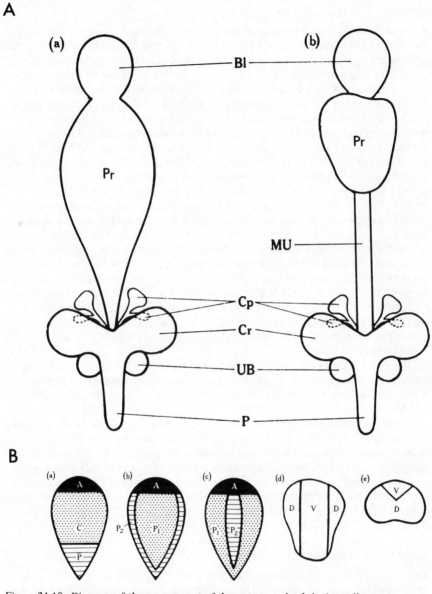

Figure 24.19 Diagrams of the arrangement of the accessory glands in Australian marsupials, ventral view. (a) Carrot-shaped prostate; (b) heart-shaped prostate. Bl: urinary
bladder: Pr: prostate; MU: membranous urethra; Cp: Cowper's glands; Cr: crus penis; UB:
urethral bulb; p: penis. (B) Diagram of the segmentation of the prostate in: (a) macropods
and *Trichosurus vulpecula*; (b) dasyurids; (c) *Pseudocheirus peregrinus*; (d) peramelids,
ventral view; (e) peramelids, transverse section. (a) – (c) in frontal section, with shading to
indicate the suggested homologies. Segments: A: anterior; C: central; D: dorsal; P: posterior;
P_1: posterior 1; P_2: posterior 2; V: ventral. Reproduced, with permission of the *Australian
Journal of Zoology,* from Rodger and Hughes (1973).

Histology

A detailed study of the histology of the accessory glands of *Didelphis* was made
by Chase (1939). She divided the prostate into three sections, each with a different
histological appearance. In the most anterior there is a fairly large amount of
connective tissue, with narrow tubules lined with low columnar or cuboidal cells
with round or oval darkly stained nuclei near their bases. The second section has
less connective tissue, wider tubules and higher and wider cells. The third section
also has little connective tissue and wide tubules, but these are characterised by
wide lumina and cuboidal cells (Chase, 1939). By histochemical techniques,
segment I was characterised by an apocrine-type secretory epithelium; in the
apical parts of the cells there were vesicles which contained mucopolysaccharides,
phospholipid, acid phosphatase, aliesterase, 5'-nucleotidase and nucleoside
diphosphatase. Segment II was characterised by two types of cells in the epithelium. In Type A cells there were large secretory granules which contained acid
mucopolysaccharides but no protein. In Type B cells there were large elongated
secretory granules which contained protein but no acid mucopolysaccharide.
Segment III was characterised by the presence of large amounts of glycogen both
in the epithelial cells and in the lumen (Martan and Allen, 1965). The distinction
between the three segments was confirmed by studies with an electron microscope by Hruban *et al.* (1965), who also showed that the first segment contained
high concentrations of iron in granules in the cells and lumen. They also found
that the third segment contained unusually high concentrations of catalase. The
prostate of the possum is also unusual in having a poorly developed rough
endoplasmic reticulum.

The appearance of the three Cowper's glands also differ. Gland I usually has
low cells with large lumina; in gland II the cells are slightly higher and in gland
III the cells are columnar with oblong nuclei near the bases of the cells (Rubin,
1944).

In the Australian marsupials the prostate consists of numerous simple branched
tubular glands, lined by a single layer of columnar cells which vary in height
between species and gland segment. The wall of the gland can be divided into a
collecting duct region, the glandular tissue which makes up the bulk of the organ
and an outer coat of smooth muscle. Cowper's glands are of the branched tubular
type lined with a columnar or sometimes pseudo-stratified epithelium, and
covered with a coat of striated muscle of variable thickness (Rodger and Hughes,
1973).

Composition of secretions

The accessory glands of eutherian mammals produce specific secretions, characterised by high concentrations of fructose, citric acid, ergothioneine, inositol, etc.
(Mann, 1964). In the prostate of *Didelphis* Mann and Lutwak-Mann (1963)
found high concentrations of a resorcinol-reactive material which they assumed
to be fructose, with the highest concentrations in the posterior segment.

Rodger and White (1974a) showed that in *T. vulpecula* and *M. rufa*, in
contrast to eutheria, fructose was virtually absent from the prostate, but that
glucose was present in high concentrations in certain parts (Table 24.9). They
also found in the posterior prostate of the possum, high concentrations of a

Table 24.9 The composition of the segments of the prostate of some marsupials. All values in mg/100 g

Species	Segment	Anthrone reactive material (as glucose)	Glucose	Fructose	N-acetyl glucosamine	Resorcinol reactive material (as fructose)	Citric acid	Ergothionine	Reference
Trichosurus vulpecula	A	54	22	2	37	4			Rodger and White (1974a and unpublished)
	C	15	6	1	161	2			
	P	29	10	1	23	51			
Pseudocheirus peregrinus	P				461	4			
Megaleia rufa	A	23	10	2	97	7			
	C	46	7	1	630	9			
	P	136	92	3	10	3			
Macropus giganteus	A	58	36	3	112	4			
	C	63	22	3	465	6			
	P	224	139	3	27	—			
M. eugenii	A	72	37	—	63	—			
	C	123	34	—	580	—			
	P	100	38	—	25	—			
Perameles nasuta	V	261	57	—	507	—			
	D	186	66	—	258	—			
Isoodon macrourus	V	289	36	—	711	—			
	D	182	56	—	278	—			
Didelphis virginiana	I					3	6	28	Mann and Lutwak-Mann (1963)
	II					5	15	18	
	III					44	12	36	

resorcinol-reactive substance which was not fructose; this has not yet been identified, but it does not react with anthrone and may be non-carbohydrate. In the central prostate of the possum they found high concentrations of N-acetyl-glucosamine (Rodger and White, 1974b), which does not occur in the semen of most eutheria (Rodger and White, 1974c). From histochemical evidence the secretion appears to be rich in 'muco-substances', except in most posterior segments; in *T. vulpecula* and *Ps. peregrinus,* 'lipid' makes a significance contribution to the secretion (Rodger and Hughes, 1973).

No studies appear to have been made of the composition of the secretion of the Cowper's glands, except that Rodger and Hughes (1973) suggest from histochemical evidence that it secretes a 'mucus material'.

'Semen' has been collected from several marsupials by electroejaculation (Howarth, 1950; Sadleir, 1965; Gilmore, 1969; Rodger and White, 1975). This is effective only anaesthetised or recently dead animals and the low and very variable concentrations of spermatozoa in the ejaculate suggests that the 'semen' collected in this way consists mainly of the secretions of the accessory glands.

Function of accessory glands

Howarth (1950) removed the prostate surgically from six adult *T. vulpecula* and found that this reduced the volume of semen produced by electroejaculation from between 5 and 31 ml to less than 1 ml. He therefore suggested 'that the main function of the prostate gland in the marsupial is to provide a fluid vehicle for the spermatozoa', a conclusion not very different from the opinion first expressed by de Graaf (1668) for the function of the prostate in man.

Conclusion

I hope that this review will serve to emphasise how little we know about the reproductive processes of male marsupials, but how many interesting lines of work are suggested by the present scanty knowledge. The real significance of the vascular rete in the spermatic cord, the curious relationship between the rete testis and the blood supply, the peculiar shape of the head of the spermatozoa, and the composition of the secretion of prostate and Cowper's glands are all worthy of attention. It will be great interest to discover the identity of the androgens in those marsupials in which testosterone does not appear to be important. The mechanism of control of the testis by gonadotrophins and the reason for the descent of the testis into a scrotum are of more general interest but data on marsupials should be added to data on eutherian species until there is sufficient information for valid generalisations to become possible.

It is perhaps too much to hope that this review will convert many people to the opinion expressed by Tyson (1704) that 'the organs of generation in the male (marsupial) are no less surprising and remarkable than in the female, and in both, they are different from any other animal that I have met with'.

Acknowledgements

I am very grateful to the authors and publishers who have allowed me to reproduce their figures and plates; Dr. A. E. Newsome and *Journal of Reproduction and Fertility* for Figure 24.1; *Comparative Biochemistry and Physiology* (Pergamon Press) for Figure 24.3; *Journal of Reproduction and Fertility* for Figures 24.4 and 24.5; Academic Press for Figure 24.7; Dr. P. Woolley and *Journal of Reproduction and Fertility* for Figure 24.8; Drs. Sapsford, Rae and Cleland and *Australian Journal of Zoology* for Figures 24.10, 24.11 and 24.12; Drs. Fawcett, Neaves and Flores and *Biology of Reproduction* for Figure 24.16c; Drs. Rodger and Hughes and *Australian Journal of Zoology* for Figure 24.19; I am especially grateful to Drs. F. N. Carrick, R. I. Cox, D. W. Fawcett, H. P. Godinho, M. T. A. Costa, F. M. Cardoso and J. C. Rodger who allowed me to use their unpublished pictures and data and to Drs. F. N. Carrick and C. H. Tyndale-Biscoe for reading the manuscripts and making helpful comments. I am also grateful to the University Library, and the Balfour Library, Department of Zoology, Cambridge for copying or allowing me to copy the pictures in Figure 24.18.

References

Adams, J. (1847-9). Prostate gland. In: *Cyclopedia of Anatomy and Physiology*, Vol. IV, Pt. 1 (ed. R. B. Todd), Longman, Brown, Green Longmans and Roberts, London, pp. 146-62.

Agar, W. E. (1923). The male meiotic phase in two genera of marsupials (Macropus and Petauroides). *Quart. J. Microscop. Sci.*, **67**, 183-202.

Azzali, G. and DiDio, L. J. A. (1965). The lymphatic system of *Didelphys azarae* and *Didelphys marsupialis. Am. J. Anat.*, **116**, 449-70.

Baldwin, J. and Temple-Smith, P. (1973). Distribution of LDH-X in mammals: presence in marsupials and absence in the monotremes Platypus and Echidna. *Comp. Biochem. Physiol.*, **46B**, 805-11.

Bancroft, B. J. (1973). Embryology of *Schoinobates volans* (Kerr) (Marsupialia: Petauridae). *Aust. J. Zool.*, **21**, 33-52.

von Bardeleben, K. (1896). Uber Spermatogenese bei Monotremen und Beutel-thieren. *Verh. Anat. Ges. Jena.* **10**, 38-43.

Bardin, C. W. and Peterson, R. E. (1967). Studies of androgen production by the rat: testosterone and androstenedione content of blood. *Endocrinology*, **80**, 38-44.

Barnett, C. H. and Brazenor, C. W. (1958). The testicular rete mirabile of marsupials. *Aust. J. Zool.*, **6**, 27-32.

Beddard, F. E. (1891). On the pouch and brain of the male thylacine. *Proc. Zool. Soc. London*, 138-45.

Benda, C. (1906). Die Spermiogenese der Marsupialier. *Denkschr. Med. Naturw. Ges. Jena.* **6**, 441-58.

Biggers, J. D. (1964). Reproduction in male marsupials. *Symp. Zool. Soc. London*, **15**, 251-80.

Binder, S. (1927). Spermatogenese von *Macropus giganteus. Z. Zellforsch*, **5**, 293-346.

Boardman, W. (1945). Some points in the external morphology of the pouch young of the marsupial *Thylacinus cynocephalus* Harris. *Proc. Linnean Soc. N. S. Wales.* **70**, 1-8.

Bolliger, A. (1942). Spermatorrhoea in marsupials with special reference to the action of sex hormones on spermatogenesis of *Trichosurus vulpecula. J. Proc. Roy. Soc. N. S. Wales.,* **76**, 86 - 92.

Bolliger, A. (1943). Functional relations between scrotum and pouch and the experimental production of a pouch-like structure in the male of *Trichosurus vulpecula. J. Proc. Roy. Soc. N. S. Wales,* **76**, 283 - 93.

Bolliger, A. (1944). An experiment on the complete transformation of the scrotum into a marsupial pouch in *Trichosurus vulpecula. Med. J. Aust.,* **11**, 56 - 8.

Bolliger, A. (1946). Some aspects of marsupial reproduction. *J. Proc. Roy. Soc. N. S. Wales,* **80**, 2 - 13.

Bolliger, A. (1954). Organ transformation induced by oestrogen in an adolescent marsupial (*Trichosurus vulpecula*) *J. Proc. Roy. Soc. N. S. Wales,* **88**, 33 - 9.

Bolliger, A. and Carrodus, A. L. (1938). Spermatorrhoea in *Trichosurus vulpecula* and other marsupials. *Med. J. Aust.,* **2**, 118 - 9.

Bolliger, A. and Carrodus, A. (1939). Experimental testicular ascent as produced in a marsupial (*Trichosurus vulpecula*) *Aust. N. Z. J. Surg.,* **9**, 155 - 63.

Bolliger, A. and Carrodus, A. (1940). The effect of testosterone propionate on pouch, scrotum, clitoris and penis of *Trichosurus vulpecula. Med. J. Aust.,* **II**, 368 - 73.

Bolliger, A. and Tow, A. J. (1946). Late effects of castration and administration of sex hormones on the male *Trichosurus vulpecula. J. Endocrinol.,* **5**: 32 - 42.

van den Broek, A. J. P. (1910). Untersuchungen über den Bau der männlichen Geschlechtsorgane der Beuteltiere. *Morphol. Jahrb.,* **41**, 347 - 436.

van den Broek, A. J. P. (1933). Gonaden und Ausführungsgänge. In: *Handbuch der vergleichenden Anatomie der Wirbeltiere* (ed. L. Bolk, E. Goppart, E. Kallius and W. Lubosch), Urban and Schwarzenburg, Berlin, pp. 1 - 154.

Broom, R. (1898). A contribution to the development of the common phalanger. *Proc. Linnean Soc. N. S. Wales,* **23**, 705 - 29.

Buchanan, G. and Fraser, E. A. (1919). The development of the urogenital system in the Marsupialia, with special reference to *Trichosurus vulpecula. J. Anat.,* **53**, 35 - 95.

de Burlet, H. M. (1921). Zur Entwicklung und Morpholigie des Säugerhodens II Marsupialer. *Z. Anat. Enwickl. Abt. 1* **61**, 19 - 31.

Burns, R. K. Jr (1939a) The differentiation of sex in the opossum (*Didelphys virginiana*) and its modification by the male hormone testosterone propionate. *J. Morphol.,* **65**, 79 - 119.

Burns, R. K. Jr (1939b). Sex differentiation during the early pouch stages of the opossum (*Didelphys virginiana*) and a comparison of the anatomical changes induced by male and female sex hormones. *J. Morphol.,* **65**, 497 - 547.

Burns, R. K. Jr (1941). The origin of the rete apparatus in the opossum. *Science* **94**, 142 - 4.

Carrick, F. N. (1969). Spermatogenesis in the brush-tailed possum. *Trichosurus vulpecula* (Kerr, 1792) with special reference to the effects of high ambient temperature. B.Sc. Honours Thesis, University of N. S. W., Sydney, Australia.

Carrick, F. N. (1972). Short range bio-telemetry of temperatures in the *Euro (Macropus robustus). Aust. Mammal.,* **1**, 65.

Carrick, F. N. and Cox, R. I. (1973). Testosterone concentrations in the spermatic vein plasma of marsupials. *J. Reprod. Fert.*, **32**, 338 - 9.

Carrodus, A. and Bolliger, A. (1939). The effect of oestrogenic hormone on the prostate of the marsupial *Trichosurus vulpecula. Med. J. Aust.*, **II**, 633 - 41.

Chase, E. B. (1939). The reproductive system of the male opossum, *Didelphis virginiana* Kerr and its experimental modification *J. Morphol*, **65**, 215 - 39.

Christensen, A. K. and Fawcett, D. W. (1961). The normal fine structure of opossum testicular interstitial cells. *J. Biophys. Biochem. Cytol.*, **9**, 653 - 70.

Cleland, K. W. (1956). Acrosome formation in bandicoot spermiogenesis. *Nature*, **177**, 387 - 8.

Clermont, Y. (1972). Kinetics of spermatogenesis in mammals: seminiferous epithelium cycle and spermatogonial renewal. *Physiol. Rev.*, **52**, 198 - 236.

Cook, B., Sutterlin, N. S., Graber, S. W. and Nalbandov, A. V. (1974). Gonadal steroid synthesis in the Virginian opossum, *Didelphis marsupialis. J. Endocrinol.*, **61**, ix.

Costa, M. T. A., Godinho, H. P. and Cardosa, F. (1975). Unpublished observations.

Courot, M., Hochereau-de Reviers, M-T., and Ortavant, R. (1970). Spermatogenesis. In: *The Testis*, Vol. 1 (ed. A. D. Johnson, W. R. Gomes and N. L. Van Demark), Academic Press, New York., pp. 339 - 432.

Cowper, W. (1699). An account of two glands and their excretory ducts lately discovered in human bodies. *Phil. Trans. Roy Soc.* **21**, 364 - 9.

Cowper, W. (1704). A letter to Dr. Edward Tyson giving an account of the anatomy of those parts of a male opossum that differ from the female. *Phil Trans. Roy, Soc.*, **24**, 1576 - 90.

Cunningham, D. J. (1882). Report on some points in the anatomy of the Thylacine (*Thylacinus cynocephalus*) Cuscus (*Phalagista maculata*) and *Phascogale* (*Phascogale calura*) collected during the voyage of H. M. S. Challenger in the years 1873 - 1875. *Rep. Sci. Res. Voyage H. M. S. Challenger, Zoology*, Vol. 5, Part 16, 1 - 192.

Curtis, G. M. (1918). The morphology of the mammalian seminiferous tubule. *Am. J. Anat.*, **24**, 339 - 94.

Cuvier, G. (1805). *Lecons d'Anatomée Comparée*, Tome V, Baudouin, Paris.

Disselhorst, R. (1904). Die männlichen Geschlechtsorgane der Monotremen und einiger Marsupialen. *Denkschr. Med. Naturw. Ges. Jena*, **6**, 122 - 50.

Duesberg, J. (1920). Cytoplasmic structures in the seminal epithelium of the opossum. *Cont. Embryol.*, **9**, 49 - 84.

Einer-Jensen, N. (1974). Local recirculation of 133 Xenon and 85 Krypton to the testes and the caput epididymidis in rats. *J. Reprod. Fert.*, **37**, 55 - 60.

Fawcett, D. W., Neaves, W. B. and Flores, M. N. (1974). Comparative observations on intertubular lymphatics and the organization of the interstitial tissue of the mammalian testis. *Biol. Reprod.*, **9**, 500 - 32.

Finkel, M. P. (1945). The relation of sex hormones to pigmentation and to testis descent in the opossum and ground squirrel. *Am. J. Anat.*, **76**, 93 - 151.

Fordham, M. G. C. (1928). The anatomy of the urogenital organs of the male *Myrmecobius fasciatus. J. Morphol. Physiol.*, **46**, 563 - 81.

Fraser, E. A. (1919). The development of the urogenital system in the Marsupialia, with special reference to *Trichosurus vulpecula. J. Anat.*, **53**, 97 - 129.

Frith, H. J. and Sharman, G. B. (1964). Breeding in wild populations of the Red Kangaroos *Megaleia rufa. C.S.I.R.O. Wildlife Res.*, **9**, 86 - 114.

452   The biology of marsupials

Fürst, C. M. (1880). Ueber die Entwicklung der Samenkörperchen bei den
    Beuteltheiren, Arch. Mikrosk. Anat., 30, 336 - 65.
Gilmore, D. P. (169). Seasonal reproductive periodicity in the male Australian
    Brush-tailed possum (Trichosurus vulpecula). J. Zool. London, 157, 75 -98.
de Graaf, F. (1668). Tractatus de virorum organis generationi inservientibus Lugd.
    Batav et Roterod. Hackiana. Translated by H. D. Jocelyn and B. P. Setchell.
    J. Reprod. Fert., Suppl., 17, 1 - 76 (1972).
Green. L. M. A. (1963). Interstitial cells in the testis of an Australian phalanger
    (Trichosurus vulpecula). Aust. J. Exp. Biol., 41, 99 - 104.
Greenwood, A. W. (1923). Marsupial spermatogenesis. Quart. J. Microscop. Sci.,
    67, 203 - 18.
Guiler, E. R. (1970). Observations on the Tasmanian Devil Sarcophilus harrisii
    (Marsupialia: Dasyuridae). II. Reproduction, breeding and growth of pouch
    young. Aust J. Zool., 18, 63 - 70.
Guiler, E. R. and Heddle, R. W. L. (1970). Testicular and body temperature in
    the Tasmanian Devil and three other species of marsupial. Comp. Biochem.
    Physiol., 33, 881 - 91.
Guraya, S. S. (1971). Morphological and histochemical changes in the mito-
    chondria during spermiogenesis in the opossum. Acta Anat., 79, 120 - 5.
Hamilton, D. W. (1972). The mammalian epididymis. In: Reproductive Biology
    (Ed. H. Balin and S. Glasser), Excerpta Medica, Amsterdam, pp. 268 - 337.
Harrison, R. G. (1948). Vascular patterns in the testis, with particular reference
    to Macropus. Nature, 161, 399.
Harrison, R. G. (1949). The comparative anatomy of the blood supply of the
    mammalian testis. Proc. Zool. Soc. London, 119, 325 - 44.
Harrison, R. G. (1951). Applications of microradiography. In: Microarteriography
    (ed. A. E. Barclay). Blackwell, Oxford, pp. 89 - 90.
Hartmann, C. G. (1952). Possums, University of Texas Press, Austin.
Hearn, J. P. (1975). The role of the pituitary in the reproduction of the male
    tammar wallaby, Macropus eugenii. J. Reprod. Fert., 42, 399 - 402.
Heddle, R. M. L. and Guiler, E. R. (1970). The form and function of the
    testicular rete mirabile of marsupials. Comp. Biochem. Physiol., 35, 415 - 25.
Heller, C. G. and Clermont, Y. (1964). Kinetics of the germinal epithelium in
    man. Recent Progr. Hormone Res., 20, 545 - 75.
Hill, J. P. and Hill, W. C. O. (1955). The growth-stages of the pouch-young of
    the Native Cat (Dasyurus viverrinus) together with observations on the
    anatomy of the new born young. Trans. Zool. Soc. London. 28, 349 - 452.
Holmes, R. S. (1972). Evolution of lactate dehydrogenase genes. FEBS Lett.,
    28, 51 - 5.
Holmes, R. S., Cooper, D. W. and Vanderberg, J. L. (1973). Marsupial and mono-
    treme lactate dehydrogenase isoenzymes: phylogeny, ontogeny and
    homology with eutherian mammals. J. Exp. Zool., 184, 127 - 48.
Home, E. (1795). Some observations on the mode of generation of the Kanguroo,
    with a particular description of the organs themselves. Phil. Trans. Roy.
    Soc. 85, 221 - 38.
Home, E. (1808). As account of some peculiarities in the anatomical structure
    of the wombat, with observations on the female organs of generation. Phil.
    Trans. Roy Soc., 98, 304 - 12.

Howarth, V. S. (1950). Experimental prostatectomy in a marsupial (*Trichosurus vulpecula*). *Med. J. Aust.*, **II** 325 - 30.

Hruban, Z., Martan, J., Slesers, A., Steiner, D. F., Lubran, M. and Rechcigl, M. Jr (1965). Fine structure of the prostatic epithelium of the opossum (*Didelphis virginiana* Kerr). *J. Exp. Zool.*, **160**, 81 - 106.

Huckins, C. (1971). The spermatogonial stem cell population in adult rats. I. Their morphology, proliferation and maturation. *Anat. Rec.*, **169**, 533 - 58.

Jacks, F. and Setchell, B. P. (1973). A technique for studying the transfer of substances from venous to arterial blood in the spermatic cord of wallabies and rams. *J. Physiol. London.* **233**, 17 - 18P.

de Jongh, F. H., Hey, A. H. and van der Molen, H. J. (1973). Effect of gonadotrophins on the secretion of oestradiol - 17β and testosterone by the rat testis. *J. Endocrinol.*, **57**, 277 - 84.

Jordan, H. E. (1911). The spermatogenesis of the opossum. (*Didelphys virginiana*) with special reference to the accessory chromosome and the chondriosomes. *Arch. Zellforsch.*, **7**, 41 - 86.

Jost, A. (1953). Problems of fetal endocrinology: The gonadal and hypophyseal hormones. *Recent Progr. Hormone Res.*, **8**, 379 - 418.

Jost, A., Vigier, B., Prépin, J. and Perchellet, J-P. (1973). Studies on sex differentiation in mammals. *Recent Prog. Hormone Res.*, **29**, 1 - 41.

Kaiser, W. (1931). Die Entwicklung des Scrotums bei Didelphis Aurita Wied *Morph. Jb*, **68**, 391 - 433.

von Korff, K. (1902). Zur Histogenese der Spermien von *Phalangista vulpina*. *Arch. Mikrosk. Anat. Entw. Mech.*, **60**, 232 - 60.

Lacy, D. (1962). Certain aspects of testis structure and function. *Brit. Med. Bull.*, **18**, 205 - 8.

Lacy, D. (1967). The seminiferous tubule in mammals. *Endeavour*, **26**, 101 - 6.

Ladman, A. J. (1967). Fine structure of the ductuli efferentes of the opossum. *Anat. Rec.*, **157**, 559 - 76,

Lipsett, M. B. (1970). Steroid secretion by the human testis. In: *The Human Testis* (ed. E. Rosemberg and C. A. Paulsen), Plenum Press, New York, pp. 407 - 21.

Lyne, A. G. and Verhagen, A. M. W. (1957). Growth of the marsupial *Trichosurus vulpecula* and a comparison with some higher mammals. *Growth*, **21**, 167 - 95.

McCrady, E. (1938). The embryology of the opossum. *Am. Anat. Mem.*, **16**, 1 - 233.

Mackenzie, W. C. and Owen, W. J. (1919). The genito-urinary system in monotremes and marsupials. In: *The Comparative Anatomy of Australian Mammals*, Part IV, Jenkins, Buxton, Melbourne.

Mann, T. (1964). *The Biochemistry of Semen and of the Male Reproductive Tract*, Methuen, London.

Mann, T. and Lutwak - Mann, C. (1963). Comparative biochemical aspects of animal reproduction. *Bull. Acad. R. Med. Belg.*, 7e series, **3**, 563 - 97.

Marlow, B. J. (1961). Reproductive behaviour of the marsupial mouse *Antechinus flavipes* (Waterhouse) (Marsupialia) and the development of the pouch young. *Aust. J. Zool.*, **9**, 203 - 18.

Martan, J. and Allen. J. M. (1965). The cytological and chemical organization of the prostatic epithelium of *Didelphis virginiana* Kerr. *J. Exp. Zool.*, **159**, 209 - 30.

Martin, W. and Jones, R. (1834). Notes on the dissection of Azara's opossum (*Didelphis azarae*, Temm). *Proc Zool. Soc. London.*, II, 101 - 4.

Mason, K. and Blackshaw, A. W. (1973). The spermatogenic cycle of the bandicoot, *Isoodon macrourus. J. Reprod. Fert.*, 32, 307 - 8.

Maynes, G. M. (1973). Reproduction in the parma wallaby, *Macropus parma* Waterhouse, *Aust. J. Zool.*, 31, 331 - 51.

Moore, C. R. (1941). Embryonic differentiation of opossum prostate following castration and responses of the juvenile gland to hormones. *Anat. Rec.*, 80, 315 - 27.

Moore, C. R. (1943). Sexual differentation in the opossum after early gonadectomy. *J. Exp. Zool.*, 94, 415 - 62.

Moore, C. R. (1950). Studies on sex hormones and sexual differentiation in mammals. *Arch. Anat. Microsc. Morphol. Exp.*, 39, 484 - 98.

Moore, C. R. and Morgan, C. F. (1943). First response of developing opossum gonads to equine gonadotrophin treatment. *Endocrinology*, 32, 17 - 26.

Morehead, J. R. and Setchell, B. P. Unpublished observations.

Mykytowycz, R. and Nay, T. (1964). Studies of the cutaneous glands and hair follicles of some species of Macropodidae. *C.S.I.R.O. Wildlife Res.*, 9, 200 - 17.

Newsome, A. E. (1965). Reproduction in natural populations of the red kangaroo *Megaleia rufa* (Desmarest) in Central Australia. *Aust. J. Zool.*, 13, 735 - 49.

Newsome, A. E. (1973). Cellular degeneration in the testis of red Kangaroos during hot weather and drought in central Australia. *J. Reprod. Fert.*, Suppl., 19, 191 - 201.

Ortavant, R., Orgebin, M. C. and Singh, G. (1962). Etude comparative de la durée des phenoménes spermatogénétiques chez les animeau domestiques. In: *Symp. Use Radioisotope Anim. Biol. Med. Sci. Mexico 1961.* Academic Press, New York, pp. 321 - 7.

Osgood, W. H. (1921). A monographic study of the American marsupial, Caenolestes. *Field Mus. Nat. Hist. Publ. 207, Zool. Ser.*, 14, 1 - 162.

Oudemans, J. T. (1892). Die Accessorischen Geschlechtsdrüsen der Saügetiere. *Natuur. Verh. Holland Maatsch. Wet. Haarlem*, 3 Verg. Deel 5, 2de Stuk, 1 - 96.

Owen, R. (1839 - 47). Marsupialia. In: *Cyclopedia of Anatomy and Physiology* (ed. R. B. Todd), Longman, Brown, Green, Longmans and Roberts, London, pp. 257 - 330.

Owen, R. (1868). *On the Anatomy of Vertebrates,* Vol. III, Longmans, Green, London.

Painter, T. S. (1922). Studies in mammalian spermatogenesis. I. The spermatogenesis of the opossum (*Didelphys virginiana*). *J. Exp. Zool.*, 35, 13 - 38.

Pavaux, C. (1962). L'appareil urinaire et genital mâle de *Macropus ruficollis Benetti.* Mammalia, 26, 72 - 83.

Phillips, D. M. (1970). Development of spermatozoa in the Wooly Opossum with special reference to the shaping of the sperm head. *J. Ultrastruct. Res, 33*, 369 - 380.

Pocock, R. I. (1926). The external characters of Thylacinus, Sarcophilus and some related marsupials. *Proc. Zool. Soc. London*, 1926, 1037 - 84.

Poole, W. E. (1973). A study of breeding in grey Kangaroos, *Macropus giganteus* Shaw and *M. fulginosus* (Desmarest) in central New South Wales. *Aust. J. Zool.*, 21, 183 - 212.

Regaud, C. (1901). Etudes sur la structure des tubes séminifères et sur la spermatogenése chez les mammifères. *Arch. Anat. Microscop*, 4, 101 - 155, 291 - 380.

Rodger, J. C. and Hughes, R. L. (1973). Studies of the accessory glands of male marsupials. *Aust. J. Zool.*, 21, 303 - 20.

Rodger, J. C. and White, I. G. (1974a). Carbohydrates of the prostate of two Australian marsupials, *Trichosurus vulpecula* and *Megaleia rufa*. *J. Reprod. Fert.*, 38, 267 - 73.

Rodger, J. C. and White, I. G. (1974b). Sugars of the accessory sexual secretions of males of marsupial and eutherian species. *Proc. Aust. Soc. Reprod. Biol. 6th Annual Conference*, Abstract No. 1.

Rodger, J. C. and White, I. G. (1974c). Free N-acetylglucosamine in marsupial semen. *J. Reprod. Fert.*, 39, 383 - 6.

Rodger, J. C. and White, I. G. (1975). Electroejaculation of Australian marsupials and analysis of the sugars in the seminal plasma from three macropod species. *J. Reprod. Fert.*, 43, 233 - 9.

Roosen-Runge, E. C. and Giesel, C. O. (1950). Quantitative studies on spermatogenesis in the albino rat. *Am. J. Anat.*, 17, 1 - 30.

Rotenberg, D. (1929). Notes on the male generative apparatus of *Tarsipes spenserae*. *J. Roy. Soc. W. Australia*, 15, 9 - 14.

Rubin, D. (1943). Embryonic differentiation of Cowper's and Bartholin's glands of the opossum following castration and ovariotomy. *J. Exp. Zool.*, 94, 463 - 75.

Rubin, D. (1944). The relation of hormones to the development of Cowper's and Bartholin's glands in the opossum (*Didelphys virginiana*). *J. Morphol.*, 74, 213 - 85.

Sadleir, R. M. F. S. (1965). Reproduction in two species of Kangaroo (*Macropus robustus* and *Megaleia rufa*) and in the arid Pilbara region of Western Australia. *Proc. Zool. Soc. London*, 145, 239 - 61.

Sapsford, C. S. and Rae, C. A. (1969). Ultrastructural studies on Sertoli cells and spermatids in the Bandicoot and ram during the movement of mature spermatids into the lumen of the seminiferous tubule. *Aust. J. Zool.*, 17, 415 - 45.

Sapsford, C. S., Rae, C. A. and Cleland, K. W. (1967). Ultrastructural studies on spermatids and Sertoli cells during early spermiogenesis in the bandicoot *Perameles nasuta* Geoffroy (Marsupialia). *Aust. J. Zool*, 15, 881 - 909.

Sapsford, C. S. Rae, C. A. and Cleland, K. W. (1969a). Ultrastructural studies on maturing spermatids and on Sertoli cells in the bandicoot *Perameles nasuta* Geoffroy Marsupialia. *Aust. J. Zool.*, 17, 195 - 292.

Sapsford, C. S., Rae, C. A. and Cleland, K. W. (1969b). The fate of residual bodies and degenerating germ cells and the lipid cycle in Sertoli cells in the bandicoot *Perameles nasuta* Geoffroy (Marsupialia). *Aust. J. Zool.*, 17, 729 - 53.

Sapsford, C. S., Rae, C. A. and Cleland, K. W. (1970). Ultrastructural studies on the development and form of the principal piece sheath of the bandicoot spermatozoon. *Aust. J. Zool.,* **18**, 21 -48.

Setchell, B. P. (1970a). Testicular blood supply, lymphatic drainage and secretion of fluid. In: *The Testis,* Vol. 1 (ed. A. D. Johnson, W. R. Gomes, and N. L. Van Demark), Academic Press, New York, pp. 101 - 239.

Setchell, B. P. (1970b). Fluid secretion by the testes of an Australian marsupial *Macropus eugenii. Comp. Biochem. Physiol.,* **36**, 411 - 14.

Setchell, B. P. (1973). Venous to arterial transfer in the spermatic cord. *Res. Reprod.,* **5** (5), 3 - 4.

Setchell, B. P. (1974). Secretions of the testis and epididymis. *J. Reprod. Fert.,* **37**, 165 - 77.

Setchell, B. P. and Carrick, F. N. (1973). Spermatogenesis in some Australian marsupials. *Aust. J. Zool.,* **21**, 491 - 9.

Setchell, B. P. and Hinks, N. T. (1969). Absence of countercurrent exchange of oxygen, carbon dioxide, hydrogen ions or glucose between the arterial and venous blood in the spermatic cords of rams and two marsupials (*Macropus eugenii* and *Megaleia rufa*). *J. Reprod. Fert.,* **20**, 179 - 81.

Setchell, B. P. and Hinton, B. T. (1973). Action of gonadotrophins on the testis in mammals. *Bibliog. Reprod.,* **21**, 817 - 26, 959 - 67.

Setchell, B. P. and Thorburn, G. D. (1969). The effect of local heating on blood flow through the testes of some Australian marsupials. *Comp. Biochem. Physiol.,* **31**, 675 - 7.

Setchell, B. P. and Thorburn, G. D. (1970). The effect of artificial cryptorchidism on the testis and on testicular blood flow in an Australian marsupial, *Macropus eugenii. Comp. Biochem. Physiol.,* **38**, 705 - 8.

Setchell, B. P. and Waites, G. M. H. (1969). Pulse attenuation and countercurrent heat exchange in the internal spermatic artery of some Australian marsupials *J. Reprod. Fert.,* **20**, 165 - 9.

Setchell, B. P., Waites, G. M. H. and Lindner, H. R. (1965). Effect of undernutrition on testicular blood flow and metabolism and the output of testosterone in the ram. *J. Reprod. Fert.,* **9**, 149 - 162.

Sharman, G. B. (1959a). Marsupial reproduction. *Monogr. Biol.,* **8**, 332 - 68.

Sharman, G. B. (1959b). Some effects of X-rays on dividing cells in the testis and bone marrow of the marsupial *Potorous tridactylus. Int. J. Rad. Biol.,* **2**, 115 - 30.

Sharman, G. B. (1961). The mitotic chromosomes of marsupials and their bearing on taxonomy and phylogeny. *Aust. J. Zool.,* **9**, 38 - 60.

Sharman, G. B. and Calaby, J. H. (1964). Reproductive behaviour in the red kangaroo, *Megaleia rufa* in captivity. *C.S.I.R.O. Wildlife Res.,* **9**: 58 - 85

Sharman, G. B., Frith, H. J. and Calaby, J. H. (1964). Growth of the pouch young, tooth eruption and age determination in the red kangaroo, *Megaleia rufa. C.S.I.R.O. Wildlife Res.,* **9**, 20 - 49.

Sharman, G. B. and Pilton, P. E. (1964). The life history and reproduction of the red kangaroo (*Melgaeia rufa*). *Proc. Zool. Soc. London,* **142**, 29 - 48.

Sharman, G. B., Robinson, E. S., Walton, S. M. and Berger, P. J. (1970). Sex chromosomes and reproductive anatomy of some intersexual marsupials. *J. Reprod. Fert.,* **21**. 57 - 68.

Smith, M. J. Brown, B. K. and Frith, H. J. (1969). Breeding of the brush-tailed possum (*Trichosurus vulpecula* Kerr) in New South Wales. *C.S.I.R.O. Wildlife Res.*, **14**, 181 – 93.

Smith, R. F. C. (1969). Studies on the marsupial glider *Schoinobates volans* (Kerr) *Aust. J. Zool.*, **17**, 625 – 36.

Steinberger, E. (1971). The hormonal control of spermatogenesis. *Physiol. Rev.*, **51**, 1 – 22.

Steinberger, E. and Steinberger, A. (1974). Hormonal control of testicular function. In: *Handbook of Physiology*, Section 7, *Endocrinology*, Vol. IV, Part 2 (ed. E. Knobil and W. H. Sawyer), American Physiological Society, Washington, D.C., pp. 325 – 45.

Stirling, E. C. (1891). Description of a new genus and species of Marsupialia 'Notoryctes typhlops'. *Roy. Soc. S. Australia Trans. Proc. Rep.*, **14**, 254 – 87.

Suzuki, Y. and Eto, T. (1962). Andogens in testicular venous blood in the adult rat. *Endocrinol Japon.*, **9**, 277 – 83.

Sweet, G. (1897). The skin, hair and reproductive organs of Notoryctes. *Quart. J. Microscop. Sci.*, **51**, 325 – 44.

Swierstra, E. E. (168). Cytology and duration of the cycle of the seminiferous epithelium of the boar, duration of spermatozoon transit through the epididymis. *Anat. Rec.*, **161**, 272 – 85.

Tyndale-Biscoe, C. H. (1968). Reproduction and post-natal development in the marsupial *Bettongia lesueur* (Quoy and Gaimard). *Aust. J. Zool.*, **16**, 577 – 602.

Tyson, E. (1704). Carigueya seu marsupiale Americum masculum. *Phil. Trans. Roy. Soc.*, **24**, 1565 – 75.

Vinson, G. P. and Phillips, J. G. (1969). Formation of testosterone by a special zone in the adrenal cortex of the Brush Possum (*Trichosurus vulpecula*). *Gen. Comp. Endocrinol.*, **13**, 538 – 9.

Waites, G. M. H. (1970). Temperature regulation and the testis. In: *The Testis*, Vol. 1. (ed. A. D. Johnson, W. R. Gomes and N. L. Van Demark), Academic Press, New York, pp. 241 – 79.

Waites, G. M. H. and Moule, G. R. (1960). Blood pressure in the internal spermatic artery of the ram. *J. Reprod. Fert.*, **1**, 223 – 9.

Waites, G. M. H., Setchell, B. P. and Quinlan, D. (1973). Effect of local heating of the scrotum, testes and epididymides of rats on cardiac output and regional blood flow. *J. Reprod. Fert.*, **34**, 41 – 49.

Wilkens, H., Regelson, W. and Hoffmeister, F. S. (1962). The physiologic importance of pusatile bloodflow. *New Eng. J. Med.*, **267**, 443 – 6.

Woolley, P. (1964). Reproduction in Antechinus spp. and other Dasyurid marsupials *Symp. Zool. Soc. London*, **15**, 281 – 94.

Woolley, P. (1975). The seminiferous tubules in Dasyurid marsupials. *J. Reprod. Fert.*, **45**, 255 – 61.

Young, A. H. (1879). On the male generative organs of the Koala (*Phascolarctos cinereus*). *J. Anat. Physiol.*, **13**, 305 – 17.

Author Index

459

Subject Index

471